Asian
Functional
Foods

NUTRACEUTICAL SCIENCE AND TECHNOLOGY

Series Editor

FEREIDOON SHAHIDI, PH.D., FACS, FCIC, FCIFST, FRSC
University Research Professor
Department of Biochemistry
Memorial University of Newfoundland
St. John's, Newfoundland, Canada

1. Phytosterols as Functional Food Components and Nutraceuticals, *edited by Paresh C. Dutta*
2. Bioprocesses and Biotechnology for Functional Foods and Nutraceuticals, *edited by Jean-Richard Neeser and Bruce J. German*
3. Asian Functional Foods, *John Shi, Chi-Tang Ho, and Fereidoon Shahidi*

ADDITIONAL VOLUMES IN PREPARATION

Asian Functional Foods

John Shi
Chi-Tang Ho
Fereidoon Shahidi

CRC Press
Taylor & Francis Group
Boca Raton London New York

CRC Press is an imprint of the
Taylor & Francis Group, an **informa** business
A TAYLOR & FRANCIS BOOK

CRC Press
Taylor & Francis Group
6000 Broken Sound Parkway NW, Suite 300
Boca Raton, FL 33487-2742

First issued in paperback 2019

ISBN-13: 978-0-8247-5855-4 (hbk)
ISBN-13: 978-0-367-39323-6 (pbk)

Library of Congress Cataloging-in-Publication Data

Shi, John.
  Asian functional foods / by John Shi, Chi-Tang Ho, and Fereidoon Shahidi
    p. cm.
    Includes bibliographical references and index.
    ISBN 0-8247-5855-2 (alk. paper)
    1. Functional foods. 2. Diet—Asia. I. Ho, Chi-Tang, 1944- II. Shahidi, Fereidoon, 1951- III. Title.

QP144.F85S535 2005
613.2'095--dc22   2004057109

Preface

Health and "healing" foods have a long history in Asian cultures. The Asians such as Chinese and Indians have long known that food and medicine are from the same source and can treat illnesses and build up a healthy life. The Chinese as well as other Asians are proud of their heritage and the ingenuity of their early scientific and cultural accomplishments. One of their most remarkable contributions to civilization was the wealth of information they collected on the uses of natural substances, plants, chemicals, and animals in treating illnesses. Many unique traditional Asian functional foods were developed by combining food with herbal medicines. It appears that both East and West are in agreement with these concepts. As early as 2,500 years ago, Hippocrates, the father of the Western medicine, said "Let your food be your medicine and medicine be your food."

Traditional Asian functional foods derived from cereals, vegetables, and fruits are consumed on a regular basis and can be considered as nutritious supplements, while the special functional foods such as herbs (ginseng, *Lingzhi* [Ganoderma]), meat (black-bone chicken, duck skin) and some seafoods (sea cucumber) are consumed less frequently. Many plants such as medicinal herbs have been used for thousands of years to maintain health and treat disease. They can be used either as a single herb or as multiple formulated herbs in herbal foods, teas, wines, congees, and pills (or powder). Many of these herbal products have been shown to improve immunopotentiation, increase systemic circulation, assist disease prevention, and slow down the aging process.

The health benefits of muscle foods, including seafoods, have also been known for thousands of years. The inhabitants of Asian countries, especially the Chinese, Japanese, and Indians, have for thousands of years considered certain meat and meat products as special health-healing foods. In addition, a number of nontraditional animal-derived foods, such as sea cucumber, shark cartilage and the like are also found in Asian supermarkets and are considered as health-enhancing food items.

Most Asian regions are located in tropical and subtropical and monsoon zones. Thus many kinds of cereals, oil seeds, and nuts can be cultivated. The traditional edible oils have been extracted from seeds or nuts (groundnut, rapeseed, sesame, perilla, walnut, and torreya) since ancient times. These oils have desirable flavor and color as well as fat soluble antioxidative substances that possess radical scavenging and antioxidant properties. Today more and more people believe that Asian functional foods can prevent diseases, maintain health, and make their dream of living longer and healthy come true. The long history of Asian functional foods, where herbal products are used as traditional medicines, and health care based on natural products has given a new meaning to functional food in the world. As traditional Asian functional foods gain the attention of the general public, manufacturers will try to fill a growing consumer appetite for these health-promoting products derived from foods.

Traditionally, Asian functional foods were produced on a small scale with manual operations, and then consumed locally. In the last few decades, mass production of functional foods with modern equipment and technologies has begun to play an increasingly important role in the production of many Asian functional food products. The production of functional foods, however, requires maximizing the retention of biologically active components that are usually heat sensitive and susceptible to process-induced changes as well as oxidative reactions.

During the past decade the consumption of functional foods has emerged as a major consumer-driven trend, serving the needs of an aging population that wants to exercise greater control over its health and well being. This trend is expected to continue, and the need for scientific information on all aspects of functional foods is vital to the advancement of this emerging sector. The increase in consumer demand for functional food has prompted international health organizations and governmental agencies to develop specific guidelines for their production and use. Accordingly, the scientific community must apply modern technologies to ensure the efficacy and safety of these traditional functional foods before developing them into first-class dietary supplements.

In order to gain a better understanding and to disseminate the latest developments in this rapidly expanding field, this book, *Asian Functional Foods*, in the Series of Nutraceutical Science and Technology, was developed. The 21 chapters in this book cover a wide range of traditional Asian functional foods, including the source of the traditional Asian functional foods, their history, functionality, the chemical, physical and physiological properties, health benefits, mechanisms of antioxidant action, anticancer, antiaging properties, as well as clinical and epidemiological evidence. The processing technology and process systems, equipment, material preparation, food preparation, and quality control during processing are also discussed. The stability, shelf life and storage technology (including packaging technology) of traditional functional food products, industrial production, homemade products, consumer and marketing issues, and social and economical impact are also presented in these chapters.

While Asian functional foods steadily gain in popularity in the Western world, food cultures from the Western countries are also being widely accepted in Asia. People around the world are accepting the concept of functional foods as more than just a source of simple nutrition. This book discusses the theoretical and practical aspects of functional foods, from fundamental concepts of biochemistry, nutrition, and physiology to food technology. The information in this volume may initiate communications between East and West, and open up areas of common interest. This in turn may generate opportunities for greater utilization of traditional Asian functional food in the Western world.

The production of this book was made possible by the efforts of international experts, and different areas are presented. This book will be of interest to a wide spectrum of food scientists and technologists, nutritionists, biochemists, engineers, and entrepreneurs worldwide. It will also serve to further stimulate the development of functional foods and nutraceuticals, and contribute to providing consumers worldwide with products that prevent diseases and maintain health.

About the Editors

John Shi is a research scientist in Federal Department of Agriculture and Agri-Food Canada, adjunct professor in University of Guelph, Canada and South China Institute of Botany, Chinese Academy of Sciences, China. He is coeditor of *Functional Foods II*. He graduated from Zhejiang University, China, and received an M.A. degree in 1985, and Ph.D. degree in 1994 from the Polytechnic University of Valencia, Spain. Dr Shi is an international editor of *Journal of Food Science and Nutrition* and *Nutraceuticals and Foods*, a member of the editorial board of *Journal of Medicinal Foods*, and *Journal of Agriculture, Food and Environment*. He is a visiting professor and has done international collaborative research at the Norwegian Institute of Fishery and Aquaculture, and Lleida University, Spain. His current research interests focus on separation technologies for health-promoting components from natural products to develop functional foods.

Professor Chi-Tang Ho received his B.S. degree in chemistry from National Taiwan University in Taipei, Taiwan in 1968. He then went on to receive both his M.A. in 1971 and his Ph.D. in 1974 in organic chemistry from Washington University in St. Louis. After completing two years as a postdoctorate fellow at Rutgers University, he joined the Rutgers faculty as an assistant professor in the Department of Food Science. He was promoted to associate professor in 1983. In 1987 he was promoted to Professor I, and in 1993 he was promoted to Professor II. He has published over 480 papers and scientific articles, edited 27 professional books and is an editorial board member for a variety of publications, including the Journal of Agricultural and Food Chemistry. He has also won numerous awards including the Stephen S. Chang Award in Lipid and Flavor Science from the Institute of Food Technology and two honorary professorships, and has served in the Division of Agricultural and Food Chemistry of the American Chemical Society in various positions including as division chair. His current research interests focus on flavor chemistry and the antioxidant and anti-cancer properties of natural products.

Fereidoon Shahidi, Ph.D., FACS, FCIC, FCIFST, FRSC, has reached the highest academic level, university research professor, in the Department of Biochemistry at Memorial University of Newfoundland (MUN). He is also cross-appointed to the Department of Biology, Ocean Sciences Centre, and the aquaculture program at MUN. Dr. Shahidi is the author of nearly 500 research papers and book chapters and has authored or edited over 30 books and made over 300 presentations at scientific conferences. His research contributions have led to several industrial developments around the globe.

Dr. Shahidi's current research interests include different areas of nutraceuticals and functional foods as well as marine foods and natural antioxidants, among others. Dr. Shahidi serves as the editor-in-chief of the *Journal of Food Lipids* and is an editorial board member of *Food Chemistry, Journal of Food Science, Journal of Agricultural and Food Chemistry, Nutraceuticals and Food*, and the *International Journal of Food Properties*. He was the recipient of the 1996 William J. Eva Award from the Canadian Institute of Food Science and Technology in recognition of his outstanding contributions to food science in Canada through research and service, and also the 1998 Earl P. McFee Award from the Atlantic Fisheries Technological Society in recognition of his exemplary contributions in the seafood area and their global impact. He has also been recognized as one of the most highly cited authors in the world in the discipline of agriculture, plant and animal sciences and was the recipient of the 2002 ADM Award from the American Oil Chemists' Society.

Dr. Shahidi is the immediate past chairperson of the nutraceuticals and functional foods division of the Institute of Food Technologists and the past chair of Lipid Oxidation and Quality of the American Oil Chemists' Society. He is also the chair of the agricultural and food chemistry division of the American Chemical Society. Dr. Shahidi serves as a member of the Expert Advisory Panel of Health Canada on Standards of Evidence for Health Claims for Foods, Standards Council of Canada on Fats and Oils, Advisory Group of Agriculture and Agri-Food Canada on Plant Products and the Nutraceutical Network of Canada. He is a member of the Washington-based Council of Agricultural Science and Technology on Nutraceuticals.

Series Introduction

The Nutraceutical Science and Technology series provides a comprehensive and authoritative source of the most recent information for those interested in the field of nutraceuticals and functional foods. There is a growing body of knowledge, sometimes arising from epidemiological studies and often substantiated by preclinical and clinical studies, demonstrating the relationship between diet and health status. Many of the bioactives present in foods, from both plant and animal sources, have been shown to be effective in disease prevention and health promotion. The emerging findings in the nutrigenomics and proteomics areas further reflect the importance of diet in a deeper sense, and this, together with the increasing burden of prescription drugs in treatment of chronic diseases such as cardiovascular ailments, certain types of cancer, diabetes, and a variety of inflammatory diseases, has raised interest in functional foods and nutraceuticals to a new high. The interest is quite widespread from producers to consumers, regulatory agencies, and health professionals.

In this series, particular attention is paid to the most recent and emerging information on a range of topics covering the chemistry, biochemistry, epidemiology, nutrigenomics and proteomics, engineering, formulation, and processing technologies related to nutraceuticals, functional foods, and dietary supplements. Quality management, safety, and toxicology, as well as disease prevention and health promotion aspects of products of interest, are addressed. The series also covers relevant aspects of preclinical and clinical trials, as well as regulatory and labeling issues.

This series provides much needed information on a variety of topics. It addresses the needs of professionals, students, and practitioners in the fields of food science, nutrition, pharmacy, and health, as well as leads conscious consumers to the scientific origin of health-promoting substances in foods, nutraceuticals, and dietary supplements. Each volume covers a specific topic of related foods or prevention of certain types of diseases, including the process of aging.

Fereidoon Shahidi

Contributors

Suad N. Al-Hooti Biotechnology Department, Kuwait Institute for Scientific Research, Safat, Kuwait

Toyohiko Ariga Laboratory of Nutrition and Physiology, Department of Biological and Agricultural Chemistry, Nihon University College of Bioresource Sciences, Fujisawa, Japan

Karen J. Auborn North Shore—Long Island Jewish Research Institute, Manhasset, New York, Departments of Otolaryngology and Microbiology and Immunology, Albert Einstein College of Medicine, Bronx, New York, U.S.A.

C. J. Carriere Cereal Products and Food Science Research Unit, National Center for Agricultural Utilization Research, ARS, USDA, Peoria, Illinois, U.S.A.

Chung Chi Chou New Orleans, Louisiana, U.S.A.

Yi Dang Beijing University of Chinese Medicine, Beijing, China

Hao Feng Department of Food Science and Human Nutrition, University of Illinois at Urbana-Champaign, Urbana, Illinois, U.S.A.

Yasuko Fukuda   Nagoya Women's University, Nagoya, Japan

Asbjørn Gildberg   Norwegian Institute of Fisheries and Aquaculture Research, Tromsø, Norway

Chi-Tang Ho   Department of Food Science, Rutgers University, New Brunswick, New Jersey, U.S.A.

G.E. Inglett   Cereal Products and Food Science Research Unit, National Center for Agricultural Utilization Research, ARS, USDA, Peoria, Illinois, U.S.A.

Hideji Itokawa   Natural Products Laboratory, School of Pharmacy, University of North Carolina, Chapel Hill, North Carolina, U.S.A.

Yi Jin   Department of Food Science, Rutgers University, New Brunswick, New Jersey, U.S.A.

Keun-Ok Jung   Pusan National University, Busan, Korea

Kenji Koge   Planning and Research Section, Chigasaki Laboratory, Shin Mitsui Sugar Co., Ltd., Kanagawa-ken, Japan

Mutsuo Kozuka   Natural Products Laboratory, School of Pharmacy, University of North Carolina, Chapel Hill, North Carolina, U.S.A.

Tai-Wan Kwon   Department of Food and Nutrition, Inje University, Kimhae, Korea

Cherl-Ho Lee   Graduate School of Biotechnology, Korea University, Seoul, Korea

Kuo-Hsiung Lee   Natural Products Laboratory, School of Pharmacy, University of North Carolina, Chapel Hill, North Carolina, U.S.A.

Su-Rae Lee   Division of Agricultural and Fishery Sciences, Korean Academy of Science and Technology, Gyeonggi-Do, Korea

S. Maneepun   Institute of Food Research and Product Development, Kasetsart University, Bangkok, Thailand

Teruo Miyazawa   Food and Biodynamic Chemistry Laboratory, Graduate School of Agriculture Science, Tohoku University, Sendai, Japan

Scott A. Morris   Department of Food Science and Human Nutrition, University of Illinois at Urbana-Champaign, Urbana, Illinois, U.S.A.

James H. Moy   Department of Molecular Biosciences and Biosystems Engineering, University of Hawaii at Manoa, Honolulu, Hawaii, U.S.A.

W.J. Mullin   Agriculture and Agri-Food Canada, Ontario, Canada

Mayumi Nagashima   Department of Culture and Domestic Science, Nagoya Keizai University Junior College, Aichi, Japan

Kiyotaka Nakagawa   Food and Biodynamic Chemistry Laboratory, Graduate School of Agriculture Science, Tohoku University, Sendai, Japan

Kun-Young Park   Pusan National University, Busan, Korea

Sook-Hee Rhee   Pusan National University, Busan, Korea

Michael Saska   Audubon Sugar Institute, Louisiana State University Agricultural Center, Baton Rouge, Louisiana, U.S.A.

Taiichiro Seki   Laboratory of Nutrition and Physiology, Department of Biological and Agricultural Chemistry, Nihon University College of Bioresource Sciences, Fujisawa, Japan

Fereidoon Shahidi   Department of Biochemistry, Memorial University of Newfoundland, St. John's, Newfoundland, Canada

John Shi   Guelph Food Research Center, Agriculture and Agri-Food Canada, Guelph, Ontario, Canada

Jiwan S. Sidhu   Department of Family Sciences, College for Women, Kuwait University Safat, Kuwait

James E. Simon New Use Agriculture and Natural Plant Products Program, Department of Plant Biology and Pathology, Rutgers University, New Brunswick, New Jersey, U.S.A.

Chaufah Thongthai Mahidol University, Bangkok, Thailand

Mark L. Wahlqvist Monash Asia Institute, Monash University, Melbourne, Australia

Mingfu Wang New Use Agriculture and Natural Plant Products Program, Department of Plant Biology and Pathology, Rutgers University, New Brunswick, New Jersey, U.S.A.

Naiyana Wattanapenpaiboon Monash Asia Institute, Monash University, Melbourne, Australia

Qing-Li Wu New Use Agriculture and Natural Plant Products Program, Department of Plant Biology and Pathology, Rutgers University, New Brunswick, New Jersey, U.S.A.

Youling L. Xiong Department of Animal Sciences, University of Kentucky, Lexington, Kentucky, U.S.A.

Table of Contents

1

Functional Foods and Their Impact on Nutrition and Health: Opportunities in the Asia Pacific

MARK L. WAHLQVIST and
NAIYANA WATTANAPENPAIBOON

PEOPLE IN ASIA

Countries of the Asian Pacific region have a wide range of nutritional status, with some countries wrestling with problems of undernutrition (like Cambodia, Myanmar, and Papua New Guinea), and others suffering more from health problems associated with overnutrition (such as Australia and Singapore). In between are countries (such as China and Malaysia), where the transition from a closed economy to an open economy takes place, the so-called "diseases of affluence" (obesity, diabetes, cardiovascular disease, osteoporosis, and certain cancers) may coexist with those of "undernutrition" (protein-energy malnutrition, micronutrient deficiency, and food-borne illness). This phenomenon of the double burden of disease is now increasingly common.[1] This requires the application of a new paradigm, arguably one that more effectively relates the human condition to the environment.[2] There is great potential to address both sets of problems through nutritional means.

KEY PUBLIC HEALTH NUTRITION ISSUES IN THE ASIA PACIFIC REGION

Transitional Health

The emerging nutrition problems in the Asia Pacific region are probably partly explained by the Barker hypothesis, also referred to as the thrifty phenotype hypothesis.[3,4] It is postulated that nutritional transition begins *in utero* where the mother is exposed to nutritional inadequacy. In response to maternal and fetal malnutrition, there are adaptive changes, in order to survive, in fetal organ development. These adaptations may permanently alter adult physiology and metabolism in ways that are beneficial to survival under continued conditions of malnutrition, but detrimental when the food supply is abundant. This fetal "programming" is thought to lead to increased rates of coronary heart disease, hypertension, insulin resistance syndrome, noninsulin-dependent diabetes, obesity, osteoporosis, and some cancers later in life.

Urbanization and its Effects

This social change has a remarkable effect on diet in the developing countries. Urbanization increases labor-force participation of women and it indirectly affects the diet of the family.[5] Whereas the food supply of rural populations comes from its own food production, the food supply of urban populations has to be purchased, and more processed, rather than represent fresh animal products and garden produce.[6]

Changing Demography with an Aging Population

There are increasing numbers of people in communities that age to 70, 80, and beyond. Thus, it is increasingly important to understand the relationship between food and health in later life.

Food Security

Without the ability to stay near the place of food production there is not the same scrutiny of the food supply. People become increasingly dependent on others to produce food and become more vulnerable from a health point of view.

Loss of Traditional Food Culture

One of the great threats to human civilization is that food diversity may progressively diminish.

OPPORTUNITIES FOR FUNCTIONAL FOOD DEVELOPMENT

Thousands of products with supposed health benefits, ranging from the nutritionally beneficial to the fraudulent, are already available in the world market, and the number of products is soaring. Science and technology, agriculture, food manufacture and markets are driven by the belief and actuality that food characteristics are relevant to health.[7]

It is proposed that food-health relationships could be categorized for the purposes of functional food development.[8] The extensive list of general and particular possibilities and roles for foodstuffs (Table 1.1) was presented to and discussed at a Joint FAO/WHO Workshop on Novel Foods in Nutrition Health and Development: Benefits, Risks and Communication.[9] Some of these relationships are pertinent to the key public health issues in the Asia Pacific region.

Food Shortage and Malnutrition

A 1997 report by the International Food Policy Research Institute predicted that by 2010, every 20th person in East Asia is likely to have an insecure food situation.[10] Pervasive poverty rather than food shortage is frequently the underlying cause of hunger, limiting an individual's access to food. Poverty also exacerbates access to education, health care services, and a clean living environment, while malnutrition coupled with a poor education often impairs employment, in turn earning prospects, and ultimately there is less money to buy food.

TABLE 1.1   Categorizing Food–Health Relationships
for the Purposes of Food Product Development (8)

Health Category	Food Characteristics
Disease related to environmental degradation and methods of food production	Eco-sensitive foods (e.g., produced in sustainable ways; biodegradable or edible packaging; identifiable biosecurity for animal-derived foods; nature of genetic material)
Food shortage and PEM (protein-energy malnutrition)	Technologies which minimize postharvest loss, increase shelf life and maintain palatability
Disease related to protein quality, fat quality and micronutrient status	Nutrient-dense foods; fish or its plant or microbial food surrogates
Physical inactivity and health (especially over fatness; also loss of lean mass, particularly muscle)	Food of low energy density and high nutrient density
Phytochemical deficiency disorders including menopause, macular degeneration, osteopenia	Greater emphases on plant-derived foods and their variety
Diseases of changing demography	
• Aging	Anti-aging food, especially ones to delay body compositional change (bone, muscle and fat); loss of sensory function; decline in immune function; proneness to neoplastic disease; decline in cardio-respiratory function; and decline in cognitive function; and anti-inflammatory foods
• Rapid loss of traditional food culture and acquisition of new food cultures	Maintenance of traditional foods in convenient, affordable and recognizable form
New psychosocial stressors and mood change	Food which favorably affects mood
Food-borne illness and the microbiological safety of foods	• Pre- and probiotic foods • Immune system enhancing foods
Illness related to the chemical safety of foods (e.g., pesticide residues)	Regional origin and certification of foods

Biotechnology can greatly advance staples farming, and undoubtedly, the production of functional foods is a ray of hope in the fight against malnutrition. Genetically improved foods with a higher yield and more resistance to diseases and insects have the capacity to alleviate food insecurity and malnutrition in the world's underfed populations. More resources and research need to be directed toward resolving these problems in developing countries.

Food fortification has been a major and integral food-based strategy to eliminate micronutrient deficiencies, especially iodine deficiency disorders, vitamin A deficiency and iron deficiency anaemia, in developing countries and an increasing number of developed countries. Recently, biofortification has been proposed to combat micronutrient malnutrition. Supplying micronutrients to vulnerable populations by using conventional plant breeding and biotechnology is low cost and sustainable. "Golden Rice" is an example of biofortification through gene manipulation to improve nutritional value, vitamin A in particular. Genes from the daffodil and a bacterium were introduced to the rice line to complete the biosynthetic pathway to β-carotene, a provitamin A carotene.[11] The insertion of these genes into rice to express β-carotene was necessary because parts of the pathway had been lost, although the downstream parts of the pathway were still expressed. The resultant transgenic rice line synthesizes enough β-carotene in the endosperm to meet part of the vitamin A requirements of people dependent on rice as a primary food staple.

At the same time, the rediscovery of hundreds of natural "yellow" rice cultivars, which may have the missing genes, has encouraged the study and propagation of rice varieties that can be used as natural food sources of provitamin A carotenoids.[12] Production of added value grains of this type is likely to be their provision of a wider array of carotenoids with wider health benefits than the prevention of vitamin A deficiency.[13] As grains richer in other micronutrients, like iron and zinc, are identified in seed banks and libraries and in traditional communities, the prospects of safe and effective biofortification in the food supply would increase.[14]

Overweight and Obesity

At the other end of the malnutrition scale, obesity threatens to become the leading cause of chronic disease in the world. The health consequences of obesity are well known. For more than a decade, dietary guidelines advising a reduced fat intake to no more than approximately 30% of total calories have been issued in developed countries. This triggered a proliferation of low-fat, fat-reduced, and fat-free food products targeted for consumers seeking a more healthy diet.

However, the role these foods have to play on a population basis is less clear. In many countries, total fat consumption as a percentage of calories has slightly declined on a population basis, while total calorie intake has risen, along with the prevalence of obesity.[15] There are too few prospective studies to conclude that dietary fat plays a role in the development of obesity. Epidemiological analysis of nutrient intake and obesity provide inconclusive results that high fat/low carbohydrate diets lead to the development of obesity.[16]

Body fatness may be altered by a number of food factors; they are either physico-chemical or chemical, with mechanisms of action that alter energy or food intake, energy expenditure, or the deposition of fat in various anatomical sites. Food factors could alter energy intake by suppressing or stimulating appetite, or by interfering with or enhancing absorption or utilization of energy (Figure 1.1). Simply using a wide range of foods, rather than bland food, can stimulate appetite without excessive energy consumption, and it can be associated with lower body fat, provided that the foods have low energy density.[17] This can be achieved by changes in the organoleptic properties of food, this being the taste, smell, and texture, and even the sound of chewing, which influence food intake. In addition, there may be various anorectic properties of food, operating in the central nervous system at cortical or subcortical levels, notably at the level of the hypothalamus and pituitary. Examples of compounds in food with such properties include xanthines (caffeine in tea and coffee, and theobromine in cocoa), and small peptides with opioid activity derived from gluten, β-casein, and compounds found in coffee

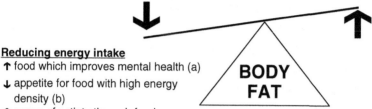

Increasing energy expenditure
↑ basal metabolic rate through
↑ increased lean mass
   thermogenesis
↑ physical activity

Reducing energy intake
↑ food which improves mental health (a)
↓ appetite for food with high energy
   density (b)
↑ sense of satiety through food
   with low energy density (c)
↓ digestion & absorption

BODY
FAT

(a) breakfast food and fish are candidates
(b) food variety for taste, not energy
(c) food with high water and dietary fiber content

Figure 1.1 Ways for food factors to alter energy balance safely.

with roasting or brewing.[18] Supplementation with garlic (*Allium sativum*) for 12 weeks was found to reduce fat and carbohydrate intakes in mild or moderate hypercholesterolemia.[19] The choice of food may also affect energy expenditure.

Food could potentially affect basal metabolic rate, thermogenesis, and physical activity (not only exercise, but also fidgeting, gesture, and other self-paced activity). Certainly one component of energy expenditure can be attributed directly to food; this is the difference in thermic response to various foods which may, in part, offer a basis for preference in macronutrient choice, with fat being least and protein most wasteful of energy in this respect. Capsaicin found in chilli (*Capsicum frutescens*) or red pepper (*Capsicum annuum*) may increase thermogenicity, principally by way of effects on the vasculature,[20,21] and consumption of red pepper has recently been reported to decrease energy intake as well.[22] While caffeine has the potential to reduce energy intake, it also stimulates the central nervous system and cardiac function, resulting in an increase in spontaneous activity and consequently energy expenditure. Additionally, caffeine is known to enhance physical performance, so that food and beverages containing it may, in this way, increase energy expenditure (Figure 1.1); this may be more in evidence when the food provides liberal carbohydrate for glycogen stores in muscle and liver. The combination of red pepper and caffeine in foods is now attracting attention in relation to energy balance.[23] Precaution should, however, be taken with the ingestion of large doses of caffeine present in certain foods and diet pills, as it can induce cardiac arrhythmia.[24]

It is generally agreed that fat intake is a significant factor in allowing energy imbalance. In addition, reduction in fat intake together with total energy intake can favorably influence body fat. The best indicator of energy balance in humans is the measure of body storage fat. But not only is the total amount of body fat important for human health, so too is the distribution of body fat. An increase in the polyunsaturated (principally n-6) and saturated fat ratio in serum phospholipids in overweight men has been found to predict a decrease in body mass index,[25] and n-3 fatty acids are less prone than other fat to increase truncal fatness in rodents.[26]

Phytochemical Deficiency Disorders

The use of plants for health promotion has been known in food cultural folklore for many centuries. However, only a limited number of biologically active compounds, known as phytochemicals, have been identified in plant food. For most of these compounds, their mechanisms of action are not clearly understood, and sometimes it has been difficult to reproduce their therapuetic benefits. In the mid-1980s, the link between coronary heart disease and diet focused exclusively on the lipid hypothesis or the theory that high blood lipids were a significant risk factor in cardiovascular disease (CVD) morbidity and mortality.

A considerable number of epidemiological studies in recent decades suggest that the high consumption of fruit and vegetables (as a collective term), or in some studies, of specific vegetables, is associated with low morbidity and mortality from CVD and certain cancers.[27–29] New analytical technology and advances in molecular biology have identified phytochemical components of popular foods used in health promotion that may explain traditional health benefits and help quantify their effects. A problem in considering the place of phytochemicals in human health is that they are numerous, alongside a few known essential nutrients, and, therefore, their net interactive effect ultimately requires a study of food itself and food patterns,[30] or that the intake of food components be subjected to sophisticated mathematical modeling. However, the advent of advanced informatics may help resolve this dilemma[31].

While evidence continues to emerge showing that the intake of certain foods and their constituents can have profound physiological effects, including the potential to prevent or delay the onset of chronic diseases such as CVD and certain types of cancer, the role of specific food phytochemicals in the prevention and/or treatment of health conditions is rapidly unfolding.[32] Compounds found in fruits and vegetables, such as polyphenols in fruits, isoflavones in the legume soy and β-carotene in vegetables, have been considered to be the responsible active compounds. However, several intervention studies have shown that ingestion of some of these isolated compounds, in tablet or capsule form, cannot confer similar health benefits to those observed with the intact food from which they come.[33–35] Studies of intact foods on health outcomes, like those with whole grains on diabetes and CVD, support these findings.[36–38] The similar dichotomy between ingredient-based phytochemical-rich foods and the administration of isolated phytochemicals is illustrated by the field of phytoestrogens and menopause. It was demonstrated in the late 1980s that certain foodstuffs (soy flour, flaxseed, and red clover as sprouts) could exhibit estrogenic effects including improving vaginal health and decreasing pituitary follicle stimulating hormone production in postmenopausal women.[39]

Thereafter, there have been those who have favored food ingredient-based studies and applications, especially bone health and breast cancer protection, to women's health[40-42] and those who have pursued a more nutraceutical or pharmaceutical approach.[43,44] The risk–benefit ratio in each case is quite different, with more checks and balances on intake in the phytochemical-rich food ingredient approaches than with encapsulated phytochemicals.

When particular bodily functions or disease processes are considered alongside the phytochemical or other nonnutrient food components, which may modulate them, the potential for making better use of basic food commodities and of relevant functional foods becomes apparent. The nutritional trend is increasingly shifting to use protective foods to prevent and manage diseases rather than to depend on exclusion of detrimental items. This may equally be the focus of prevention and management of phytochemical deficiency.

Disease of Changing Demography

Aging

There is a worldwide increase in the proportion of people in the older age groups. It is predicted that by the year 2020 people over 65 years in some countries, such as Japan, will comprise nearly one-quarter of the population. Socioeconomic progress and advances in medical care have apparently underpinned this increase in longevity, but it is also likely that it has been facilitated, in part, through food availability and quality. With the help of micronutrients, phytochemicals, and probably other compounds, it is hoped that people will live longer and remain healthier.

Although aging appears to be an inevitable, natural process programmed into the genes, many of the changes are partly the result of lifestyle or environmental factors. As people grow older, a decline in muscle mass and increase in body fat tends to occur. A major contributor to these changes is the increasingly sedentary nature of lifestyle. Appropriate health promotion strategies that encourage desirable food habits and other lifestyle factors need to be implemented to maximize the quality of life for elderly populations and reduce the cost of health care.

Particular food patterns have not only been able to predict overall mortality,[30] favorable patterns have also been shown to reduce mortality. Studies have shown that consuming a wide variety of foods, especially plant foods,[45] and having a proportionately higher intake of plant foods relative to animal foods is associated with longevity.[30,46,47]

Loss of Traditional Food Culture

Economic development together with recent technological innovation and modern marketing techniques have modified dietary preferences, and consequently, led to major changes in the composition of diet. There was a shift toward high fat, refined carbohydrate, and low fiber diets. The accelerating factors for the rapid transition include globalization, e.g., exposure to the global mass media, shift in occupational structure including the trend from labor-intensive occupations and leisure time activities toward more capital-intensive, less strenuous work and leisure. The globalization of human diet is very much a combined outcome of cultural sensory preferences coupled with the greater availability of cheap fats in the global economy and rapid social changes in the lower-income world. As people move into cities, their food supplies change, and therefore, so do their diets and body composition. Traditional staples are often more expensive in urban areas than in rural areas, whereas processed foods are less expensive. This favors the consumption of new processed foods. The shift from traditional staples to processed foods in urban areas is also strongly enhanced by the advocacy of western culture through mass media, commercial marketing and through other channels related to globalization.

Food acculturation normally occurs when a population has migrated and established itself amongst a majority food culture, and consequently, the loss of traditional food culture.

An example can be seen from the Asian migration to Australia. There has been a decrease in energy expenditures, an increase in food energy density (through increased fat and sugary drink consumption), and a decrease in certain health protective foods (lentils, soy, green leaf vegetables) and beverages (tea), amongst Asians in Australia. However, food acculturation with migration is generally bidirectional. The process of "Asianification" of Australian eating patterns has been evident through fresh food markets and groceries, restaurants, and the development of household cooking skills. The great advantage of Asian migration to Australia, for the majority population, has been the increased diet diversification, most of which has been with rice, soy, green leafy and root vegetables, and various "exotic fruits."[48]

Depression and Mood Change

The World Health Organization (WHO) estimates that depression is one of the major single causes of disability worldwide. Loss of appetite and loss of interest in surroundings and social relationships occur in depression. The risk of being depressed is rising rapidly in developed countries, for reasons that are unknown. There is a growing body of evidence to suggest that n–3 fatty acids may play an important role in the etiology of depression.[49–51] The variation in the prevalence of depression found in different countries appears to relate to fish consumption.[52] For example, in Japan the prevalence is 0.12%, whereas in New Zealand the prevalence is 5.8%; fish consumption is 67 and 18 kg per person per year, respectively.

Several nutritional factors appear to be associated with mood. Carbohydrates have been shown to influence brain serotonin levels, and in individuals under emotional stress like depression, a preference for sweet simple carbohydrates has been demonstrated. One particularly striking observation is that total long-chain polyunsaturated fatty acid levels in plasma are positively related to cerebrospinal fluid 5-hydroxyindolactic acid, the main metabolite of serotonin, a neurotransmitter involved in mood control.[53]

Immunodeficiency

Nutritional immunoenhancement is relevant to chronic infectious conditions that are characterized by multiple nutrient deficiencies. In some situations like inadequate food or nutrient intake, excessive nutrient loss or increased nutrient demand, immunodeficiency may be reversible by nutritional means, to the extent that it has been nutritionally caused.[54,55] However, impairment of immune function can be caused not only by deficiencies of various nutrients, but also by imbalances of nutrients, such as iron excess and changes in the ratios of n-3 to n-6 fatty acids.[56,57] There are many potential therapeutic applications of specific nutrients for immune defence. The n-3 fatty acids are of interest as anti-inflammatory agents, acting at least partly by influencing leukotriene and prostaglandin balance.

Another area of nutritional immunoenhancement would be fermented foods. The link between fermented foods and longevity received interest over a century ago, and the idea of beneficial gut microflora was introduced. Fermented milks and yogurts are consumed worldwide and research has shown them to be associated with various physiological benefits such as regulation of intestinal peristalsis and alleviation of the symptoms of lactose intolerance. Study of the organisms contained in fermented foods led to the concept of probiotics, which can be defined as a live microbial food ingredient that, when consumed in sufficient quantities, exerts health benefits on the host beyond basic nutrition.[58] Probiotic therapy has been implicated in enhancing the colonization resistance of the gut microflora against intestinal pathogens such as *Helicobacter pylori*, *Salmonella sp.*, *Clostrolidium difficile* and rotavirus infection, gut and systemic immunity, as well as colonocyte nutrition from the lumen. However, there is still divided opinion about the general effectiveness of lactic acid bacteria in promoting human health. Differences in strains, dose, model systems, and stringency of data interpretation have led to some of the inconsistencies in conclusions. In the meantime, although research support is lacking for many claims about live bacterial culture-induced promotion of intestinal and human health, products containing such cultures are marketed successfully worldwide.

HEALTH CLAIMS AND SAFETY CONSIDERATIONS

The social, food, nutrition and biomedical sciences allow for increasingly rational product development to fit consumers' health needs. The development of new food products brings with it elements of the unknown and therefore risk. However, when it comes to foodstuffs and beverages, it is expected that the risk will be negligible. Monitoring and surveillance is a requisite of any community that is exposed to products launched for purported physiological or health reasons.[59]

In order to optimize the risk–benefit relationships of functional foods with definable health properties, the following approach has been recommended by a Joint WHO/FAO Working Group on Novel Foods in Nutrition Health and Development: Benefits, Risks and Communication, and published in the Metung Report.[60]

1. Consider the health outcome in question
2. Select a plant food or foods, which confer these characteristics, preferably with an established food cultural base
3. Formulate a food for trial
4. Carry out a risk evaluation
5. Conduct a food trial using biomarkers and/or health outcomes
6. Develop an appropriate monitoring and surveillance strategy
7. Seek regulatory approach as novel food for safety
8. Formulate a food-based educational and informational framework, with or without health claims (depending on regulatory regime)

In all cases, affordability and sustainability should be considered. These considerations will be increasingly important as food novelty and related health opportunities become more evidence.

REFERENCES

1. Gillespie, S. and Haddad, L., Attacking the Double Burden of Malnutrition in Asia and the Pacific. Asian Development Bank and the International Food Policy Research Institute, Washington, D.C., 2001.

2. Wahlqvist, M.L. and Specht, R.L., Food variety and biodiversity: econutrition, *Asia Pacific J. Clin. Nutr.*, 1998, 7(3/4), 314–319.

3. Hales, C.N. and Barker, D.J.P., Type 2 (noninsulin-dependent) diabetes mellitus: the thrifty phenotype hypothesis, *Diabetologia*, 1992, 35, 595–601.

4. Barker, D.J., Martyn, C.N., Osmond, C., Hales, C.N., Fall, C.H. Growth *in utero* and serum cholesterol concentrations in adult life. *BMJ* 1993, 307(6918), 1524–1527.

5. Ruel, M.T., Haddad, L., Garrett, J.L., Rapid urbanization and challenges of obtaining food and nutrition security, in Semba, R.D., Bloem, M.W., Eds., *Nutrition and Health in Developing Countries*, Totowa, NJ, Humana Press Inc., 2001.

6. Solomons, N.W. and Gross, R., Urban nutrition in developing countries, *Nutr. Rev.* 1995, 369, 39–48.

7. Wahlqvist, M.L. and Wattanapenpaiboon, N., Can functional foods make a difference to disease prevention and control? Yach, D. and Puska, P., *Globalization, Diets and Noncommunicable Diseases*, World Health Organization, Geneva, Switzerland, 2002.

8. Wahlqvist, M.L., Principles for safety and efficacy in the development of novel and functional foods, *Scand. J. Nutr.* 2000, 44, 128–129.

9. Wahlqvist, M.L., Focusing on novel foods: their role, potential and safety, *Asia Pacific J. Clin. Nutr.* 2002, 11(Suppl. 6), S98–S99.

10. Pinstrup-Anderson, P. and Pandya-Lorch, R., World food needs towards 2000, *Am. J. Agr. Econ.* 1997, 5, 1465–1466.

11. Ye, X., al-Babili, S., Klöti, A., Zhang, J., Lucca, P., Beyer, P., Potrykus, I., Engineering the provitamin A (β-carotene) biosynthetic pathway into (carotenoid-free) rice endosperm, *Science* 2000, 287, 303–305.

12. Graham, R.D. and Rosser, J.M., Carotenoids in staple foods: their potential to improve human nutrition, *Food Nutr. Bulletin* 2000, 21, 405–409.

13. Cooper, D.A., Eldridge, A.L., Peters, J.C., Dietary carotenoids and certain cancers, heart disease, and age-related macular degeneration: a review of recent research, *Nutr. Rev.* 1999, 57(7), 201–214.

14. Bouis, H.E., Plant breeding: a new tool for fighting micronutrient malnutrition, *J. Nutr.* 2002, 132, 491S–494S.

15. Grundy, S.M., Multifactorial causation of obesity: implication for prevention, *Am. J. Clin. Nutr.* 1998, 67(Suppl.), 563S–572S.

16. Prentice, A. and Jebb, S., Aetiology of obesity: dietary factors, in *Obesity. The Report of the British Nutrition Foundation Taskforce*, Oxford, Blackwell Science, 1999.

17. McCrory, M.A., Fuss, P.J., McCallum, J.E., Yao, M., Vinken, A.G., Hays, N.P., Roberts, S.B., Dietary variety within food groups: association with energy intake and body fatness in men and women, *Am. J. Clin. Nutr.* 1999, 69(3), 440–407.

18. Wahlqvist, M.L., Functional foods in the control of obesity, in Goldberg, I., Ed., *Functional Foods: Designer Foods, Pharmafoods, Nutraceuticals*. Chapman & Hall, New York, 1994 pp. 71–86.

19. Kannar, D., Wattanapenpaiboon, N., Savige, G.S., Wahlqvist, M.L., Hypocholesterolemic effect of an enteric-coated garlic supplement, *J. Am. Coll. Nutr.* 2001, 20(3), 225–231.

20. Henry, C.J. and Emery, B., Effect of spiced food on metabolic rate, *Hum. Nutr. Clin. Nutr.* 1986, 40(2), 165–168.

21. Colquhoun, E.Q., Eldershaw, T.P., Bennett, K.L., Hall, J.L., Dora, K.A., Clark, M.G., Functional and metabolic evidence for two different vanilloid (VN1 and VN2) receptors in perfused rat hindlimb, *Life Sci.* 1995, 57(2), 91–102.

22. Yoshioka, M., St. Pierre, S., Drapeau, V., Dionne, I., Doucet, B., Suzuki, M., Tremblay, A., Effects of red pepper on appetite and energy intake, *Br. J. Nutr.* 1999, 82, 115–123.

23. Yoshioka, M., Doucet, E., Drapeau, V., Dionne, I., Tremblay, A., Combined effects of red pepper and caffeine consumption on 24 h energy balance in subjects given free access to foods, *Br. J. Nutr.* 2001, 85, 203–211.

24. Cannon, M.E., Cooke, C.T., McCarthy, J.S., Caffeine-induced cardiac arrhythmia: an unrecognised danger of healthfood products, *Med. J. Aust.* 2001, 174, 520–521.

25. Stewart, A.J., Wahlqvist, M.L., Stewart, B.J., Oliphant, R.C., Ireland, P.D., Fatty acid pattern outcomes of a nutritional programme for overweight and hyperlipidemic Australian men, *J. Am. Coll. Nutr.* 1990, 9(2), 107–113.

26. Parrish, C.C., Pathy, D.A., Parkes, J.G., Angel, A., Dietary fish oils modify adipocyte structure and function, *J. Cell. Physiol.* 1991, 148(3), 493–502.

27. Kant, A.K., Schatzkin, A., Ziegler, R.G., Dietary diversity and subsequent cause-specific mortality in the NHANES I epidemiologic follow-up study, *J. Am. Coll. Nutr.* 1995, 14, 233–238.

28. Kushi, L., Lew, R.A., Stare, F.J., Ellison, C.R., Lozy, M.E., Bourke, G., Daly, L., Graham, I., Hickey, N., Mulcahy, R., Kevaney, J., Diet and 20-year mortality from coronary heart disease. The Ireland-Boston Diet Heart Study, *N. Engl. J. Med.* 1985, 312, 811–818.

29. World Cancer Research Fund/American Institute for Cancer Research. Food, nutrition and the prevention of cancer: a global perspective. American Institute for Cancer Research, Washington, D.C., 1997.

30. Trichopoulou, A., Kouris-Blazos, A., Wahlqvist, M.L., Gnardellis, C., Lagiou, P., Polychronopoulos, E., Vassilakou, T., Lipworth, L., Trichopoulos, D., Diet and overall survival of the elderly, *BMJ* 1995, 311:1457-1460.

31. Kouris-Blazos, A., Setter, T.L., Wahlqvist, M.L. Nutrition and health informatics, *Nutr. Res.,* 2001; 21, 269–278.

32. Wahlqvist, M.L., Nutritional deficiencies underpin some clinical disorders, *Current Therapeutics* 1997, 38(12), 34–35.

33. MacLennan, R., Macrae, F., Bain, C., Battistutta, D., Chapuis, P., Gratten, H., Lambert, J., Newland, R.C., Ngu, M., Russell, A., Ward, M., Wahlqvist, M.L., The Australian Polyp Prevention Project. Randomized trial of intake of fat, fiber, and beta carotene to prevent colorectal adenomas, *J. Natl. Cancer Inst.* 1995, 87(23), 1760–1766.

34. Omenn, G.S., Goodman, G.E., Thornquist, M.D., Balmes, J., Cullen, M.R., Glass, A., Keogh, J.P., Meyskens, F.L. Jr., Valanis, B., Williams, J.H. Jr., Barnhart, S., Cherniack, M.G., Brodkin, C.A., Hammar, S., Risk factors for lung cancer and for intervention effects in CARET, the Beta-Carotene and Retinol Efficacy Trial, *J. Natl. Cancer Inst.* 1996, 88(21), 1550–1559.

35. Stevinson, C., Pittler, M.H., Ernst, E., Garlic for treating hyper-cholesterolemia. A meta-analysis of randomized clinical trials, *Annals Intern. Med.* 2000, 133(6), 420–429.

36. Jacobs, D.R., Meyer, K.A., Kushi, L.H., Folsom, A.R., Whole-grain intake may reduce the risk of ischemic heart disease death in postmenopausal women: the Iowa Women's Health Study, *Am. J. Clin. Nutr.* 1998; 68: 248-257.

37. Salmeron, J., Aserio, A., Rimm, E.B., Colditz, G.A., Spiegelman, D., Jenkins, D.J., Stampfer, M.J., Wing, A.L., Willett, W.C., Dietary fiber, glycemic load, and risk of NIDDM in men, *Diabetes Care* 1997, 20, 545–550.

38. Salmeron, J., Manson, J.E., Stampfer, M.J., Colditz, G.A., Wing, A.L., Willett, W.C., Dietary fiber, glycemic load, and risk of noninsulin-dependent diabetes mellitus in women, *JAMA* 1997, 277, 472–477.

39. Wilcox, G., Wahlqvist, M.L., Burger, H.G., Medley, G., Oestro-genic effects of plant-derived foods in postmenopausal women, *BMJ* 1990, 301, 905–906.

40. Wahlqvist, M.L. and Dalais, F., Phytoestrogens — the emerging multifaceted plant compounds. (Editorial) *Med. J. Aust.* 1997, 167 (3), 119–120.

41. Murkies, A., Dalais, F.S., Briganti, E.M., Burger, H.G., Healy, D.L., Wahlqvist, M.L., Davis, S.R., A study of phytoestrogen levels in postmenopausal Australian women diagnosed with breast can-cer compared with controls, *Menopause* 2000, 7, 289–296.

42. Worsley, A., Wahlqvist, M.L., Dalais, F.S., Savige, G.S., Charac-teristics of soy bread users and their beliefs about soy products, *Asia Pacific J. Clin. Nutr.* 2002, 11(1), 51–55.

43. Hodgson, J.M., Puddey, I.B., Beilin, L.J., Mori, T.A., Burke, V., Croft, K.D., Rogers, P.B., Effects of isoflavonoids on blood pres-sure in subjects with high-normal ambulatory blood pressure levels: a randomised controlled trial, *Am. J. Hypertension* 1999, 12(1 pt. 1), 47–53.

44. Howes, J.B., Sullivan, D., Lai, N., Nestel, P., Pomeroy, S., West, L., Eden, J.A., Howes, L.G., The effects of dietary supplemen-tation with isoflavones from red clover on the lipoprotein pro-files of post menopausal women with mild to moderate hypercholesterolaemia, *Atherosclerosis* 2000, 152, 143–147.

45. Kant, A.K., Schatzkin, A., Harris, T.B., Ziegler, R.G., Block, G., Dietary diversity and subsequent mortality in the First National Health and Nutrition Examination Survey Epidemiologic Follow-up Study, *Am. J. Clin. Nutr.* 1993, 57, 434–440.

46. Kouris-Blazos, A. and Wahlqvist, M.L., The traditional Greek food pattern and overall survival in elderly people, *Aust. J. Nutr. Dietetics* 1998, 55(4 Suppl.), S20–S23.

47. Walter, P., Effects of vegetarian diets on aging and longevity, *Nutr. Rev.* 1997, 55, S61–S68.

48. Wahlqvist, M.L., Asian migration to Australia: its food and health consequences, *Asia Pacific J. Clin. Nutr.* 2002, 11 (Suppl. 3), S562–S568.

49. Adams, P.B., Lawson, S., Sanigorski, A., Sinclair, A., Arachidonic acid to eicosapentaenoic acid ratio in blood correlates positively with clinical symptoms of depression, *Lipids* 1996, 31, S157–S161.

50. Hibbeln, J.R. and Salem, N. Jr., Dietary polyunsaturated fatty acids and depression: when cholesterol does not satisfy, *Am. J. Clin. Nutr.* 1995, 62, 1123–1129.

51. Peet, M., Edwards, R.W., Lipids, depression and physical disease, *Curr. Opin. Psychiatr.* 1997, 10, 477–480.

52. Hibbeln, J.R., Fish consumption and major depression, *Lancet* 1998, 351, 1213.

53. Hibbeln, J.R., Linnoila, M., Umhau, J.C., Rawlings, R., George, D.T., Salem, N. Jr., Essential fatty acids predict metabolites of serotonin and dopamine in cerebrospinal fluid among healthy control subjects, and early- and late-onset alcoholics, *Biol. Psychiatry* 1998, 44, 235–242.

54. Chandra, R.K., Effect of vitamin and trace-element supplementation on immune responses and infection in elderly subjects, *Lancet* 1992, 340, 1124–1127.

55. Chandra, R.K. and Kumari, S., Effects of nutrition on the immune system, *Nutrition* 1994, 10(3), 207–210.

56. Ayala, A. and Chaudry, I.H., Dietary n-3 polyunsaturated fatty acid modulation of immune cell function before or after trauma, *Nutrition* 1995, 11, 1–11.

57. DeMarco, D.M., Santoli, D., Zurier, R.B., Effects of fatty acids on proliferation and activation of human synovial compartment lymphocytes, *J. Leukocyte Biol.* 1994, 56, 612–615.

58. Mitsuoka, T., Intestinal flora and human health, *Asia Pacific J. Clin. Nutr.* 1996, 5 (1), 2–9.

59. Mahoney, D., Food component safety: risk–benefit analysis in developing countries, *Asia Pacific J. Clin. Nutr.* 2002, 11 (Suppl. 6), S212–S214.

60. Clugston, G., Lupien, J.R., Savige, G.S., Winarno, F.G., Wahlqvist, M., Okada, A., Eds., Novel foods in nutrition health and development: benefits, risks and communication, Metung, Australia, 11–14 November 2001, *Asia Pacific J. Clin. Nutr.* 2002, 11 (Suppl.), S97–S229.

2

Asian Herbal Products: The Basis for Development of High-Quality Dietary Supplements and New Medicines

KUO-HSIUNG LEE, HIDEJI ITOKAWA, and
MUTSUO KOZUKA

Univerity of North Carolina,
Chapel Hill, NC, USA

INTRODUCTION

Asian herbal products have been used for thousands of years to maintain health and treat disease. Now, these same ancient remedies should be reassessed in our modern era for development as high-quality dietary supplements and new medicines in the 21st century, particularly as medicinal herb use is increasing rapidly worldwide. Private health industries are marketing and popularizing herbal products, including in the U.S., comfrey (*Symphytum officinale*) (Boraginaceae), echinacea (*Echinacea purpurea*) (Asteraceae), feverfew (*Tanacetum parthenium*) (Asteraceae), garlic (*Allium sativum*) (Lilliaceae), ginkgo (*Ginkgo biloba*) (Ginkgoaceae), ginseng (*Panax ginseng*) (Araliaceae), saw palmetto (*Serenea repens*) (Arecaceae), and St. John's wort (*Hypericum perforatum*) (Hypericaceae). The increasing consumer use and demand have prompted international health organizations and governmental agencies to publish guidelines for herbal medicine use. Accordingly, the scientific community must apply modern technologies to assure the efficacy and safety of these traditional remedies and develop them as first-class dietary supplements and new medicines.

Asian herbal products and traditional Chinese medicine (TCM) are widely used in China, Japan,* Korea, Taiwan, and Southeast Asia. The former are used as dietary supplements, including both daily foods (cereals, vegetables, fruits) and "functional" foods or "Yao Shan," which are TCM-based dietary supplement dishes. Yao Shan is a Chinese eating culture, which combines TCM and food for replenishment and medical purposes. It is connected with immunopotentiation, improving systemic circulation, disease prevention, and aging control. Either a single herb or multiple, formulated herbs are used in herbal foods, teas, wines, congees, and pills (or powder). Two examples of herbal wines are Ginseng Wine (*Panax ginseng*), which is used for revitalization, immunoregulation, and stimulation, and Schisandra Wine (*Schisandra chinensis*, Wu Wei Tzu), which is taken for chronic cough and asthma.

Natural products, mainly (>80%) plants, are the basis of TCM. Approximately 5,000 plants species have been identified to have therapeutic value. Many of these medicinal plants are used as folk drugs (Min Chien Yao) and ca. 500 plant species are commonly prescribed by doctors of TCM as Chinese Materia Medica or traditional drugs (Chung Yao). In China, TCM holds a predominant position in medicine with ca. 2,500 hospitals, 360,000 medical doctors, 30 advanced schools, and 60 research institutes established specifically for TCM practice. In addition, TCM makes up a large proportion of the over $15 billion annual worldwide sales of natural medicines, as

---

* Kampo is the word used to describe traditional medicine in Japan, and Chinese prescriptions are the main source.

consumers look for ways to improve their quality of life and for alternative disease treatments to Western medicine.

Western and Chinese medicine vary in practice, theory, and thus, drugs of choice. Western medicine uses pure natural or synthetic compounds aimed at a single target, while Chinese medicine uses processed crude multicomponent natural products, in various combinations and formulations aimed at multiple targets, to treat a totality of different symptoms. In ancient Chinese literature, 110 (Chin Kuei Yao Lueh, Summaries of Household Remedies) to over 100,000 (I Fang Chi Chieh) different formulas were recorded. Even today, commonly used contemporary formulas number ca. 365, 1,200, and 200 in Taiwan, China, and Japan, respectively.

The main principle of TCM treatment is to establish a holistic balance of such forces in the body and, thus, promote health. Specific TCM treatments are chosen based on careful diagnostic observations and systematic principles, which originate from the Yin-Yang theory. TCM doctors believe that disease is caused by an imbalance of Yin and Yang. The main purpose of drugs, then, should be to restore equilibrium of Yin and Yang to the body. Accordingly, if a patient suffers from Yang-fever, a Yin-cool drug is prescribed, but conversely, if the patient has a Yin-cool problem, a Yang-warm drug is advised.[1]

Chinese herbal pharmacology studies the composition, actions, indications, dosages, and clinical uses of herbal formulations. The component herbs fall into four categories, according to their importance and role in the formulations.[2] The supporting herbs aid the effectiveness of the principle herb.

- Imperial Herb — the chief herb (main ingredient) of a formula; it is tonic and nontoxic
- Ministerial Herb — ancillary to the imperial herb, it augments and promotes the action of the chief herb
- Assistant Herb — reduces the side effects of the imperial herb
- Servant Herb — harmonizes or coordinates the actions of the other herbs

Any change in composition (for example, two different four-herb formulations with three identical but one different ingredient and with different proportions and different imperial/ministerial herbs) induces different pharmacological actions.[3] For example, Mahuang combination and Mahuang apricot seed combination both contain Mahuang, apricot seed, and licorice (in different proportions); however, the former formulation contains Cassia (ministerial herb), which assists Mahuang (imperial herb) in liberating heat while the latter formulation contains gypsum (assistant/Imperial herb), which is antagonistic with Mahuang (imperial/assistant). Overall, Mahuang combination suppresses coughing and induces sweating, while Mahuang apricot seed combination suppresses coughing and suppresses sweating. TCM principles are based on therapeutic effects and the body's reactions to herbal products. Effective and safe herbal formulations are developed through diligent attention to these principles.

Over China's long history, 33 kinds (2,088 volumes) of Chinese Materia Medica books and 572 volumes of Food Recipe books have been published. Three classics are *Shen Nung Pen Ts'ao Ching* (The Book of Herbs by Shen Nung), which is the earliest known description of Chinese herbal folk medicine, *Huang Ti Nei Ching* (The Yellow Emperor's Classic on Internal Medicine), and *Shang Han Tsa Ping Lun* (Treatise on Febrile and Miscellaneous Diseases). The first governmental Chinese Pharmacopoeia was *Sin Siu Pen Taso* in 659 BC. *Pen Tsao Kan Mu* (A General Catalog of Herbs) was published in 1590 and has been translated into many languages, including Korean, Japanese, Latin, English, French, and German. Currently, the 34-volume Chung Hua Pen Tsao (Chinese Materia Medica) published in 1999 is the most comprehensive and detailed publication on TCM to date.

A prevailing concept in TCM is that "Therapy by Food is Better than Therapy by Medicine" — maintaining balance and health through daily diet and functional ("Yao Shan") foods to achieve the balance of Yin and Yang. Accordingly, in the classic work, Shen Nung Pen Tsao Chung, Shen Nung recorded 365 herbs classified as Upper Class, Middle Class, and Lower Class. These three classifications were based on herbal toxicities and encompass the four categories mentioned above, in general. In the first of these classes, the Upper Class

TABLE 2.1 Examples of Upper, Middle, and Lower Herbal Species.

Upper Class Herbs	Middle Class Herbs	Lower Class Herbs
Ling Chih (*Ganoderma lucidum*)	Tang Kuei (*Angelica sinensis*)	Kan Sui (*Euphorbia kansui*)
Ginseng (*Panax ginseng*)	Ko Ken (*Pueraria lobata*)	Hsia Ku Tsao (*Prunella vulgaris*)
Tan Shen (*Salvia miltiorrhiza*)	Kan Chiang (*Zingiber officinale*)	Kuei Chiu (*Podophyllum emodi*)
Tu Chung (*Eucommia ulmoides*)	Huang Ching (*Scutellaria baicalensis*)	Da Huang (*Rheum palmatum*)
Wu Wei Tzu (*Schisandra chinensis*)	Ma Huang (*Ephedra sinica*)	Fu Tzu (prepared *Aconitum carmichaelii*)
Huang Chi (*Astragalus membranaceus*)		Ban Tsia (*Pinellia ternata*)
Kou Chi (*Lycium barbarum*)		
Ta Tsao (*Zizyphus jujuba*)		
Gan Tsao (*Glycyrrhiza uralensis* or *glabra*)		

herbs are nontoxic and rejuvenating. They can be taken continuously for a long period and form the main components of the dietary supplement dishes (Yao Shan). The Middle Class herbs promote mental stability and have nontoxic or toxic effects and, thus, may need more precaution regarding potential toxicity. The Lower Class herbs cannot be taken for extended periods because of toxic effects and are used in TCM formulations for treating various diseases only after proper processing to reduce their toxicity. Examples of all three classes are found in Table 2.1.

ASIAN HERBAL PRODUCTS

The following section will supply details concerning specific Asian herbal products, including folkloric use, chemical composition, and currently identified biological activities. The herbs and information discussed are not intended to be comprehensive, but will supply an introduction to the promising resource of Asian herbal products.

Upper Class Herbs

Medicinal Mushrooms

The medicinal properties of mushrooms have made a major contribution to human life. For thousands of years, many kinds of mushrooms have been recognized for their healing effects together with few side effects and used as folk medicine throughout the world. Mushrooms contain several kinds of bioactive compounds from small molecules to polysaccharides. Five major medicinal mushrooms are discussed below, including "Ling Chih," which was classified as an Upper Class herb in the Chinese classic *Shen Nung Pen Ts'ao Ching*.

*Ling Chih (*Ganoderma lucidum *Karst, Polyporaceae)*

Ling Chih, Reishi mushroom, or "Spirit Plant" is the common Chinese fungus *G. lucidum*. It is mentioned in Chinese medical classics, including *Shen Nung Pen Ts'ao Ching* (the Book of Herbs by Shen Nung) and *Pen Tsao Kang Mu*. The latter work contains the following description, "Continued use of Ling Chih will lighten weight and increase longevity." The fruit body and mycelia of *G. lucidum* are now cultured, making Ling Chih the leading cultivated medicinal mushroom. It is used as a tonic and sedative and to treat hyperlipidemia, angina pectoris, chronic bronchitis, hepatitis, leukopenia, and autoimmune disease.[4–6]

Chemical composition and pharmacological properties. Like many other mushrooms, Ling Chih spores contain polysaccharides, including $(1\rightarrow3)$-β-, $(1\rightarrow4)$-β-, $(1\rightarrow6)$-β-D-glycans and linear $(1\rightarrow3)$-α-D-glucans.[7] These compounds are immunostimulating in both animals and humans. Ling Chih also has antimicrobial, antiviral, hypoglycemic, antitumor, free radical scavenging, and antioxidative (i.e., antiaging) activities.[7]

Other bioactive compounds found in the spores of *G. lucidum* are triterpenoids, both ganoderic acids (**1–4**, Figure 2.1) and ganolucidic acid A. These compounds have anti-tumor and anti-HIV-protease activities, as well as analgesic effects on the CNS.[8] Oxygenated triterpenes (**5–7**, Figure 2.2) in *G. lucidum* also exhibit hypolipidemic activity by blocking cholesterol absorption and inhibiting HMG-CoA reductase.[6]

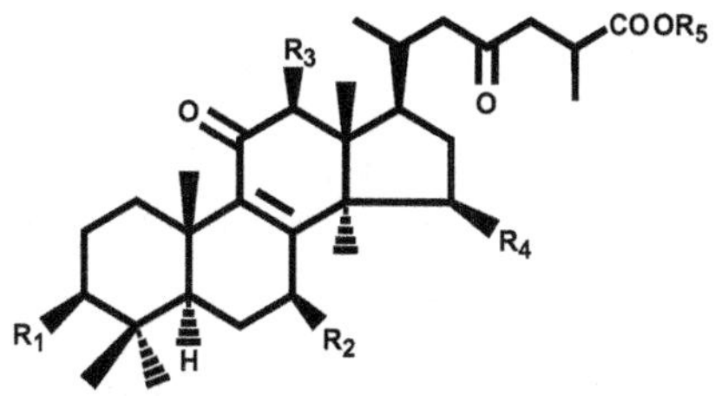

Ganoderic Acid A (1): R_1 = O=, R_2 = β-OH, R_3 = R_5 =H, R_4 = α-OH
Ganoderic Acid B (2): R_1 = R_2 = β-OH, R_3 = R_5 = H, R_4 = O=
Ganoderic Acid G (3): R_1 = R_2 = R_3 = β-OH, R_4 = O=, R_5= CH_3
Ganoderic Acid H (4): R_1 = β-OH, R_2 = R_4 = O=, R_3 = β-OAc, R_5 = CH_3

Figure 2.1

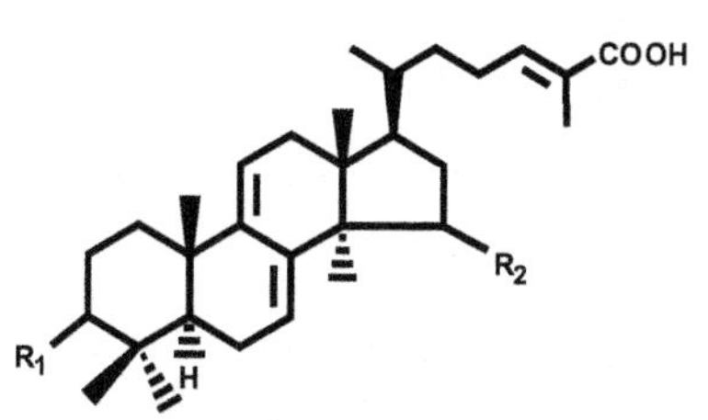

Oxygenated Triterpenes in *Ganoderma lucidum*
5: R_1 = R_2 = α-OAc
6: R_1 = α-OAc, R_2 = α-OH
7: R_1 = β-OH, R_2 = α-OH

Figure 2.2

Yun-Chih (Coriolus versicolor Quél, Polyporaceae)

The mushroom Yun-Chih is known as "Turkey Tail" in the U.S. It has been used in traditional East Asian folk medicine to modern medicine as an immunostimulating, antitumor, and hypoglycemic agent.[9] It contains a protein-bound polysaccharide (PSK, trade name: Krestin). PSK enhances the immune system and consequently has a tumor-retarding effect.[10] It is used as an immune enhancement agent in modern Japanese therapy.

Hsian Ku (China) or Shiitake (Japan) [Lentinus edodes (Berk.) Sing, Tricholomataceae]

This edible fungus is native to Asian forests and is the second most commonly cultivated edible mushroom also used for medicinal purposes.[11] It contains free radical scavenging and immunostimulating polysaccharides, including linear $(1\rightarrow3)$-β-D-glucans.[12] Lentinan, a polysaccharide (β-D-glucan) fraction from *L. edodes* has been studied more extensively than other similar substances. It exhibits notable antitumor activity and is significantly more potent than polysaccharides from many other fungi or from higher plants. Lentinan appears to be active in certain animals against various tumor types.[13] Recently, a chemopreventive effect for polysaccharides from this mushroom was reported.[14]

Shen Ku (Agaricus blazei Murill, Agaricaceae)

Other names for this basidiomycetous fungus are "mushroom of God," Hime Matsutake (Japan), and Pa Hsi Mo Ku (Brazil). It originates in the Pier Date mountain area of Sao Paolo suburbs, Brazil, and has attracted attention because the inhabitants of that area suffer very little from cancer and other age-related diseases. This rare mushroom is now cultivated in Brazil, Japan, China, and the U.S., and is widely popular in Japan, Korea, and China as a dietary supplement.

In the 1980s, numerous research groups published reports on the anticancer activity of *A. blazei*. However, not only antitumor effects, but also antiviral, tumor chemopreventive,[15] immunostimulating, blood sugar lowering, and cholesterol reducing activities are known for this mushroom.[13] It contains immunostimulating polysaccharides, including (1→6)-β- and (1→3)-β-D-glycans,[16] in addition to an antitumor glycoprotein complex, cytotoxic ergosterol derivatives, antimutagenic compounds, and bactericidal substances.[17]

Ginsengs

*Asian Ginseng (*Panax ginseng, *Araliaceae)*

Ginseng is the root of *Panax ginseng* found in China and Korea. The genus name Panax is from the Greek pan (all) and akos (remedy). This panacea (panakeia) was believed to be a universal remedy. Correspondingly, in Oriental medicine, ginseng has been known since ancient times for being tonic, regenerating, and rejuvenating. Wild ginseng is scarce and has been replaced by cultivated ginseng or "true" ginseng. Other species include:

> American ginseng (*P. quinquefolium*); cultivated in North America.
> Japanese ginseng (*P. japonicus*); widely distributed in Japan.
> San-chi ginseng (*P. notoginseng*); reputed as a tonic and hemostatic in China.

Chemical composition[18-20] ***and pharmacological properties.*** Many compounds have been isolated from the roots of ginseng. These compounds include polysaccharides, glycopeptides (panaxanes), vitamins, sterols, amino acids and peptides, essential oil, and polyalkynes (panaxynol, panaxytriol). Approximately 30 saponins have been isolated from the root, including oligoglycosides of tetracyclic dammarane aglycones, more specifically a 3β,12β,20(S)-trihydroxylated type (protopanaxadiol) and a 3β,6α,12β,20(S)-tetrahydroxylated type (protopanaxatriol). The saponins (**ginsenosides, 8–19**) differ in the mono-, di-, or trisaccharide nature of the two sugars attached at the C-3 and C-20 or the C-6 and C-20 hydroxyl groups (Table 2.2). In exceptional cases, all three hydroxyl groups at C-3, C-6, and C-12 of protopanaxatriol can form glycosidic bonds (e.g., ginsenoside 20-gluco-Rf). Malonyl-ginsenosides have also been characterized in white ginseng. Traditionally, ginseng is used to restore normal pulse, remedy collapse, benefit the spleen and liver, promote production of body fluid, calm nerves, and treat diabetes and cancer. A recent report has discussed ginseng's effects on quality of life.[21] Ginseng and its congeners have also been studied for cancer prevention effects, including:

- Anticarcinogenic effects against chemical carcinogens: Ginsenosides Rg3 and Rg5 significantly reduced lung tumor incidence, and Rg3, Rg5 and Rh2 (red ginseng) showed anticarcinogenic activity.[22] Ginsenoside Rg_3 inhibited cancer cell invasion and metastasis, Rb_2 inhibited tumor angiogenesis, and Rh_2 inhibited human ovarian cancer growth in nude mice.[19] After oral administration, ginsenosides Rb_1 and Rb_2 are metabolized by intestinal bacteria to compound K, also known as M1, which induces apoptosis of tumor cells. Compound K was shown to affect nucleosomal distribution.[23]
- Expression of cyclooxygenase-2 (COX-2) in TPA-stimulated mouse skin was markedly suppressed by ginsenoside Rg_3 pretreatment.[24]
- A case-control study on the relationship between cancer and ginseng intake was reported.[25]

American Ginseng (Panax quinquefolium, Araliaceae)

P. quinquefolium (American ginseng) and *P. ginseng* (Asian ginseng) contain many identical components.[26] Thus, American ginseng could be used for the same medical conditions as Asian ginseng. However, according to TCM theory, American ginseng is somewhat cool and, accordingly, is mainly used to reduce internal heat and promote secretion of body fluids.

TABLE 2.2 Structures of Ginsenosides

Structure	Ginsenoside	R_1	R_2
Protopanaxadiol, $R_1 = R_2 = H$	Rb1 (**8**)	Glc(2-1)Glc	Glc(6-1)Glc
	Rb2 (**9**)	Glc(2-1)Glc	Glc(6-1)Ara(p)
	Rc (**10**)	Glc(2-1)Glc	Glc6Ara(f)
	Rd (**11**)	Glc(2-1)Glc	Glc
	Rg3 (**12**)	Glc(2-1)Glc	H
	Rh2 (**13**)	Glc	H
Protopanaxatriol $R_1 = R_2 = H$	Re (**14**)	Glc(2-1)Rha	Glc
	Rf (**15**)	Glc(2-1)Glc	H
	Rg1 (**16**)	Glc	Glc
	Rg2 (**17**)	Glc(2-1)Rha	H
	Rh1 (**18**)	Glc	H
	Rg5 (**19**)	Glc(2-1)Glc	—

Biological[27] and chemical[28–31] differences between American ginseng and Asian ginseng (**Figure 2.3 and Figure 2.4**). It is generally accepted that American ginseng stimulates proliferation of human lymphocytes. Meanwhile, Asian ginseng does not significantly alter the proliferative response *in vitro*, and Siberian ginseng enhances proliferation

Asian and American ginsengs differ in the presence or absence of ginsenoside Rf(**15**) and 24(R)-pseudoginsenoside F_{11}(**20**). Ginsenoside Rf is found in Asian ginseng, but not in American ginseng, while 24(R)-pseudoginsenoside F_{11} is abundant in American ginseng, but found only in trace amounts in Asian ginseng.

Sanchi ginseng (Panax notoginseng, Araliaceae)

This ginseng exerts a major effect on the cardiovascular system. It dilates the coronary vessels and reduces vascular resistance, resulting in increased coronary flow and decreased blood pressure. In TCM, this ginseng is used to arrest bleeding, remove blood stasis, and relieve pain. Recent studies have shown that a preparation (Sanqi Gaunxin Ning) from this herb can produce a 95.5% improvement in symptoms of angina pectoris and 83% improvement in the electrocardiogram (ECG) pattern. The herb can usually stop bleeding in cases of hemoptysis and hematemesis. *P. notoginseng* contains similar saponins as *P. ginseng*. In addition, three new ginsenoside type saponins were recently isolated, together with 11 known saponins[32] and two small-molecular weight compounds, dencichine [$H_2NCH(COOH)CH_2NHCOCOOH$](**21**) (which arrests bleeding)[33,34] and a pyrazine derivative.[35]

Siberian ginseng (Eleutherococcus senticosus, Araliaceae)

Although Siberian ginseng is not a true ginseng such as *Panax ginseng* or *P. quinquefolia*, it also belongs to the Araliaceous plant family and has its own bioactive ingredients with unique and proven medicinal values. Siberian ginseng is harvested naturally from Russia and northeast China, where it

Ginsenoside Rf(15)
(in Oriental Ginseng)

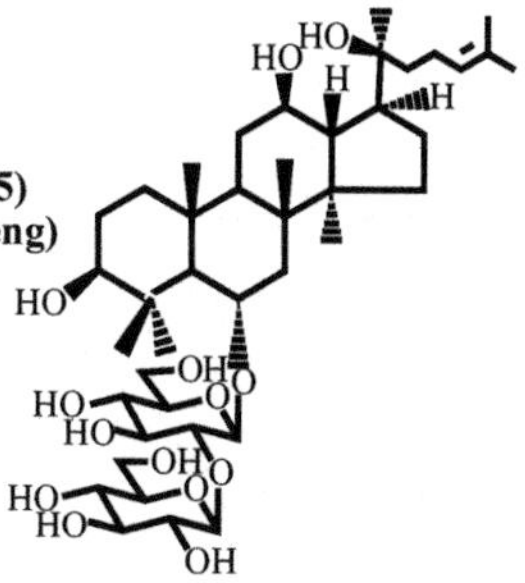

Figure 2.3

24(R)-PseudoginsenosideF$_{11}$(20)
(in American Ginseng)

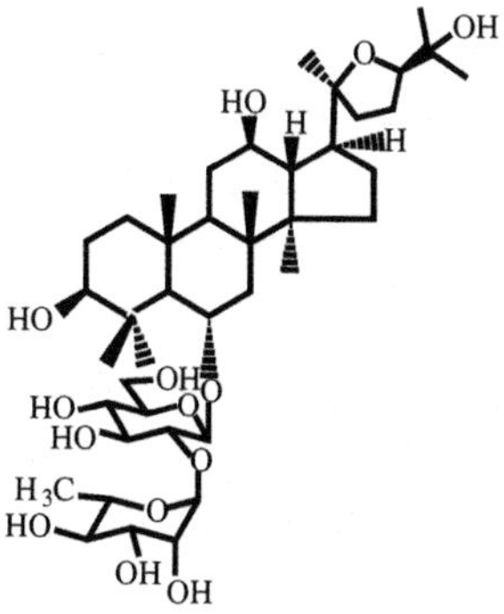

Figure 2.4

has been used for over 2,000 years. The root contains polysaccharides, phenolics (coumarins, lignans, phenylpropionic acids),
and eleuterosides. Of the latter compounds, some are triterpenoid in nature (eleuterosides I–M) and others belong in miscellaneous series, including isofraxoside (eleutheroside B$_1$),
syringaresinol glycosides (eleutherosides D–E), sinapyl alcohol
glycosides, and the methyl ester of galactose (eleutheroside C).

Two major glycosides, eleutheroside B (syringin) and
eleutheroside E (**22**, syringaresinol 4′,4″-di-O-β-D-glycopyra-
noside) are usually used as marker compounds. Recently, a
new lignan glycoside, eleutheroside E$_2$ (**23**) was isolated
(Figure 2.5).[36]

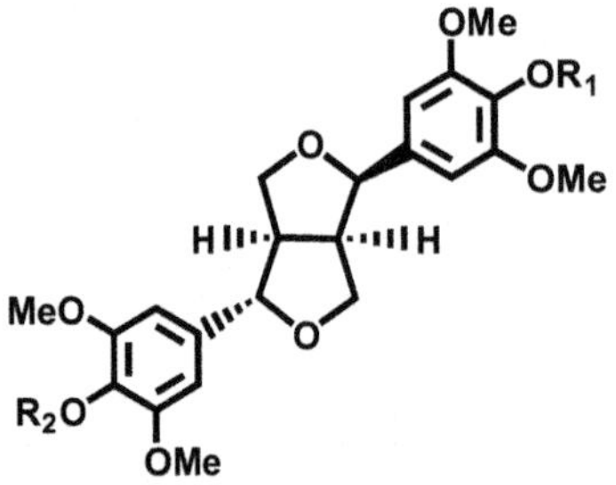

Eleutheroside   E (22): $R_1 = R_2 = $ -O-β-D-Glc
Eleutheroside E_2 (23): $R_1 = H$, $R_2 = $ -O-β-D-Glc

Figure 2.5

Siberian ginseng possesses significant adaptogenic action and is recommended as a general tonic. Because of its nonspecific mechanism of action, Siberian ginseng has a broad range of clinical applications. Recently, it was observed to reduce cardiovascular stress.[37]

Other Herbs

Tan Shen or Sage (Salvia miltiorrhiza, Labiatae)

The rhizome and roots of *S. miltiorrhiza* have been widely used in TCM to treat various cardiovascular diseases. This herb exhibits hypotensive and positive inotropic effects, causes coronary artery vasodilation, and inhibits platelet aggregation. In Europe and America, *S. officinalis* is available.

Chemical composition and pharmacological properties. A high-speed counter-current chromatography (HSCCC) method was developed for preparative separation and purification of six diterpenoids, (dihydrotanshinone I, cryptotanshinone, methylenetanshiquinone, tanshinone I, tanshinone IIA, and danshenxinkun B) from *S. miltiorrhiza*.[38] Sodium tanshinone II-A sulfonate (**24**, a water soluble sulfonate, Figure 2.6) is used to treat angina pectoris and myocardial infarction. It exhibits a strong membrane-stabilizing effect on red blood corpuscles and may also act similarly to the clinically used verapamil. Accordingly, an intravenously applicable *S. miltiorrhiza/Dalbergia*

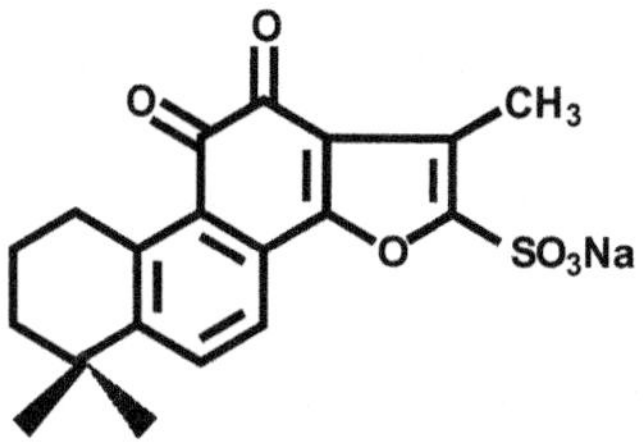

Sodium Tanshinone II-A
Sulfonate (24)

Figure 2.6

odorifera TCM mixture may have potential as an antianginal drug.[39] Chronotropic, inotropic, and coronary vasodilator actions of the available ampoule preparation were examined using canine-isolated, blood-perfused heart preparations. *S. militiorrhiza* also exerts clear cytotoxic effects and strongly inhibits the proliferation of HepG(2) cells.[40]

Tu Chung (Eucommia ulmoides, Eucommiaceae)

The dried bark has been used to supplement the liver and kidney, strengthen muscles and bones, and stabilize the fetus, while the leaves can be used as a tea (Tu Chung tea). The plant is said to be a longevity and antistress herb, and exhibits antihypertensive effects in animal and clinical experiments.[41]

Chemical composition and pharmacological properties. The active compounds present belong to the families of (1) iridoids, including geniposidic acid, geniposide, asperulosidic acid, deacetyl asperulosidic acid, and asperuloside; (2) phenols, including pyrogallol, protocatechuic acid, and *p*-trans-coumaric acid; and (3) triterpenes and lignans.

Their pharmacological activities are mainly due to lignans and iridoid glycosides.[41] The main component of Tu Chung, geniposidic acid (**25**, Figure 2.7), stimulates the parasympathetic nervous system through the muscarinic ACh receptor agonist.[42] The administration of geniposidic acid or

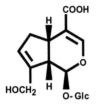

Geniposidic Acid (25)

Figure 2.7

aucubin stimulates collagen synthesis in aged model rats.[43] *E. ulmoides* extract could possibly act as a prophylactic agent to prevent free radical-related diseases.[44]

> *Wu Wei Tzu* (Schisandra chinensis, Schisandraceae)

Wu Wei Tzu is the dried fruit of *S. chinensis* (northern China) or *S. spenanthera* (southern China). It astringes the lungs, nourishes the kidneys, promotes secretion of fluids, reduces hyperhidrosis, and controls seminal emission and diarrhea. Traditionally, it is used for dyspnea and cough, dry mouth and thirst, spontaneous diaphoresis, night sweats, insomnia, and amnesia.

Chemical composition and pharmacological properties. The plant contains essential oils, lignans, and citric, malic, and tartaric acids. Lignans, including dibenzocyclooctadienes such as schisandrin, deoxyschisandrin, pregomisin, and gomisins A–D, F–H, and J, show various biological activities. Schisanhenol (**26**, Figure 2.8) completely inhibited peroxidative damage of brain mitochondria and rat membrane,[45] and schisandrin B (**27**) protected against hepatic oxidative damage in mice.[46] Gomisin A (**28**) inhibits tumor promotion, probably due to its anti-inflammatory activity.[47] Gomisin G (**29**) shows potent anti-HIV activity against HIV replication in H9 lymphocytes with an EC_{50} of 0.006 µg/mL and a therapeutic index of 300.[48]

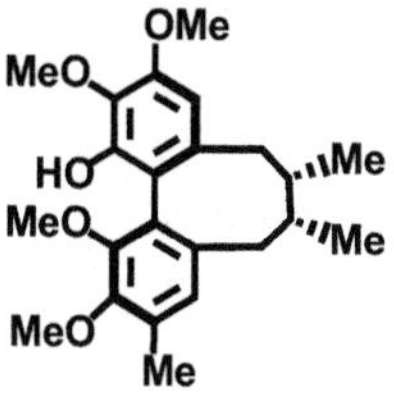

Schisanhenol (26) 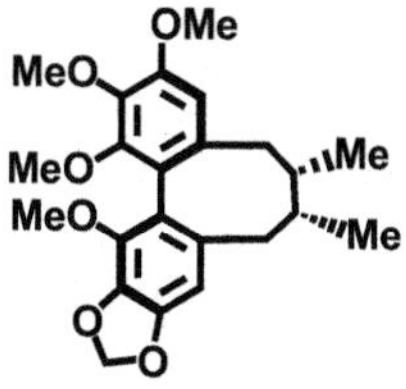   Schisandrin B (27)

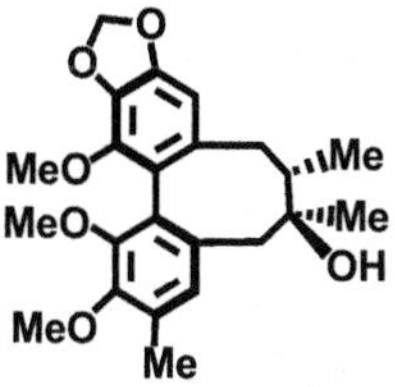

Gomisin A (28) 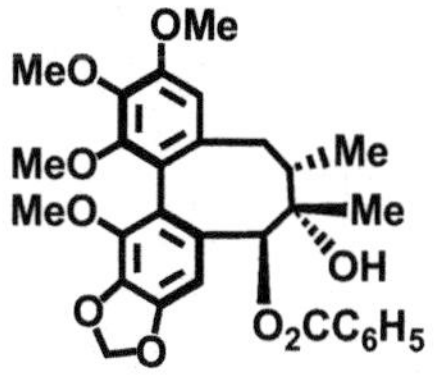   Gomisin G (29)

Figure 2.8

Astragalus or Huang-qi (Astragalus membranaceus, Leguminosae)

The name Astragalus serves both as a botanical generic name and an English common name for the Chinese medicinal plant known as "Huang-qi." The Chinese consider cultivated roots to be of superior quality to wild-harvested roots. Generally, astragalus is considered to have immunostimulant, antioxidant, antiviral, and antitumor activities. Huang-qi is used in TCM prescriptions for ch'i (energy) deficiency and general weakness and specifically for shortness of breath and palpitation, collapse, spontaneous perspiration, night sweats, edema due to physical deficiency, chronic nephritis, pulmonary diseases, lingering diarrhea, rectal and uterine prolapse, nonfestering boils, and hard-to-heal sores and wounds.

Chemical composition and pharmacological properties. Astragalus contains flavonoids, polysaccharides, and triterpenoids. Various biological activities have been attributed to various triterpenoid saponin, called astragalosides (**30–35**), including acetylastragaloside I, astragaloside I, astragaloside

III, and astragaloside IV.[49] In particular, astragaloside IV increased T,B lymphocyte proliferation and antibody production *in vivo* and *in vitro*, and inhibited production of IL-1 and TNF-α from peritoneal macrophages *in vitro*.[50] In addition, astragalosides I, VII, and VIII have shown antiviral suppression.[51]

Structures of astragalosides.

Compound	R_1	R_2	$R_{2'}$	$R_{3'}$	$R_{4'}$
Acetylastragaloside I (**30**)	β-D-Glu	H	Ac	Ac	Ac
Astragaloside I (**31**)	β-D-Glu	H	Ac	Ac	H
Astragaloside III (**32**)	H	H	β-D-Glu	H	H
Astragaloside IV (**33**)	β-D-Glu	H	H	H	H
Astragaloside VI (**34**)	β-D-Glu	H	β-D-Glu	H	H
Astragaloside VII (**35**)	β-D-Glu	Glc	H	H	H

*Kou Chi Tzu (*Lycium barbarum, *Solanaceae)*

Traditionally, *L. barbarum* is used to supplement the liver and kidney meridians in deficiency of liver and kidney yin and to treat debility of loins and knees, vertigo, excessive tearing, cough due to consumption, and diabetes. Kou chi is the dried fruit, and Ti-ku-pi is the root of this plant. The dried red berries (Barbary wolfberry fruits) are similar to raisins.

   Chemical composition and pharmacological properties. Polysaccharides and arabinogalactan-proteins are among the active components. The glycan backbone consists of $(1{\rightarrow}6)$-β-galactosyl residues, about 50% of which are substituted at

C-3 by galactosyl groups. The major nonreducing end is formed from arabinofuranosyl substituents.[52,53] The glycan (LbGp4-OL) and, to a lesser extent, its glycoconjugate (LbGp4), enhanced splenocyte proliferation in normal mice, most likely by targeting B-lymphocyte cells. The immunostimulatory effect was associated with activated expression of nuclear factor kappa B (NF-kappa B) and activator protein 1 (AP-1).[54] Other active components include betaine, vitamins, and zeaxanthin.

Tung Chung Hsia Tsao

Tung Chung Hsia Tsao (Winter Worm, Summer Grass) originates from the larvae of the caterpillar *Hepialus armoricanus* or Sphinx moth, together with the parasitic fungus *Cordyceps sinensis* (Berk.) Sacc. (Hypocreaceae). In winter, the fungus attacks the caterpillar during its underground hibernation and slowly eats it away. By the end of the winter, the fungus has killed the infected host and continues to grow throughout the dead body. In the summer, a rodlike fungal stroma grows out from mummified shell of the dead host. This stroma looks like a plant among the leaves and is then harvested as Tung Chung Hsia Tsao. It is found naturally only in the highlands of the Himalayan region, Sichuan, Qinghai, Tibet, and Yunnan, but is now cultured as *Codyceps mycellia*. Tung Chung Hsia Tsao has been traditionally used to treat chronic cough, asthma and impotence, promote longevity, relieve exhaustion, and increase athletic prowess.

Chemical composition and pharmacological properties. This product contains immunopotentiating polysaccharides, such as galactomannan, and antitumor polysaccharides, sterols, and adenosine derivatives.[55,56]

Coix Seeds (Coix lachryma-jobi *var. ma-yuen Stapf*)

The dried ripe kernels of *Coix lachryma-jobi* constitute the traditional Chinese medicine, coix seed, which is used to ease arthritis, control diarrhea, and eliminate edema by invigorating spleen function and promoting diuresis.[57] Two Chinese formulations containing coix seed are Szu Shen Tan (Dioscorea Combination or Four Wonders Soup) and Lo Shih Shu. In addition to coix seeds, the former formulation contains *Dioscorea opposita*, *Nelumbo nucifera* (lotus seed), *Poria cocos* (Hoelen) and *Euryale ferax*, and the latter contains *Wisteria floribunda*, *Trapa bispinosa*, *Terminalia chebula*, and Coix; hence it is also known as WTTC. The former prescription is well known in Taiwan for treating indigestion, especially in asthenic children. The latter is formulated as a water soluble ointment and used to treat gastric and rectal cancers, particularly to inhibit cancer cell growth and metastasis after cancer surgery. Kang Lai Te is a new anticancer drug from the active principles of coix seed. It is used as an i.v. injection and is effective in lung, liver, and bone cancer, particularly in reducing toxic side effects of chemotherapy. Coix seed can also effectively treat verrucas caused by the human papilloma virus and other tumorous diseases.

Chemical composition and pharmacological properties. Various components of coix seed show pharmacologically different activities (Figure 2.9). For example, an acidic fraction composed of four free acids, palmitic, stearic, oleic, and linoleic acids, shows antitumor activity;[58] α-monolinolein inhibits tumorigenesis[59]; three glycans, namely, coixans A, B, and C, show hypoglycemic activity[60]; and a benzoxazinone (**36**) and a benzoxazolinone (**37**) exhibit anti-inflammatory activity.[61] Isolation of the anti-HIV ellagitannins chebulinic (**38**) and chebulagic acids (**39**) from *Terminalia chebula* has also been reported.[62]

Ziziphus (Ziziphus jujuba, Rhamnaceae)

Dried ripe fruits of *Ziziphus* are used as supplements to tonify the spleen and stomach, nourish and pacify the spirit, smooth herbal action, and harmonize all drugs. The roots have been used for their hypotensive effects, and the leaves to decrease the intake of sweets (as a taste-modifier and antiobesity agent).[63]

Chemical composition and pharmacological properties (Figure 2.10). Among eight flavonoids isolated from the seeds, swertisin (**40**) and spinosin (**41**) possessed significant sedative

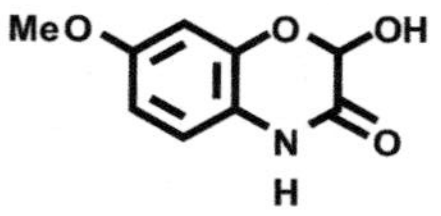

Chebulinic Acid (38): R = ‒C:O

Chebulagic Acid (39): R + R = HO

Benzoxazinone (36)

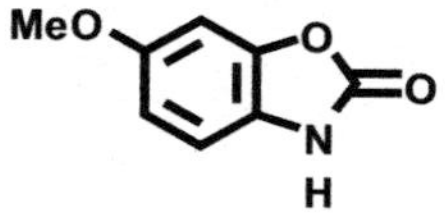

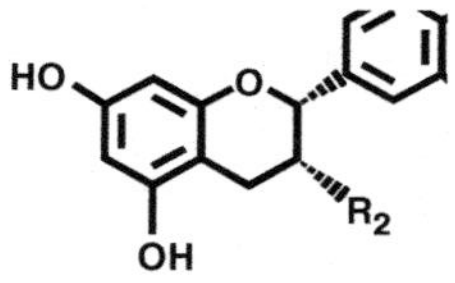

(-)-Epicatechin(EC) (61): R₁ = H, R₂ = O
(-)-Epigallocatechin (62) (EGC): R₁ = OH
(-)-Epicatechin-3-gallate (ECG) (63): R₁

(-)-Epigallocatechin-3-gallate (EGCG) (6

Benzoxazolinone (Coixol) (37)

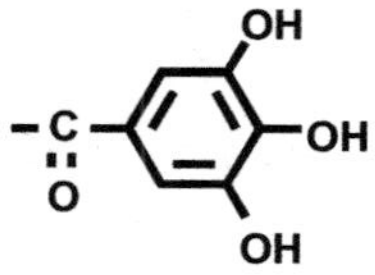

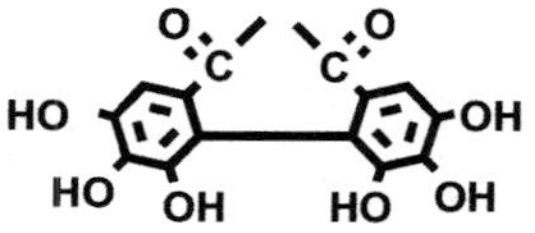

Figure 2.9

Swertisin (40): R = H
Spinosin (41): R = β-D-glu

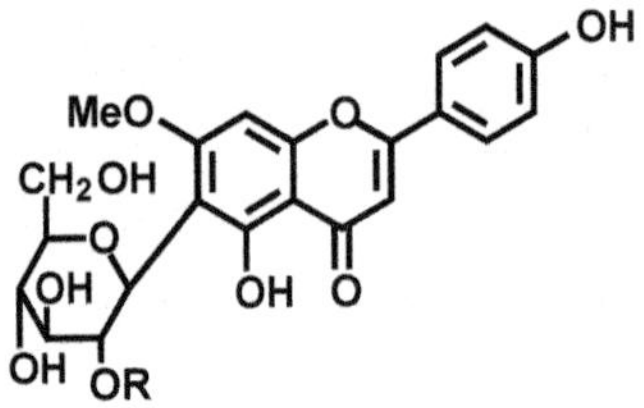

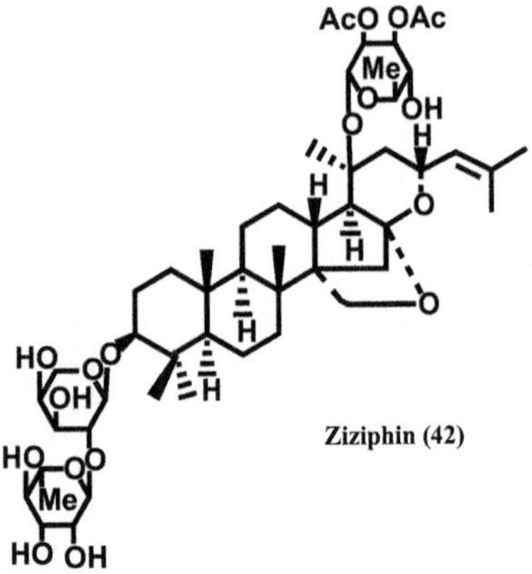

Ziziphin (42)

Figure 2.10

activity.[64] The fresh leaves contain jujubasaponins, including ziziphin (**42**), which shows sweetness-inhibiting activity (its original structure has been revised).[65,66] An ethanolic extract of *Z. jujuba* also possessed an anxiolytic effect at lower doses and a sedative effect at higher doses.[67].

Middle Class Herbs

Tang Kuei or Dong Quai (Angelica sinensis, Umbelliferae)

The dried root of *Angelica sinensis* is used to activate blood circulation, regulate menstruation, relieve pain, relax bowels, and treat anemia, menstrual disorders, rheumatic arthralgia, and traumatic injuries.

Chemical Composition and Pharmacological Properties (Figure 2.11).

- Essential oils: ligustilide (**43**), n-butylidenephthalide, n-butylphthalide, and safrole
- Fatty acids: palmitic acid, linoleic acid, stearic acid, and arachidonic acid
- Coumarins: bergapten, scopoletin, and umbelliferone
- Polysaccharides

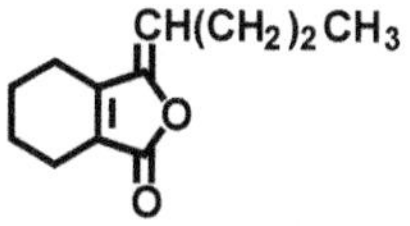

Ligustilide (43) 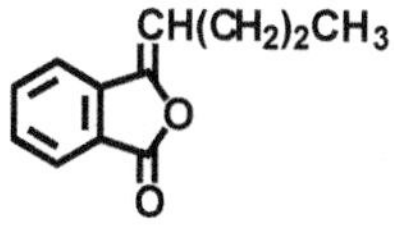   n-Butylidenephthalide (44)

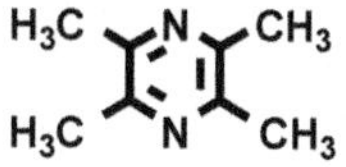

Tetramethylpyrazine (45) 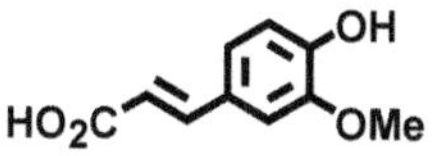   Ferulic acid (46)

Figure 2.11

A. *sinensis* has a direct mucosal healing effect on gastric epithelial cells.[68] *n*-Butylidenephthalide (**44**) modulates performance deficits induced by drugs; these effects are related to activating the central but not the peripheral cholinergic neuronal system via muscarinic and nicotinic receptors.[69] An immunostimulating, low molecular weight polysaccharide, which was isolated from A. *sinensis*, showed strong antitumor activity in Ehrlich ascites tumor bearing mice.[70] Tetramethylpyarazine (**45**) and ferulic acid (**46**) exert analgesic and anti-inflammatory effects.[71]

*Chuan Chiung (*Cnidium officinale, *Umbelliferae)*

The rhizomes of *Ligusticum wallichii* (China) or *Cnidium officinale* (Japan) are used in TCM to treat female genital inflammatory diseases. The rhizome invigorates blood circulation, promotes the flow of ch'i, and controls pain. Traditionally, it is used for headache, abdominal pain, arthralgia due to cold, tendon spasms, amenorrhea, and other menstrual disorders.

Chemical composition and pharmacological properties. Ligustilide, butylidenephthalide, butylphthalide, senkyunolide, and cnidilide are found in the essential oil of Chuan Chiung.[72] Tetramethylpyrazine is also present.[71]

This herb increases myocardial contractility and coronary circulation, decreases heart rate and oxygen consumption,

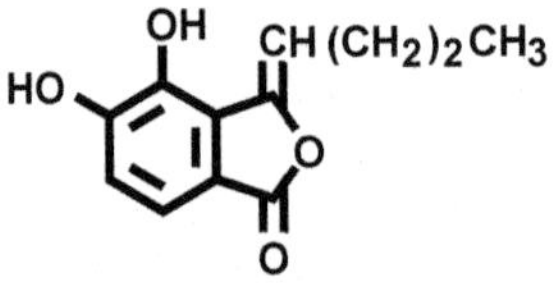

4,5-Dihydroxy-butylidenephthalide (47)

Figure 2.12

causes vasodilation, and lowers blood pressure.[73] Ligustilide and butylidenephthalide contribute to pentobarbital sleep effects in mice.[74] Among tested synthetic butylidenephthalide derivatives, BP-42 (4,5-dihydroxy-butylidenephthalide, **47**, Figure 2.12) showed the greatest antiproliferative effects in primary cultures of vascular mouse aorta smooth muscle cells, and thus, may become a trial antiatherosclerotic drug.[75]

Ko Ken (Pueralia lobata, Leguminosae)

Kudzu vine root is one of the most important Chinese medical herbs. Ko Ken Tang (Pueraria Combination) is used by TCM doctors to treat symptoms of fever, headache, neck pain, and back and neck stiffness. Ko Ken is also used for cardiovascular disease, angina pectoris, and hypertension. A unique application is treatment of alcoholism by decreasing alcohol craving.

Chemical composition and pharmacological properties (Figure 2.13). Ko Ken contains isoflavones, such as daidzein (**48**) (7,4′-dihydroxyisoflavone), daidzin (**49**), puerarin, and other derivatives.[76–78] In addition to antiarrhythmic and immunostimulant activities, daidzein also inhibits aldehyde dehydrogenase II, and thus, can suppress alcohol intake.[79] Genistein (**50**) has estrogenic activity and inhibits oxidation of LDL.[80] Puerarin (**51**) is a β-adrenergic blocker and inhibits platelet aggregation; it also shows a cardioprotective effect when given after long periods of heart arrest and reperfusion.[81] Four compounds inhibit PGE(2) production with the relative potencies: tectorigenin (**52**) > genistein > tectoridin (**53**) > daidzein.[82]

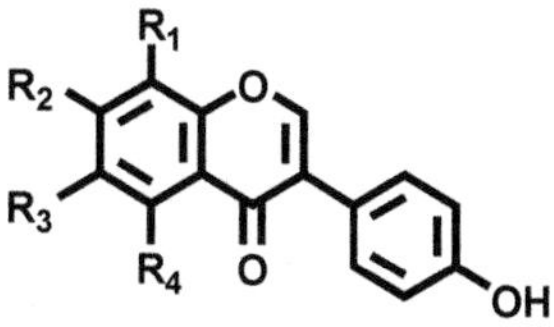

Daidzein (**48**):$R_1 = R_3 = R_4 = H$, $R_2 = OH$
Daidzin (**49**): $R_1 = R_3 = R_4 = H$, $R_2 = O\text{-}\beta\text{-D-Glucose}$
Genistein (**50**): $R_1 = R_3 = H$, $R_2 = R_4 = OH$
Puerarin (**51**): $R_1 = O\text{-Glu}$, $R_2 = OH$, $R_3 = R_4 = H$
Tectorigenin (**52**): $R_1 = H$, $R_2 = R_4 = OH$, $R_3 = OMe$
Tectoridin (**53**): $R_1 = H$, $R_2 = O\text{-Glu}$, $R_3 = OMe$, $R_4 = OH$

Figure 2.13

Shi Liu Pi (Punica granatum, *Punicaceae)*

Punica granatum (pomegranate) is native to Asia and has been cultivated for centuries for its flavorful red fruit. The edible seeds are also made into jams and jellies and the juice into the drink "grenadine." The root bark is used traditionally as a taenifuge to purge intestinal parasites and the fruit husks are used as an antiseptic for gum, tonsil, and throat inflammation and infection. It is also associated with fertility/contraception and menopausal disorders, as well as antitumor[83,84] and antioxidant[85,86] activities.

Chemical composition and pharmacological properties (Figure 2.14). The toxic alkaloid pelletierine (**54**) is found in the root and bark of *P. granatum*. Fruits with seeds contain polyphenolic compounds and estrogen. The latter is richest in the seed (17 mg/kg) and used for menopausal disorders and associated with fertility/contraception. Antioxidant activity of *P. granatum*[83] has been associated with polyphenols, including the anthocyanidin delphinidin (**55**).[84] Ellagic acid (**56**) shows anticarcinogenic effects.[85,86] Ellagitannins, including punicalin (**57**), showed anti-HIV activity.[87]

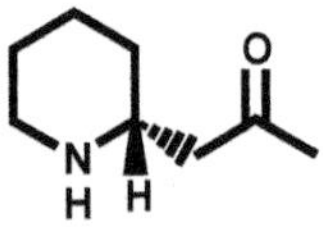

Pelletierine (54) 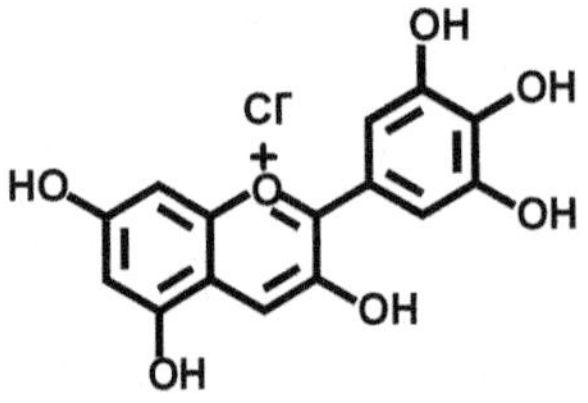   Delphinidin (55)

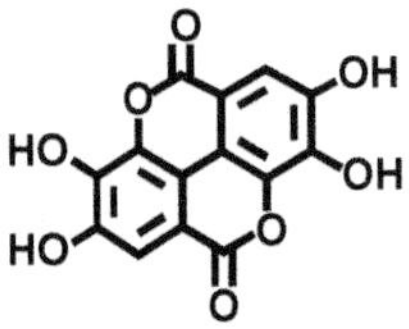

Ellagic Acid (56) 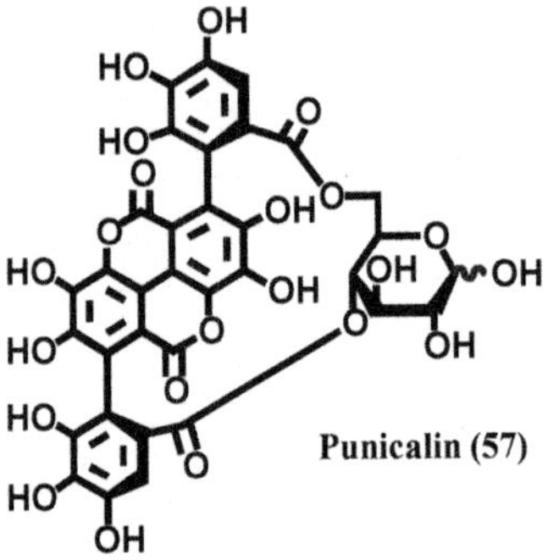   Punicalin (57)

Figure 2.14

*Ginger (*Zingiber officinale, *Zingiberaceae)*

Ginger has botanical characteristics resembling those of tur-
meric. Although originally from India, ginger is cultivated in
India, China, Southeast Asia, and tropical regions of Africa.
This spice, or rhizome, is used in Asian traditional medicines
especially for functional dyspepsia. Lately, its use as complementary and alternative medicine (CAM) by older adults has
been investigated.[88]

Chemical composition and pharmacological properties.
The rhizomes of ginger are rich in starch and contain proteins,
fats, essential oil, and a resin. The composition of the essential
oil varies as a function of the geographical origin, but the major
(30–70%) constituents are terpene hydrocarbons, including
zingiberene, ar-curcumene, and α-bisabolene. The constituents responsible for ginger's pungent taste are 1-(3′-methoxy-
4′-hydroxyphenyl)-5-hydroxyalkan-3-ones (Figure 2.15). Also
known as [3–6]-, [8]-, [10]-, and [12]-gingerols, these compounds

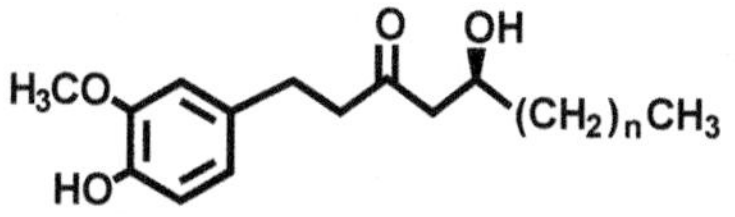

[6]-Gingerol (**58**), n = 4
[8]-Gingerol (**59**), n = 6

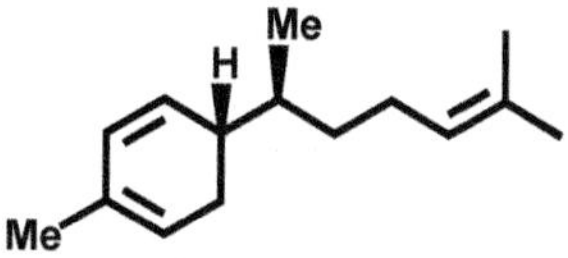

Zingiberene (**60**)

Figure 2.15

have a side chain with 7–10, 12, 14, or 16 carbon atoms, respectively, and occur alongside the corresponding ketones, and, in the dried drug, the dehydration products (shogaol). [6]-Gingerol (**58**) is a cholagogue, and [8]-gingerol (**59**) is a hepatoprotective agent.[89] Zingiberene (**60**) has antiulcer effects in rats.[90] Recently, gingerols were found to inhibit arachidonic acid-induced human platelet serotonin release and aggregation.[91] Antioxidant activity of gingerols has also been reported.[92,93]

> *Green Tea (*Thea sinensis *or* Camellia
> sinensis, *Theaceae)*

Green Tea is manufactured from the fresh leaves of *Thea sinensis*.

Chemical composition and pharmacological properties. Green tea polyphenols (GTPs) are the major bioactive constituents of green tea.[94] The major GTPs are (-)-epicatechin (EC), (-)-epigallocatechin (EGC), (-)-epicatechin-3-gallate (ECG), and (-)-epigallocatechin-3-gallate (EGCG) (**61–64**, Figure 2.16). The major biological effects of catechins are as follows.

- Cancer chemoprevention: GTPs inhibit promotion of carcinogenesis; EGCG inhibits tumor promotion (duodenal and skin cancers).[94–96]
- Antibacterial activity: Inhibit the activity of bacterial exo-toxins.[97]
- Inhibit influenza virus infection by blocking adsorption of virus to cells.[98]
- Inhibit HIV RT: ECG, EGCG: IC_{50} = 10–20 ng/mL; but EC, EGC, and gallic acid were not active.[99]

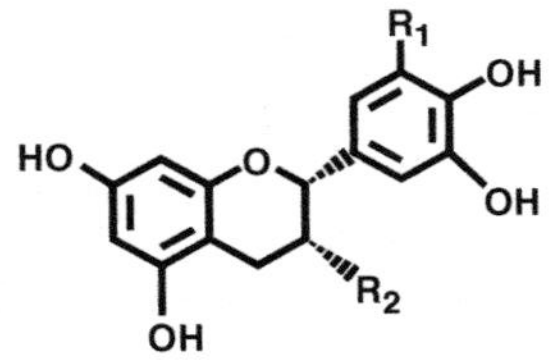

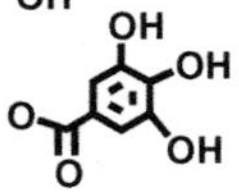

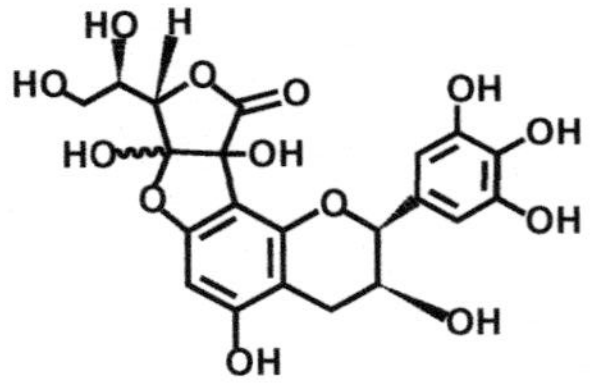

8-C-Ascorbyl
(-)-Epigallocatechin (65)

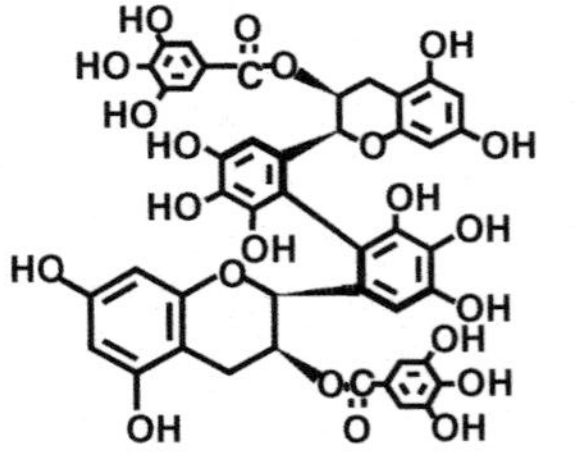

Theasinensin-D (66)

(-)-Epicatechin(EC) (61): R_1 = H, R_2 = OH
(-)-Epigallocatechin (62) (EGC): R_1 = OH, R_2 = OH
(-)-Epicatechin-3-gallate (ECG) (63): R_1 = H, R_2 =

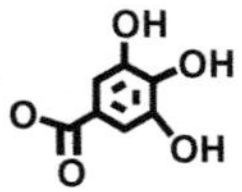

(-)-Epigallocatechin-3-gallate (EGCG) (64): R_1= OH, R_2 =

Figure 2.16

Inhibitory effects of 38 tea polyphenols for against HIV replication in H9 lymphocytic cells have been reported. 8-*C*-Ascorbyl (-)-epigallocatechin (**65**) and theasinensin-D (**66**) demonstrated anti-HIV activity with EC_{50} values of 3 and 8 µg/mL and therapeutic index values of 9.5 and 5, respectively.[100]

Ta Suan or Garlic (Allium sativum, Liliaceae)

Garlic (in Chinese: Ta Suan) is the bulb of *Allium sativum* L., which was first cultivated in ancient Egypt, Greece, Rome, India, and China for its therapeutic benefits. It contains various sulfur compounds including alliin, as well as steroid saponins and polysaccharides. Garlic was used by the first Olympic athletes as an energizer and has also been used to treat tumors, headaches, weakness and fatigue, wounds, sores, and infections. It lowers blood lipids and inhibits platelet aggregation. Garlic is safe and effective: 2 to 5 grams raw or 0.4 to 1.2 grams dried powder is equivalent to 2 to 5 mg of allicin, daily.[101]

Major chemical constituents of garlic and their biological activities (Figure 2.17). When garlic cells are crushed, alliin (**67**) is degraded and converted to allicin (**68**) by alliinase. Allicin is converted by heat to many other compounds, including diallyl sulfide. Allicin shows antiprotozoal, antibacterial, antifungal and antiviral activities, hypoglycemic effects, and decreases blood cholesterol levels.[102,103] Diallyl sulfide (**69**) reduces tumor growth in animals, and also shows anti-HIV effects. Allithiamine (**70**) is synthesized from alliin and Vitamin B_1 (VB_1) by the following route. As stated above, alliin is converted by alliinase to allicin, which then reacts with a thiol-type VB_1 to produce allithiamine, a prodrug of VB_1. Allithiamine resists the activity of aneurinase (thiaminase) and is absorbed easily in intestine to reach high blood concentration. It is converted to Vitamin B_1 in the body and, thus, can be used as an active VB_1.[104] However, taking allithiamine generates an unpleasant garlic odor; therefore, related odorless derivatives have been synthesized and used as active VB_1.

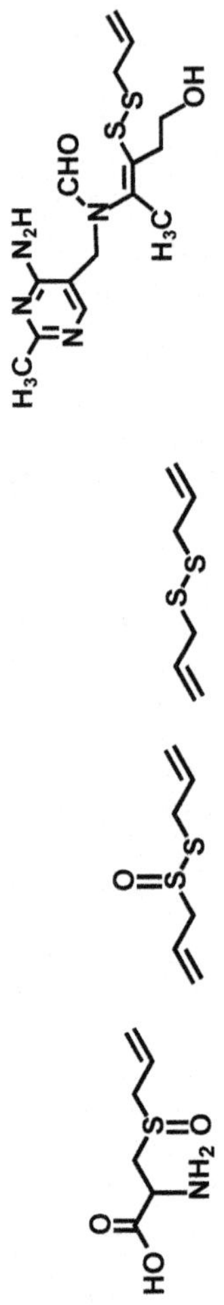

Figure 2.17

Kansuiphorin A (71)

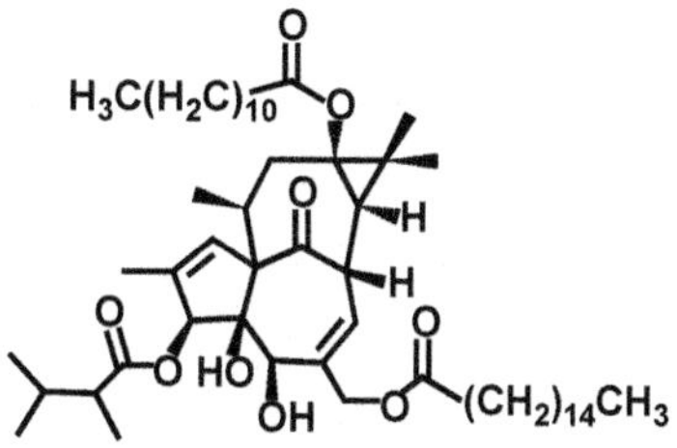

Kansuiphorin B (72)

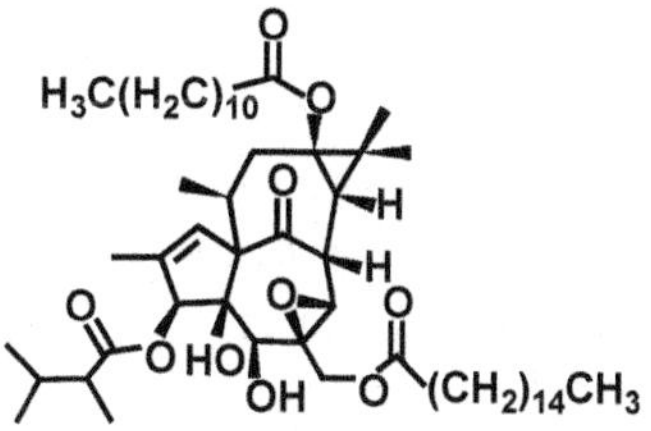

Figure 2.18

Lower Class Herbs

Kansui (Euphorbia kansui, Euphorbiaceae)

Euphorbia kansui is widely distributed in Northwest China. Its dried roots are known as Kansui and have been used as an herbal remedy for ascites and cancer in China.

Chemical composition and pharmacological properties. Ingenol derivatives, including kansuiphorins A, B, C, and D, have been isolated from Kansui.[105,106] Kansuiphorins A and B (**71** and **72**, Figure 2.18) demonstrated potent activity against P-388 leukemia in mice with T/C values of >176 and 177% at 0.1 and 0.5 mg/kg, respectively.[105] An *E. kansui* extract was found to enhance immune complex binding to macrophages, and two ingenols with dose-dependent activity were isolated.[107] Eleven compounds including one steroid, four triterpenes, and six diterpenes were assayed for their cytotoxic and antiviral activity.[108]

Hsia Ku Tsao (Prunella vulgaris L., Labiatae)

Hsia Ku Tsao (Spica Prunellae, Self-heal Spike) is the dried flowered fruit-spike of *P. vulgaris*, which is widely distributed in China and Asia. In the summer when the spike becomes

Ursolic Acid (**73**)

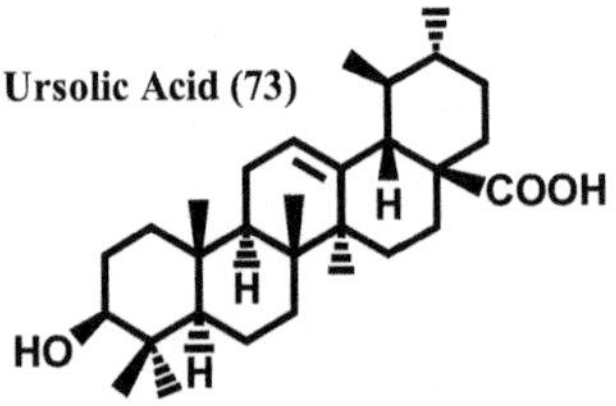

Figure 2.19

brownish red, the herb is collected and dried in the sun. This lower class herb is used for hypertension with headache, tinnitus, eye inflammation, and nocturnal eye pain. It also has hypotensive, antibacterial, and antitumor activities.[109]

  Chemical composition and pharmacological properties. Ursolic acid (**73**, Figure 2.19) was isolated from a cytotoxic extract of the fruiting spikes of *P. vulgaris* using bioassay-directed fractionation. The compound showed significant cytotoxicity in P-388 and L-1210 lymphocytic leukemia cells, as well as A-549 human lung carcinoma cells, and marginal cytotoxicity in KB and human colon (HCT-8) and mammary (MCF-7) tumor cells.[110]

  *Kuei Chiu (Podophyllum emodi,
  Berberidaceae)*

Podophyllum emodi is found in western China and grows in most of the Himalayan region. The dried roots are traditionally used as a contact cathartic. This herb has also long been used in China as an anticancer drug and to treat snakebites, periodontitis, skin disorders, coughs, and intestinal parasites.

  Chemical composition and pharmacological properties. The antimitotic lignan podophyllotoxin (**74**, Figure 2.20), as well as α- and β-peltatins, desoxypodophyllotoxin, and other close derivatives, are found in *P. emodi* and other related species, including *P. peltatum* and *P. pleianthum*. Podophyllotoxin is a mitotic spindle poison. It inhibits the polymerization of tubulin and stops cell division at the start of metaphase.[111] It can be converted into the semisynthetic antineoplastic derivatives etoposide (**75**) and teniposide (**76**), which are used

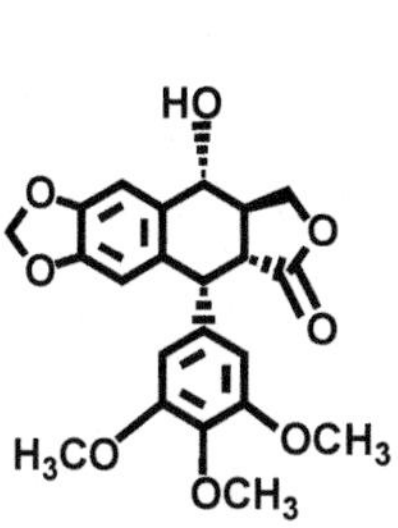

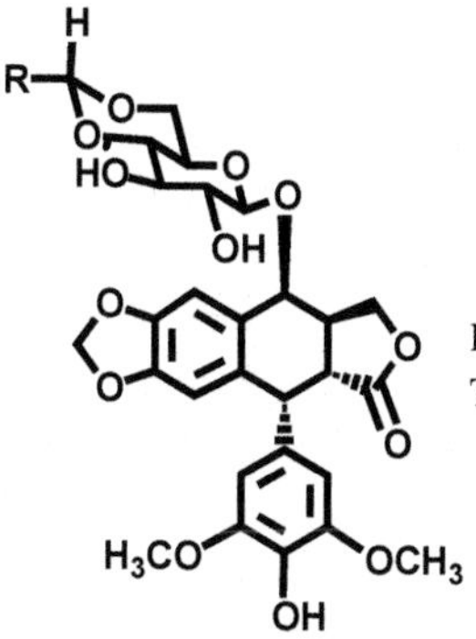

Etoposide (75): R = CH₃
Teniposide (76): R = 2-Thienyl

Podophyllotoxin (74)

Figure 2.20

to treat small cell lung cancer, testicular cancer, leukemias, lymphomas, and other cancers.[112–115] The combination of etoposide, folinic acid, and 5-fluorouracil (5-FU) provides effective chemotherapy in patients with advanced gastric cancer,[116] and etoposide with carboplatin induces remission of malignant schwannoma.[117] These semisynthetic derivatives are epipodophyllotoxins, which are demethylated at C-4″ and have the opposite stereochemistry and are glucosylated at the C-4 hydroxyl, with two of the glucose hydroxyl groups (at C-4″ and C-6″) blocked by acetalization as either a thienylidene (teniposide) or ethylidene (etoposide). In contrast with podophyllotoxin, these derivatives are inactive against microtubule assembly, but instead complex with the enzyme topoisomerase II and stop the cell cycle at the end of the S phase or at the beginning of the G2 phase.[111] Problems associated with the use of etoposide as an anticancer drug include myelosuppression, drug resistance, and poor bioavailability.[118]

DRUG DEVELOPMENT RESEARCH IN THE AUTHOR'S LABORATORY

Using the principles of lead improvement, extensive structure-activity relationship, enzyme interaction, and computational studies have been performed to generate new compounds to overcome existing limitations. In particular, several series of

4-alkylamino and 4-arylamino epipodophyllotoxin analogues were synthesized from the natural product podophyllotoxin.[119] New computational strategies continue to play an important role in the rational design of improved etoposide analogs.[120,121]

Compared with etoposide, several synthetic compounds have shown similar or increased percent inhibition of DNA topo II activity and percent protein-linked DNA breakage.[122] Even more notable is the increased cytotoxicity of these derivatives in etoposide-resistant cell lines. GL-331 (**77**, Figure 2.21),[123] which contains a *p*-nitroanilino moiety at the 4β position of etoposide, has emerged successfully from this preclinical development to proceed further along the drug development pathway.

GL-331 is a topo II inhibitor and, thus, causes DNA double strand breakage and G2-phase arrest. Formulated GL-331 shows desirable stability and biocompatibility and similar pharmacokinetic profiles to those of etoposide.[123] It has been patented by Genelabs Technologies, Inc. and has completed phase I clinical trials as an anticancer drug at the M.D. Anderson Cancer Center. Initial results from these trials[124] in four tumor types (nonsmall and small cell lung, colon, and head/neck cancers) showed good antitumor efficacy. The major toxicity was cytopenias, but side effects were minimal. Maximum tolerated dose (MTD) was declared at 300 mg/m^2, while that for etoposide was 140 mg/m^2. GL-331 has other advantages over etoposide, including (a) greater activity both *in vitro* and *in vivo*, (b) a shorter synthesis and, thus, easier manufacture, and (c) evidence of activity in refractory tumors as it overcomes drug-resistance in many cancer cell lines (KB/VP-16, KB/VCR, P388/ADR, MCF-7/ADR, L1210/ADR, HL60/ADR, and HL60/VCR).[124]

In summary, GL-331 is an exciting chemotherapeutic candidate, which exemplifies successful preclinical drug development from Asian herbal products.

RESEARCH ON NEW MEDICINES FROM HERBAL PRODUCTS

As the source of a potential new drug, an herb or herbal prescription is chosen based on folk or clinical experiences. The

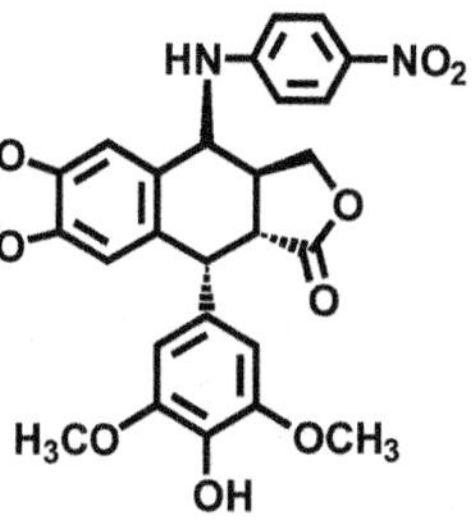

GL-331 (77)

Figure 2.21

initial research (new lead discovery) focuses on the isolation of a bioactive natural lead compound(s). After extraction of the target herbal medicine, the activity is verified by pharmacological testing. The active portions are subjected to bioactivity-directed fractionation and isolation (BDFI). After the new lead has been discovered, chemical modification or lead improvement is aimed at increasing activity, decreasing toxicity, or improving other pharmacological profiles. Active compounds are further studied in mechanism of action (e.g., enzymatic or antimitotic) and other appropriate assays. For example, with cytotoxic compounds, preclinical screening in the National Cancer Institute's (NCI) *in vitro* human cell line panels and selected *in vivo* xenograft systems is used to select the most promising drug development targets. Efficacy and toxicity must be evaluated, and production, formulation, and toxicological studies are performed prior to clinical trials.

Importantly, this same process can and should be applied to both single and formulated herbs leading to single active principles (single herbal-derived compounds), active fractions (herbal extracts), and effective and safe prescriptions (multiple herbal products) to be developed both as dietary supplements and new medicines.

By using these approaches, numerous drugs have been discovered from active principles of TCM herbs, as described in this contribution. Ephedrine and artemisinin exemplify drugs identified from single herbs, while indirubin demonstrates a drug discovered from an herbal formula.

Ephedras (*Ephedra* spp., Ephedraceae)

Ephedras are dioecious subshrubs similar to horsetails with slender, angular, strait branches and membraneous scales for leaves. *E. equisetina* and *E. sinica* are found in China, *E. intermedia* and *E. gerardiana* in India and Pakistan, and about ten species, including *E. nevadensis* or Mormon, in North America.

In TCM, ephedra is the chief drug for treatment of asthma and bronchitis. It has been used for thousands of years as a primary component of multiherb formulas prescribed to treat bronchial asthma, cold and flu, cough and wheezing, fever, chills, lack of perspiration, headache, and nasal congestion. The history of ephedrine, the main bioactive principle, includes the following milestones: 1882 discovery (Nagayoshi Nagai), 1920 synthesis (Späth, Gohring), 1928 pharmacology (K.K. Chen), and 1932 stereochemistry (Freudenberg).

Ephedras contain various alkaloids, primarily (-)-ephedrine (**78**), which occurs alongside (+)-pseudoephedrine (**79**) and the corresponding nor and N,N-dimethyl derivatives (Figure 2.22). Ephedrine is structurally similar to adrenaline and physiologically is an indirect sympathomimetic. It triggers the release of endogenous catecholamines from the postganglionic sympathetic fibers, stimulates cardiac automaticity, has a positive inotropic activity, and accelerates and intensifies respiration. Ephedrine also is a bronchodilator, stimulates the brain stem respiration center, and decreases bladder contractility. Recently, ephedrine and phenylephedrine were evaluated in randomized controlled trials to control hypotension during spinal anesthesia for cesarean delivery.[125] Dietary supplements containing ephedras are available through websites and in dietary stores in the U.S. They are widely promoted and used for weight reduction and energy enhancement.[126]

Qinghao (*Artemisia annua* L., Asteraceae)

Qinghao (Sweet Wormwood) is the dried aerial parts of *A. annua*, originally indigenous in Asia. The herb has been traditionally

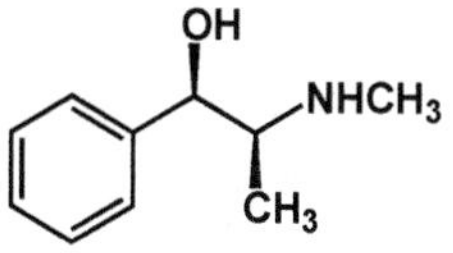

(-)-Ephedrine (78) 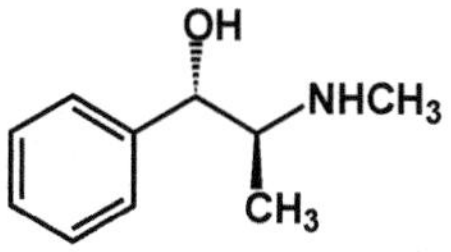   (+)-Pseudoephedrine (79)

Figure 2.22

used in China for over a thousand years to treat fever and malaria.

Chinese scientists isolated artemisinin (**80**) (Qing Hao Su), as an active principle (Figure 2.23). This novel antimalarial compound has an endoperoxide linkage, which is essential for the antimalarial activity of the molecule. Artemisinin directly kills malaria parasites in the erythrocytic stage, and acts mainly on the membranes of *Plasmodium berghei*. The curative rate is 100%, as compared with 95% for chloroquine, with low toxicity to both animal and human organs; therefore, artemisinin was introduced clinically as a new type of antimalarial agent with safe and rapid action against chloroquine-resistant *Plasmodium falciparum*.[127] Many synthetic derivatives including artemether (**81**) and arteether (**82**) have been studied. Artemether is in clinical use in China and arteether is in phase II clinical trials in the U.S. as an antimalarial drug.[128] Several antimalarial analogs related to artemisinin have been synthesized.[129–131]

Artemisinin (80)

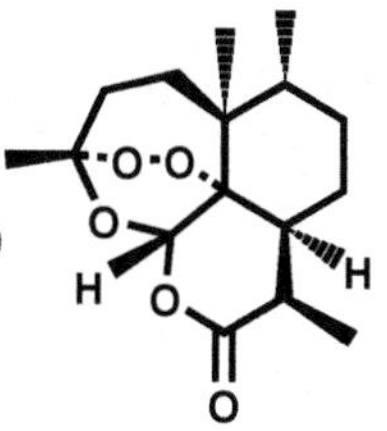

Artemether (81): R = CH$_3$
Arteether (82): R = CH$_2$CH$_3$

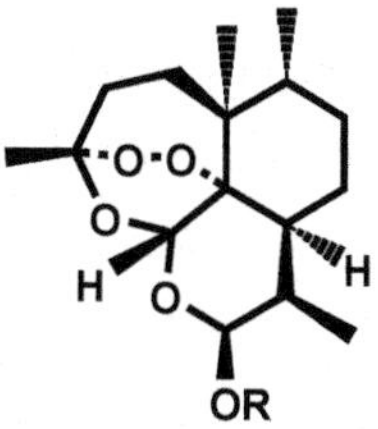

Figure 2.23

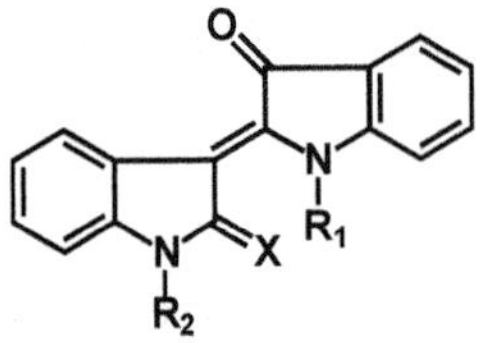

Indirubin (83): $R_1 = R_2 = H$, $X = O$
N,N'-Dimethylindirubin (84): $R_1 = R_2 = CH_3$, $X = O$
N-Methylindirubin Oxime (85): $R_1 = H$, $R_2 = CH_3$, $X = N\text{-}OH$

Figure 2.24

Indirubin (from Dang Gui Lu Hui Wan)

A traditional remedy for chronic myelocytic leukemia (diagnosed according to its symptoms) is the prescription Dang Gui Lu Hui Wan, which contains *Angelica sinensis* (root), *Aloe vera* (dried juice), *Gentiana scabra* (root), *Gardenia jasminoides* (fruit), *Scutellaria baicalensis* (root), *Phellodendron amurense* (stem-bark), *Coptis chinensis* (rhizom), *Rheum palmatum* (root), *Aucklandia lappa* (root), and Indigo naturalis, which is a product made from leaves of *Baphicacanthus cusia*, *Indigofera tinctoria*, or *Isatis indigotica*.

Chinese researchers have identified the active ingredient of this prescription as Indigo naturalis, and further study established indirubin (**83**) as the antileukemic agent[132] (Figure 2.24). The antitumor activities of various indirubin derivatives were further examined against rat carcinosarcoma W256 and mouse leukemia L7212. Among the compounds tested, N,N'-dimethyl-indirubin (**84**) and N-methylindirubin oxime (**85**) were more potent than the parent compound.[133]

MODERN DIRECTIONS BASED ON AN ANCIENT AND LONG-LASTING LEGACY

TCM has a clinical history of over 4,000 years. Its founder, Shen Nung, authored the classical and longest surviving description of herbs *Shen Nung Pen Ts'ao Ching*. The 365 herbs of *Shen Nung Pen Ts'ao Ching* and many others are still in use today as part of the traditional Chinese healing arts. Shen Nung and other herbalists realized that each herb can affect the human body to help or hinder the body's processes. Herbs can strengthen and balance the system, tonify the organs, and optimize the flow and use of energy (Chi; Qi), making them ideal as dietary supplements; however, they can also treat acute illness. Relatively nontoxic herbal products are especially attractive for the following chronic health effects: antioxidant and antiaging activity, blood pressure-lowering effects, hypolipidemic action, blood sugar-lowering effect, antiallergic functions, and antiarthritis properties. However, rigorous study will be necessary to prove efficacy and safety of these products. In this century, herbal products should continue to be excellent sources of new drugs. Although the research described above was based on the discovery and development of bioactive, pure, lead compounds, new drugs and dietary supplements can arise from three herbal-related sources:

1. Active pure compounds
2. Active fractions
3. Validated or improved effective and safe herbal formulations

CONCLUSIONS

Asian herbal products, which originate from TCM, use processed single or formulated herbal products as dietary supplements or prescriptions based upon unique TCM principles and herbal pharmacology to prevent, relieve, and cure many diseases. As recorded in both ancient and current literature, TCM has long been used for human disease prevention and treatment, and will undoubtedly provide a strong base for continued development of modern high-quality dietary supplements and modern medicines in the 21st century. As validated by modern scientific studies, including those in the author's laboratory, single herbs contain numerous bioactive compounds. Highly efficient bioactivity-directed fractionation and isolation, characterization, analog synthesis and mechanistic studies are prerequisites for the development of these new single active principles as clinical candidates for world-class new drug development. Herbal products are also being increasingly developed worldwide as dietary supplements in the 21st century. Bioactive lead compounds, active herbal fractions, and active TCM prescriptions will all be sources of new, effective, and safe world-class new medicines and dietary supplements.

ACKNOWLEDGMENT

This investigation is supported in part by grants CA-17625 and AI-33055 from the National Cancer Institute and the National Institute of Allergy and Infectious Disease, NIH, respectively, awarded to K. H. Lee.

REFERENCES

1. HY Hsu, WG Peacher. *Chinese Herb Medicine and Therapy.* Nashville, TN: Aurora Publishers, 1976, p. 18.

2. KC Huang. *The Pharmacology of Chinese Herbs.* Boca Raton, FL: CRC Press, 1993, pp 3–7.

3. E Hosoya. The pharmacology of kampo prescriptions, in T Takemi, M Hasegawa, A Kumagai, Y Otsuka, Eds. *Herbal Medicine: Kampo, Past and Present.* Tokyo, Life Science Publishing, 1985, pp 52–65.

4. TT Tsai, SN Liao. *The Application and Therapy of Ling Chih.* Taichung, Taiwan: San Yun Press, 1982.

5. J Boik. *Natural Compounds in Cancer Therapy.* Princeton, MN: Oregon Medical Press, 2001, pp 205–206.

6. MS Siao, KR Lee, LJ Li, CT Wang. Natural products and biological activities of the Chinese medicinal fungus Ganoderma lucidum, in CT Ho, T Osawa, MT Huang, RT Rosen, Eds. *Food Phytochemicals for Cancer Prevention.* ACS Symposium Series 547. Washington, D.C., American Chemical Society, 1994, pp 342–354.

7. X Bao, J Duan, X Fang, J Fang. Chemical modifications of the $(1{\to}3)$-α-D-glucan from spores of Ganoderma lucidum and investigation of their physicochemical properties and immunological activity. *Carbohyd Res* 336:127–140, 2001.

8. BS Min, JJ Gao, M Hattori, HK Lee, YH Kim. Anticomplement activity of terpenoids from the spores of Ganoderma lucidum. *Planta Medica* 67:811–814, 2001.

9. American Cancer Society, *Guide to Complementary and Alternative Cancer Methods*. Atlanta, GA: American Cancer Society, 2000, pp 314–315.

10. H Sakagami, K Sugaya, A Utsumi, S Fujinaga, T Sato, M Takeda. Stimulation by PSK of interleukin-1 production by human peripheral blood mononuclear cells. *Anticancer Res* 13:671–675, 1993.

11. *American Cancer Society's Guide to Complementary and Alternative Cancer Methods*. Atlanta, GA: American Cancer Society, 2000, pp 345–346.

12. F Liu, VEC Ooi, ST Chang. Free radical scavenging activities of mushroom polysaccharide extracts. *Life Sciences* 60:763–771, 1997.

13. SP Wasser, AL Weis. Medicinal properties of substances occurring in higher basidiomycetes mushrooms: current perspectives. *Int J Med Mushrooms* 1:31–62, 1999.

14. HS Kim, S Kacew, BM Lee. In vitro chemopreventive effects of plant polysaccharides (Aloe barbadensis miller, Lentinus edodes, Ganoderma lucidum and Coriolus versicolor). *Carcinogenesis* 20:1637–1640, 1999.

15. KH Lee, M Kozuka, H Tokuda. Unpublished Data.

16. N Ohno, AM Akanuma, NN Miura, Y Adachi, M Motoi. $(1\rightarrow3)$-β-D-Glucan in the fruit bodies of Agaricus blazei. *Pharm Pharmacol Lett* 11:87–90, 2001.

17. M Hirotani, K Sai, S Hirotani, T Yoshikawa. Blazeispirols B, C, E and F, des-A-ergostane-type compounds, from the cultured mycelia of the fungus Agaricus blazei. *Phytochemistry* 59:571–577, 2002.

18. S Shibata, M Fujita, H Itokawa, O Tanaka. Studies on the constituents of Japanese and Chinese crude drugs. XI Panaxadiol, a sapogenin of Ginseng roots. *Chem Pharm Bull* 11:759, 1963.

19. S Shibata. Chemistry and cancer preventing activities of ginseng saponins and some related triterpenoid compounds. *J Korean Med Sci Suppl* 16:28–37, 2001.

20. O Tanaka. Ginseng and its congeners, in CT Ho, T Osawa, MT Huang, RT Rosen, Eds. *Food Phytochemicals for Cancer Prevention*. ACS Symposium Series 547. Washington, D.C.: American Chemical Society, 1994, pp 335–341.

21. JM Ellis, P Reddy. Effects of Panax ginseng on quality of life. *Ann Pharmacother* 36:373–379, 2002.

22. TK Yun, YS Lee, YH Lee, SI Kim, HY Yun. Anticarcinogenic effect of Panax ginseng CA Meyer and identification of active compounds. *J Korean Med Sci* 16:1–18, 2001.

23. AS Attele, JA Wu, CS Yuan. Ginseng pharmacology: multiple constituents and multiple actions. *Biochem Pharmacol* 58:1685–1693, 1999.

24. YJ Surh, HK Na, JY Lee, YS Keum. Molecular mechanisms underlying antitumor-promoting activities of heat-processed Panax ginseng CA Meyer. *J Korean Med Sci* 16:38–41, 2001.

25. TK Yun, SY Choi. A case-control study of ginseng intake and cancer. *Intl J Epidemiol* 19:871–876, 1990.

26. H Besso, R Kasai, JF Wang, Y Saruwatari, T Fuwa, O Tanaka. Further studies on dammarane-saponins of American ginseng, roots of Panax quiquefolium L. *Chem Pharm Bull* 30:4534–4538, 1982.

27. AT Borchers. Comparative effects of three species of ginseng on human peripheral blood lymphocyte proliferative responses. *Intl J Immun* 14:143–152, 1998.

28. W Li, C Gu, H Zhang, DVC Awang, JF Fitzloff, HHS Fong, RB van Breemen. Use of high-performance liquid chromatography-tandem mass spectrometry to distinguish Panax ginseng CA Meyer (Asian ginseng) and Panax quinquefolius L. (North American ginseng). *Anal Chem* 72:5417–5422, 2000.

29. TWD Chan, PPH But, SW Cheng, IMY Kwok, FW Lau, HX Xu. Differentiation and authentication of Panax ginseng, Panax quinquefolius, and ginseng products by using HPLC/MS. *Anal Chem* 72:1281–1287, 2000.

30. DQ Dou, WB Hou, YJ Chen. Studies on the characteristic constituents of Chinese ginseng and American ginseng. *Planta Medica* 64:585, 1998.

31. WK Li, JJ Fitzloff. HPLC analysis of ginsenosides in the roots of Asian ginseng (Panax ginseng) and North American ginseng (Panax quinquefolius) with in-line photodiode array and evaporative light scattering detection. *Liquid Chromat Related Technol* 25:29–41, 2002.

32. WG Ma, M Mizutani, KE Malterrud, SL Lu, F Ducrey, S Tahara. Saponins from the roots of Panax notoginseng. *Phytochemistry* 52:1133–1139, 1999.

33. FY Gang, GZ Zheng. Chemical compounds from Panax notoginseng. *Zhongguo Yaoxue Zazhi* 27:138–143, 1992.

34. T Kosuge, M Yokota, A Ochiai. Studies on antihemorrhagic principles in the crude drugs for hemostatics. II. On antihemorrhagic principle in Sanchi Ginseng Radix. *Yakugaku Zasshi* 101:629–632, 1981.

35. Q Li, YH Ye, AX Yan, YW Zhou, H Shen, QY Xing. Isolation, identification and physiological activities of 2-(1′,2′,3′,4′-tetrahydroxybutyl)-6-(2″3″,4″trihydroxybutyl)-pyrazine from Panax notoginseng. *Chem J* Chinese Univ (Chinese) 22:1824–1828, 2001.

36. XC Li, DL Barnes, IA Khan. A new lignan glycoside from Eleutherococcus senticosus. *Planta Medica* 67:776–778, 2001.

37. F Facchinelli, L Neri, M Larabusi. Eleutherococcus senticosus reduces cardiovascular stress response in healthy subjects: a randomized, placebo controlled trial. *Stress and Health* 18:11–17, 2002.

38. HB Li, F Chen. Preparative isolation and purification of six diterpenoids from the Chinese medicinal plant Salvia miltiorrhiza by high-speed countercurrent chromatography. *J Chromatogr* A 925:109–114, 2001.

39. A Sugiyama, BM Zhu, A Takahara, Y Satoh, K Hashimoto. Cardiac effects of salvia miltiorrhiza/dalbergia odorifera mixture, an intravenously applicable Chinese medicine widely used for patients with ischemic heart disease in China. *Circulation* J 66:182–184, 2002.

40. J Liu, HM Shen, CN Ong. Salvia miltiorrhiza inhibits cell growth and induces apoptosis in human hepatoma HepG(2) cells. *Cancer Lett* 153:85–93, 2000.

41. T Deyama, S Nishibe, Y Nakazawa. Constituents and pharmacological effects of Eucommia and Siberian ginseng. *Acta Pharmac Sinica* 22:1057–1070, 2001.

42. T Nohara. Scientific Study on Tochu-tea, The 3rd Symposium on Medicinal Foods, Abstract, 2000, pp 67–76.

43. YM Li, T Sato, K Metori, K Koike, QM Che, S Takahashi. The promoting effects of geniposidic acid and aucubin in Eucommia ulmoides Oliver leaves on collagen synthesis. *Biol Pharm Bull* 21:1306–1310, 1998.

44. GC Yen, CL Hsieh. Reactive oxygen species scavenging activity of Du-zhong (Eucommia ulmoides oliv.) and its active compounds. *J Agricul Food Chem* 48:3431–3436, 2000.

45. JY Xue, GT Liu, HL Wei, Y Pan. Antioxidant activity of two dibenzocyclooctene lignans on the aged and ischemic brain in rats. *Free Radical Biol Med* 12:127–135, 1992.

46. SP Ip, HY Yiu, KM Ko. Schisandrin B protects against menadione-induced hepatotoxicity by enhancing DT-diaphorase activity. *Mol Cell Biochem* 208:151–155, 2000.

47. K Yasukawa, Y Ikeya, H Mitsuhashi, M Iwasaki, M Aburada, S Nakagawa, M Takeuchi, M Takido. Gomisin A inhibits tumor promotion by 12-O-tetradecanoylphorbol-13-acetate in two-stage carcinogenesis in mouse skin. *Oncology* 49:68–71, 1992.

48. DF Chen, SX Zhang, L Xie, JX Xie, K Chen, Y Kashiwada, BN Zhou, F Wang, LM Cosentino, KH Lee. Structure–activity correlations of gomisin-G-related anti-HIV lignans from Kadsura interior and of related synthetic analogs. *Bioorg Med Chem* 5:1715–1723, 1997.

49. M Hirotani, Y Zhou, H Lui, T Furuya. Cycloartane triterpene glycosides from the hairy root cultures of Astragalus membranaceus. *Phytochemistry* 36:665–670. 1994.

50. YP Wang, XY Li, CQ Song, ZB Hu. Effect of astragaloside IV on T,B lymphocyte proliferation and peritoneal macrophage function in mice. *Acta Pharmacol Sinica* 23:263–266, 2002.

51. Hong Kong Monograph of Chinese Materia Medica, 2000.

52. XM Peng, G Tian. Structural characterization of the glycan part of glycoconjugate LbGp2 from Lycium barbarum L. *Carbohyd Res* 331:95–99, 2001.

53. XM Qin, R Yamauchi, R Aizawa, T Inakuma, K Kato. Structural features of arabinogalactan-proteins from the fruit of Lycium chinense Mill. *Carbohyd Res* 333:79–85, 2001.

54. XM Peng, LJ Huang, CH Qi, YX Zhang, GY Tian. Studies on chemistry and immunomodulating mechanism of a glycoconjugate from Lycium barbarum βL.*Chinese J Chem* 19:1190–1197, 2001.

55. *Chinese Herbs*. SRA "Chinese Herbs" Editorial Committee, Ed. Shanghai: Shanghai Science and Technology Publishing Company, 1999, p 494.

56. JW Bok, L Lermer, J Chilton, HG Klingeman, GH Towers. Antitumor sterols from the mycelia of Cordyceps sinensis. *Phytochemistry* 51:891–898, 1999.

57. *A Coloured Atlas of the Chinese Materia Medica*. Pharmacopoeia Commission of the Ministry of Public Health, P.R. China. Guang Zhou, China: Guangdong Science & Technology Press, 1995, p 486.

58. M Numata, A Yamamoto, A Moribayashi, H Yamada. Antitumor components isolated from the Chinese herbal medicine Coix lachryma-jobi. *Planta Medica* 60:356–359, 1994.

59. T Matsumoto, H Ichikawa, T Sei, H Tokuda. α-Monolinolein as inhibitor against carcinogen promoters. *Jpn Kokai Tokkyo Koho* JP2240019, 1990.

60. M Takahashi, C Konno, H Hikino. Isolation and hypoglycemic activity of coixans A, B and C, glycans of Coix lachryma-jobi var. ma-yuen seeds. *Planta Medica* 52:64–65, 1986.

61. H Otsuka, Y Hirai, T Nagao, K Yamasaki. Anti-inflammatory activity of benzoxazinoids from roots of Coix lachryma-jobi var. ma-yuen. *J Nat Prod* 51:74–79, 1988.

62. G Nonaka, I Nishioka, M Nishizawa, T Yamagishi, Y Kashiwada, GE Dutschman, AJ Bodner, RE Kilkuskie, YC Cheng, KH Lee. Inhibitory effects of tannins on HIV reverse transcriptase and HIV replication in H9 lymphocyte cells. *J Nat Prod* 53:587–595, 1990.

63. R Suttisri, IR Lee, AD Kinghorn. Plant-derived triterpenoid sweetness inhibitors. *J Ethnopharm* 47:9–26, 1995.

64. G Cheng, Y Bai, Y Zhao, J Tao, Y Liu, G Tu, L Ma, N Liao, X Xu. Flavanoids from Ziziphus jujuba Mill var. spinosa. *Tetrahedron* 56:8915–8920, 2000.

65. Y Kurihara, K Ookubo, H Tasaki, H Kodama, Y Akiyama, A Yagi, B Halpern. Purification and structure determination of sweetness inhibiting substance in leaves of Ziziphus jujuba. *Tetrahedron* 44:61–66, 1988.

66. K Yoshikawa, N Shimono, S Arihara. Antisweet substances, jujubasaponins I-III from Zizyphus jujuba. Revised structure of ziziphin. *Tetrahedron Lett* 32:7059–7062, 1991.

67. WH Peng, MT Hsieh, YS Lee, YC Lin, J Liao. Anxiolytic effect of seed of Ziziphus jujuba in mouse models of anxiety. *J Ethnopharmacol* 72:435–441, 2000.

68. YN Ye, ESL Liu, VY Shin, MWL Koo, Y Li, H Matsui, CH Cho. A mechanistic study of proliferation induced by Angelica sinensis in a normal gastric epithelial cell line. *Biochem Pharmacol* 61:1439–1448, 2001.

69. MT Hsieh, CR Wu, LW Lin, CC Hsieh, CH Tsai. Reversal caused by n-butylidenephthalide from the deficits of inhibitory avoidance performance in rats. *Planta Med* 67:38–42, 2001.

70. YM Choy, KN Leung, CS Cho, CK Wong, PK Pang. Immunopharmacological studies of low molecular weight polysaccharide from Angelica sinensis. *Am J Chin Med* 22:137–145, 1994.

71. Y Ozaki. Anti-inflammatory effect of tetramethylpyrazine and ferulic acid. *Chem Pharm Bull* 40:954–956, 1992.

72. HS Choi, MSL Kim, M Sawamura. Constituents of the essential oil of Cnidium officinale Makino, a Korean medicinal plant. *Flav Frag J* 17:49–53, 2002.

73. H Itokawa. *High Technology Information for Specialist*. Tokyo: CMC Press, 2001, p 266.

74. K Matsumoto, S Kohno, K Ojima, Y Tezuka, S Kadota, H Watanabe. Effects of methylenechloride-soluble fraction of Japanese angelica root extract, ligustilide and butylidenephthalide, on pentobarbital sleep in group-housed and socially isolated mice. *Life Sci* 62:2073–2082, 1998.

75. Y Mimura, S Kobayashi, T Naitoh, I Kimura, M Kimura. The structure–activity relationship between synthetic butylidenephthalide derivatives regarding the competence and progression of inhibition in primary cultures proliferation of mouse aorta smooth muscle cells. *Biol Pharm Bull* 18:1203–1206, 1995.

76. T Murakami, Y Nishikawa, T Ando. Constituents of Japanese and Chinese crude drugs. IV. Constituents of Pueraria root. *Chem Pharm Bull* 8:688–691, 1960.

77. HJ Rong, JF Stevens, ML Deinzer, L DeCooman, D De Keukeleire. Identification of isoflavones in the roots of Pueraria lobata. *Planta Medica* 64:620–627, 1998.

78. G Chen, JX Zhang, JN Ye. Determination of puerarin, daidzein and rutin in Pueraria lobata (Wild.) Ohwi by capillary electrophoresis with electrochemical detection. *J Chromat A* 923:255–262, 2001.

79. WM Keung, BL Vallee. Kudzu root: an ancient Chinese source of modern antidipsotropic agents. *Phytochemistry* 47:499–506, 1998.

80. G Grynkiewicz, O Achmatowicz, W Pucko. Bioactive isoflavone — genistein. Synthesis and prospective applications. *Herba Polonica* 46:151–160, 2000.

81. KC Huang. *The Pharmacology of Chinese Herbs*. 2nd ed. Boca Raton, FL: CRC Press, 1999, p 87.

82. K Yamaki, DH Kim, N Ryu, YP Kim, KH Shin, K Ohuchi. Effects of naturally occurring isoflavones on prostaglandin E2 production. *Planta Medica* 68:97–100, 2002.

83. Y Noda, T Kaneyuki, A Mori, L Packer. Antioxidant activities of pomegranate fruit extract and its anthocyanidins: delphinidin, cyanidin, and pelargonidin. *J Agric Food Chem* 50:166–171. 2002.

84. KN Chidambara Murthy, GK Jayaprakasha, RP Singh. Studies on antioxidant activity of pomegranate (Punica granatum) peel extract using *in vivo* models. *J Agric Food Chem* 50:4791–4795, 2002.

85. A Constantinou, GD Stoner, R Mehta, K Rao, C Runyan, R Moon. The dietary anticancer agent ellagic acid is a potent inhibitor of DNA topoisomerases *in vitro*. *Nutr Cancer* 23:121–130, 1995.

86. ND Kim, R Mehta, W Yu, I Neeman, T Livney, A Amichay, D Poirier, P Nicholls, A Kirby, W Jiang, R Mansel, C Ramachandran, T Rabi, B Kaplan, E Lansky. Chemopreventive and adjuvant therapeutic potential of pomegranate (Punica granatum) for human breast cancer. *Breast Cancer Res Treat* 71:203–217, 2002.

87. G Nonaka, I Nishioka, M Nishizawa, T Yamagishi, Y Kashiwada, GE Dutschman, AJ Bodner, RE Kilkuskie, YC Cheng, KH Lee. Inhibitory effects of tannins on HIV reverse transcriptase and HIV replication in H9 lymphocyte cells. *J Nat Prod* 53:587–595, 1990.

88. RJ Cohen, K Ek, CX Pan. Complementary and alternative medicine (CAM) use by older adults: a comparison of self-report and physician chart documentation. *J Gerontol A Biol Sci Med Sci* 57:223–227, 2002.

89. H Hikino, Y Kiso, N Kato, Y Hamada, T Hiori, R Aiyama, H Itokawa, F Kiuchi, U Sankawa. Antihepatotoxic action of gingerols and diarylheptanoids. *J Ethnopharm* 14:31–39, 1985.

90. J Yamahara, M Mochizuki, HQ Rong, H Matsuda, H Fujimura. The antiulcer effect in rats of ginger constituents. *J Ethnopharmacol* 23:299–304, 1988.

91. KL Koo, AJ Ammit, VH Tran, CC Duke, BD Roufogalis. Gingerols and related analogues inhibit arachidonic acid-induced human platelet serotonin release and aggregation. *Thrombosis Res* 103:387–397, 2001.

92. N Nakatani. Phenolic antioxidants from herbs and spices. *Biofactors* 13:141–146, 2000.

93. BL Halvorsen, K Holte, MC Myhrstad, I Barikmo, E Hvattum, SF Remberg, AB Wold, K Haffner, H Baugerod, LF Andersen, O Moskaug, DR Jacobs Jr, R Blomhoff. A systematic screening of total antioxidants in dietary plants. *J Nutrition* 132:461–471, 2002.

94. CT Ho, T Ferraro, Q Chen, RT Rosen, MT Huang. Phytochemicals in teas and rosemary and their cancer-preventive properties. In: CT Ho, T Osawa, MT Huang, RT Rosen, Eds. *Food Phytochemicals for Cancer Prevention. ACS Symposium Series 547*. Washington, D.C.: American Chemical Society: Washington, D.C., 1994, pp 2–19.

95. M Demente, J Michaud-Levesque, B Annabi, D Gingras, D Boivin, J Jodoin, S Lamy, Y Bertrand, R Beliveau. Green tea catechins as novel antitumor and antiangiogenic compounds. *Curr Med Chem Anticancer Agents* 2:441–463, 2002.

96. JD Lambert, CS Yang. Cancer chemopreventive activity and bioavailability of tea and tea polyphenols. *Mut Res* 523:201–208, 2003.

97. N Caturla, E Vera-Samper, J Villalain, C Reyes Mateo, V Micol. The relationship between the antioxidant and the antibacterial properties of galloylated catechins and the structure of phospholipid model membranes. *Free Radic Biol Med* 6:648–662, 2003.

98. T Shimamura. Inhibition of influenza virus infection by tea polyphenols. In: CT Ho, T Osawa, MT Huang, RT Rosen, Eds. *Food Phytochemicals for Cancer Prevention.* ACS Symposium Series 547. Washington, D.C.: American Chemical Society, 1994, pp 101–104.

99. H Nakane, Y Hara, K Ono. Tea polyphenols as a novel class of inhibitors for human immunodeficiency virus reverse transcriptase. In: CT Ho, T Osawa, MT Huang, RT Rosen, Eds. *Food Phytochemicals for Cancer Prevention.* ACS Symposium Series 547. Washington, D.C.: American Chemical Society, 1994, pp 56–64.

100. F Hashimoto, Y Kashiwada, G Nonaka, I Nishioka, T Nohara, LM Cosentiono, KH Lee. Evaluation of tea polyphenols as anti-HIV agents. *Bioorg Med Chem Lett* 6:695–700, 1995.

101. http//www.alternative-medicines.com/herbdesc1/1garlic.htm

102. JC Harris, SL Cottrell, S Plummer, D Lloyd. Antimicrobial properties of Allium sativum (garlic). *Appl Microbiol Biotechnol* 57:282–286, 2001.

103. BS Reddy, CV Rao. Chemoprevention of colon cancer by thiol and other organosulur compounds. In: CT Ho, T Osawa, MT Huang, RT Rosen, Eds. *Food Phytochemicals for Cancer Prevention*. ACS Symposium Series 546. Washington, D.C.: American Chemical Society, 1994, pp 164–172.

104. MR Doyle, MF Webster, LD Erdmann. Allithiamine ingestion does not enhance isokinetic parameters of muscle performance. *Intl J Sport Nutri* 7:39–47, 1997.

105. TS Wu, YM Lin, M Haruna, DJ Pan, T Shingu, YP Chen, HY Hsu, T Nakano, KH Lee. Kansuiphorins A and B, two novel antileukemic diterpene esters from Euphorbia kansui. *J Nat Prod* 54:823–829, 1991.

106. DJ Pan, CQ Hu, JJ Chang, TTY Lee, YP Chen, HY Hsu, DR McPhail, AT McPhail, KH Lee. Kansuiphorin-C and -D, cytotoxic diterpenes from Euphorbia kansui. *Phytochemistry* 30:1018–1020, 1991.

107. T Matsumoto, JC Cyong, H Yamada. Stimulatory effects of ingenols from Euphorbia kansui on the expression of macrophage Fc receptor. *Planta Medica* 58:255–258, 1992.

108. WF Zheng, Z Cui, Q Zhu. Cytotoxicity and antiviral activity of the compounds from Euphorbia kansui. *Planta Medica* 64:754–756, 1998.

109. YP Zhu. *Chinese Materia Medica, Chemistry, Pharmacology and Applications*. The Netherlands: Harwood Academic Publishers, 1998, pp 120–122.

110. KH Lee, YM Lin, TS Wu, DC Zhang, T Yamagishi, T Hayashi, IH Hall, JJ Chang, RY Wu, TH Yang. The cytotoxic principles of *Prunella vulgaris, Psychotria serpens* and *Hyptis capitata*: ursolic acid and related derivatives. *Planta Medica* 54:308–311, 1988.

111. G Cragg, M Suffness. Metabolism of plant-derived anticancer agents. *Pharmac Ther* 37:425–461, 1988.

112. C Keller-Juslen, M Kuhn, A von Wartburg, H Stahelin. Synthesis and antimitotic activity of glycosidic lignan derivatives related to podophyllotoxin. *J Med Chem* 14:936–940, 1971.

113. PJ O'Dwyer, MT Alonso, B Leyland-Jones, S Marsoni. Teniposide: a review of 12 years of experience. *Cancer Treat Rep* 68:1455–1466, 1984.

114. VePesid Product Information Overview, Bristol Lab, 1983.

115. BF Issell, FM Muggia, SK Carter, Eds. *Etoposide (VP-16) Current Status and New Developments*. Orlando, FL: Academic Press, pp 5–8, 1984.

116. H Schulze-Bergkamen, I Zuna, A Teufel, W Stremmel, J Rudi. Treatment of advanced gastric cancer with etoposide, folinic acid, and fluorouracil in the clinical setting: efficacy of therapy and value of serum tumor markers. *Med Oncol* 19:43–53, 2002.

117. MB Steins, H Serve, M Zuhldorf, N Senninger, M Semik, WE Berdel. Carboplatin/etoposide induces remission of metastasised malignant peripheral nerve tumours (malignant schwannoma) refractory to first-line therapy. *Oncol Rep* 9:627–630, 2002.

118. JM van Maanen, J Retel, J deVries, HM Pinedo. Mechanism of action of antitumor drug etoposide: a review. *J Natl Cancer Inst* 80:1526–1533, 1988.

119. KH Lee, Y Imakura, M Haruna, SA Beers, LS Thurston, HJ Dai, CH Chen, SY Liu, YC Cheng. New cytotoxic 4-alkylamino analogues of 4'-demethyl-epipodophyllotoxin as inhibitors of human DNA topoisomerase II. *J Nat Prod* 52:606–613, 1989.

120. SJ Cho, A Tropsha, M Suffness, YC Cheng, KH Lee. Three-dimensional quantitative structure–activity relationship study of 4'-O-demethylepipodophyllotoxin analogs using the modified CoMFA/q^2-GRS approach. *J Med Chem* 39:1383–1395, 1996.

121. Z Xiao, YD Xiao, J Feng, A Golbraikh, A Tropsha, KH Lee. Modeling of epipodophyllotoxin derivatives using variable selection k nearest neighbor QSAR method. *J Med Chem* 45:2294–2309, 2002.

122. JY Chang, FS Han, SY Liu, HK Wang, KH Lee, YC Cheng. Effect of 4 β-arylamino derivatives of 4'-O-demethylpodophyllotoxin on human DNA topoisomerase II, tubulin polymerization, KB cells, and their resistant variants. *Cancer Res* 51:1755–1759, 1991.

123. ZW Wang, YH Kuo, D Schnur, JP Bowen, SY Liu, FS Han, JY Chang, YC Cheng, KH Lee. New 4 β-arylamino derivatives of 4'-O-demethylepipodophyllotoxin and related compounds as potent inhibitors of human DNA topoisomerase II. *J Med Chem* 33:2660–2666, 1990.

124. SY Liu, R Soikes, J Chen, T Lee, G Taylor, KM Hwang, KH Lee, YC Cheng. GL331: a Novel Compound which Can Overcome Multiple Drug Resistance. 84th AACR Annual Conference, Orlando, FL, 1993.

125. A Lee, WD Ngan Kee, T Gin. A quantitative, systematic review of randomized controlled trials of ephedrine versus phenylephrine for the management of hypotension during spinal anesthesia for cesarean delivery. *Anesthesia Analgesis* 94:920–926, 2002.

126. J Arditti, JH Bourdon, M Spadari, L de Haro, N Richard, M Valli. Ma Huang, from dietary supplement to abuse. *Acta Clin Belg Suppl* 57:34–36, 2002.

127. SR Meshnick. In: PJ Rosenthal, Ed. *Antimalarial Chemotherapy*. Totowa, NJ: Human Press, 2001, pp 191–201.

128. MA Avery, G McLean, G Edwards, A Ager. Structure-activity relationships of peroxide-based artemisinin antimalarials. In: SJ Cutler, HG Cutler, Eds. *Biologically Active Natural Products: Pharmaceuticals*. Boca Raton, FL: CRC Press, 2000, pp 121–132.

129. Y Imakura, T Yokoi, T Yamagishi, J Koyama, H Hu, DR McPhail, AT McPhail, KH Lee. Synthesis of desethanoqinghaosu, a novel analogue of the antimalarial qinghaosu. *J Chem Soc Chem Commun* 372–374, 1988.

130. Y Imakura, K Hachiya, T Ikemoto, S Yamashita, M Kihara, T Shingu, WK Milhous, KH Lee. Acid degradation products of qinghaosu and their structure–activity relationships. *Heterocycles* 31:1011–1016, 1990.

131. Y Imakura, K Hachiya, T Ikemoto, S Kobayashi, S Yamashita, J Sakakibara, FT Smith, KH Lee. Antimalarial artemisinin analogs: synthesis of 2,3-desethano-12-deoxoartemisinin-related compounds. *Heterocycles* 31:2125–2129, 1990.

132. PK Hsiao. Traditional experience of Chinese herb medicine: its application in drugs research and new drug searching. In: JL Beal, E Reinhard, Eds. *Natural Products as Medicinal Agents*. Stuttgart, Germany: Hippokrates Verlag, 1981, pp 351–394.

133. Q Zeng, D Du, D Xie, X Wang, C Ran. Antitumor activities of indirubin derivatives. *Chin Trad Herb Drugs* 13:24–30, 1982.

3

Chemistry, Pharmacology, and Quality Control of Selected Popular Asian Herbs in the U.S. Market

MINGFU WANG, QING-LI WU, and
JAMES E. SIMON

New Use Agriculture and Natural Plant
Products Program, Department of Plant
Biology and Pathology, Rutgers University,
New Brunswick, NJ, USA

YI JIN and CHI-TANG HO

Department of Food Science, Rutgers
University, New Brunswick, NJ, USA

INTRODUCTION

Herbs have been used as medicines and functional foods in the Asian world for thousands of years. Before western medicines were introduced into Asia, herbs had been the main method for the treatment of diseases and remain a main source of drugs in primary healthcare. Currently, over half of the

Chinese population use traditional herbal remedies, particularly when western medicines do not appear to be as effective, as in the case of chronic ailments such as age-related diseases. The traditional Chinese medicine in China is undergoing a renaissance, as the current Chinese government has recognized that traditional Chinese medicine is a treasure for the nation and can provide both improved health care for its own citizens as well as serving as an excellent source of phytomedicines for the international export market when it is combined with modern science. The Chinese government is now attempting to modernize traditional Chinese medicine (TCM). The push toward modernization in this field has led to an increase in the use of modern pharmacological experiments, standardization of the active components, identification and use of marker compounds in herbal prescription, establishing fingerprinting profiles (chemically and genetically) for single and blended herbs, and a wide range of other quality-related issues that now face the international acceptance and use of TCM. The long history of TCM, and its associated assumption of safety, if not efficacy, has attracted much interest for the European and North American marketplace to examine and use Asian herbs for disease prevention and treatment. Given the recent demographic trends in the U.S. (the latest U.S. national census), there is a strong rise in the number of first and second generation Chinese and Asian families. This demographic development has contributed to an increased demand and interest to have the same herbs available in China, now available for use in their new country. Consequently, due to perceived consumer demand for these products, many western companies have been importing, promoting, and distributing a wide range of Asian herbs in the western market. Popular Asian herbs including ginseng (*Panax ginseng* C. A. May), ginkgo (*Ginkgo biloba*), Dong Quai (*Angelica sinensis* [Oliv.] Diels) and Siberian ginseng (*Eleutherococcus senticosius*) are among the top 20 selling herbs in the U.S. market.

Herbal products are marketed in the U.S. as dietary supplements and not as medicinal plants or medicines. The U.S. Food and Drug Administration (FDA) defines dietary supplements as:

A dietary supplement is a product taken by mouth that contains a "dietary ingredient" intended to supplement the diet. The "dietary ingredients" in these products may include: vitamins, minerals, herbs or other botanicals, amino acids, and substances such as enzymes, organ tissues, glandulars, and metabolites. Dietary supplements can also be extracts or concentrates, and may be found in many forms such as tablets, capsules, softgels, gelcaps, liquids, or powders. They can also be in other forms, such as a bar, but if they are, information on their label must not represent the product as a conventional food or a sole item of a meal or diet (FDA website).

The dietary supplements are regulated by Dietary Supplement Health and Education Act (DSHEA), a branch of the FDA. This passage of DSHEA has opened up opportunity for a plethora of new herbs to enter the U.S. marketplace. While this provides consumers with a wider range of options, the main issues relating to the safety, efficacy, and quality control of herbs clearly sold and marketed as medicinal plants are never really addressed. As a result of the ambiguity of the law, the lack of strict guidelines, and regulations, many problems have been identified in herbal products, including those related to Asian herbs. In this chapter, we present general information on a number of popular Asian herbs in the U.S. market, discuss their current applications, chemistry, pharmacology, and quality control. We focus our comments on four popular Asian herbs: *Angelica sinensis* (Oliv.) Diels, *Rhodiola rosea, Pueraria lobata,* and *Panax ginseng* C. A. May.

ANGELICA SINENSIS (OLIV.) DIELS (DONG QUAI OR DANG GUI, OR TANG KUEI)

The whole root of this Chinese plant, *Angelica sinensis* (Family: Apiaceae) is used as an herbal medicine in China and is a well-recognized tonic herb for women. Traditionally, *Dong Quai* is used to treat obstetric and gynecological problems; the herb drug acts as a mild laxative, a uterine tonic, antispasmodic, and alterative (blood purifying), it increases blood circulation, relaxes the uterus, stabilizes pregnancy, and is used for regulating the menstrual cycle. Currently, in the U.S., it is marketed towards alleviating female disorders such as premenstrual syndrome, menstrual cramps, and to ease discomfort associated with menopause. *Dong Quai* is sold as a single herb or herbal extract, or combined with other herbs, such as black cohosh (*Cimicifuga racemosa*), chase tree berries (*Vitex agnus castus*), blue cohosh (*Caulophyllum thalictroides*), and astragalus (*Astragalus membranaceous*) in complex formulas in the U.S. market.

Chemical Components

The investigation of the phytochemistry of *Angelica sinensis* root has revealed the presence of several distinct groups of chemical compounds.[1-9]

1. Amino acids: some 20 amino acids have been reported in *Angelica sinensis* with arginine and glutamic acid as the major ones.
2. Essential oils: 49 volatile aromatic compounds have been identified by gas chromatography–mass spectrometry (GC-MS) with ligustilides (Figure 3.1) reported as the major constituents.
3. Sterols: β-sitosterol, stigmasterol, and β-sitosterol-D-glucoside have been identified.
4. Fatty acids and organic acids: palmitic acid, linoleic acid, stearic acid, arachidonic acid, ferulic acid, and vanillic acid were present.
5. Coumarins: bergaptene, imperatorin, psoralen, osthole, oxypeucedanin, scopoletin, and umbelliferone were present.

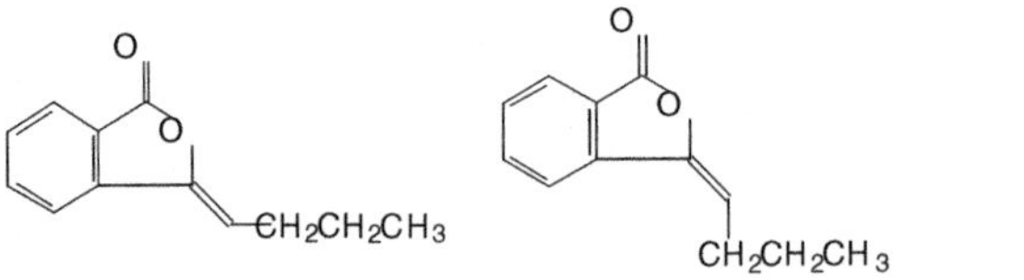

ligustilide   *E*-ligustilide

Figure 3.1 Structures of ligustilide and E-ligustilide.

6. Polysaccharides were identified.
7. Other components: E232, angelicide, brefeldin A, tetradecan-1-ol, tetramethylpyarazine were present.

Pharmacological Activity

The extracts of *Angelica sinensis* showed antiarrhythmic effects on adrenaline induced arrhythmia in cats,[10] while promoting melanocytic proliferation, melanin synthesis, and tyrosinase activity.[11] *Dong Quai* was also found to improve the blood circulation of the injured nerve,[12] to regulate lipopolysaccharide (LPS)-induced elevation of Ca^{2+} intracellular level of alveolar macrophages, and may inhibit nonspecific inflammation of airways in chronic bronchitis,[13] to protect the human vascular endothelial cell from the effects of oxidized low-density lipoprotein *in vitro*,[14] to enhance gastric ulcer healing in rats and promote wound repair in RGM-1 cells[15] and to stimulate the proliferation, alkaline phosphatase (ALP) activity, protein secretion, and particularly type I collagen synthesis of human osteoprecursor cells (OPC-1).[16] *Dong Quai* and its constituent ferulic acid have been reported to potentiate the phagocytic activity of macrophages and inhibit blood platelet aggregation and serotonin (5HT) release by blood platelets of rat *in vivo* and *in vitro*,[17] to stimulate murine spleen lymphocytes and their proliferation, and to increase Con-A stimulated DNA and protein synthesis and interleukin-2 production.[18] Ferulic acid was found to have anti-inflammatory effects, significantly inhibiting the edema induced by carrageenin.[19] The *Angelica sinensis* polysaccharides were found to decrease colony formation in spleen hematopoietic tissue of irradiated mice,[20] to increase the proliferation of several types of precursor cells in healthy and anemic mice and increase hematopoisis[21] and showed protective effects on gastrointestinal damage induced by ethanol or indomethacin in rats,[22] protective effects on hepatic injury induced by acetaminophen in rodents,[23] antianemic and immunofunction-regulating activities,[24] an extensive effect on immunocompetence,[25] promoting gastric ulcer healing[26] and augmenting mice splenocyte proliferation, released interferon-γ (IFN-γ) and increased IFN-γ bioactivity.[27] Sodium ferulate and ethanol sediments from Angelica were also found to have protective effects on the immunological liver injury induced by lipopolysaccharide in bacillus calmette-guerin primed mice.[28]

Analysis and Quality Control of Dong Quai Products

Various analytical methods have been used to analyze *Dong Quai* products including GC-MS analysis of the essential oils,[29–31] analysis of ferulic acid content in angelica root and its preparation by high-pressure liquid chromatography (HPLC),[32,33] quantitative determination of ligustilide in *Dong Quai* using HPLC with fluorometric detection,[34] liquid chromatography–mass spectrometry (LC–MS) analysis of phthalides including ligustilide in extracts,[35] analysis of the chemical components of angelica and related umbelliferous drugs by thin layer chromatography, HPLC, and LC-MS,[36] quantitative analysis of ferulic acid in *Dong Quai* by high-performance capillary electrophoresis[37,38] and RAPD (random amplified polymorphic DNA) analysis of angelica.[39,40]

The *Pharmacopoeia of the People's Republic of China* (English edition 2000, Chemical Industry Press, Beijing) has a monograph on *Dong Quai*, where it defines this drug as the dried root of *Angelica sininsis* (Oliv.) Diels. The drug must contain not more than 7.0% of total ash, not more than 2.0% acid-insoluble ash, and extractives in 70% ethanol should not be less than 45%. In the U.S. market, the contents of ligustilide and *E*-ligustilide (Structures are illustrated in Figure 3.1) are used as quality control standards.

We have developed a robust and reliable HPLC method to analyze ligustilide and *E*-ligustilide contents in *Dong Quai* using the following conditions and representative HPLC chromatogram as shown in Figure 3.2.

Column, Waters Nova-Pak C18, 3.9*150 mm; mobile phase, 0.1% phosphoric acid solution/methanol isocratic (45:55); extraction solvent, methanol; flow rate, 1.0 mL/min; temperature, ambient; detection wavelength, 274 nm; injection volume, 10 µL; running time, 20 minutes; retention time, *E*-ligustilide, 8.1 min; and ligustilide, 9.9 min.

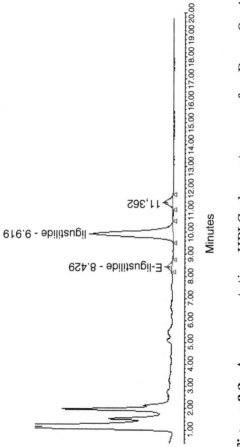

Figure 3.2 A representative HPLC chromatogram of a Dong Quai extract (1% ligustilide).

RHODIOLA ROSEA (GOLDEN ROOT, ROSEROOT)

The genus of Rhodiola consists of over 200 species, most found in Asia, and many used in TCM. In TCM, Rhodiola is used as a general tonic in the treatment and invigoration of the human body. In the old Chinese pharmacopoeias such as the *Ben Cao*, Rhodiola was documented to prolong human life, to enhance the *Qi*, and keep the body "light." In the U.S. market, the best-known Rhodiola species is *Rhodiola rosea* L., also known as golden root or roseroot. *Rhodiola rosea* grows primarily in dry sandy ground at high altitudes in the Arctic areas of Europe and Asia. The plant is perennial with a thick rhizome, which is used as the herbal drug. Russians were the first to introduce Rhodiola into the U.S. market. Extracts of the *Rhodiola Rosea* root has been greatly researched in Russia and found to contain powerful adaptogens. In the U.S., Rhodiola is marketed and sold as a product to improve mental health, with adaptogenic, antistress and cardioprotective agents. Rhodiola is sold as single herbal extract though it can be found combined with other well-known adaptogenic herbs such as ginseng and Siberian ginseng.

Chemical Components

The investigation of the phytochemistry of *Rhodiola rosea* root has revealed the presence of several distinct groups of chemical compounds.[41–43]

1. Phenylpropanoids: rosavin, rosin, and rosarin were present.
2. Phenylethanol derivatives: salidroside and tyrosol were present.
3. Flavanoids: acetylrodalgin, kaempferol, kaempferol 7-rhamnoside, rodiolin, rodionin, rodiosin, tricin; tricin 5-glucoside, and tricin 7-glucoside have been identified.
4. Monoterpenes: such as rosiridol and rosaridin were present.
5. Sterols: daucosterol and -sitosterol were present.

6. Phenolic acids: chlorogenic acid, hydroxycinnamic acids, and gallic acid have been identified.
7. Polysaccharides.
8. Tannins.

Pharmacological Studies

Extracts of *Rhodiola rosea* exhibited adaptogenic effects in mice and rabbits,[44] cardio-protective and antiaderenergic effects during stress.[45] One standardized extract SHR-5 was reported to significantly relieve stress-induced fatigue in a double blind cross-over study.[46] An alcohol-aqueous extract (1:1) was found to improve learning and long-term memory in mice.[47] Rhodiola extracts were also reported to prevent ischemic brain damage development,[48] to scavenge free radicals,[49] to show antimicrobial effects against some strains of *Staphylococcus aureus*,[50] antitumor effect in experiments on inbred and noninbred mice and rats with transplantable NK/Ly tumor, Ehrlich's adenocarcinoma, melanoma B16, and Lewis lung carcinoma,[51] to decrease cyclophosphamide haematotoxicity in mice with Ehrlich and Lewis transplantable tumors[52] and to increase the resistance of experimental animals to adrenalin and $CaCl_2$-induced arrhythmias.[53] In a small clinical trial with 12 superficial bladder patients, the oral administration of Rhodiola extract was found to improve the characteristics of the erothelial tissue integration, parameters of leukocyte integrins and T-cell immunity, and average frequency of relapses for these patients has been found to fall twice.[54]

Analysis and Quality Control of Rhodiola

Analytical methods including GC, thin-layer chromatography (TLC), HPLC and colorimetric methods have been used to analyze the chemical components in *Rhodiola*. These reported methods include GC and GC-MS analysis of volatiles with decanol, geraniol and 1,4-p-menthadien-7-ol as the major volatile,[55] reverse phase HPLC analysis of salidroside and tyrosol with a Nova-Pak C_{18} column and 6.5% methanol as mobile phase,[56] analysis of rosavin by reverse phase HPLC,[57] TLC analysis of rosin, rosavin and rosarin with $CHCl_3$-MeOH-H_2O (26:14:3) as mobile phase[58] and a photometric method for quantification of salidroside.[59] The best reported analytical method was developed by Ganzera et al with reverse phase HPLC simultaneously determining five marker compounds (salidroside, rosavin, rosin, rosarin, and rosiridin) in Rhodiola.[60] In general, salidroside, rosavin, rosarin, and rosin (structures are shown in Figure 3.3) were used as marker compounds to control the quality of Rhodiola extract. In the U.S. market, there are different graded products of Rhodiola, including 4% salidroside and 4% total rosavins extract and 1% salidroside and 4% total rosavins extract. However, from a quality perspective, *Rhodiola rosea* extract must contain all four marker compounds. In our screening of this herb, we have developed an analytical HPLC method to simultaneously analyze all four marker compounds in *Rhodiola rosea*. This was achieved using the following conditions, and the representative HPLC chromatogram was shown in Figure 3.4.

Column, Phenomenex Phenyl-hexyl, 4.6×150 mm, 3 μ*M*; mobile phase, 0.2% phosphoric acid (A)-acetonitrile (B), initial

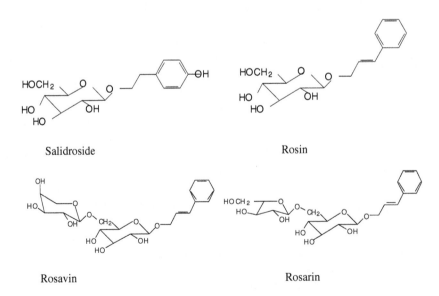

Figure 3.3 Structures of marker compounds in *Rhodiola rosea*.

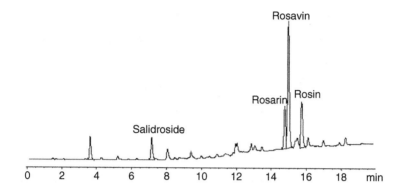

Figure 3.4 Representative HPLC chromatogram of *Rhodiola rosea* extract.

4% B, linear gradient to 30% B in 20 min; extraction solvent, 40% methanol aqueous solution; flow rate, 1.2 mL/min; temperature, ambient; detection wavelength, 225 nm for 0 to 10 min and 254 nm for 10 to 20 min; injection volume, 10 µL; running time, 20 min; retention time, salidroside, 7.2 min; rosarin, 14.7 min; rosavin, 15.0 min; and rosin, 15.7 min.

PUERARIA LOBATA (KUDZU, GE GENG)

Kudzu is one of the earliest medicinal plants used in traditional Chinese herbal medicine. The roots of kudzu have been used as antidiarrhetic, antipyretic, diaphoretic, and antiemetic agents, as well as to treat alcohol-related problems (intoxication and alcohol abuse). Currently, kudzu is marketed as a rich source of isoflavone, as an herb for women's health, and as an antialcohol abuse plant material. As a plant for women's health, this herbal drug is formulated together with soy isoflavone extract, red clove isoflavone extract, and chase tree berries.

Chemistry

The chemistry of kudzu has been studied extensively. The major components were isoflavones and saponins[61–68] including:

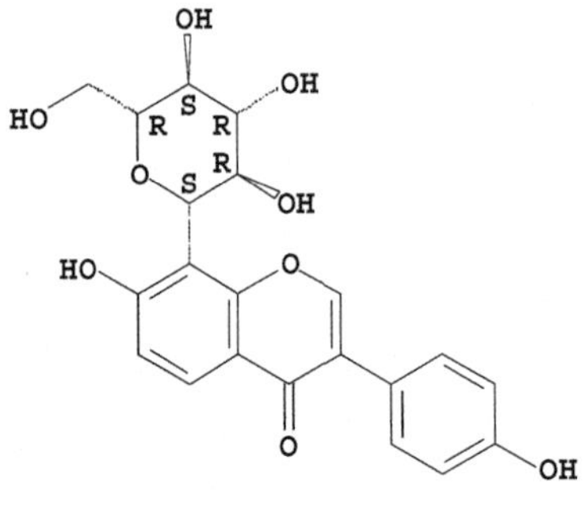

Figure 3.5 Structure of puerarin.

1. Isoflavones: Puerarin (Figure 3.5), daidzin, daidazein-4,7-diglucoside, 6,7-dimethoxy-3′,4′-methylenedioxy-isoflavone, formononetin, mirificin, 3′-methoxy puerarin, genistein 8-C-glucoside, genistin, genistein, 6″-O-malonyldaidzin, 3′-hydroxy-4′-O-D-glucosylpuerarin and 3′-methoxydaidzin have been identified.
2. Chalcones: isoliquiritigenin was present.
3. Aromatic glycoside: pueroside-A and -B, but-2-enolides, sophoroside A.
4. Sterols: β-sitosterol and daucosterol have been identified.
5. Saponins: kudzusaponins A1, A2, A3, A4, A5, SA1, SA2, SA3, SA4, SB1 and C1, soyasaponins I SA3, and I have been identified.
6. Tryptophan derivatives: PF-P was present.
7. Volatile compounds: such as methyl palmitate, dimethyl suberate, and furfuryl alcohol were present.

Pharmacological Activities

Kudzu extracts have shown antimutagenic activity,[69] antidipsotropic activity[70] and suppressed alcohol preference in a pharmacogenetic rat model of alcoholism.[71] The flavone extracts of kudzu affect coronary circulation, cardiac hemodynamics and myocardial metabolism in dogs. The extracts resulted in hypotensive effects on anesthetized dogs and unanesthetized hypertensive dogs, decreased vascular resistance in anesthetized dogs, and increased peripheral and cerebral circulation.[72,73] Puerarin showed stimulatory effect on α_{1A}-adrenoceptor to increase glucose uptake into cultured C_2Cl_2 cells of mice,[74] antioxidant activity.[75] Puerarin, daidzin, and daidzein showed anti-inebriation and the antidipsotropic effects.[76] The saponins from Kudzu also showed protective effects on *in vitro* immunological liver injury of rat primary hepatocyte cultures.[77,78]

Analysis and Quality Control of Kudzu

HPLC and TLC methods have both been widely used to analyze the isoflavones, including puerarin in kudzu and kudzu extracts.[79–81] Other analytical methods include determination of puerarin, daidzein, and rutin in *Pueraria lobata* (Wild.) Ohwi by capillary electrophoresis with electrochemical detection[82] and HPLC analysis of hepatoprotective oleanene-glucuronides in *Puerariae Flos*.[83] *Pharmacopoeia of the People's Republic of China* includes a monograph for Kudzu in which it defines Kudzu as the dried root of *Pueraria lobata* (Wild.) Ohwi or *Pueraria thomsonii* Benth. (Fam. Leguminosae). In *Pueraria lobata*, the herbal product must contain not less than 2.4% of puerarin, not more than 14% moisture, and not more than 7% of total ash. In the U.S. market, Kudzu is usually sold as an extract (ca. 40% isoflavones). We have developed an HPLC method to analyze isoflavones in Kudzu under the following conditions:

HPLC column, Water Symmetry C_{18}, 5 µm, 3.90 × 150 mm; column temperature, ambient; mobile phase, the mobile phase consisted of solvent A (0.1% formic acid solution) and solvent B (acetonitrile) with the following gradient solvent system:

Time (min)	%A	%D
0.0	90	10
40	65	35

flow rate, 0.8 mL/min; injection volume, 10 μL; detection wavelength, 255 nm; running time, 40 min, and postrun time, 15 min; retention times of puerarin, daidzin, glycitin, genistin, daidzein, glycitein, and genistein are about 6.5, 9.5, 10.4, 14.7, 23.5, 24.7, and 31.7 min, respectively.

GINSENG (PANAX GINSENG C. A. MAY)

Ginseng (*Panax ginseng* C. A. May) is the most popular traditional Chinese medicinal herb. It is not only widely used in Asian countries, such as China, Korea, and Japan, but also widely used in many western countries. For thousands of years, ginseng has been used as a tonic to increase nonspecific resistance against a wide array of various stress agents, to prevent and cure many health conditions, and has been used as an emergency medicine to save dying patients. In the U.S., it is marketed to improve mental performance in times of stress, to enhance overall health and vitality, to improve resistance to the damaging effects of stress, and to increase endurance. Ginseng is used and marketed as a major tonic, stimulant, and immune booster. Ginseng is available in the U.S. market in a myriad of products as a powder, extract (7% ginsenosides and 15% ginsenosides extract), or combined with other herbs such as American ginseng, rhodiola, Siberian ginseng and various vitamins. The dosage varies based upon each of its formulations.

Chemical Components

The chemistry of ginseng has been extensively studied, and the main components include saponins and polysaccharides.[84–88]

1. Saponins: about 30 saponins have been purified from the root of ginseng with ginsenoside, Rg_1, Re, Rb_1, Rc, Rd, and Rf as the major saponins (structures are shown in Figure 3.6) have been identified in root extracts.
2. Sterols: β-sitosterol, stigmasterol, and campesterol were present.

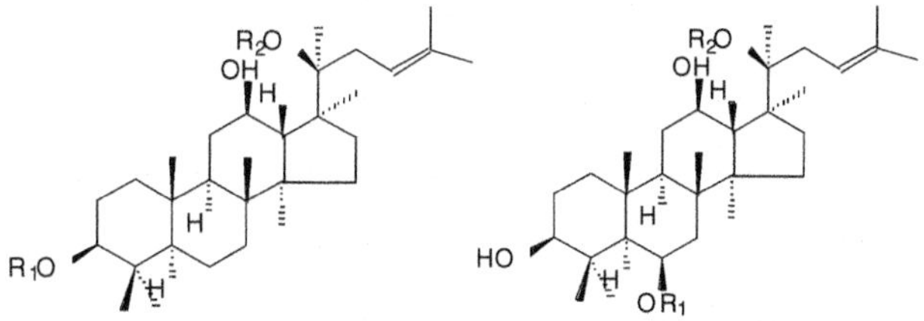

Rb1	Glc(2-1)Glc	Glc(6-1)Glc	Re	Glc(2-1)Rha	Glc
Rb2	Glc(2-1)Glc	Glc(6-1)Ara(p)	Rf	Glc(2-1)Glc	H
Rc	Glc(2-1)Glc	Glc6Ara(f)	Rg1	Glc	Glc
Rd	Glc(2-1)Glc	Glc			

Figure 3.6 Structures of the major ginsenoside in Asian ginseng root.

3. Polyalkynes: heptadeca-1-en-4,6-diyne-3,9,10-triol, panaxynol, panaxynol, panaxydol, and panaxytriol, ginsenoynes A, B, C, D, and E have been identified.
4. Fatty acids: linoleic acid, palmitic acid, oleic acid, and linolenic acid were present.
5. Amino acids.
6. Peptides.
7. Polysaccharides.

Pharmacological Studies

Extensive pharmacological studies have been reported on ginseng powder, ginseng extracts, and ginseng components.[89,90] Ginseng was found to improve different aspects of cognitive performance of healthy young adults,[91] and to result in a reduction in the bile flow and bile secretion of total lipids and cholesterol, while increasing the secretion of proteins in a dose-dependent manner.[92] The butanol fraction of ginseng was found to inhibit gastric damage.[93] Ginseng saponins were discovered to contain components potentiating the apoptosis of MMS-exposed NIH3T3 cells via p53 and p21 activation, accompanied with down-regulation of cell cycle-related protein expression,[94] protecting hippocampal CA1 and CA3 cells against KA-induced neurotoxicity[95] and inhibiting EGF-induced cell proliferation via decrease of c-fos and c-jun gene expression in primary cultured rabbit renal proximal tubular cells.[96] Topical application of ginsenosides significantly attenuated ear edema induced by 12-O-tetradecanoylphorbol-13-acetate (TPA) and ginsenosides also suppressed expression of cyclooxygenase-2 (COX-2) and activation of NF-B in the TPA-treated dorsal skin of mice.[97] The total ginseng saponins and ginsenoside Rb1 and Rg1 showed neuroprotective effects on spinal cord neurons, with Rb1 and Rg1 protecting spinal neurons from excitotoxicity induced by glutamate and kainic acid, as well as oxidative stress induced by H_2O_2.[98] Rb1 showed nootropic properties,[99] ginsenoside Rb2 showed epidermis proliferative effect[100] and ginsenoside Rg2 blocked the nicotinic acetylcholine receptors in bovine chromaffin cells.[101] Ginsenoside Rg3 was found to modulate Ca^{2+} channel currents in rat sensory neurons[102] and to inhibit N-methyl-D-aspartate (NMDA) receptors.[103] Panaxytriol, a polyalkyne from ginseng was found to inhibit tumor cell proliferation and induct G2/M cell cycle arrest.[104] The polysaccharides from *Panax ginseng* showed antisepticaemic activities[105] and immunostimulating effect.[106]

Analysis and Quality Control of Ginseng

Various methods have been applied to analyze ginseng, especially ginsenosides. These methods include, but are not limited to, colorimetric, TLC, GC, HPLC and LC-MS. HPLC with UV detection at 203 nm is the most popular and accepted method for analysis of ginsenosides. The content of ginsenosides is being used in industry quality control standards. USP 25 (U.S. pharmacopoeia National Formulary, 2002 edition) has official monographs for ginseng, powdered ginseng, and powder ginseng extract. In this monograph, ginseng root must contain at least 0.2% of ginsenoside Rg1 and 0.1% Rb1, less than 12% moisture, less than 2% foreign organic matter, less than 8.0% total ash, and not less than 14% alcohol soluble extractives. In the *Pharmacopoeia of the People's Republic of*

China (English edition, 2000, Chemical Industry Press) ginseng is defined as the dried root of *Panax Ginseng* C. A. May. (Fam. Araliaceae), and it must contain not less than 0.25% of the sum of ginsenoside Rg1 and Re. We have developed a robust HPLC method to analyze ginsenosides in ginseng and American ginseng. The representative HPLC chromatograms for ginseng main root, ginseng powder, and ginseng leaves (Figure 3.7) were characterized using the following conditions:

Column, Phenomenex Phenyl-hexyl 4.6 × 150 mm, 3 µM; mobile phase, water (A)-acetonitrile (B) gradient as described:

Time (min)	%B	%A
0.0	20	80
10.0	20	80
45.0	30.0	70.0
55.0	33.0	67.0
70	50.0	50.0
71	100	0
75	100	0
76	20	20
90	20.0	80.0

Extraction solvent, 60% methanol aqueous solution; flow rate, 0.8 mL/min; temperature, ambient; detection Wavelength, 203 nm; injection volume, 25 µL; running time, 90 min; ginsenoside Rg1 elution time, approximately 16.5 min; ginsenoside Re elution time, approximately 17.5 min; ginsenoside Rb1 elution time, approximately 45.3 min; ginsenoside Rc elution time, approximately 47.5 min; ginsenoside Rb2 elution time, approximately 50.3 min; ginsenoside Rd elution time, approximately 54.5 min.

A study was recently carried out to evaluate the quality of ginseng products in the U.S. market. Six ginseng (*Panax ginseng*) finished products either in the capsule or in the tables form were selected at random and purchased from local health food stores and supermarkets. Based on the label claims, each product was claimed as ginseng root or root extract. Four ginseng extracts (15% ginsenosides) were also acquired from local

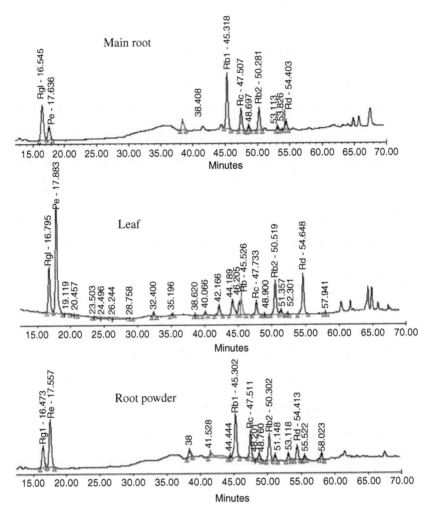

Figure 3.7 HPLC chromatogram of ginseng main root, leaf and powder.

or oversea botanical raw material suppliers. Among the six commercial ginseng products, four commercial sources reported the total amount of ginsenosides and claimed the product as a ginseng extract. Based on our testing results, three met their own product label claims, one failed containing only 50% of the reported minimum and actually appeared to be a leaf extract

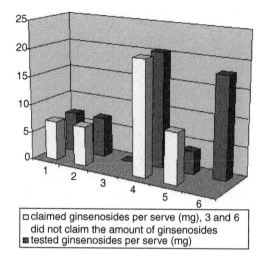

Figure 3.8 Contents of ginsenosides in commercial products.

by its HPLC chromatogram rather than a root extract as it was labeled. The other two commercial products were labeled as ginseng powders, with one source appearing to be a real ginseng main root powder, containing 1.4% ginsenosides. The other source, in contrast, did not contain any ginsenosides (0%) although it claimed to be ginseng (see Figure 3.8).

For the four 15% ginsenosides extracts (all claimed as root extract), it was concluded that none of these products were pure root extract and each appeared to be spiked with ginseng leaf extracts as based on the HPLC chromatograms (in which we observed increased levels of ginsenoside Re, Rg1 and Rd, and decreased levels of Rb1) (Figure 3.9).

CONCLUDING REMARKS ON QUALITY

As Chinese herbal medicines move into U.S. markets, and into the dietary supplement mainstream, the value of these products to the health care industry, and to consumers, will be predicated largely on their proper use, additional scientific studies using both animal and human studies to evaluate efficacy, and ensuring that a quality product reaches the

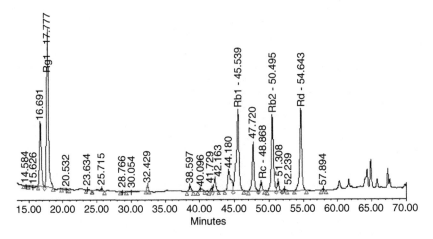

Figure 3.9 Representative HPLC chromatogram of ginseng root extract spiked with leaves extract.

consumer. Issues and problems that now surround these medicinal plants include lack of botanical authentication; lack of natural product standardization, whether for a single, blended or complex herbal mix; and the spiking or adulteration of final products. Other product problems in Asian herbs, such as the presence of undesired heavy metals, pesticides and nontarget plant debris in the final products, are also part and parcel of such a quality-control imperative, though not addressed in this brief overview. All of these issues can be minimized or eliminated with a strong scientific-driven quality-control program that if implemented can further promote Chinese Traditional Medicines to the mantle it richly deserves within an integrated western health care system.

REFERENCES

1. P Shang, T Yang, M Jia, Q Mei, W Zhao, Z Cao, D Zhao. Purification and analysis of polysaccharides of *Angelica sinensis*. *Disi Junyi Daxue Xuebao* 22: 1311–1314, 2001.

2. LW Zhang, RD Huang. Purification, characterization and structure analysis of polysaccharide As-III a and As-III b from *Angelica sinensis*. *Jiguang Shengwu Xuebao* 8(2): 123–126, 1999.

3. H Wang, R Chen, H Xu. Study on chemical constituents of radix *Angelicae sinensis* Diels. *Zhongguo Zhongyao Zazhi* 23(3): 167–168, 1998.

4. H Zhang, Z Li, Y Chen. Studies of the active principles of *Angelica sinensis* (Oliv.) Diels — Isolation, characterization and biological effect of its polysaccharides. *Lanzhou Daxue Xuebao, Ziran Kexueban* 25(4): 78–81, 1989.

5. P Hon, C Lee, TF Choang, K Chui, HNC Wong. A ligustilide dimer from *Angelica sinensis*. *Phytochemistry* 29: 1189–1191, 1990.

6. Y Chen, H Zhang, W Cai. Study on the analysis of chemical constituents of *Angelica sinensis*. III. Determination of amino acids in Min-Gui. *Lanzhou Daxue Xuebao, Ziran Kexueban* 19 (Huaxue Jikan): 194–195 and 199, 1983.

7. Y Chen, N Chen, X Ma, H Li. Analysis of the composition of *Angelica sinensis* — Determination of the essential oil composition by capillary column GC/MS. *Gaodeng Xuexiao Huaxue Xuebao* 5(1): 125–128, 1984.

8. Y Chen, Z Duan, H Zhang, J Tao, Y Ruan, Q Mei, S Liu, Q Tian, F Xie, Y Yu. Chemical composition and pharmacological effects of *Angelica sinensis* (Oliv.) Diels. *Lanzhou Daxue Xuebao, Ziran Kexueban* 20(1): 158–160, 1984.

9. Y Chen, H Zhang. Analysis of the chemical ingredients of *Angelica sinensis* — Analysis of nonvolatile constituents of roots. *Gaodeng Xuexiao Huaxue Xuebao* 5(4): 515–520, 1984.

10. L Cha, CC Chien, FH Lu. Antiarrhythmic effect of *Angelica sinensis* root, tetrandrine and *Sophora flavescens* root. *Yaoxue Tongbao* 16(4): 53–54, 1981.

11. Y Deng, L Yang. Effect of *Angelica sinensis* (Oliv.) on melanocytic proliferation, melanin synthesis and tyrosinase activity *in vitro*. *Di Yi Jun Yi Da Xue Xue Bao* 23(3): 239–241, 2003.

12. W Yang, W Liao, J Tian, S Zhu, B Zhao. An experimental study on the effects of radix *Angelica* on oxygen free radicals and lipid peroxidation (LPO) of injured peripheral nerves. *Zhonghua Chuangshang Zazhi* 12(5): 300–302, 1996.

13. Z Peng, ZX Zhang, YJ Xu, ZL Liu, MJ Song. Regulation of LPS-induced elevation of Ca^{2+} intracellular level of alveolar macrophages in chronic bronchitis by *Angelica sinensis* and nifedipine. *Zhongguo Bingli Shengli Zazhi* 16(8): 738–740, 2000.

14. X Yan, JP Ou-Yang, S Tu. Angelica protects the human vascular endothelial cell from the effects of oxidized low-density lipoprotein *in vitro*. *Clin Hemorheology Microcirculation* 22: 317–323, 2000.

15. YN Ye, ESL Liu, VY Shin, MWL Koo, Y Li, EQ Wei, H Matsui, CH Cho. A mechanistic study of proliferation induced by *Angelica sinensis* in a normal gastric epithelial cell line. *Biochem Pharmacology* 61: 1439-1448, 2001.

16. Q Yang, SM Populo, J Zhang, G Yang, H Kodama. Effect of *Angelica sinensis* on the proliferation of human bone cells. *Clin Chim Acta* 324(1–2): 89–89, 2002.

17. ZZ Yin, LY Zhang, LN Xu. Effect of Dang-Gui (*Angelica sinensis*) and its ingredient ferulic acid on rat platelet aggregation and release of 5-HT. *Yaoxue Xuebao* (6): 321–326, 1980.

18. L Xu, R Ouyang, Z Yin, L Zhang, L Ji. Effect of Dang Gui (*Angelica sinensis*) and its constituent ferulic acid on phagocytosis in mice. *Yaoxue Xuebao* 16(6): 411–414, 1981.

19. Y Ozaki. Anti-inflammatory effect of tetramethylpyrazine and ferulic acid. *Chem Pharm Bull* 40: 954–956, 1992.

20. Q Mei, J Tao, H Zhang, Z Duan, Y Chen. Effects of *Angelica sinensis* polysaccharides on hemopoietic stem cells in irradiated mice. *Zhongguo Yaoli Xuebao* 9(3): 279–282, 1988.

21. Y Wang, B Zhu. The effect of Angelica polysaccharide on proliferation and differentiation of hematopoietic precursor cells. *Zhonghua Yixue Zazhi* 76(5): 363–366, 1996.

22. CH Cho, QB Mei, P Shang, SS Lee, HL So, X Guo, Y Li. Study of the gastrointestinal protective effects of polysaccharides from *Angelica sinensis* in rats. *Planta Med* 66: 348–351, 2000.

23. YN Ye, ESL Liu, Y Li, HL So, CCM Cho, HP Sheng, SS Lee, CH Cho. Protective effect of polysaccharides-enriched fraction from *Angelica sinensis* on hepatic injury. *Life Sci* 69: 637–646, 2001.

24. S Wang, Y Wang, Q Dai, M Zheng, R Jiang. Study on effect of Angelica polysaccharide on modulation of human CFU-GM. *Jiepou Xuebao* 32(3): 241–245, 2001.

25. X Xia, R Peng, Z Wang. Effect of *Angelica sinensis* polysaccharides and its ingredients on immunocompetence of mice. *Wuhan Daxue Xuebao, Yixueban* 22(3): 199–201 and 207, 2001.

26. YN Ye, HJ So, ESL Liu, VY Shin, CH Cho. Effect of polysaccharides from *Angelica sinensis* on gastric ulcer healing. *Life Sci* 72: 925–932, 2003.

27. J Shan, Y Wang, S Wang, D Liu, Z Hu. Effect of *Angelica sinensis* polysaccharides on lymphocyte proliferation and induction of IFN-γ. *Yaoxue Xuebao* 37(7): 497–500, 2002.

28. Y Li, R Peng. Protective effects of sodium ferulate and ethanol sediments from Danggui (*Angelica sinensis*) on immunological liver injury. *Zhongcaoyao* 31(4): 274–276, 2000.

29. Y Chen, H Zhang, N Chen, T Zhao, M Wang. Analysis of the ingredients of *Angelica sinensis* — Determination of the structure of angelicide. *Kexue Tongbao* (Foreign Language Ed.) 29(4): 560–562, 1984.

30. I Takano, I Yasuda, N Takahashi, T Hamano, T Seto, K Akiyama. Analysis of essential oils in various species of Angelica root by capillary gas chromatography. *Tokyo-toritsu Eisei Kenkyusho Kenkyu Nenpo* 41: 62–69, 1990.

31. C Zhou, J Zou, Y Chen, Y Zhao. Determination of ligustilide in essential oil of *Angelica sinensis* by GC-MS. *Yaowu Fenxi Zazhi* 22(4): 290–292, 2002.

32. H Chen, S Liu, G Li, Q Li. Determination of ferulic acid in the Chinese angelica (*Angelica sinensis*) and its preparations by HPLC. *Zhongcaoyao* 19: 447–448, 1988.

33. RM Lu, LI Ho, SY Lo. Determination of ferulic acid in danggui (*Angelica sinensis*). *Zhongcaoyao* 11: 395–398, 1980.

34. N Kikuchi, HL Lay, T Tanabe, T Miki. High-performance liquid chromatographic separation and quantitative determination of ligustilide in the Angelica plant using fluorometric detection. *Acta Chromatogr* 1: 23–33, 1992.

35. LZ Lin, XG He, LZ Lian, W King, J Elliott. Liquid chromatographic-electrospray mass spectrometric study of the phthalides of *Angelica sinensis* and chemical changes of Z-ligustilide. *J Chromatogr A* 810: 71–79, 1998.

36. S Zschocke, JH Liu, H Stuppner, R Bauer. Comparative study of roots of *Angelica sinensis* and related umbelliferous drugs by thin layer chromatography, high-performance liquid chromatography, and liquid chromatography — mass spectrometry. *Phytochem Anal* 9(6): 283–290, 1998.

37. Y Chen, Z Cheng, F Han, X Yang. Quantitative analysis of ferulic acid in Danggui by high-performance capillary electrophoresis. *Fenxi Huaxue* 27(12): 1424–1427, 1999.

38. Z Chen, M Zhang, J Mo, P Cai, H Wu, K Zhang. Determination of ferulic acid in *Angelica sinensis* roots. *Zhongcaoyao* 31(7): 506–508, 2000.

39. KT Cheng, CP June, HC Chang. RAPD analysis of *Angelica sinensis* medicine in Taiwan. *Chin Pharm J* (Taipei) 51(5): 307–312, 1999.

40. W Gao, E Qin, X Xiao, H Yu, G Gao, S Chen, Y Zhao, S Yang. Analysis on genuineness of *Angelica sinensis* by RAPD. *Zhongcaoyao* 32(10): 926–929, 2001.

41. RP Brown, PL Gerbarg, Z Ramazanov. *Rhodiola rosea:* A phytomedicinal overview. *HerbalGram* 56: 40–52, 2002.

42. GG Zapesochnaya, VA Kurkin, AN Shchavlinskii. The chemical study of *Rhodiola rosea* L. Int. Conf. Chem. Biotechnol. Biol. *Act Nat Prod* [Proc.], 3rd 4: 404–408, 1987.

43. LS Wang, L Wang, JJ Lu, ZY Liu. Microwave technique extraction and content determination of polysaccharide in *Rhodiola rosea*. *Shihezi Daxue Xuebao, Ziran Kexueban* 6(1): 18–19 and 22, 2002.

44. RA Aksenova, MI Zotova, MF Nekhoda, SG Cherdyntsev. Stimulating and adaptogenic effects of a refined preparation of *Rhodiola rosea*; rhodosin. *Stimulatory Tsent Nerv Sist* 77–79, 1966.

45. LV Maslova, BY Kondratyev, LN Maslov, YB Lishmanov. On the cardioprotective and antiadrenergic activity of *Rhodiola rosea* extract during stress. Eksperimental'naya i Klinicheskaya *Farmakologiya* 57(6): 61–63, 1994.

46. V Darbinyan, A Kteyan, A Panossian, E Gabrielian, G Wikman, H Wagner. *Rhodiola rosea* in stress induced fatigue — a double blind cross-over study of a standardized extract SHR-5 with a repeated low-dose regimen on the mental performance of healthy physicians during night duty. *Phytomed* 7(5): 365–371, 2000.

47. VD Petkov, D Yonkov, A Mosharoff, T Kambourova, L Alova, VV Petkov, I Todorov. Effects of alcohol aqueous extract from *Rhodiola rosea* L. roots on learning and memory. *Acta Physioligica et Pharmacologica Bulgarica* 12(1): 3–16, 1986.

48. VE Pogorelyi, LM Makarova. *Rhodiola rosea* extract for prophylaxis of the ischemic cerebral circulation disorders. *Eksperimental'naya i Klinicheskaya Farmakologiya* 65(4): 19–22, 2002.

49. Z Yuan, Z Ma. Studies on the scavenging effect of *Rhodiola rosea* L. and Radix *Salvia miltiorrhiza* on O_2-.bul. and .bul. OH by electrochemistry method. *Fenxi Huaxue* 27(6): 626–630, 1999.

50. M Furmanowa, B Starosciak, J Lutomski, J Kozlowski, N Urbanska, A Krajewska-Patan, A Pietrosiuk, W Szypula. Antimicrobial effect of *Rhodiola rosea* L. roots and callus extracts on some strains of Staphylococcus aureus. *Herba Polonica* 48(1): 23–31, 2002.

51. LA Dement'eva, KV Iaremenko. Effect of a Rhodiola extract on the tumor process in an experiment. *Voprosy Onkologii* 33(7): 57–60, 1987.

52. SN Udintsev, VP Schakhov. Decrease of cyclophosphamide haematotoxicity by *Rhodiola rosea* root extract in mice with Ehrlich and Lewis transplantable tumors. *European J Cancer* 27(9): 1182, 1991.

53. IB Lishmanov, LV Maslova, LN Maslov, EN Dan'shina. The antiarrhythmia effect of *Rhodiola rosea* and its possible mechanism. *Bull Eksperimentaalnoi Biologii I Meditsiny* 116(8): 175–176, 1993.

54. OA Bocharova, BP Matveev, AI Baryshnikov, KM Figurin, RV Serebriakova, NB Bodrova. The effect of a *Rhodiola rosea* extract on the incidence of recurrences of a superficial bladder cancer (experimental clinical research). *Urologiia i Nefrologiia* (2): 46–47, 1995.

55. J Rohloff. Volatiles from rhizomes of *Rhodiola rosea* L. *Phytochemistry* 59: 655–661, 2002.

56. PT Linh, YH Kim, SP Hong, JJ Jian, JS Kang. Quantitative determination of salidroside and tyrosol from the underground part of *Rhodiola rosea* by high-performance liquid chromatography. *Arch Pharm Res* 23(4): 349–352, 2000.

57. AG Dubichev, VA Kurkin, GG Zapesochnaya, ED Vorontsov. HPLC study of the composition of *Rhodiola rosea* rhizomes. *Khimiya Prirodnykh Soedinenii* (2): 188–193, 1991.

58. AA Kir'yanov, LT Bondarenko, VA Kurkin, GG Zapesochnaya, AA Dubichev, ED Vorontsov. Determination of biologically active constituents of *Rhodiola rosea* rhizome. *Khimiya Prirodnykh Soedinenii* (3): 320–323, 1991.

59. RI Peshekhova, VD Gol'tsev, LA Khnykina. Determination of salidroside in *Rhodiola rosea* extracts. *Usp Izuch Lek Rast Sib, Mater Mezhvuz Nauch Konf* 83–84, 1973.

60. M Ganzera, Y Yayla, IA Khan. Analysis of the marker compounds of *Rhodiola rosea* L. (golden root) by reversed phase high-performance liquid chromatography. *Chem Pharm Bull* 49: 465–467, 2001.

61. M Miyazawa, H Kameoka. Volatile flavor components of crude drugs. Part IV. Volatile flavor components of *Puerariae* Radix (*Pueraria lobata* Ohwi). *Agric Biol Chem* 52: 1053–1055, 1988.

62. K Hirakura, M Morita, K Nakajima, K Sugama, K Takagi, K Nitsu, Y Ikeya, M Maruno, M Okada. Phenolic glucosides from the root of *Pueraria lobata*. *Phytochemistry* 46(5): 921–928, 1997.

63. J Kinjo, J Furusawa, T Nohara. Two novel aromatic glycosides, pueroside-A and -B from *Puerariae radix*. *Tetra Lett* 26: 6101–6102, 1985.

64. J Kinjo, J Kurusawa, J Baba, T Takeshita, M Yamasaki, T Nohara. Studies on the constituents of *Pueraria lobata*. III. Isoflavonoids and related compounds in the roots and the voluble stems. *Chem Pharma Bull* 35: 4846–4850, 1987.

65. J Kinjo, T Takeshita, T Nohara. Constituents of *Pueraria lobata*. V. A tryptophan derivative from *Puerariae flos*. *Chem Pharma Bull* 36: 4171–4173, 1988.

66. T Arao, J Kinjo, T Nohara, R Isobe. Leguminous plants. III. Oleanene-type triterpene glycosides from Puerariae Radix. IV. Six new saponins from Pueraria lobata. *Chem Pharm Bull* 45: 362–366, 1997.

67. Z Zhang, X Wang, Q Liu, Z Chen, Z Gao. Studies on isoflavonoid constituents of roots of Qin mountain Taibai *Pueraria lobata*. *Zhongguo Yaoxue Zazhi* (Beijing) 34(5): 301–302, 1999.

68. S Li, J Deng, X Liu, S Zhao. Chemical constituents of *Pueararia lobata*. *Zhongcaoyao* 30(6): 416–417, 1999.

69. M Miyazawa, K Sakano, S Nakamura, H Kosaka. Antimutagenic activity of isoflavone from *Pueraria lobata*. *J Agric Food Chem* 49: 336–341, 2001.

70. WM Keung, BL Vallee. Kudzu root: an ancient Chinese source of modern antidipsotropic agents. *Phytochemistry* 47: 499–506, 1997.

71. RC Lin, S Guthrie, CY Xie, K Mai, DY Lee, L Lumeng, TK Li. Isoflavonoid compounds extracted from *Pueraria lobata* suppress alcohol preference in a pharmacogenetic rat model of alcoholism. *Alcoholism Clin Exp Res* 20: 659–663, 1996.

72. L Fan, G Zeng, Y Zhou, L Zhang, Y Cheng. Pharmacologic studies on Radix *Pueraria*. Effects of Pueraria flavones on coronary circulation, cardiac hemodynamics, and myocardial metabolism in dogs. *Chin Med J* 95(2): 145–150, 1982.

73. KY Tseng, YP Chou, LY Chang, LL Fan. Pharmacologic studies on *Radix puerariae*. I. Effects on dog arterial pressure, vascular reactivity, cerebral and peripheral circulation. *Zhonghua Yixue Zazhi* 54(5): 265–270, 1974.

74. HH Hsu, CK Chang, HC Su, IM Liu, JT Cheng. Stimulatory effect of puerarin on α_{1A}-adrenoceptor to increase glucose uptake into cultured C_2C_{12} cells of mice. *Planta Med* 68: 999–1003, 2002.

75. R Cervellati, C Renzulli, MC Guerra, E Speroni. Evaluation of antioxidant activity of some natural polyphenolic compounds using the Briggs-Rauscher reaction method. *J Agric Food Chem* 50: 7504–7509, 2002.

76. RC Lin, TK Li. Effects of isoflavones on alcohol pharmacokinetics and alcohol-drinking behavior in rats. *Am J Clin Nutr* 68(6 Suppl): 1512S–1515S, 1998.

77. T Arao, M Udayama, J Kinjo, T Nohara, T Funakoshi, S Kojima. Preventive effects of saponins from puerariae radix (the root of *Pueraria lobata* Ohwi) on *in vitro* immunological injury of rat primary hepatocyte cultures. *Biol Pharm Bull* 20: 988–991, 1997.

78. T Arao, M Udayama, J Kinjo, T Nohara. Preventive effects of saponins from the *Pueraria lobata* root on *in vitro* immunological liver injury of rat primary hepatocyte cultures. *Planta Med* 64: 413–416, 1998.

79. Z Zhang, B Yang, Q Liu, P Liu. Isolation of puerarin from *Pueraria lobata* and determination by HPTLC. *Zhongguo Yiyao Gongye Zazhi* 32(7): 291–293, 2001.

80. B Chen, H Zhao, Y Yan. Rapid HPLC determination of puerarin and daidzin in *Pueraria lobata. Shipin Kexue* (Beijing) 22(4): 64–66, 2001.

81. H Rong, D De Keukeleire, L De Cooman, WRG Baeyens, G Van Der Weken. Narrow-bore HPLC analysis of isoflavonoid aglycons and their O- and C-glycosides from *Pueraria lobata. Biomed Chromatogr* 12(3): 170–171, 1998.

82. G Chen, J Zhang, J Ye. Determination of puerarin, daidzein and rutin in *Pueraria lobata* (Wild.) Ohwi by capillary electrophoresis with electrochemical detection. *J Chromatog A* 923: 255–262, 2001.

83. J Kinjo, K Aoki, M Okawa, Y Shii, T Hirakawa, T Nohara, Y Nakajima, T Yamazaki, T Hosono, M Someya, Y Niiho, T Kurashige. Constituents of leguminous plants. Part LX1. Studies on hepatoprotective drugs. Part IX. HPLC profile analysis of hepatoprotective oleanene-glucuronides in *Puerariae Flos. Chem Pharm Bull* 47: 708–710, 1999.

84. YS Ko. Studies on the oil soluble constituents of Korean Ginseng. Part 1. On the composition of ginseng sterols. *Han'guk Sikp'um Kwahakhoechi* 8(4): 201–206, 1976.

85. KJ Choi, DH Kim. Studies on the lipid components of fresh ginseng, red ginseng and white ginseng. *Saengyak Hakhoechi* 16(3): 141–150, 1985.

86. K Hirakura, M Morita, K Nakajima, Y Ikeya, H Mitsuhashi. Polyacetylenes from the roots of *Panax ginseng. Phytochemistry* 30: 3327–3333, 1991.

87. SC Khim, HY Koh, BH Han. Polyacetylene compounds from *Panax ginseng* C.A. Meyer. *Bull Korean Chem Soc* 4(4): 183–188, 1983.

88. F Soldati. *Panax Ginseng*: Standardization and biological activity. In *Biologically Active Natural Products: Pharmaceuticals* (SJ Cutler, HG Cutler, Eds), CRC Press, Boca Raton, FL, pp. 209–232, 1999.

89. DD Kitts. Chemistry and pharmacology of ginseng and ginseng products. In *Herbs, Botanicals & Teas* (G Mazza, BD Oomah, Eds), Technomic Publishing Co., Inc., Lancaster, PA, pp. 23–44, 2000.

90. GB Mahady, HHS Fong, NR Farnsworth. Panax ginseng. In *Botanical Dietary Supplements: Quality, Safety and Efficacy.* Swets & Zeitlinger Publ., Lisse, pp. 207–224, 2001.

91. DO Kennedy, AB Scholey, KA Wesnes. Modulation of cognition and mood following administration of single doses of *Ginkgo biloba*, ginseng, and a ginkgo/ginseng combination to healthy young adults. *Physiology and Behavior* 75(5): 739–751, 2002.

92. OMES Salam, SA Nada, MS Arbid. The effect of ginseng on bile-pancreatic secretion in the rat. Increase in proteins and inhibition of total lipids and cholesterol secretion. *Pharmacol Res* 45(4): 349–353, 2002.

93. CS Jeong. Effect of butanol fraction of *Panax ginseng* head on gastric lesion and ulcer. *Arch Pharm Res* 25(1): 61–66, 2002.

94. SJ Hwang, JY Cha, SG Park, GJ Joe, HM Kim, HB Moon, SJ Jeong, JS Lee, DH Shin, SR Ko, JK Park. Diol- and triol-type ginseng saponins potentiate the apoptosis of NIH3T3 cells exposed to methyl methanesulfonate. *Toxicol Appl Pharm* 181(3): 192–202, 2002.

95. JH Lee, SR Kim, CS Bae, D Kim, HN Hong, SY Nah. Protective effect of ginsenosides, active ingredients of *Panax ginseng*, on kainic acid-induced neurotoxicity in rat hippocampus. *Neurosci Lett* 325(2): 129–133, 2002.

96. HJ Han, BC Yoon, SH Lee, SH Park, JY Park, YJ Oh, YJ Lee. Ginsenosides inhibit EGF-induced proliferation of renal proximal tubule cells via decrease of c-fos and c-jun gene expression *in vitro*. *Planta Med* 68: 971–974, 2002

97. YJ Surh, JY Lee, KJ Choi, SR Ko. Effects of selected ginsenosides on phorbol ester-induced expression of cyclooxygenase-2 and activation of NF-kappaB and ERK1/2 in mouse skin. *Ann NY Acad Sci* 973: 396–401, 2002.

98. B Liao, H Newmark, R Zhou. Neuroprotective effects of ginseng total saponin and ginsenosides Rb1 and Rg1 on spinal cord neurons *in vitro*. *Exp Neurology* 173(2): 224–234, 2002.

99. JD Churchill, JL Gerson, KA Hinton, JL Mifek, MJ Walter, CL Winslow, RA Deyo. The nootropic properties of ginseng saponin Rb1 are linked to effects on anxiety. *Integrative Physiol Behavioral Sci* 37(3): 178–187, 2002.

100. S Choi. Epidermis proliferative effect of the *Panax ginseng* ginsenoside Rb2. *Arch Pharm Res* 25(1): 71–76, 2002.

101. F Sala, J Mulet, S Choi, SY Jung, SY Nah, H Rhim, LM Valor, M Criado. Effects of ginsenoside Rg2 on human neuronal nicotinic acetylcholine receptors. *J Pharm Exp Therapeutics* 301(3): 1052–1059, 2002.

102. H Rhim, H Kim, DY Lee, TH Oh, S Nah. Ginseng and ginsenoside Rg3, a newly identified active ingredient of ginseng, modulate Ca^{2+} channel currents in rat sensory neurons. *Euro J Pharm* 436: 151–158, 2002.

103. S Kim, K Ahn, TH Oh, SY Nah, H Rhim. Inhibitory effect of ginsenosides on NMDA receptor-mediated signals in rat hippocampal neurons. *Biochem Biophys Res Commun* 296: 247–254, 2002.

104. JY Kim, KW Lee, SH Kim, JJ Wee, YS Kim, HJ Lee. Inhibitory effect of tumor cell proliferation and induction of G2/M cell cycle arrest by panaxytriol. *Planta Med* 68: 119–122, 2002.

105. DS Lim, KG Bae, IS Jung, CH Kim, YS Yun, JY Song. Antisepticaemic effect of polysaccharide from Panax ginseng by macrophage activation. *J Infection* 45(1): 32–38, 2002.

106. JY Shin, JY Song, YS Yun, HO Yang, DK Rhee, S Pyo. Immunostimulating effects of acidic polysaccharides extract of *Panax ginseng* on macrophage function. *Immunopharm Immunotoxicol* 24(3): 469–482, 2002.

4

Health-Related Fat Replacers Prepared from Grain for Improving Functional and Nutritive Values of Asian Foods

GEORGE E. INGLETT and CRAIG J. CARRIERE

Cereal Products and Food Science Research
Unit, National Center for Agricultural
Utilization Research, ARS, USDA,
Peoria, IL, USA

SAIPIN MANEEPUN

Institute of Food Research and Product
Development, Kasetsart University,
Bangkok, Thailand

INTRODUCTION

It is well recognized that the health benefits of cereal oat products are associated with prevention of hypercholesterolemia, one of the primary factors contributing to atherosclerosis and heart disease.[1,2] Dietary fibers from oat bran have been primarily reported to have hypocholesterolemic activities in both animal and human subjects.[3] Oat bran is presently not only the source of this key ingredient, but beta-glucans, (1-3)(1-4)-beta-D-glucans, a soluble fiber from oat also plays a potential role in these health improvements.[4-6] Dietary fat-reduction and controlled caloric intake are important for maintaining good health.

Recently, the U.S. Food and Drug Administration[7,8] has authorized a health claim related to oat products, stating that oat products, in conjunction with a diet low in saturated fat and cholesterol, may reduce heart disease and must contain 0.75 g of beta-glucans per serving. Nutrim OB derived from oat bran is a new beta-glucan-rich hydrocolloid for increasing textural qualities and health benefits of functional foods. It acts as a fat substitute with a high concentration of about 10% beta-glucan soluble fibers. Preparation in high yields involved a natural extraction process that removes most of the cellulose material from the oat bran, making it easier for the body to absorb beta-glucans.[9] Nutrim OB produces soft and smooth textured material of fat-like gel. It can provide more nutritious functional foods along with large reductions in calories and fat depending upon the amount of fat replacement. In laboratory studies by Yokoyama et al.,[10] they reported that Nutrim OB lowered cholesterol levels in hamsters by 27%, substantially more than unprocessed oats did, and low-density lipid (LDL) cholesterol was reduced by 36%.

Many Asian people are concerned with their dietary fat intakes related to health risks. It is well known that excessive fat and calorie consumption lead to several diseases such as obesity, hypertension, hypercholesterolemia, hypertriglyceridemia, diabetes mellitus, and heart disease, among others. Data from a nutritional survey on daily food consumption of people living in Bangkok showed fat consumption to be about 38.3% of total calorie intake, which was more than the Thai Recommended Dietary Allowances[11] at a recommended level of 30%. As an alternative, reduction of fat and calorie intake could be achieved by using fat substitutes.

Traditionally, a fat-rich ingredient commonly used in both Thai meals and desserts is coconut milk or cream. It contains 35.5% of fat, 31.2% of total saturated fat, and only

2.82% of polyunsaturated fat.[12] Large consumption of coconut milk could cause health problems leading to numerous diseases. The consumption of saturated fat was recommended to be as low as 10% of total calorie intake.[13]

Therefore, research on the modification of low-fat or low-calorie Thai foods, particularly desserts containing a high proportion of coconut milk, should be conducted with fat replacers such as Nutrim OB. Hence, the influence of Nutrim OB on increasing the nutritional value of Thai foods that use substantial amounts of coconut fat in their preparation was examined.

In order to determine Nutrim OB's suitability as a replacer, it was necessary to determine whether Nutrim OB had similar rheological and sensory properties that were comparable to coconut cream. Since coconut cream is a principal source of saturated fat in the Thai diet, this fat could contribute to health problems in Southeastern Asian countries. These conditions also exist in the Western cultures that consume diets rich in saturated fats.

PHYSICAL AND CHEMICAL PROPERTIES

Nutrim OB Preparation

Nutrim OB hydrocolloid was prepared as follows: To 5,100 mL deionized water in a 5-gal (19 L) container, 900 g of oat bran (OB) concentrate (Quaker Oats Company, Chicago, IL) were added and mixed at about 10,000 rpm with a dispersator (Premier Mill Corp, Reading, PA; PMC Model 90, high viscosity head) to generate a temperature in the range of 80 to 95°C. Continuous shear force was applied to maintain this temperature for 30 min before adding 6 L of boiling water. The slurry was steam jet-cooked at 138 to 141°C and 40 to 45 psi. The hot slurry from the cooker was immediately passed into a Sweco separator (Sweco International, Florence, KY) with 50 and 80 steel mesh sieves to recover the hydrocolloid liquid. The wet fiber solids from the sieves were collected, reslurried with boiling water, and recollected on the sieves. The liquid wash was combined with the hydrocolloid liquid before drum drying the liquid to give oat bran hydrocolloid, 536 g. The combined wet fiber solids were oven dried to give 175 g of product. The Nutrim OB hydrocolloid composition in percent are: moisture, 6.7; ash, 2.2; fat (diethyl ether extraction), 1.1; protein (nitrogen $\times$ 6.25), 9.7; crude fiber, 0.25; and beta-glucan, 8.6. The pH of a 10% slurry was 5.5 to 6.5. A commercial coconut cream sample (D'Best Coconut Cream Kakang Gata, Simplex International, South San Francisco, CA, 69% fat by weight) was purchased and used as a reference for the rheological studies. The sample was used as received without any additional modification.

Nutrim OB Rheological Measurements

Nutrim OB suspensions were produced at a concentration of 10% by weight in deionized water.[14] The solid Nutrim OB product was initially slurried in deionized water and then introduced into a colloidal mill (Polytron PT6000, Kinematica GmbH, Kriens-Luzen, Switzerland) and sheared at 2,000 rpm for 5 min to ensure thorough suspension of the material. The sample was allowed to cool from 100°C to room temperature and the resulting suspension was used in the rheological experiments. New samples were produced daily to avoid any possible problems with sample degradation.

Dynamic rheological properties were measured using a CarriMed TM CSL^2 500 (Dorking, England) controlled-stress rheometer using a cone-and-plate fixture. All the rheological studies were conducted using a 6-cm diameter plate and a 4° cone. The rheometer is capable of measuring torques from 2 to 500,000 g-cm. The temperature of the sample was controlled using a Peltier plate, which enabled the chamber of the viscometer to be controlled to within ± 0.1°C. The transition from linear to nonlinear viscoelastic behavior for each of the materials was investigated using a stress sweep experiment at a fixed frequency of 1 s^{-1}.

Preparation of Thai Desserts

The preparation of the eight popular Thai desserts containing a high proportion of coconut milk was selected in this study.

They were grouped into four categories according to their moisture contents. A dry or low-moisture product was represented by crispy pancake (tong-pup). Semimoistened products were taro conserve (puek-kuan), steamed banana cake (khanom-kuay), coconut pudding (tako saku), and steamed glutinous rice with coconut cream (kaoniew moon), whereas moistened products were coconut jelly (vunsangkaya) and coconut-cantaloupe ice cream. A liquid product such as pumpkin in coconut milk (fuktong kaengbuad) was included.

The standard formulas of eight desserts were developed and represented as control samples as shown in Table 4.1. Coconut milk was substituted with Nutrim OB at 60, 80, and 100% of coconut milk on a weight basis for crispy pancake, taro conserve, steamed banana cake, and coconut jelly. Whereas coconut pudding, steamed glutinous rice with coconut cream, coconut cantaloupe ice cream, and pumpkin in coconut milk, containing high amounts of coconut milk as a major ingredient in the formulas, would be replaced with Nutrim OB at levels of 40, 60, and 80%, respectively. The Nutrim OB dry powder was converted into a soft gel by blending with hot water containing 5% Nutrim OB followed by refrigeration overnight. The coconut milk was prepared by mixing 2 kg of grated coconut with 500 g of water and extracting the coconut milk using an electric press.

Coconut Jelly (Vun-sangkaya)

Ingredients for the agar suspension included 12 g agar powder, 45 g sugar, and 530 g water. Ingredients for the coconut cream were 200 g coconut cream, 125 g duck eggs, 250 g palm sugar, 1 g salt, and 5 g vanilla flavor. Preparation of the jelly started with mixing agar powder, sugar, and water thoroughly in a bowl. The mixture was brought to a boil and then set aside and allowed to cool. The coconut cream, eggs, palm sugar, and salt were combined and then mixed thoroughly. The mixture was poured into the agar suspension, stirred, and boiled again. The suspension was transferred into the desired mold and allowed to set at room temperature.

TABLE 4.1 Standard Formulas of Thai Desserts for 100 g

Products	Coconut cream	Sugar	Egg	Flour	Salt	Other ingredients	Water	Others[a]
Coconut jelly	17.12	25.25	10.70	—	0.09	1.03 (agar powder)	45.38	0.43
Taro conserve	25.61	23.05	—	—	0.13	51.22 (mashed taro)	—	—
Crispy pancake	37.40	20.33	4.07	36.58	0.40	1.22 (sesame seed)	—	—
Steamed banana cake	20.67	23.85	—	11.13	0.60	39.75 (mashed banana)	—	4.00
Pumpkin in coconut milk	44.40[b]	11.10	—	—	0.09	44.40 (pumpkin)	—	—
Coconut pudding	36.10	11.91	—	3.97	0.36	10.83 (sago)	28.88	7.95
Steamed glutinous rice with coconut cream	33.71	14.61	—	—	1.12	50.56 (glutinous rice)	—	—
Coconut cantaloupe ice cream	43.23**	12.97	—	—	0.29	43.23 (cantaloupe)	—	0.29

[a] Including vanilla powder, shredded coconut, lotus seed, pandan leaf juice, and gelatin powder, respectively.
[b] Addition of water to coconut cream in the ratio of 1:1.

Taro Conserve (Puck-kaoon)

Ingredients were 400 g mashed, steamed taro, 200 g coconut cream, 180 g sugar, and 1 g salt. To prepare the conserve the ingredients were mixed together and cooked in a sauce pan until thickened. The mixture was then placed into the desired mold.

Crispy Pancake (Tong-pup)

Ingredients were 200 g rice flour, 250 g cassava flour, 150 g palm sugar, 100 g white sugar, 460 g coconut cream, 50 g egg, 5 g salt, and 15 g black sesame seeds. The pancake was prepared by mixing all of the ingredients together except the sesame seeds. The mixture was kneaded until it softened and became a homogenous batter. The mixture was set aside for 15 min and then sesame seeds were added. The mixture was baked in a tong-pup electric heater.

Steamed Banana Cake (Kanom-kuay)

Ingredients contained 500 g mashed banana, 70 g rice flour, 70 g cassava flour, 150 g sugar, 260 g coconut cream, 7.5 g salt, and 50 g shredded coconut. Methods used included mixing mashed banana, flour, sugar, and salt together and kneading it until softened. Then adding coconut cream and mixing thoroughly. The mixture was spooned into 20 small cups, sprinkled with shredded coconut and then steamed over boiling water for 15 min.

Pumpkin in Coconut Milk

Ingredients were 500 g small pieces pumpkin, 500 g coconut milk (60% coconut cream in water), 100 g palm sugar, 25 g white sugar, and 1 g salt. Methods included combining all the ingredients together in a pot. Bringing the mixture to a boil for 2 min and then letting it cool.

Coconut Pudding

Ingredients for sago included 150 g small size sago, 400 g water, 140 g sugar, 70 g boiled, lotus seed, and 40 g condensed pandan leaf water. The pudding was prepared by first washing the sago and draining off the water. The sago was then cooked in water with continuous stirring until the material boiled and thickened. To this was added sugar and pandan leaf water. This mixture was cooked until sticky and transferred into a small plastic mold holding 3/4 of portion size. The mixture was allowed to cool. Ingredients for coconut topping were 500 g coconut cream, 5 g salt, 55 g rice flour, and 25 g sugar. The topping was prepared by mixing all the ingredients together and stirring until they dissolved. The mixture was then cooked with continuous stirring until it had thickened. The topping was then poured over the sago mixture and allowed to set.

Steamed Glutinous Rice with Coconut Cream

Ingredients were 450 g glutinous rice, 300 g coconut cream, 130 g sugar, and 10 g salt. The glutinous rice was steamed over boiling water for 30 min. At the same time, the coconut cream, sugar, and salt were mixed together and brought to a boil. The coconut cream mixture was poured into the cooked glutinous rice while still hot. The rice was then mixed thoroughly and heated with a cover for 30 min. The dish was served with ripened mango.

Coconut–Cantaloupe Ice Cream

Ingredients were 600 g coconut cream, 600 g blended fresh cantaloupe, 180 g sugar, 4 g salt, and 4 g gelatin powder. To prepare the ice cream, coconut cream, sugar, salt, and gelatin powder were combined and brought to a boil in a container. The boiling mixture was stirred gradually and mixed with the cantaloupe suspension. The ice cream was cooled in a refrigerator and then transferred into an ice cream freezer.

Physical and Chemical Evaluation

A sensory panel with 25 members with acceptability and preference training evaluated the eight Thai desserts for the following characteristics: color, appearance, odor, taste, and texture using a 9-point hedonic scale. Samples with acceptability scores of more than six were analyzed for proximate composition[15] and total saturated fats.[16] The texture was detected using a TA.XT2i/25 Texture Analyzer (Stable Micro System Ltd, Scarsdale, NY 10583) using 2-mm diameter cylinder stainless probe and HDP/BSK blade set with knife edge and viscosity of the product was measured using a Brookfield TC 500 viscometer (Brookfield Company, Middleboro, MA 02346) operating under Rheocale V 1.0. The statistical analysis was assessed by analysis of variance (ANOVAS) and Duncan's multiple range (DMRT) tests.[17]

RESULTS AND DISCUSSION

Rheological Properties

Nutrim OB

The rheological responses [viscosity versus shear rate] of samples produced from OB (oat bran) and Nutrim made from OB subjected to thixotropic loop experiments are presented in Figure 4.1.[14] The OB data display a clockwise thixotropic loop with shear-thinning behavior evident throughout the shear rate range studied. The viscosity of the OB material after the thixotropic loop is lower than the initial starting viscosity indicating that OB may have experienced some slight shear degradation during the experiment. The observed drop in viscosity could also be due to disruption of some intermolecular associations that have occurred during the preparation of the suspension rather than actual shear degradation of the components; however, allowing the suspension to stand for one hour and repeating the thixotropic loop experiment (data not shown) produced a curve, which followed the downward cycle of the initial loop. The initial starting viscosity values could not be recovered. This result indicates that the drop in viscosity is due to a permanent change in the material (such as shear degradation) rather than the temporary disruption of transient associations.

In comparison, Nutrim produced from OB displayed a lower viscosity than the starting OB throughout the shear

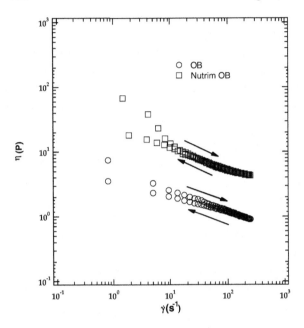

Figure 4.1   The response of OB and Nutrim OB to the imposition
of a thixotropic loop experiment conducted at 25°C. Both materials
display shear-thinning behavior across the shear-rate range studied.
The arrows indicate the direction of the applied shear for each of
the loops.

rate range studied (Figure 4.1). This is indicative of the loss
of insoluble components in the OB during conversion to
Nutrim as well as the mechanical shear degradation of the
components of OB. Nutrim materials, in general, are observed
to possess lower viscosities than the starting materials.
Nutrim produced from OB displayed a clockwise thixotropic
loop with shear-thinning behavior throughout the studied
shear rate range. Nutrim OB also showed a slight degree of
shear degradation during the experiment, in that the viscosity
at the end of the thixotropic loop was slightly lower than the
viscosity at the beginning of the experiment. The dependence
of the viscosity on shear rate for both OB and Nutrim OB was
comparable throughout the shear rate range studied. In general, the rheological response of Nutrim OB mimicked the

observed behavior of OB. It is interesting to note that the Nutrim OB viscosities were many times higher than Oatrim, the enzyme-hydrolyzed flour used in many food products.[18]

During the application of a second thixotropic loop experiment (data not displayed), Nutrim OB again displayed a clockwise hysteresis loop as was observed in the initial thixotropic loop experiment. Throughout the second thixotropic loop, only shear-thinning behavior was observed. The upward and downward cycles of the second loop followed the downward cycle of the first loop indicating that the application of subsequent shear did not cause further shear degradation of the material.

Coconut Cream

The response of the commercial coconut cream sample to a thixotropic loop experiment is illustrated in Figure 4.2. Two thixotropic loop experiments were conducted; the second loop was initiated immediately after the first loop had been completed. During the initial upward cycle of the first thixotropic loop, a marked region of shear thickening was observed from shear rates of 1.7-2.5 s^{-1}. At shear rates above 2.5 s^{-1}, the sample exhibited shear-thinning behavior. The ending viscosity at the completion of the first thixotropic loop was slightly lower than the starting viscosity. Repetition of the thixotropic loop experiment (Figure 4.2) followed the viscosity versus shear rate curve that was generated during the second half of the initial experiment. Repeating the thixotropic loop experiment with a fresh sample of coconut cream again displayed the shear-thickening behavior shown in Figure 4.2. This data indicates that the coconut cream sample initially possessed some internal structure or aggregation that was broken by the initial application of the shear field. Using a fresh sample of coconut cream and executing a thixotropic loop experiment starting at 250 s^{-1} and decreasing to 0.8 s^{-1} also did not show any evidence of the shear-thickening behavior shown in Figure 4.2 (data not shown). This result again supports the conclusion that the applied shear field disrupted some structure or aggregation originally present in the sample.

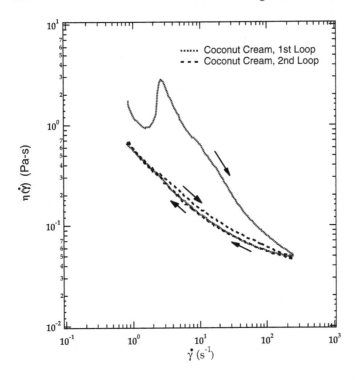

Figure 4.2   The response of a commercial coconut cream sample to a thixotropic loop experiment conducted at 25°C. A shear-thickening region is clearly observed during the initial thixotropic loop experiment starting at approximately 1.7 s⁻¹. The arrows indicate the direction of the applied shear for each of the loops.

Constitutive Analysis of the Rheological Response

The rheological behaviors of the Nutrim OB suspension and coconut cream were characterized using a power law constitutive equation.[19] The power law equation may be written as

$$\eta = K\dot{\gamma}^{n-1} \qquad (4.1)$$

where η is the shear viscosity, K is the front factor, $\dot{\gamma}$ is the shear rate, and n is the power law exponent. Equation (4.1) was fit to both cycles of the thixotropic loop for the Nutrim OB suspensions. For the coconut cream sample, the model

TABLE 4.2 Power Law Model Parameters

Material	K (Pa-s^n)	n
Nutrim OB 1st	2.4 ± 0.2	0.68 ± 0.05
Nutrim OB 2nd	2.3 ± 0.2	0.68 ± 0.06
Coconut cream 1st	8.0 ± 0.2	−0.13 ± 0.02
Coconut cream 2nd	0.553 ± 0.004	0.392 ± 0.007

Note: ± values represent two standard deviations
Note: 1st: initial part of the thixotropic loop from 0 to 250 s^{-1}
Note: 2nd: second part of the thixotropic loop from 250 to 0 s^{-1}

was fitted to the data in the upward cycle in the shear-thinning region after the shear-thickening regime. The results of the data fits are summarized in Table 4.2. The majorities of fluids exhibit pseudoplastic behavior and have values of n between 0.15 and 0.6. From the data presented in Table 4.2 it is evident that Nutrim OB suspensions behave as pseudoplastic fluids.

Fat Replacer in Prepared Thai Foods

Fat substitutes can be broadly grouped into either lipid-based materials or carbohydrate- and protein-based materials. For the most part, lipid-based fat substitutes have functional and sensory properties similar to those of the fats, but their applications are limited due to the toxicology and metabolism of these compounds, while most of the carbohydrate- and protein-based substitutes are either approved or near approval. In particular carbohydrate-based fat substitutes such as Oatrim or Nutrim OB, prepared by incorporating water into a gel-type structure, can result in rheological properties similar to those of fats; however, they are rather limited in their ability to carry lipid-based flavors. In addition, the high water content of the gel may reduce the shelf life of the product.[6,20] The fat replacer, Nutrim OB, in this study was used as a coconut milk substitute and nutrifying ingredient in eight Thai desserts.

Eight Thai desserts were chosen because they were expected that their specific components from cereals, roots, fruits, or others (Table 4.1) could be suitable for fat replacement of their principal fat source, coconut fat, with the fat replacer, Nutrim OB.

Substitution of Nutrim OB for coconut milk in taro conserve, crispy pancake, steamed banana cake, and coconut jelly could be achieved by complete substitution, whereas steamed glutinous rice with coconut cream and coconut cantaloupe ice cream were accepted at 80% of the substitution. Coconut pudding and pumpkin in coconut milk were acceptable at 60% substitution. This was based on the acceptability scores equal to or more than six. The results of this study were associated with a recent report on use of Oatrim as fat replacer for coconut milk in three desserts.[4]

Oatrim is also a soluble fiber derived from oat flour and bran, containing 4.5 to 5.5% beta-glucans possessing similar functional and nutritional qualities to Nutrim OB.[21] It was found that Oatrim could totally replace coconut milk in mungbean converse and coconut cream spread with good acceptability. When compared to the standard formulas, these two desserts with 60% of Oatrim replacement showed no significant difference ($p > .05$) in color, taste, texture, and acceptability. Beside that maximum use of Oatrim for butter replacer was at 70% in banana cake and brownie. Another study of using Oatrim for fat substitute in shortbread cookies showed substitution at 35% having the least negative effects on the physical attributes and had breaking force and toughness values most similar to the traditional full-fat shortbread cookie.[22]

The sensory evaluation data of the substitution of Nutrim OB for coconut milk are shown in Table 4.3. Preparation of the desserts could be substituted with Nutrim OB up to 100% level in some products. The results showed that levels of Nutrim OB substitution at 60, 80 and 100% in either taro conserve, steamed banana cake, or coconut jelly were not significantly different in color and appearance. Also odor and taste of taro conserve and steamed banana cakes had no significant difference among 60, 80, and 100% of Nutrim OB substitution. When compared to the control, steamed banana cake with all levels of Nutrim OB substitution were not significantly different in appearance, odor, texture, and acceptability, but showed some

TABLE 4.3 Sensory Evaluation of Thai Desserts Using Nutrim OB as Coconut Cream Replacer[a]

Products	Color	Appearance	Odor	Taste	Texture	Acceptability
Coconut jelly						
Control	6.70[b]	6.72[b]	7.38[a]	7.28[a]	7.22[a]	7.38[a]
60%	7.32[a]	7.32[a]	6.68[b]	6.98[ab]	6.98[ab]	6.82[b]
80%	7.34[a]	7.54[a]	6.24[c]	6.74[bc]	6.64[b]	6.66[b]
100%	7.24[a]	7.36[a]	6.48[bc]	6.58[c]	6.62[b]	6.54[b]
Taro conserve						
Control	7.62[a]	7.65[a]	7.77[a]	7.77[a]	7.56[a]	7.63[a]
60%	7.62[a]	7.62[a]	7.53[ab]	7.56[ab]	7.40[a]	7.38[a]
80%	7.58[a]	7.50[a]	7.33[b]	7.50[b]	7.21[ab]	7.27[a]
100%	7.48[a] ($p > .01$)	7.50[a] ($p > .01$)	7.17[b]	7.31[b]	6.85[b]	6.88[b]
Crispy pancake						
Control	7.98[a]	7.92[a]	8.18[a]	8.04[a]	8.06[a]	8.22[a] 0[c]
60%	7.54[b]	7.62[b]	7.42[b]	7.50[b]	7.72[b]	7.46[b]
80%	7.12[c]	7.52[b]	7.02[b]	7.36[b]	7.52[b]	7.16[a]
100%	6.76[c]	6.92[c]	6.02[c]	6.68[c]	7.00[c]	6.20[c]
Steamed banana cake						
Control	7.00[b]	7.26[a]	7.46[a]	7.66[a]	7.42[a]	7.52[a]
60%	7.56[a]	7.44[a]	7.28[a]	7.50[ab]	7.20[a]	7.30[a]
80%	7.50[a]	7.44[a]	7.14[a]	7.44[ab]	7.26[a]	7.24[a]
100%	7.52[a]	7.44[a] ($p > .01$)	7.20[a] ($p > .01$)	7.22[b]	6.94[a] ($p > .01$)	7.02[a] ($p > .01$)
Pumpkin in coconut milk						
Control	7.62[a]	7.46[a]	7.48[a]	7.56[a]	7.46[a]	7.60[a]
40%	7.56[a]	7.30[a]	7.20[ab]	7.30[ab]	7.12[ab]	7.30[ab]
60%	7.34[a]	7.30[a]	6.94[b]	7.10[b]	6.78[b]	6.88[b]
80%	6.78[b]	6.52[b]	6.10[c]	5.94[c]	5.94[c]	5.86[c]

TABLE 4.3 (CONTINUED) Sensory Evaluation of Thai Desserts Using Nutrim OB as Coconut Cream Replacer[a]

Products	Color	Appearance	Odor	Taste	Texture	Acceptability
Coconut pudding						
Control	7.94[a]	7.74[a]	7.76[c]	7.68[a]	7.44[a]	7.60[a]
40%	7.06[b]	6.96[b]	6.94[b]	7.10[b]	6.66[b]	6.88[b]
60%	6.18[c]	6.44[c]	6.62[b]	6.84[b]	6.42[b]	6.42[c]
80%	5.80[d]	5.94[d]	5.78[d]	6.02[c]	5.46[c]	5.54[d]
Steamed glutinous rice with coconut cream						
Control	7.92[a]	7.70[a]	7.60[a]	7.70[a]	7.58[a]	7.72[a]
40%	7.52[b]	7.44[a]	7.06[b]	7.16[b]	7.24[b]	7.28[b]
60%	7.00[c]	6.88[b]	6.46[c]	6.78[c]	6.44[d]	6.52[c]
80%	7.10[c]	6.96[b]	6.60[c]	6.68[c]	6.86[c]	6.70[c]
Coconut cantaloupe ice cream						
Control	7.92[a]	7.66[a]	7.26[a]	7.54[a]	7.34[a]	7.40[a]
40%	7.56[b]	7.22[b]	7.30[a]	7.46[a]	7.26[a]	7.34[a]
60%	7.36[bc]	7.02[bc]	7.10[a]	7.48[a]	6.94[a]	7.24a
80%	7.22[c]	6.86[c]	7.00[a] $(p > .01)$	7.12[a] $(p > .01)$	7.00[a] $(p > .01)$	6.94[a] $(p > .01)$

Note: In a column, means followed by same superscript are not significantly different at $p > .05$ and at $p > .01$, shown with parenthesis by ANOVA and DMRT.

[a] Prepared by blending 5% Nutrim OB in hot water (by weight) and refrigerated overnight before use.

difference in taste at a 100% level (Table 4.3). Crispy pancake with 60 and 80% of Nutrim OB substitution revealed no significant difference in appearance, odor, taste, texture, and acceptability. Coconut pudding and pumpkin in coconut milk acquired quite low scores of characteristics and acceptability when substituted at a level of 80% Nutrim OB. Steamed glutinous rice with coconut cream substituted with Nutrim OB at 60 and 80% were not significantly different in color, appearance, odor, taste, and acceptability. In particular, steamed banana cake and coconut–cantaloupe ice cream was most acceptable and satisfactory with no difference in acceptability score of all substitution when compared to the controls. Apparently, Nutrim OB and mashed banana possessed similarity in texture and color, therefore Nutrim OB substituted steamed banana cake could not be differentiated by the panelists from the control. Likewise, it is expected that cantaloupe ice cream produced with Nutrim OB should give good characteristics due to the viscous texture imparted by Nutrim OB.

Textures of the desserts were obtained by measuring shear force, crispness, penetration force, and viscosity as shown in Table 4.4. The taro conserve (semimoistened products) became tougher than steamed banana cake when the substitution with Nutrim OB was increased. In contrast, the acceptability showed a very good score with little change from the controls. The coconut jelly, as a moistened product, resulted in a very soft texture. The crispy pancake, as a dry product, showed no significant difference in the maximum force of shearing in all samples with Nutrim OB substitution relative to the control. The force (g/s), which represents the crispness of the crispy pancake, decreased distinctively at levels of 80 and 100% Nutrim OB substitution. The product with 100% substitution was described as having the toughest texture, although the data showed no significant difference from the lower substitution. The pumpkin in coconut milk as a liquid product became more viscous and formed a thick suspension when the Nutrim OB substitution was increased. It appeared that the poor characteristics of both the coconut milk suspension from pumpkin in coconut milk, and the coconut topping from coconut pudding, resulted from the separated

TABLE 4.4 Measurement of Force in Compression and Viscosity[a]

Products	Nutrim OB substitution level			
	Control	60%	80%	100%
Penetration force (g)**				
Coconut jelly	35.94[a]	26.91[b]	26.54[b]	25.01[c]
Taro conserve	20.57[b]	21.84[b]	24.33[ab]	27.60[a]
Steamed banana	31.38[a]	32.02[a]	32.54[b]	22.03[a]
Shearing force (g)[b]				
Crispy pancake	805.56[a]	696.98[a]	664.86[a]	702.03[a]
Fracturability (g/sec)[b]				
Crispy pancake	744.31[a]	710.47[a]	454.53[b]	359.31[b]
Viscosity (cps)[c]	Control	40%	60%	80%
Pumpkin in coconut milk	182.4	280.8	418.4	818.7

[a] In a row measurements followed by same superscript are not significantly different at $p > .01$ and $p > .05$ by ANOVA and DMRT.
[b] Performed by TA.ST2/5 Texture Analyzer, Stable Micro Systems using: 2 mm diameter cylinder stainless probe and 11DP/BSK blade set with knife edge.
[c] Detected by Brookfield TC 500 with Rheocale V1.0, controlling temperature of water bath at 28°C.

preparation of coconut milk. To improve the process for achieving the 100% substitution, Nutrim OB as a coconut milk substitute should be homogeneously incorporated with the other ingredients.

Samples with acceptability sensory panel scores of more than six and the controls were analyzed for proximate composition (Table 4.5) and total saturated fat (Table 4.6). Total fats and energy contents of Thai desserts with Nutrim OB substitution were reduced from their controls. Namely, these reductions were 83.4, 97.4, 85.4, 55.4, 73.1, 75.1, 74.6, and 49.9% and 13.7, 22.9, 7.1, 21.4, 15.5, 27.5, 45.4, and 22.4% in crispy pancake, taro conserve, steamed banana cake, coconut pudding, steamed glutinous rice with coconut cream, coconut jelly, coconut–cantaloupe ice cream, and pumpkin in coconut milk, respectively. The analysis data indicated a decrease in energy content, ranging from a low value of 7.1% in steamed banana cake, to a high of 45.4% in coconut–cantaloupe ice cream. Despite the distinctive reduction in total fat content

TABLE 4.5 Proximate Composition of Thai Desserts with Nutrim OB Substitution for Coconut Cream (per 100 g sample)

Thai desserts	Moisture (g)	Fat (g)	Protein (g)	Carbohydrate (g)	Ash (g)	Fiber (g)	Energy (kcal)
Coconut jelly							
Control	62.54	6.98	1.96	27.57	0.76	0.19	180.94
100% Nutrim OB	67.82	1.74	1.67	27.22	0.75	0.80	131.22
Taro conserve							
Control	37.81	9.81	2.70	47.12	1.11	1.45	287.57
100% Nutrim OB	42.70	0.26	2.24	52.62	1.01	1.17	221.78
Crispy pancake							
Control	2.11	15.81	4.23	75.31	1.27	1.27	460.45
100% Nutrim OB	1.84	2.62	3.70	89.81	1.15	0.88	397.62
Steamed banana cake							
Control	46.29	7.05	1.12	43.27	1.09	1.18	241.01
100% Nutrim OB	42.05	1.03	0.48	53.16	0.93	2.35	223.83
Pumpkin in coconut milk							
Control	67.58	8.29	1.43	21.16	0.90	0.64	164.97
60% Nutrim OB	72.11	4.15	0.88	21.78	0.76	0.32	127.99

TABLE 4.5 (CONTINUED) Proximate Composition of Thai Desserts with Nutrim OB
Substitution for Coconut Cream (per 100 g sample)

Thai desserts	Moisture (g)	Fat (g)	Protein (g)	Carbohydrate (g)	Ash (g)	Fiber (g)	Energy (kcal)
Coconut pudding							
Control	57.45	10.92	1.76	28.13	0.79	0.94	217.84
60% Nutrim OB	61.42	4.87	1.22	30.64	0.60	1.25	171.27
Steamed glutinous rice with coconut cream							
Control	39.91	7.63	3.33	46.92	1.15	1.06	269.67
80% Nutrim OB	43.40	2.05	2.92	49.44	1.00	1.19	227.89
Coconut–cantaloupe ice cream							
Control	67.39	11.24	1.81	18.21	0.69	0.66	181.24
80% Nutrim OB	77.17	2.85	0.96	17.38	0.55	1.09	99.01

TABLE 4.6   Reduction of Total Fats and Saturated Fats in Thai
Desserts with Nutrim OB Substitution for Coconut Cream

Thai desserts	Total fats (g/100 g)	% Reduction	Total saturated fats (g/100 g)	% Reduction
Coconut jelly				
Control	6.98	75.1	5.11	87.7
100% Nutrim OB	1.74		0.63	
Taro conserve				
Control	9.81	97.4	8.39	99.4
100% Nutrim OB	0.26		0.05	
Crispy pancake				
Control	15.81	83.4	12.76	95.5
100% Nutrim OB	2.62		0.58	
Steamed banana cake				
Control	7.05	85.4	5.93	86.3
100% Nutrim OB	1.03		0.81	
Pumpkin in coconut milk				
Control	8.29	49.9	7.14	58.3
60% Nutrim OB	4.15		2.98	
Coconut pudding				
Control	10.92	55.4	7.94	46.2
60% Nutrim OB	4.87		4.27	
Steamed glutinous rice with coconut cream				
Control	7.63	73.1	6.68	80.4
80% Nutrim OB	2.05		1.31	
Coconut–cantaloupe ice cream				
Control	11.24	74.6	9.66	79.8
80% Nutrim OB	2.85		1.95	

in all Nutrim OB-substituted products, the desserts such as
crispy pancake, steamed banana cake, and steamed glutinous
rice with coconut milk showed little decrease in energy content when compared to the others. This might be due to the
amount of flour or other carbohydrate sources used in the
formulas. The high-carbohydrate constituent could lead to a
high-energy content, even though Nutrim OB was used as a

fat replacer. Interestingly, all of the products, i.e., crispy pancake, taro conserve, steamed banana cake, coconut pudding, steamed glutinous rice with coconut cream, coconut jelly, coconut–cantaloupe ice cream, and pumpkin in coconut milk exhibited an overall decrease in the total saturated fat content of 95.5, 99.4, 86.3, 46.2, 80.4, 87.7, 79.8, and 58.3% from the controls, respectively. These reductions would enable the desserts to be designated as low saturated fat Thai dessert products.

CONCLUSIONS

Nutrim OB proved to be a suitable fat substitute for coconut milk in some Thai desserts prepared with high levels of coconut fat. Carbohydrate-based fat substitutes such as Nutrim OB from oats can result in lubricant or flow properties similar to the texture of coconut fat. Nutrim OB had similar rheological properties compared to coconut cream as determined by the results of thixotropic loop experiments for the Nutrim OB suspension and a commercial coconut cream sample. At high substitution levels, dramatic reduction of energy contents, total fats, and saturated fats were obtained. A decrease in energy content, ranging from a low of 7.13% in pumpkin in coconut milk, to a high of 44.82% in coconut–cantaloupe ice cream were observed. Crispy pancake, taro conserve, steamed banana cake, and coconut jelly were acceptable at the 100% substitution level, and the steamed glutinous rice with coconut cream and coconut–cantaloupe ice cream could possibly be replaced at 80%, whereas coconut pudding and pumpkin in coconut milk were acceptable at 60% substitution levels. The principal effects of Nutrim OB used as a coconut milk substitute depended on the moisture content of the products and component interactions that produce unfavorable characteristics. When the texture of the highest substituted desserts were compared to the controls, the crispy pancake as a dry product increased in hardness, the taro conserve and steamed banana cake as semimoistened products increased in toughness, the coconut jelly became too soft and the pumpkin in coconut milk became too viscous. Total fat percentage reductions for coconut jelly, taro conserve, crispy pancake, steamed banana cake, pumpkin in coconut milk, coconut pudding, steamed glutinous rice with coconut cream, and coconut–cantaloupe ice cream were 75.1, 97.4, 83.4, 85.4, 49.9, 74.6, 55.40, and 73.13, respectively. Saturated fat percentage reductions were 87.7, 99.4, 95.5, 86.3, 58.3, 79.8, 46.22, and 80.39, respectively.

ACKNOWLEDGMENTS

We wish to express my sincere thanks to Steven A. Lyle and Mary P. Kinney for technical assistance.

REFERENCES

1. TS Kahlon, FI Chow. Hypocholesterolemic effects of oat, rice, and barley dietary fibers and fractions. *Cereal Food World* 42:86–92, 1997.

2. CM Ripsin, JM Keenan, DR Jacobs. Oat products and lipid lowering. *JAMA* 267: pp 3317–3325, 1992.

3. AP DeGroot, R Luyken, NA Pikaar. Cholesterol-lowering effect of rolled oats. *Lancet* 2: pp 303–304, 1963.

4. GE Inglett, S Maneepun, N Vatanasuchart. Evaluation of hydrolyzed oat flour as a replacement for butter and coconut cream in bakery products. *Food Sci Technol Int* 6: pp 457–462, 2000.

5. GE Inglett, CJ Carriere, S Maneepun, P Tungtrakul. Nutritional and functional properties of a hydrocolloidal composite from processing rice bran and barley flour for improving the functional and nutritive values of Asian foods. *Int J Food Sci Technol.* 39: 1–10, 2004.

6. GE Inglett, CJ Carriere, S Maneepun, T Boonpunt. Nutritional value and functional properties of a hydrocolloidal soybean and oat blend for use in Asian foods. *J Sci Food Agri.* 83: 86–92, 2002.

7. Food and Drug Administration. Food Labeling: Health Claims; Soluble fiber from whole oats and risk of coronary heart disease. *Federal Register* 62: pp 15343–15344, 1997a.

8. Food and Drug Administration. Food Labeling: Health Claims; Oats and coronary heart disease. *Federal Register* 62: pp 3584–3601, 1997b.

9. GE Inglett. Soluble hydrocolloid food additives and method of making. U. S. Patent 6,060,519, 2000.

10. WH Yokoyama, BE Knuckles, D Wood, GE Inglett. Food processing reduces the size of soluble cereal beta-glucan polymers without the loss of cholesterol-reducing properties. In: *Bioactive Compounds in Foods. Effects of Processing and Storage.* Tung-Chin Lee, Chi-Tang Ho, Eds. ACS Symposium Series 816. American Chemical Society, Washington, D.C. 2002. pp 105–116.

11. Committee on Recommended Daily Dietary Allowances. Department of Health. Ministry of Public Health. Recommended Daily Dietary Allowances for Healthy Thais. Bangkok, Thailand. 1989, pp 159.

12. Nutrition Division, Thai Food Composition Tables. Department of Health. Ministry of Public Health Thais. Bangkok, Thailand, 1989, p 97.

13. Food Act. B.E. 2541. Nutrition Labeling. Department of Health. Ministry of Public Health. Notification of the Ministry of Public Health. Bangkok, Thailand. No.182: pp 46, 1998.

14. CJ Carriere, GE Inglett. Effect of processing conditions on the viscoelastic behavior of Nu-TrimX: an oat-based beta-glucan-rich hydrocolloidal extractive. *J Text Stud* 31, 123–140, 2000.

15. *Official Methods of Analysis*, 15th ed. Association of Official Analytical Chemists, AOAC, Arlington, VA, 1990, pp 777–782, 1103-1106.

16. GN Jham, F Teles, LG Campas. Use of aqueous HCl/MeOH as esterification reagent for analysis of fatty acids derived from soybean lipids. *JAOCS* 59: pp 132–133, 1992.

17. Irristat version 90-1, Department of Statistics, International Rice Research Institute, Los Banos, Laguna, Philippines, 1990.

18. CJ Carriere, AJ Thomas, and GE Inglett. Prediction of the nonlinear transient and oscillatory rheological behavior of flour suspensions using a strain-separable integral constitutive equation. *Carbohydrate Polymers*, 47, pp 219–231, 2002.

19. RB Bird, RC Armstrong & O Hassager. Dynamics of polymeric liquids, in Vol 1. *Fluid Mechanics*. John Wiley and Sons: New York, 1977. pp 208.

20. CA Hassel. Nutritional implications of fat substitutes. *Cereal Foods World* 38: pp 142–144, 1993.

21. GE Inglett, CJ Carriere. Oatrim and NutrimX: Technological development and nutritional properties, In: BV McCleary, L Prosky, Eds. *Advanced Dietary Fibre Technology*. Oxford, UK: Blackwell Science, Chapter 24, 2001, pp 270–276.

22. C Sanchez, cf. Klopfenstein, CE Walker. Use of carbohydrate-based fat substitutes and emulsifying agents in reduced-fat shortbread cookies. *Cereal Chem* 72: pp 25–29, 1995.

5

Antiaging Properties of Asian Functional Foods: A Historical Topic Closely Linked to Longevity

YI DANG

Beijing University of Chinese Medicine, China

INTRODUCTION

Human beings have always aspired to live longer lives. More and more people nowadays believe that Oriental functional foods with antiaging activities can prevent diseases, maintain the health of people and make their dream of living long lives come true.

Aging is a natural physiological process of multisystematic functions of the human body that decline gradually along with the increase of age. Quality of life is the main concern of the elderly and those who are approaching the later stages of their life. To this end, it was found that many antiaging

Chinese materia medica and functional foods could regulate immune systems, thus playing certain crucial roles in the antiaging processes.

CONCEPT AND DEFINITION OF ANTIAGING FUNCTIONAL FOODS

The proper regulations on application, approval, production management, labeling, specification, advertising, and supervision of functional foods have been in place in China since the Measures of Functional Food Administration (MFFA) was enforced on June 1, 1996. In spite of the difference in nomenclature and classification of functional foods in different countries, the consensus is a focus on the function of health care. The MFFA, on the other hand, provides a unified concept of functional food as a guideline. As stipulated in Article 2 of Chapter 1 by MFFA, functional food is defined as the food with specific function for health care. Moreover, it should be suitable for use by a specific group of people to improve their body functions, but the use of it might not be aimed at treating diseases.[1]

More than 2,000 functional food items have so far been approved by the Ministry of Public Health and are entitled to the label of "Functional Food" on their packages. The authorized institutions usually perform 30 different analyses for testing functional food. These include: immune regulation, postponement of senility, memory improvement, promotion of growth and development, antifatigue, body weight reduction, oxygen deficit tolerance, radiation protection, antimutation, antioxidation, blood lipid regulation, sex potency improvement, blood glucose regulation, digestion function improvement, sleep improvement, improvement of nutritional anemia, protection of liver from chemical damages, relief of the side-effects in radiotherapy or chemotherapy, lactation improvement, dispelling acne, dispelling chloasma, skin moisture or oil improvement, vision improvement, promotion of lead removal, removal of "intense heat" from the throat and moistening of the throat, blood pressure regulation, enhancement of bone calcification, intestinal bacterium regulation, moistening the bowels to relieve constipation, and stomach mucous membrane protection.

Differentiation between antiaging functional food and nonantiaging functional food is not an easy task. Generally speaking, most functional foods exhibit antiaging activities, such as those for immune regulation, postponement of senility, memory improvement, blood lipid regulation, blood glucose regulation, digestion function improvement, sleep improvement, skin moisture or oil improvement, vision improvement, blood pressure regulation, enhancement of bone calcification, intestinal bacterium regulation, and moistening the bowels to relieve constipation, etc. In principle, all of the herbs or foods that have been recognized officially as both food and Chinese medicine by the Ministry of Health of the People's Republic of China are safe and suitable for use as antiaging functional food. Up to now, there are 77 items of dietetic Chinese medicine that have been recognized formally as both food and Chinese medicine by the Ministry of Health of the People's Republic of China[2] (see Table 5.1).

CHINESE MATERIA MEDICA AS ANTIAGING FOODS AND FOR PREVENTING DISEASES

The following two aspects should be considered in defining the concept of preventing diseases through using functional foods:

1. Strengthening of the body constitution and promotion of immunity against disease through scientific food structure in general.
2. Prevention of certain diseases by eliminating pathogenic factors or by supplying specific nutrition such as diarrhea and cancer with garlic.

Chinese medicine has always placed its emphasis on prevention. For example, it was pointed out in *The Inner Canon of the Yellow Emperor* that "action should be taken before a disease arises." In addition, the remark that "Great

TABLE 5.1   77 Items of Dietetic Chinese Medicine that Have Been Recognized Formally as Both Food and Chinese Medicine

Drug Latin name	Chinese name	English name	Latin name
Agkistrodon	*Fushe*	Pallas Pit Viper	*Agkistrodon halys*
Arillus Longan	*Longyanrou*	Longan Aril	*Dimocarpus longan*
Bulbus Allii Macrostemi	*Xiebai*	Longstamen Onion Bulb	*Allium macrostemon*
Bulbus Lilii	*Baihe*	Lily Bulb	*Lilium lancifolium;* *L. brownii var. viridulum;* *L. pumilum*
Concha Ostreae	*Muli*	Oyster Shell	*Ostrea gigas;* *O. talienwhanensis;* *O. rivularis*
Cortex Cinnamomi	*Rougui*	Cassia Bark	*Cinnamomum cassia*
Endothelium Corneum Gigeriae Galli	*Jineijin*	Chicken's Gizzard-skin	*Gallus gallus domesticus*
Exocarpium Citri Rubrum	*Juhong*	Red Tangerine Peel	*Citrus reticulata*
Flos Carthami	*Honghua*	Safflower	*Carthamus tinctorius*
Flos Caryophylli	*Dingxiang*	Clove	*Eugenia caryophylata*
Flos Chrysanthemi	*Juhua*	Chrysanthemum Flower	*Chrysanthemum morifolium*
Folium Mori	*Sangye*	Mulberry Leaf	*Morus alba*
Folium Nelumbinis	*Heye*	Lotus Leaf	*Nelumbo nucifera*
Folium Perillae	*Zisuye*	Perilla Leaf	*Perilla frutescens*
Fructus Amomi	*Sharen*	Villous Amomum Fruit	*Amomum villosum;* *A. villosum var. Xanthioides;* *A. longiligulare*
Fructus Anisi Stellati	*Bajiaohuixiang*	Chinese Star Anise	*Illcium verum*
Fructus Aurantii	*Daidaihua*	Orange Fruit	*Citrus aurantium 'Daidai'*
Fructus Canarii	*Qingguo*	Chinese White Olive	*Canarium album*
Fructus Cannabis	*Huomaren*	Hemp Seed	*Cannabis sativa*

TABLE 5.1 (CONTINUED)   77 Items of Dietetic Chinese Medicine that Have Been Recognized Formally as Both Food and Chinese Medicine

Drug Latin name	Chinese name	English name	Latin name
Fructus Chaenomelis	*Mugua*	Common Floweringqin ce Fruit	*Chaenomeles speciosa*
Fructus Citri	*Xiangyuan*	Citron Fruit	*Citrus medica;* *C. wilsonii*
Fructus Citri Sarcodactylis	*Foshou*	Finger Citron	*Citrus medica var. sarcodactylis*
Fructus Crataegi	*Shanzha*	Hawthorn Fruit	*Crataegus pinnatifida var. major;* *C. pinnatifida*
Fructus Foeniculi	*Xiaohuixiang*	Fennel	*Foeniculum vulgare*
Fructus Gardeniae	*Zhizi*	Cape Jasmine Fruit	*Gardenia jasminoides*
Fructus Hippophae	*Shaji*	Seabuckthorn Fruit	*Hippophae rhamnoides*
Fructus Hordei Germinatus	*Maiya*	Germinated Barley	*Hordeum vulgare*
Fructus Jujubae	*Dazao*	Chinese Date	*Ziziphus jujuba*
Fructus Lycii	*Gouqizi*	Barbary Wolfberry Fruit	*Lycium barbarum*
Fructus Momordicae	*Luohanguo*	Grosvenor Momordica Fruit	*Momordica grosvenori*
Fructus Mori	*Sangshen*	Mulberry Fruit	*Morus alba*
Fructus Mume	*Wumei*	Smoked Plum	*Prunus mume*
Fructus Piperis	*Hujiao*	Pepper Fruit	*Piper nigrum*
Herba Cichorii	*Juju*	Chicory Herb	*Cichorium glandulosum;* *C. intybus*
Herba Moslae	*Xiangru*	Haichow Elsholtzia Herb	*Mosla chinensis*
Herba Menthae	*Bohe*	Peppermint	*Mentha haplocalyx*
Herba Pogostemonis	*Guanghuoxiang*	Cablin Patchouli Herb	*Pogostemon cablin*

TABLE 5.1 (CONTINUED)   77 Items of Dietetic Chinese Medicine that
Have Been Recognized Formally as Both Food and Chinese Medicine

Drug Latin name	Chinese name	English name	Latin name
Herba Portulacae	*Machixian*	Purslane Herb	*Portulaca oleracea*
Mel	*Fengmi*	Honey	*Apis cerana; A. mellifera*
Pericappium Citri Reticulatae	*Chenpi*	Dried Tangerine Peel	*Citrus reticulata*
Pericarpium Zanthoxyli	*Huajiao*	Pricklyash Peel	*Zanthoxylum schinifolium; Z. bungeanum*
Poria	*Fuling*	Indian Bread	*Poria cocos*
Radix Angelicae Dahuricae	*Baizhi*	Dahurian Angelica Root	*Angelica dahurica; A. var. formosana*
Radix Glycyrrhizae	*Gancao*	Liquorice Root	*Glycyrrhiza uralensis; G. inflata; G. glabra*
RhizomaApiniae Officinarum	*Gaoliangjiang*	Lesser Galangal Rhizome	*Alpinia officinarum*
Rhizoma Dioscoreae	*Shanyao*	Common Yam Rhizome	*Dioscorea opposita*
Rhizoma Imperatae	*Baimaogen*	Lalang Grass Rhizome	*Imperata cylindrica var. major*
Rhizoma Phragmitis	*Lugen*	Reed Rhizome	*Phragmites communis*
Rhizoma Zingiberis Recens	*Shengjiang*	Fresh Ginger	*Zingiber officinale*
Semen Armeniacae Amarum	*Xingren*	Apricot Seed	*Prunus armeniaca var. ansu ; P. sibirica; P. mandshurica; P. armeniaca*
Semen Canavaliae	*Daodou*	Jack Bean	*Canavalia gladiata*
Semen Cassiae	*Juemingzi*	Cassia Seed	*Cassia obtusifolia; C. tora*
Semen Coicis	*Yiyiren*	Coix Seed	*Coix lacrymajobi var. Ma-yuen*
Semen Euryales	*Qianshi*	Gordon Euryale Seed	*Euryale ferox*

TABLE 5.1 (CONTINUED) 77 Items of Dietetic Chinese Medicine that Have Been Recognized Formally as Both Food and Chinese Medicine

Drug Latin name	Chinese name	English name	Latin name
Semen Ginkgo	Baiguo	Ginkgo Seed	Ginkgo biloba
Semen Lablab Album	Baibiandou	White Hyacinth Bean	Dolichos lablab
Semen Myristicae	Roudoukou	Nutmeg	Myristica fragrans
Semen Nelumbinis	Lianzi	Lotus Seed	Nelumbo nucifera
Semen Persicae	Taoren	Peach Seed	Prunus persica; P. davidiana
Semen Phaseoli	Chixiaodou	Rice Bean	Phaseolus calcaratus; P. angularis
Semen Pruni	Yuliren	Chinese Dwarf Cherry Seed	Prunus humilis; Prunus japonica; P. pedunculata
Semen Raphani	Laifuzi	Radish Seed	Raphanus sativus
Semen Sesami Nigrum	Heizhima	Black Sesame	Sesamum indicum
Semen Brassicae Junceae	Huangjiezi	Yellow Mustard Seed	Brassica juncea
Semen Sojae Preparatum	Dandouchi	Fermented Soybean	Glycine max;
Semen Torreyae	Feizi	Grand Torreya Seed	Torreya grandis
Semen Ziziphi Spinosae	Suanzaoren	Spine Date Seed	Ziziphus jujuba var. spinosa
Thallus Laminariae	Kunbu	Kelp Or Tangle	Laminaria japonica
Zaocys	Wushaoshe	Black Snake	Zaocys dhumnades
Fructus Phyllanthi	Yuganzi	Emblic Leafflower Fruit	Phyllanthus emblica L.
Flos Lonicerae	Jinyinhua	Honeysuckle Flower	Lonicera japonica Thunb.; Lonicera hypoglauca Miq.; Lonicera confusa DC.; Lonicera dasystyla Rehd.

TABLE 5.1 (CONTINUED)   77 Items of Dietetic Chinese Medicine that
Have Been Recognized Formally as Both Food and Chinese Medicine

Drug Latin name	Chinese name	English name	Latin name
Herba Houttuyniae	*Yuxingcao*	Heartleaf Houttuynia Herb	*Houttuynia cordata Thunb.*
Fructus Alpiniae Oxyphyllae	*Yizhi*	Sharpleaf Glangal Fruit	*Alpinia oxyphylla Miq.*
Semen Sterculiae Lychnophorae	*Pangdahai*	Boat-fruited Sterculia Seed	*Sterculia lychnophora Hance*
Herba Lophatheri	*Danzhuye*	Lophatherum He	*Lophatherum gracile Brongn.*
Radix Puerariae	*Gegen*	Kudzuvine Root	*Pueraria lobata (Willd.) Ohwi.; Pueraria thomsonii Benth.*
Herba Taraxaci	*Pugongying*	Dandelion	*Taraxacum mongolicum Hand.-Mazz.; Taraxacum sinicum Kitag.*

doctors give treatment before diseases occur and prevent disorders before they rise" was provided in the historical book
of *The Inner Canon of the Yellow Emperor*. Moreover, *Sun
Simiao*, a famous doctor in the *Tang* dynasty of China, stated
"Doctors should understand the pathogenesis of the disease
at first, and then treat it with food while prescription drugs
should be used only if food therapy fails."[3]

It is known now that some foods can be utilized to prevent
diseases. Radish, for example, not only is prepared into very
delicious dishes, but is also used as medicine for treating
symptoms such as indigestion caused from overeating and
abdominal distension, because there is a type of oil with
pungent and hot taste present in radish that can promote the
movement of stomach and small intestine, and consequently
improve appetite. In addition, fried shredded radish with beef
functions to reinforce the spleen and stomach, to strengthen
tendon and bone, to promote the circulation of blood, and to
eliminate sputum, etc.

It should also be pointed out that Chinese practitioners and physicians over the centuries have made efforts to promote longevity and ensure the well-being and vitality of elderly patients via tonics, restoratives, and strengthening agents.

In addition, certain herbal medicine or foods themselves have the function of prolonging life. Let us take the herbal medicine in *Shen Nong's Materia Medica* as an example. *Shen Nong's Materia Medica* was one of the earliest books to review Chinese Materia Medica in China. This book, completed in the early part of the first century A.D., classified 365 Chinese Materia Medica into three categories: superior-grade, intermediate-grade, and low-grade. The Chinese Materia Medica in the superior-grade mainly are the nourishing and strengthening types and are nontoxic, and suitable for long-term administration, which include ginseng and Chinese jujuba. On the other hand, at least half of the Chinese Materia Medica reported in this book can be used as medicine or food, such as lotus root, grape, Chinese yam, honey, orange, sesame, some of which can prolong life according to the written record reported in this book. Some of the most commonly used Chinese Materia Medica used as antiaging food in China are cited as follows: barbary wolfberry (*Lycium babarum, fr.*), black ear mushroom (*Auricularia auricula, fr-bd*), black sesamum (*Sesamum indicum, sd.*), Chinese date (*Zizyphus jujuba, fr.*), chrysanthemum (*Chrysanthemum morifolium, fl*), euryale (*Euryale ferox, sd*), fragrant mushroom (*Lentinus edodes, fr-bd*), garlic (*Allium sativum, bul*), hawthorn (*Crataegus pinnatifida, fr*), Job's tears (*Coix lachryma-johi var. ma-yuan*), lily (*Lilium brownii var, viridulum, bul*), longan aril (*Euphoria longon, fr*), lotus (*Nelumbo nucifera, sd*), mulberry (*Morus alba, fr*), seabuckthom (*Hippophae rhamnoides, fr*), silver ear mushroom (*Tremella fusiformis, fr-bd*), walnut (*Juglans regia, fr*), and yam (*Dioscorea opposita, rhz*).

THE FUNCTIONS RELATED TO ANTIAGING FUNCTIONAL FOODS

Modern functional foods in China have been developed on the basis of the food therapy of Chinese medicine, which has attracted considerable international attention in the past few years owing in part to its long medical history, unique food therapy theory, extensive Chinese Materia Medica resources and splendid Oriental culture.

The antiaging activities of functional food are exemplified as follows:

1. **Immune regulation** — American ginseng (*Panax quinquefolius*), ginseng (*Panax ginseng*), barbary wolfberry fruit (*Lycium barbarum*), astragalus (*Astragalus membranaceus; A. Membranaceus var. Mongholicus*), Chinese caterpillar fungus (*Cordyceps sinensis*), gingko leaf (*Ginkgo biloba*), walnut (*Juglans regia*), and Chinese date (*Ziziphus jujuba*).

2. **Life extension** — Green tea (*Camellia sinensis*), fleece flower root (*Polygonum multiflorum*), black sesame (*Sesamum indicum*), mulberry fruit (*Morus alba*), and barbary wolfberry fruit (*Lycium barbarum*).

3. **Blood lipid regulation** — Hawthorn fruit (*Crataegus pinnatifida*), soybean (*Glycine max*), peach seed (*Prunus persica*), chrysanthemum flower (*Chrysanthemum morifolium*), spine date seed (*Ziziphus jujuba var. spinosa*), corn oil (*Zea mays*), extract of flax seed (*Linum usitatissimum*), and safflower seed (*Carthamus tinctorius*).

4. **Blood glucose regulation** — Five leaf Gynostemma (*Gynostemma pentaphyllum*), hawthorn fruit (*Crataegus pinnatifida*), Chinese yam (*Dioscorea opposita*), buckweat poria (*Fagopyrum esculentum*), cocos poria (*Poria cocos*), extract of pumpkin (*Cucurbita moschata*), and powder of spirulina (*Spirulina princeps*).

CATEGORIES OF CHINESE MATERIAL MEDICAL FOR ANTIAGING ACTIVITY

Chinese Materia Medica for antiaging activity can be categorized into nourishing *Yin, Yang, Qi,* and *Xue* (blood) according to their functions.

Yin tonics are used for replenishing the vital essence and fluid, and in the treatment of vital essence and fluid deficiency marked by dry mouth and throat, constipation, vertigo, tinnitus, feverishness of the palms and soles, night sweating, and insomnia. Since many aged people are always suffering with *Yin*-deficiency, the *Yin* tonics have been widely used in geriatrics. The dried fruit of *Lycium barbarum* is one of the most commonly used *Yin* tonic in Chinese medicine, for example, it (at a dosage of 100 mg/d) was reported to treat 194 cases of aged people with insufficiency syndromes for 2 months. The corresponding results showed that the symptoms of senility such as dizziness, fatiguing easily, sleeplessness, and poor appetite were improved significantly. In addition, it was found out that the functions of immune as well as the critical blood lipids were enhanced.

1. **Chinese Materia Medica for *Yang* tonics** — *Yang* tonics are the drugs for reinforcing the vital function, chiefly for the kidney function. According to the theory of Chinese medicine, most of the kidney function deteriorates with the increase of a person's age. Modern scientific experiments revealed that many of these drugs exhibit excitatory effects on the endocrine system and stimulation and modulating effects on immunological functions.

2. **Chinese Materia Medica for *Qi* tonics** — *Qi* tonics are the drugs that reinforce or invigorate the vital energy. They are usually used for the treatment of *Qi* deficiency and commonly exist in elderly people with syndromes such as pale complexion, low and weak voice, indigestion, anemia, and palpitation, etc.

3. **Chinese Materia Medica for blood-tonics** — "Blood" here should be interpreted as blood in the ordinary sense including all of the physical elements of the body and living cells, such as cytoplasmic substances, enzymes, DNA, RNA, and hormones. Blood tonics are mainly used for the treatment of blood deficiency and for regulating the menstrual flow, etc., which are beneficial to the elderly.

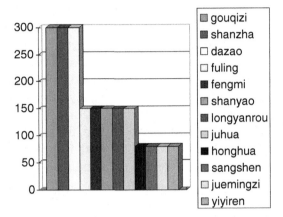

Figure 5.1 Frequency of appearance of *Fructus Lycii* (*Gouqizi*) in functional food products.

LYCIUM (*GOUQIZI*), ONE OF THE MOST COMMONLY USED ANTIAGING HERBS

Lycium (*Gouqizi*), is the dried ripe fruit of *Lycium barbarum L.* (*Fam. Solanaceae*). The application of Lycium in Chinese medicine has a long history. Lycium is now used not only as an herb for treatment of diseases but also as an antiaging functional food. Among the 77 items of dietetic Chinese medicine that were recognized officially as both food and Chinese medicine by the Ministry of Health of the People's Republic of China, Lycium (*Gouqizi*) is one of the most commonly used herbs and functional food linked to antiaging therapy (see Figure 5.1)

MODERN RESEARCH ON LYCIUM AND ITS APPLICATION

Lycium (*Gouqizi*), barbary wolfberry fruit is often collected in the seasons of summer and autumn when the fruit turns to orange-red.

1. **Identified chemical constituents of Lycium** — The identified chemical constituents of Lycium

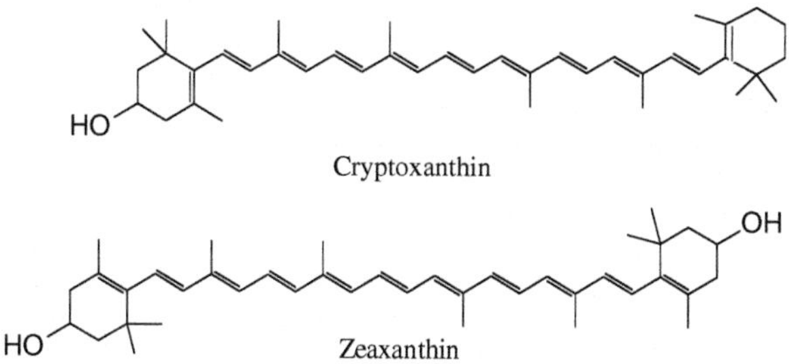

Figure 5.2 Cryptoxanthin and Zeaxanthin.

include carotene, thamine, riboflavine, nicotinic acid, ascorbic acid, β-sitosterol, linoleic acid, zeaxanthin, betanine, physalien, cryptoxanthin, astropine, hyoscuyamine, and acopoletin (Figure 5.2).

2. **Pharmacological action of Lycium** — It was reported that the aqueous extract of Lycium enhanced nonspecific immunity as well as relieved the liver damage induced by CCl_4 in mice.

3. **Attributes of Lycium** — Lycium is sweet in taste and mild in property. In addition, it acts on liver and kidney meridians.

4. **Action and indications**
   a. Lycium nourishes the liver and kidneys and is effective for treating people with yin and blood deficiency such as general debility with deficiency of vital essence manifested by aching of the loins and knees, low abdominal pain, impotence, nocturnal emission, dizziness and tinnitus; diabetes caused by internal heat; anemia; impaired vision, and consumption. Because this herb is neither hot nor cold, it is ideal and commonly used for treating liver and kidney deficiency.
   b. Lycium benefits the essence and brightens the eyes: for liver and kidney deficiency patterns where the essence and blood are unable to nourish the eyes. The manifestations of this symptom

include dizziness, blurred vision, and diminished visual acuity.

   c. The dried fruit enriches the yin and moistens the lungs for consumptive cough.

ANALYSIS OF THE FUNCTIONAL FOOD PRODUCTS THAT PASSED THE FUNCTIONAL FOOD EVALUATION

There are more than 2,000 kinds of functional foods that passed the functional food evaluation, most of which were developed by using modern techniques established on the basis of nutritional science and traditional Chinese nutrition knowledge in Chinese medicine. Novel formulation and processes were also utilized for the production of many functional food items.

As shown in Table 5.2, there are over 2,000 functional foods that were approved by the government and the functional

TABLE 5.2   Frequency of Appearance of Functions in the Approved Functional Food Products (for the 13 Most Common Functions)

No	Function	Frequency of appearance	%
1	Immune regulation	904	43.73
2	Blood lipid regulation	440	21.29
3	Antifatigue	436	21.09
4	Postponement of senility	177	8.56
5	Oxygen deficit tolerance	110	5.32
6	Digestion function improvement	95	4.60
7	Sleep improvement	82	3.97
8	Blood glucose regulation	75	3.63
9	Body weight reduction	58	2.81
10	Memory improvement	56	2.71
11	Protection of liver from chemical damages	45	2.18
12	Antimutation	34	1. 64
13	Enhancement of bone calcification	33	1. 60

TABLE 5.3   Frequency of Appearance of the Herbs or Foods
in the Approved Functional Food Products (for the Above 30
Most Common Herbs or Foods)

Frequency of appearance	Name of herbs or foods
200–470	Barbary wolfberry fruit (*Gouqizi*), hawthorn (*Shanzha*), Chinese date (*Dazao*)
100–199	India bread (*Fuling*), American ginseng (*Xiyangshen*), honey (*Fengmi*), glossy ganoderma (*Lingzhi*), ginseng (*Renshen*), astragalus (*Huangqi*), Chinese yam (*Shanyao*), longan aril (*Longyanrou*), soybean (*Dadou*), Chinese caterpillar fungus (*Dongchongxiacao*), spirulina (*Luoxuanzao*), chrysanthemum flower (*Juhua*)
70–99	Germinated barley (*Damaiya*), safflower (*Honghua*), mulberry fluid (*Sangshen*), milk (*Niunai*), green tea (*Lucha*), cassia seed (*Juemingzi*), coix seed (*Yiyiren*), fish oil (*Yuyou*), fleeceflower root (*Heshouwu*)
50–69	Fiveleaf Gynosttemma (*Jiaogulan*), dried tangerine peel (*Chenpi*), mushroom (*Xianggu*), Chinese angelica (*Danggui*), gingko leaf (*Yinxingye*), cassia bark (*Rougui*), ant (*Mayi*), chicken (*Jirou*), etc.

foods with immune regulation, blood lipid regulation, post-ponement of senility, antifatigue, etc. are the most common.[4]

Among more than 2,000 kinds of approved functional food by the government, Lycium, barbary wolfberry fruit (*Gouqizi*), hawthorn (*Shanzha*), and Chinese date (*Dazao*) are the most common herbs (Table 5.3).

The long history of traditional Chinese medicine in food therapy provides us with profound information on traditional functional foods. Barbary wolfberry fruit can be considered one of the traditional functional foods possessing antiaging functions. It is anticipated that with the development of advanced technology, a new generation of functional foods for antiaging can be discovered in the near future. It is expected that barbary wolfberry fruit will play a very important role in prolonging life and in maintaining the health of human beings sooner or later. In addition, Chinese functional food and Chinese Materia Medica as a treasure house is worthy of further exploration for antiaging agents. It is hoped that the research and development of Chinese functional food will be advanced and that the life quality of human beings will be further improved as we acquire more information and knowledge about functional foods in the future.

REFERENCES

1. The Ministry of Public Health of the People's Republic of China, Measures of Functional Food Administration, 1996.

2. Department of Hygiene Supervision of the Ministry of Public Health of the People's Republic of China, Collection of Administration Laws of the Functional Food, Jilin Science and Technology Publishing House, 1997.

3. *Sun, Simiao, Prescription Worth a Thousand Gold for Emergencies, the Tang Dynasty*, Beijing, the People's Medical Publishing House, Beijing, 1955.

4. Y Dang, Y Xiao, *The Research and Development of Chinese Functional Food*, Beijing, the People's Medical Press, 2002.

6

A New Treatment and Prevention Strategy for Human Papillomaviruses

KAREN J. AUBORN

North Shore—Long Island Jewish Research
Institute, Manhasset, NY, USA;
Departments of Otolaryngology and
Microbiology and Immunology,
Albert Einstein College of Medicine,
Bronx, NY, USA

INTRODUCTION

Approximately 100 types of cruciferous vegetables, alias Brassicas, are used in human diets. Some of the common ones include cabbage, broccoli, and kale. This group of vegetables is important in traditional Asian diets and is likely to contribute to the low rate of certain cancers, e.g., endocrine-related cancers, in Asian countries. Low consumption of these vegetables may be a factor in the increase in these cancers in

Asians after their diets become more Western. The importance of cruciferous vegetables in cancer prevention has been documented in Western countries as well. A study of postmenopausal women in Sweden — 2,832 case patients and 2,650 controls — found no correlation between a diet rich in many fruits and vegetables and breast cancer. However, in this same study, consumption of Brassicas was inversely associated with cancer risk.[1] In the case of prostate cancer, two or more servings of cruciferous vegetables per week have been reported to decrease risk.[2] No epidemiological dietary studies have indicated that cruciferous vegetables decrease the risk of tumors for which papillomaviruses are cofactors, e.g., cervical cancer. Both animal and human studies with indole-3-carbinol, a phytochemical in cruciferous vegetables, suggest that its consumption will prevent these tumors as well as being useful in treatment.[3–7]

INDOLE-3-CARBINOL

Cruciferous vegetables contain at least three phytochemicals that protect against a variety of cancers. One of these compounds is indole-3-carbinol (I3C), which is converted in the stomach to diindolylmethane (DIM) and other condensation products. Numerous animal studies (Table 6.1)[8–16] support the efficacy of I3C for the prevention of a variety of cancers. Recently, it has become clear that I3C has the potential to prevent and even to treat a number of common cancers, especially those that are estrogen-related.

Laboratory studies suggest that I3C can act in several different ways to prevent cells from becoming transformed, i.e., prevention of progression to tumors and killing transformed cells selectively. I3C is an antioxidant and could in theory protect against deleterious effects of active electrophiles and free radicals.[17,18] Much of the prophylactic effect of I3C and DIM can be ascribed to their ability to induce a variety of enzymes that detoxify carcinogens. Induction of many phase I and phase II detoxifying enzymes occurs because DIM is a weak ligand for the aryl hydrocarbon receptor.[19–21] Several of the phase I enzymes induced by I3C/DIM alter estrogen

TABLE 6.1 Chemoprevention of tumors by indole-3-carbinol in rodents

Tumor tissue	Species	Initiator	Reference
Mammary	Rat	DMBA	Wattenberg et al. 1978[8]
Forestomach	Mouse	B[a]P	Wattenberg et al. 1978[8]
Mammary	Mouse	MMTV	Bradlow et al. 1990[9]
Endometrium	Rat	None	Kojima et al. 1990[10]
Lung	Mouse	NNK	Morse et al. 1990[11]
Larynx	Mouse	HPV	Newfield et al. 1993[3]
Colon	Rat	PhIP	Guo et al. 1995[12]
Liver	Mouse	DEN	Oganesian et al. 1997[13]
Liver	Rat	AFB	Manson et al. 1998[14]
Skin	Mouse	DMBA	Srivastava et al. 1998[15]
Cervix	Mouse	HPV transgenes	Jin et al. 1999[4]
Colon	Mouse	None	Kim et al. 2003[56]

Note: DMBA, dimethybenz(a)anthracine; MMTV, mouse mammary tumor virus; HPV, human hapillomavirus; PhIp, 2-amino-1-methyl-6-phenylimidazo [4,5-b] pyridine; B[a]P, benzo(a)pyrene; NNK, nitrasamine-4-(methyl-nitrosamino)-1-(3-pyridye)-1-butanone; DEN, diethylnitrosamine; AFB, aflatoxin B1

metabolism[22–24] and thus affect the establishment and progression of estrogen-responsive tumors. In addition to inducing the battery of phase I enzymes, I3C is also a ligand for the estrogen receptor and has been shown to be a negative regulator of estrogen signaling.[25] This ability to alter estrogen effects has been the major rationale for the use of I3C/DIM in treatment of papillomavirus-induced lesions of both the larynx and the cervix.

More recently, I3C and DIM have been shown to decrease proliferation and induce apoptosis of cervical,[26] prostate,[27] and breast cancer cells.[28,29] These activities occur independently of estrogen signaling. Additionally, I3C has been shown to inhibit invasion and migration of breast cancer cells.[30,31] Thus, the evidence is that I3C/DIM affects multiple cellular pathways. Insight into the mechanisms whereby I3C/DIM changes the microenvironment of a cell is surfacing from nutritional genomics studies. Clearly, I3C or DIM alters gene expression. I3C/DIM abrogates changes in gene expression brought about by estrogen,[32] which is important since estrogen increases proliferation[33,34] and inhibits apoptosis in estrogen-sensitive cells.[35] Using microarray profiling, DIM was found to alter the expression of more than 100 genes.[36] Many of the genes whose stimulation is induced by DIM encode for transcription factors and proteins involved in signaling, stress response, and growth. I3C downregulates expression of papillomavirus oncogenes. Proteins induced by DIM, i.e., GADD153 and cEBPβ, decrease expression of these oncogenes.[36]

Together the activities of I3C/DIM support the epidemiological observations of the benefits of cruciferous vegetables in the prevention and treatment of tumors.

PAPILLOMAVIRUSES

Papillomaviruses transform cells producing proliferative lesions. These viruses are very abundant, affecting keratinocytes in humans and other species. More than 70 types of papillomaviruses have been identified. The number of types (based on DNA homology of papillomavirus genomes) keeps growing as more are being identified. Some papillomaviruses have a high oncogenic potential.[37] It is clear that some types of human papillomaviruses (HPVs) are cofactors (together with estrogen) for cervical cancer.[38] Growing evidence supports a role for other HPVs as cofactors with ultraviolet light for nonmelanoma skin cancers.[39]

Most infections with HPV result in an inactive infection. The viral DNA is present in cells, but that is the only sign of infection. However, under certain circumstances that are not clearly understood, the virus expresses its oncogenes. Viral oncogenes make a cell favorable for replication and enable the cell to avoid apoptosis.[40,41] Oncogenes, especially those from highly oncogenic types of HPVs, inactivate tumor suppressors, most notably retinoblasoma and p53. This suppression of retinoblastoma enables the virus to replicate. This is important since viral replication occurs in keratinocytes that are undergoing differentiation and not normally replicating. Inactivation of p53 increases survival of a cell that has undergone DNA damage. Hence, papillomaviruses can make a cell vulnerable to transformation especially where the regulation of the expression of the viral oncogenes goes astray.

Both benign and malignant tumors are associated with HPVs. Common skin warts and planter warts have an HPV etiology. As described above, more evidence is accumulating that certain types of HPVs are cofactors for nonmelanoma skin cancers. It is clear that HPVs infect the epithelium of the aerogenital tract, and HPVs cause both condylomas and laryngeal papillomas. The highly oncogenic HPVs are cofactors for cervical cancer. Much circumstantial evidence indicates that both HPVs and estrogen are combined cofactors for cervical cancer and noncancerous lesions in the aerogenital tract. This mouse model provides evidence that this is the case. A mouse with transgenes for HPV type 16 develops cervical cancer only when given estradiol chronically.[42]

Some HPV types are apparently ubiquitous, albeit the virus is generally latent. In the case of HPVs that affect the aero and genital tracts, about 30% of the population is infected.[43] Hence, it is important to identify ways to prevent and treat the early virus pathology so that the infection does not result in serious disease or cancers.

INDOLE-3-CARBINOL AND PAPILLOMAVIRUSES

It is apparent that I3C has the potential to prevent or ameliorate papillomavirus lesions in the aero and genital tracts. The initial rationale for a possible benefit from I3C is its ability to induce 2-hydroxylation of estradiol with an overall antiestrogen effect. Studies evaluating estrogen metabolism in cells derived from tissue explants of the larynx[3] or the cervix[44] indicate that cells infected with active HPV increase utilization of the 16α-hydroxylation pathway upstream of estrone. The importance of increased 16α-hydroxylation is that products of this metabolism have a prolonged estrogen activity and are carcinogenic.[45,46] On the other hand, 2-hydroxylation of estrone, which is an alternate pathway to 16α-hydroxylation, results in products that are not estrogenic and rapidly *O*-methylated to metabolites that are antiproliferative, proapototic, and antiangiogenic.[47–49] Both *in vitro* laboratory studies[3,24–26,33,36] and animal studies[3,4] and clinical studies[5,6,50] support the efficacy of I3C/DIM in prevention of laryngeal papillomas and cervical cancer. Other activities of I3C/DIM, e.g., inhibiting proliferation and induction of apoptosis, could increase the efficacy of these nutrients for these and other tumors. Finally, I3C directly downregulates expression of HPV type 11 and 16 genes.[36]

PREVENTION AND THERAPY WITH INDOLE-3-CARBINOL

Recurrent respiratory papillomatosis (RRP) is a rare disease having an HPV etiology.[51] RRP results in benign hyperplastic epithelial tumors that are typically found on the vocal cords, the most hormonally sensitive part of the respiratory tract. The primary symptom of the disease is hoarseness where the lesions interfere with vocal cord function, but bulky lesions can cause life-threatening airway obstruction. A hallmark of the disease is the tendency of papillomas to recur at regular intervals after their surgical removal, hence the name RRP. Malignant transformation is infrequent but irradiation and other DNA-damaging agents predispose individuals with RRP to further transformation of tumors that were benign.[52]

I3C or DIM is the most popular adjunct therapy for RRP because of its virtual lack of toxicity.[43] Studies show that a diet rich in cruciferous vegetables or supplements of I3C/DIM is a useful adjunct therapy for RRP.[5,6,50] Previous studies in mice indicate that a diet supplemented with I3C prevents papilloma cysts in xenografts of laryngeal tissue infected with HPV type 11.[3] A child having severe RRP was cured after being fed daily with a diet that included cabbage juice.[5] Other studies implementing diets that include substantial amounts of cruciferous vegetables or supplements (generally used at concentrations that people could achieve by eating vegetables) follow and support the efficacy of I3C. The Recurrent Respiratory Papillomatosis Foundation makes a concerted effort to keep statistics and evaluate the effectiveness of therapies for RRP. While a subgroup of patients appear to be refractory to the benefits of I3C/DIM, complete remission of the disease or improvement (longer intervals between needed surgeries) occur in about two-thirds of patients.[53]

Cervical cancer is prevalent worldwide, accounting for 5% of all new cancers, and is the second most common cancer in women.[54] Preneoplastic lesions (cervical intraepithelial neoplasia) can develop as a result of HPV infections. Some of the women with such lesions develop invasive cervical cancer. The preneoplastic lesions can be detected, monitored, and treated, e.g., ablatement of the transformation zone of the cervix, the most estrogen-sensitive genital site.[55] A phase II placebo-controlled study indicates that I3C supplements (the equivalent of one-third of a head of cabbage) reverse preneoplastic stage II and stage III lesions of the cervix.[7] Similar to the RRP studies, about one-third of the patients did not benefit from I3C supplements. The translational study of preneoplastic lesions of the cervix followed a study indicating that a diet supplemented with I3C not only prevents cervical cancer in a mouse model but reduces cervical dysplasia induced by estrogen.[4]

I3C is among an array of dietary compounds that are identified as natural prophylactic and anticancer agents. I3C or DIM is proving useful clinically for prevention and treatment of HPV-induced pathologies. Clinical trials are planned to test the efficacy of I3C as a preventative treatment for breast cancer.[55,56] It is becoming clear that I3C has the potential to prevent and even to treat a number of common cancers, especially those that are estrogen-sensitive.

REFERENCES

1. P Terry, A Wolk, I Persson, C Magnusson. Brassica vegetables and breast cancer risk. *JAMA*. 285:2975–2977, 2001.

2. JH Cohen, AR Kristal, JL Stanford. Fruit and vegetable intakes and prostate cancer risk. *J Natl Cancer Inst* 92:61–68, 2000.

3. L Newfield, A Goldsmith, HL Bradlow, K Auborn. Estrogen metabolism and human papillomavirus-induced tumors of the larynx: chemo-prophylaxis with indole-3-carbinol. *Anticancer Res* 13:337–341, 1993.

4. L Jin, M Qi, DZ Chen, A Anderson, GY Yang, JM Arbeit, KJ Auborn. Indole-3-carbinol prevents cervical cancer in human papilloma virus type 16 (HPV16) transgenic mice. *Cancer Res* 59:3991–3997, 1999.

5. DA Coll, Rosen CA, Auborn K, Potsic WP, Bradlow HL. Treatment of recurrent respiratory papillomatosis with indole-3-carbinol. *Am J Otolaryngol* 18:283–285, 1997.

6. CA Rosen, GE Woodson, JW Thompson, AP Hengesteg, HL Bradlow. Preliminary results of the use of indole-3-carbinol for recurrent respiratory papillomatosis. *Otolaryngol Head Neck Surg* 118:810–815, 1998.

7. MC Bell, P Crowley-Nowick, HL Bradlow, DW Sepkovic, D Schmidt-Grimminger, P Howell, EJ Mayeaux, A Tucker, EA Turbat-Herrera, JM Mathis. Placebo-controlled trial of indole-3-carbinol in the treatment of CIN. *Gynecol Oncol* 78:123–129, 2000.

8. LW Wattenberg, WD Loub. Inhibition of polycyclic aromatic hydrocarbon-induced neoplasia by naturally occurring indoles. *Cancer Res* 38:1410–1413, 1978.

9. HL Bradlow, J Michnovicz, NT Telang, MP Osborne. Effects of dietary indole-3-carbinol on estradiol metabolism and spontaneous mammary tumors in mice. *Carcinogenesis* 12:1571–1574. 1991.

10. T Kojima, T Tanaka, H Mori. Chemoprevention of spontaneous endometrial cancer in female Donryu rats by dietary indole-3-carbinol. *Cancer Res* 54:1446–1449, 1994.

11. MA Morse. Inhibition of NNK-induced lung tumorigenesis by modulators of NNK activation. *Exp Lung Res* 24:595–604, 1998.

12. D Guo, HA Schut, CD Davis, Snyderwine e.g., GS Bailey, RH Dashwood Protection by chlorophyllin and indole-3-carbinol against 2-amino-1-methyl-6-phenylimidazo[4,5-b]pyridine (PhIP)-induced DNA adducts and colonic aberrant crypts in the F344 rat. *Carcinogenesis* 16:2931–2937, 1995.

13. A Oganesian, JD Hendricks, DE Williams. Long-term dietary indole-3-carbinol inhibits diethylnitrosamine-initiated hepatocarcinogenesis in the infant mouse model. *Cancer Lett* 16:118:87–94, 1997.

14. MM Manson, EA Hudson, HW Ball, MC Barrett, HL Clark, DJ Judah, RD Verschoyle, GE Neal. Chemoprevention of aflatoxin B1-induced carcinogenesis by indole-3-carbinol in rat liver—predicting the outcome using early biomarkers. *Carcinogenesis* 19:1829–1836, 1998.

15. B Srivastava, Y Shukla. Antitumour promoting activity of indole-3-carbinol in mouse skin carcinogenesis. *Cancer Lett* Dec 11;134:91–95, 1998.

16. DJ Kim, DH Shin, B Ahn, JS Kang, KT Nam, CB Park, CK Kim, JT Hong, YB Kim, YW Yun, DD Jang, KH Yang. Chemoprevention of colon cancer by Korean food plant components. *Mutat Res* 523–524:99–107, 2003.

17. HG Shertzer, MW Tabor, IT Hogan, SJ Brown, M Sainsbury. Molecular modeling parameters predict antioxidant efficacy of 3-indolyl compounds. *Arch Toxicol* 70:830–834, 1996.

18. MB Arnao, J Sanchez-Bravo, M Acosta. Indole-3-carbinol as a scavenger of free radicals. *Biochem Mol Biol Int* 39:1125–1134, 1996.

19. CA Miller 3rd. Expression of the human aryl hydrocarbon receptor complex in yeast. Activation of transcription by indole compounds. *J Biol Chem* 272:3282–3284, 1997.

20. MM Manson, HW Ball, MC Barrett, HL Clark, DJ Judah, G Williamson, GE Neal. Mechanism of action of dietary chemoprotective agents in rat liver: induction of phase I and II drug metabolizing enzymes and aflatoxin B1 metabolism. *Carcinogenesis* 18:1729–1738, 1997.

21. CW Nho, E Jeffery. The synergistic upregulation of phase II detoxification enzymes by glucosinolate breakdown products in cruciferous vegetables. *Toxicol Appl Pharmacol* 174:146–152, 2001.

22. CP Martucci, J Fishman. P450 enzymes of estrogen metabolism. *Pharmacol Ther* 57:237–257, 1993.

23. HL Bradlow, J Michnovicz, NT Telang, MP Osborne. Effects of dietary indole-3-carbinol on estrogen metabolism and spontaneous mammary tumors in mice. *Carcinogenesis* 12:1571–1574, 1991.

24. F Yuan, DZ Chen, K Liu, DW Sepkovic, HL Bradlow, K Auborn. Antiestrogenic activities of indole-3-carbinol in cervical cells: implication for prevention of cervical cancer. *Anticancer Res* 19:1673–1680, 1999.

25. Q Meng, F Yuan, ID Goldberg, EM Rosen, K Auborn, S Fan. Indole-3-carbinol is a negative regulator of estrogen receptor-alpha signaling in human tumor cells. *J Nutr* 130:2927–2931, 2000.

26. D-Z Chen, M Qi, KJ Auborn, and TH Carter. Indole-3-carbinol and diindolymethane induce apoptosis of human cervical cells and in murine HPV16-transgenic preneoplastic cervical epithelium. *J Nutr* 131:3294–3302, 2001.

27. SR Chinni, Y Li, S Upadhyay, PK Koppolu, FH Sarkar. Indole-3-carbinol (I3C) induced cell growth inhibition, G1 cell cycle arrest and apoptosis in prostate cancer cells. *Oncogene* 20:2927–2936, 2001.

28. CM Cover, SJ Hsieh, SH Tran, G Hallden, GS Kim, LF Bjeldanes, GL Firestone. Indole-3-carbinol inhibits the expression of cyclin-dependent kinase-6 and induces a G1 cell cycle arrest of human breast cancer cells independent of estrogen receptor signaling. *J Biol Chem* 273:3838–4387, 1998.

29. X Ge, FA Fares, S Yannai. Induction of apoptosis in MCF-7 cells by indole-3-carbinol is independent of p53 and bax. *Anticancer Res* 19:3199–3203, 1999.

30. Q Meng, ID Goldberg, EM Rosen, S Fan. Inhibitory effects of indole-3-carbinol on invasion and migration in human breast cancer cells. *Breast Cancer Res Treat* 63:147–152, 2000.

31. Q Meng, M Qi, D-Z Chen, R Yuan, ID Goldberg, EM Rosen, K Auborn, S Fan. Suppression of breast cancer invasion and migration by indole-3-carbinol: associated with upregulation of BRCA1 and E-cadherin/catenin complexes. *J Mol Med* 78:155–165, 2000.

32. I Chen, T Hsieh, T Thomas, S Safe. Identification of estrogen-induced genes downregulated by AhR agonists in MCF-7 breast cancer cells using suppression subtractive hybridization. *Gene* 10:262:207–14, 2001.

33. L Newfield, HL Bradlow, DW Sepkovic, K Auborn. Estrogen metabolism and the malignant potential of human papillomavirus immortalized keratinocytes. *Proc Soc Exp Biol Med* 217:322–326, 1998.

34. C Dabrosin, K Palmer, WJ Muller, J Gauldie. Estradiol promotes growth and angiogenesis in polyoma middle T transgenic mouse mammary tumor explants. *Breast Cancer Res Treat* 78:1–6, 2003.

35. RJ Alvarez, SJ Gips, N Moldovan, CC Wilhide, EE Milliken, AT Hoang, RH Hruban, HS Silverman, CV Dang, PJ Goldschmidt-Clermont. 17beta-estradiol inhibits apoptosis of endothelial cells. *Biochem Biophys Res Comm* 237:372–381, 1997.

36. TH Carter, K Liu, W Ralph Jr, D Chen, M Qi, S Fan, F Yuan, EM Rosen, KJ Auborn. Diindolylmethane alters gene expression in human keratinocytes *in vitro*. *J Nutr* 132:3314–3324, 2002.

37. H zur Hausen, EM de Villiers. Human papillomaviruses. *Ann Rev Microbiol* 48:427–447, 1994.

38. EM de Villiers. Relationship between steroid hormone contraceptives and HPV, cervical intraepithelial neoplasia and cervical carcinoma. *Int J Cancer* 103:705–708, 2003.

39. EM de Villiers, A Ruhland, P Sekaric. Human papillomaviruses in nonmelanoma skin cancer. *Sem Cancer Biol* 9:413–422, 1999.

40. H zur Hausen. Immortalization of human cells and their malignant conversion by high-risk human papillomavirus genotypes. *Sem Cancer Biol* 9:405-411, 1999.

41. K Munger, PM Howley. Human papillomavirus immortalization and transformation functions. *Virus Res* 89:213–228, 2002.

42. JM Arbeit, PM Howley, D Hanahan. Chronic estrogen-induced cervical and vaginal squamous carcinogenesis in human papillomavirus type 16 transgenic mice. *Proc Natl Acad Sci USA* 93:2930–2935, 1996.

43. RF Rando. Human papillomavirus: implications for clinical medicine. *Ann Internal Med* 108:628–630, 1988.

44. KJ Auborn, C Woodworth, JA DiPaolo, HL Bradlow. The interaction between HPV infection and estrogen metabolism in cervical carcinogenesis. *Int J Cancer* 49:867–869, 1991.

45. J Fishman, C Martucci. Biological properties of 16 alpha-hydroxyestrone: implications in estrogen physiology and pathophysiology. *J Clin Endocrinol Metabol* 51:611–615, 1980.

46. GE Swaneck, J Fishman. Covalent binding of the endogenous estrogen 16 alpha-hydroxyestrone to estradiol receptor in human breast cancer cells: characterization and intranuclear localization. *Proc Natl Acad Sci U S A* 85:7831–7835, 1988.

47. M Gupta, A McDougal, S Safe. Estrogenic and antiestrogenic activities of 16alpha- and 2-hydroxy metabolites of 17beta-estradiol in MCF-7 and T47D human breast cancer cells. *J Steroid Biochem Mol Biol* 67:413–419, 1998.

48. BT Zhu, AH Conney. Functional role of estrogen metabolism in target cells: review and perspectives. *Carcinogenesis* 19:1–27, 1998.

49. HL Bradlow, NT Telang, DW Sepkovic, MP Osborne. 2-hydroxyestrone: the 'good' estrogen. *J Endocrinol* 150 Suppl:S259–265, 1996.

50. K Auborn, A Abramson, HL Bradlow, D Sepkovic, V Mullooly. Estrogen metabolism and laryngeal papillomatosis: a pilot study on dietary prevention. *Anticancer Res* 18:4569–4574, 1998.

51. BM Steinberg, AL Abramson. Laryngeal papillomas. *Clin Dermatol* 3:130–138, 1985

52. BM Steinberg. Human papillomaviruses and upper airway oncogenesis. *Am J Otolaryngol* 11:370–374, 1990.

53. K Auborn. Therapy for recurrent respiratory papillomatosis. *Antiviral Ther* 7:1–9, 2002.

54. WHO. The Work Health Report. Geneva, 1997.

55. K Auborn, TH Carter. Treatment of human papillomavirus gynecologic infections. *Clin Lab Med* 20:407–422, 2000.

7

Traditional Functional Foods in Korea

SU-RAE LEE

Senior Fellow, Division of Agricultural
and Fisher Sciences, Korean Academy of
Science and Technology, Gyeonggi-Do, Korea

FOOD SUPPLY AND NUTRITIONAL ADEQUACY

Early in the history of civilization, humans in their quest for food must have attempted to eat a variety of plant and animal materials. Through trial and error, they discovered two classes of materials, that is "food" and "poison." Residents of the Korean peninsula have subsisted upon a scanty food supply for 6 millennia. South Korea has a highly dense population, mountainous areas constitute two-thirds of the Korean peninsula, and the arable land is quite limited.[1] The nutrient supply of the Korean population during the past century shows that the calorie intake was minimal until the 1960s when it finally began to exceed the recommended dietary

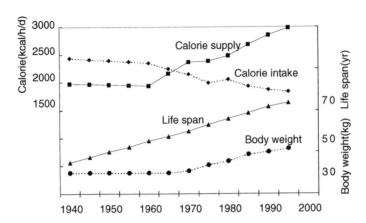

Figure 7.1 Trends in food supply and health status by year in Korea.

allowance in the 1970s. The protein supply was satisfied in the 1980s, and the animal proteins, thereafter.[2–4]

Food or diet should be satisfied with respect to nutritive value, hygienic safety, palatability, storability, and economy. An important factor in health maintenance is the nutritional adequacy of food supply or dietary intake. The statistics on food intake (calories), physique (body weight), and life span for the Korean population are shown in Figure 7.1.[4]

During the past half century, the average body weight and life span increased in proportion to the food supply. Here, a question may be raised as to how the body weight and life span are increased in spite of decreased calorie intake. A new theory on food intake and health can be postulated, i.e., "Less intake of calories elevates the health condition of humans as long as essential nutrients are supplied." There is a popular statement in the Orient that "less diet is good for health."

According to the recommended dietary allowance for Koreans, the present status of nutrient intake and its balance is appropriate on the average in the 1990s.[5,6] That is, total calorie intake, total protein, and percent of animal proteins are satisfactory. Furthermore, it is worthwhile to point out that the intake level of lipids is appropriate, the percent contribution of fats being 20% of the total calories as shown

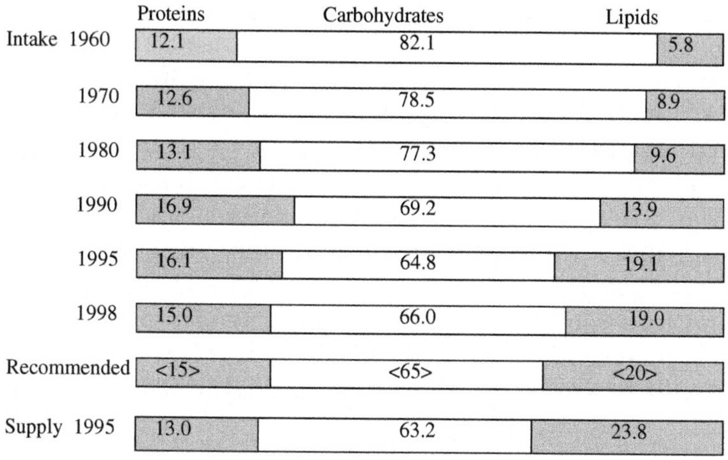

Figure 7.2 Energy contribution profile of macronutrients by Koreans (%).

in Figure 7.2. In particular, the high levels of dietary fiber/nonnutrient ingredients in the Korean diets bear a very favorable comparison with those of developed countries that are troubled with a high incidence of cardiovascular disease, obesity, and cancer through the excessive intake of fatty foods, amounting to 30% in Japan and 40% in U.S. or EU.

The proportion of nonnutrient constituents included in carbohydrates in Korean diets should be high enough to supply dietary fiber, antinutritive factors, anticarcinogenic factors, health-promoting factors and so on. These constituents should be classified as the third functionality of food materials, next to nutritive value and palatability. The low incidence of degenerative diseases in Korean population as compared with developed nations should be explained by the light food habits, which supply only necessary amounts of essential nutrients, based on a semivegetarian diet.

HEALTH CONCEPTS AND DIETARY REGIME

Human beings want to enjoy their health once they are grown. In principle, good physical health requires a balanced maintenance of diet, sports, and rest. If one deviates from this balance, he becomes unhealthy and falls down from illness or disease. Here, the diagnosis and cure of disease has been placed in the hands of medical professionals, who sometimes give warnings along with preventive measures including improved dietary habits.

Maintaining satisfactory dietary habits requires the help of dietitians, to whom nutritionists and food scientists should supply essential information and necessary food materials. They should provide information on the physiological effects of foods and food components that are capable of preventing or alleviating disease. Health is a matter of sequential phenomena and calls for expertise from different disciplines as follows:

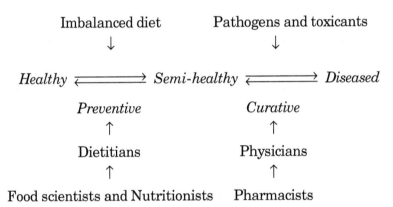

Historically, diseased persons were cared for by medical professionals who performed the diagnosis, treatment (curing), and dietary practice, and thus received the highest respect from society. Roles of health maintenance are divided among physicians (medical doctors), pharmacists, dietitians and food scientists in the Western world since the Middle Ages. The roles, however, were not divided in the Orient until 20th century. Japan, however, adopted the Western system during the later half of 19th century and modified it appropriately so as to be useful for their own social system. Japan accomplished this modernization, following economic development.

The Oriental approach toward health is different from Western countries. It is said that the Oriental approach was developed under the ultimate philosophical principles, that is, the "Yin and Yang Theory" or solar and lunar concept, and gave rise to Chinese medicine and others, including Korean, Mongolian, and Tibetan medicines.[4,7] These Oriental medicines are based on many clinical experiences and differentiate the necessary treatments/drugs/diets for health maintenance depending on the "body character" of individuals, these characters are grouped into four types according to the Korean way. They also claim the "Identical Origin of Foods and Drugs." Westerners have debated these because they were not proved by scientific method and were no more than folk remedies. The terminology and way of explanation for human physiology, pharmacology, and composition of human body or food materials in the Oriental way are not understandable by those who were trained under the Western education system.

Most food scientists and nutritionists in Korea were trained under institutional education at home or in Western countries. These educated people can not do much in teaching the common people who rely, to a great extent, on the Oriental way of thinking for their health maintenance. The common people believe that foods contain not only necessary nutrients but also some unknown health-promoting factors.

In general, the so-called "health foods" imply that they contain not only essential nutrients, but also nonnutrient constituents acting as health promoters. Recently, health foods have been produced under the label of functional foods, dietary supplements, or organic foods and sold at higher prices than traditional food commodities. Many food scientists in the Orient, and a few scientists in the Western world, are eager to undertake studies on the new functionality of uncommon food materials.

Most commercial advertisements in relation to health foods or dietetic foods are not understandable in the light of modern science. Nevertheless, many people buy these commodities at a high price and this makes the traditional food industries distorted or disturbed. It is hoped that young food scientists and nutritionists make every effort to accumulate solid scientific evidence to support the efficacy or functionality of such food products with health claims. Commercialization at the expense of unproven health claims should be avoided.

It appears that Orientals like something mysterious in their dietary life and health care even though it is not proven scientifically. The historical drama on "Dongeui-Bogum," meaning Encyclopedia of Korean Medicines, which was written by Herjun 400 years ago dealing with medical care, drugs, and diets, was broadcast on television for 6 months as a series in 2000 and showed a high viewing record of 70%. Any television program on dietary life based on modern nutrition and food science has not shown this high an interest by audiences in this country.

Overall dietary patterns may have changed in response to advanced food technology, advertising, taste, and health consciousness among other variables. It is assumed that Western civilization succeeded in emphasizing the importance of nutrients in relation to health maintenance. On the other hand, Oriental scientists emphasized the importance of non-nutritive constituents in addition to essential nutrients, without scientific evidence. It is the responsibility of future food scientists and nutritionists to provide the necessary information to interpret the unexplained function and hidden story of foods in the coming 21st century. People in the Orient should not repeat the difficult situation in Western countries that failed to control the overnutrition and subsequent health problems. A new food regimen toward good health is needed.

COMMODITIES OF FUNCTIONAL FOODS IN KOREA

The Korean food industry has succeeded in producing some specialty foods of Korean taste. Some of these are explained below in relation to their claimed functionality.

Kimchi

Kimchi is an acid-fermented vegetable of Korean taste having a 1,500-year history. There is a record that its primitive form came from China to Korea and then to Japan in early history. However, the primitive form of using mainly radish roots has gradually changed in Korea to the use of Chinese cabbage and many other flavorings including hot pepper, garlic, Welsh onion, ginger, pear, fish, and shellfish, etc. The per capita consumption of *kimchi* by Korean people amounts to 90 g per day, which is a good source of vitamin C, minerals, and dietary fiber.

The delicacy of the fermented vegetables does not last longer than a few days unless refrigerated. It was, therefore, made at household levels in the fall and stored underground in winter. Intensive investigations during the last 3 decades by Korean food scientists made it possible to clarify the chemical constituents, sensory characteristics, curing conditions, and techniques of mass production, packaging, and preservation. *Kimchi* became a Korean specialty food and it is exported to many countries in the amount of 18,000 tons, which is equivalent to U.S. $50 million annually. An international standard for *kimchi* is being advanced to step 5 of the Codex procedure.

As the *kimchi* is prepared from a variety of vegetables, it may exhibit some specific physiological activities. It was reported that *kimchi* made of Chinese cabbage showed an antimutagenic activity in the Ames test with *Salmonella typhimurium* and the SOS chromotest with *Escherichia coli*.[8,9] *Kimchi* also contains a high level of dietary fiber. It is, therefore, quite natural to anticipate that kimchi can become a functional food exhibiting health effects in terms of nutrition and third functionality.

Fermented Soybean Foods

Soybean has been used to manufacture fermented foods in Far Eastern countries for many years. The typical soy products in Korea are soy sauce, soy paste, and hot soy paste, using natural microflora. The typical daily consumption levels of these products are 20 mL of soy sauce, 20 g of soy paste, and 10 g of hot soy paste, per head. There has been a decrease of soy intake due to the use of Western seasonings such as mayonnaise, tomato ketchup, meat sauces, etc. in recent years.

The various phytochemicals and microbial metabolites in fermented soy products may exhibit health-related functionality and have attracted much attention in recent years.[10,11] Though soybean has been recognized as improper as the main food resource due to the presence of antinutritional factors, there is a tendency to reconsider the importance of soybean and its products with respect to health-related functionality such as effects toward cancer, cardiovascular disease, osteoporosis, and kidney disease, besides the nutritional aspect in Oriental diets.[12]

Korean Ginseng Products

Korean ginseng is believed to cure all diseases as indicated in its scientific name, *Panax ginseng* C.A. Meyer. It has proved its efficacy and maintained its supremacy as a king of medicinal herbs. Many scientists have studied various aspects of the plant such as its efficacy, pharmacology, and chemistry in order to reveal the hidden characteristics of this mysterious herb.

Historically, Korean ginseng has been known as an excellent health food of high biological safety. The reasons that ginseng and its products do not create any adverse side effects, despite prolonged use as medicine or food, is that ginseng's function on health is derived from its rejuvenating activity through its supplemental effect rather than allopathic effect. The true value of ginseng as a health food has been validated by a number of clinical and empirical studies..[13] Additional studies followed after the publication of *Korean Ginseng* to accumulate further evidence of its value as a health food (Chapter 16 by Hoon Park). A list of ginsengs' effects follows.

1. Antifatigue activity — proven by a swimming test of mice and 3-km race of humans; its active ingredient being the glycoside ginsenoside (Rg1). The effect was different from that of amphetamine, caffeine, and other stimulants.
2. Recovery effect of declining liver functions — liver damage can be either prevented or cured by proper administration of ginseng glycosides.
3. Effect on aging and sugar tolerance — prevents chronic degenerative diseases and has antidiabetic effect.
4. Effect on hypertension and atherosclerosis — persistent hypotensive action and transient lowering of blood

pressure, the effect differs according to constituents, doses, animal species, and physiological conditions.

5. Antineoplastic effect — nonspecifically increased resistance toward neoplastic diseases in animals and humans.

Ginseng products are produced and consumed in many different forms for the purpose of medicine and food. Products in original form are made into red ginseng or white ginseng, depending on steaming, for medicinal purposes. Processed goods are made as extracts, powders, capsules, and tablets, mainly for medicinal purposes, while candy, jelly, tea drinks, nectar, chicken soup, wine, etc. are made for food and beverage purposes.

Chemical structure of ginsenoside R_{g1}

Traditional Medicinal Foods

Korean traditional medicines are based on the works of the Great physician *Herjun* in the 16th century.[4] The book *Dongeui-Bogum*, meaning Encyclopedia of Korean Medicines, deals with medical care, drugs, and diets based on the environment, body characteristics, and available resources in the Korean peninsula. His theory was later expanded by Jema Lee in 1894 who proposed a unique addition to the theory, called *Sasang Chejil*, a system of categorizing people into four body characters; that is, *Taeyang* (bright solar), *Taeum* (bright lunar), *Soyang* (dark solar), and *Soeum* (dark lunar) characters. The body characters are determined by the body shape, behavior, physical constitutions, and dietary habits of individuals, which require the appropriate medical care as well.[14]

Various foods were recommended for certain diseases or illness depending on the body characters as shown in Table 7.1.[15] This means that foods may contain specific constituents having preventive efficacy in response to physical

TABLE 7.1 List of Health-Related Functional Foods Desirable for Different Body Characters as Recorded in the Korean Literature "*Dongeui-Bogum*"[14]

Disease	Recommended foods according to the body character[a]			
	Taeyang	Taeum	Soyang	Soeum
	(1) Respiratory diseases			
Cold	Acanthopanax root-bark extract, Chinese quince tea, pine tree leaf extractive, Korean cherry	Arrowroot ext.	Raw fruit & vegetable juices	Beefsteak plant leaf tea, Korean angelica, citrus peel tea
Aczema	Snail, Ganoderma mushroom	Omiza tea		Ginseng + walnut
Cough		Gingko seeds, Chinese bellflower root, schizandra seed, apricot kernel	Ganoderma mushroom, pumpkin, lily roots	Radish + honey mix, beefsteak plant seeds, foxtail millet
Lung diseases		Bamboo leaf, lily roots		Beefsteak leaf, ginseng-schizandra fruit tea
Tuberculosis	Pine leaf	Cattle bone jelly, sesame seed	Vinegared eggs	Royal jelly
	(2) Gastrointestinal diseases			
Gastrointestinal distress	Buckwheat, wild kiwi	Radish juice, lotus seed, cuttle fish bone	Gardenia	Cinnamon tea, ginger tea,mugwort tea, beef steak plant leaf tea

Gastro-duodenal ulcer	Kiwi juice	Job's tears		Shells, Egg shell, cabbage, potato starch, mugwort tea
Neural disorder				Mugwort tea, leek wine, ginger, ginseng, cinnamon, beefsteak plant leaf
Liver disease	Buckwheat	Snail, fruits, vegetables, honey, seaweeds, omiza tea, fresh beef fillet	Cassia tora seeds tea, Chinese matrimony vine tea, peppermint tea, gardenia tea	Artemisia capillaris herb, dry ginger
Diarrhea	Pine leaf, cooked buckwheat	Brown rice, arrowroot, glutinous rice cake	Green tea, gardenia seed extract	Leek soup, apple, persimmon punch, green onion root
Constipation	Aloe	Sweet potato, bamboo shoot, peach kernel	Green tea	Bracken, crown daisy, leaf mustard, carrot, radish, mallow
Sensitized colon syndrome	Wild kiwi, buckwheat	Lotus seed	Broad-leaved plantain	Mugwort tea, dry ginger
Foul breath	Pine tree joint + buckwheat powder	Job's tears, mulberry root peel, Chinese matrimony vine peel	Green tea	Licorice extract
Stomach ache	Chinese quince + buckwheat	Turnip seed		Cinnamon, antler
(3) Cardiovascular diseases				
Palsy	Oriental bezoar gallstone	Cornus fruit, Chinese matrimony vine seeds		Beefsteak leaf

TABLE 7.1 (CONTINUED) List of Health-Related Functional Foods Desirable for Different Body Characters as Recorded in the Korean Literature "Dongeui-Bogum"[14]

Disease	Recommended foods according to the body character[a]			
	Taeyang	Taeum	Soyang	Soeum
Facial paralysis		Arrowroot tea, Job's tears tea, schizandra fruit tea, dried yam	Raw rhemannia root ext, Chinese matrimony vine roots	Ginseng, Korean angelia tea
Hypertension	Persimmon juice, radish juice, persimmon leaf tea	Cassia seed tea, arrowroot tea, Chrysanthemum tea	Epimedium herb, red bean, Cornus fruit, Chinese matrimony vine seeds, raw rhemannia root	Mugwort, beefsteak leaf, Atractylodes rhizome white
Hypotension		Young antler	steamed rhemannia root, Cornus fruit tea, Chinese matrimony vine tea	Ginseng tea, Astragalus root tea
Arteriosclerosis	Black soybean, fruits, vegetables	Yellow soybean, fishes, fruits, vegetables	Fishes, fruits, vegetables	Fishes, fruits, vegetables
Heart disease	Raw rhemannia root, Akebia stem, bamboo leaf, Arborvitae seeds	Arrowroot tea, scallion, Trogopterorum feces tea, cattail tea	Korean angelica, steamed rhemannia root, Gardenia fruit tea, bamboo leaf tea	Hedysarum, ginseng, violet glycyrrhiza, cinnamon's sprig, Liriope tuber + ginseng + schizandra fruit
(4) Urinary diseases				
Cystitis	Buckwheat porridge, kiwi vine juice	Job's tear	Red bean, lotus root, purslane, yam, green onion roots	Citrus peels, malvae seed, green onion roots

Condition				
Incontinence of urine	Buckwheat flour, grape juice, kiwi vine	Ginko nut, apricot kernel	yam peel juice	Green onion roots
Impotence	Acanthopanax root-bark, Polygonum tuber	Yellow soybean, mulberry wine	Raspberry, Cornus fruit	Leek, onion, garlic, eel
(5) Psychoneurotic diseases				
Fatigue	Pine tree, grape	Lotus flower, arrowroot extract	Areca seed, cassia tora seeds	Quail, ginseng tea
Fret	Buckwheat	Anchovy-bean paste soup	Gardenia seeds	Roasted carrot
Nervous disorder				Triticum semen
Insomnia	Wild kiwi	Walnut	Steamed rhemania root ext	Apple
Diabetes	Vegetables, seaweeds	Mushrooms	Cuscuta seed	
(6) Women's diseases				
Irregular menstruation	Pine leaf	Job's tears	Safflower tea	Korean angelica-safflower tea
Leucorrhea				Garlic, mugwort, ginseng
Becoming thin				Pig's stomach
Obesity		Soybean, pinenut, seaweeds, lactic bacteria beverage	Barley, mung bean, green tea	Apple
Deficient lactation		Cow's leg	Pig's leg, lettuce seeds	Ginseng-chicken broth

TABLE 7.1 (CONTINUED) List of Health-Related Functional Foods Desirable for Different Body Characters as Recorded in the Korean Literature "*Dongeui-Bogum*"[14]

Disease	Recommended foods according to the body character[a]			
	Taeyang	Taeum	Soyang	Soeum
Breast fatigue		Angelica dahurica root, Fritillaria thunbergii bulb	Red bean extract	Immature citrus peel
Melena during pregnancy	Grape root	Overgrown antler jelly	Anemarrhena rhizome	Green onion roots
Postpartum gastralgia		Cattail flower	Prunella spike	Pteropus stool

[a] *Taeyang* means bright solar character, *Taeum* bright lunar character, *Soyang* dark solar character and *Soeum* dark lunar character of the body.

conditions that should be exhibited as functionality of the food commodities. Further studies should be undertaken in the future to clarify the scientific evidence to explain the efficacy of various medicinal foods, depending on the individual body characters.

FOOD MATERIALS PERMITTED LEGALLY IN KOREA

The Korea Food Standards Codex defines raw materials that can be used for manufacturing of processed foods.[16] The materials should be of good quality and freshness, and proven to be safe without toxic or hazardous substances. They may be grouped into common and rare materials. Most of the common materials were based on their nutritive value and some of them without nutritive value were chosen because of their uses as spices or herbs. The nutritional constituents of these common materials were listed in the Korean Food Composition Table.[17,18]

As the significance of health-promoting effect is recognized recently, many natural products that have not been recommended as edible foods due to the lack of nutritive value or spice effects were recently listed as raw materials applicable in food processing. Here, uncommon materials without any known toxicity may be used as major ingredients without any limitation of use level, whereas those with some pharmacological or adverse effects can be used as minor ingredients below 50% of final food products. Functionality of these rare natural products must be elucidated in the future. Scientists must be cautious in differentiating between food materials consumable by common people and medicinal materials used by diseased persons. The lists of rare materials used on the basis of nonnutrients are shown in Tables 7.2 and 7.3.[19,20]

Food materials
- Common materials
  - nutrient-based: grains, fruits, vegetables, meats, fishes, etc.
  - nonnutrient based: spices, herbs, etc.
- Rare materials
  - major ingredients: limitless use
  - minor ingredients: limited use <50%

TABLE 7.2 List of Rare Natural Products Used as Major
Ingredients in Foods[16,19]

Scientific name	Edible portion	Common name
Agastache rugosa O. Kuntze	Young leaf	Pogostemi herb
Alchemilla vulgaris L.	Leaf	Lady's mantle
Aloysia triphylla	Leaf	Lemon verbena
Althaea officinalis	Flower, root	Marshmallow
Anethum graveolens L.	Fruit	Dill
Armillariella mellea	Fruiting body	Mulberry mushroom
Aronia melanocarpa	Fruit	Black chokeberry
Artemisia capillaris Thunberg	Ground part	Art. capillaris herb
Aspalathus lineraris	Leaf	Rooibos
Astragalus membranaceus Bunge	Root	Hedysarum
Brassica campestris L.	Whole plant	Rape
Calendula officinalis	Flower	Marigold
Capparis spinosa L.	Buds	Chaparral (caper)
Carum carvi L.	Seed	Caraway
Cedrela sinensis A. Juss.	Tender shoot, young leaf	Chinaberry
Chamomilla recutita (=Matricaria recutita) *Chamaemelum nobile* (=Anthemis nobilis)	Flower, leaf	Chamomile
Chrysanthemum indicum *Chr. morifolium* Ramat.	Flower	Chrysanthemum
Cirsium maackii (=Cirsium japonicum)	Tender shoot, young leaf	Thistle
Citrus unshiu Markovich Corchorus olitorius	Peel	Citrus peel
Cordyceps militaris	Fruiting body & larva	Militaris fungus in larva
Corianthrum sativum Linnaeus	Fruit, leaf	Coriander
Cyclopia intermedia	Leaf	Honey bush
Cymbopogon citratus	Leaf, stem	Lemon grass
Cynanchum wilfordii *Polygonum multiflorum* Thunberg	Root	Polygonum tuber
Cynara scolymus L.	Tender shoot, young leaf	Artichoke
Cyperus esculentus L.	Tuber	Chufa
Dendrapolyporus umbellatus (=Grifola umbellatus)	Sclerotia	Chuling

TABLE 7.2 (CONTINUED) List of Rare Natural Products Used As Major Ingredients in Foods[16,19]

Scientific name	Edible portion	Common name
Dolichos lablab L.	Seed	White lentil
Eleocharis kuroguwai Ohwi	Seed, root	Eleocharis seed
Elsholtzia ciliata Hylander	Tender shoot, young leaf	A mint plant
Equisetum arvense L.	Leaf	Field horsetail
Erigeron canadensis L.	Tender shoot, young leaf	Horseweed
Gastrodia elata Blume	Root	Gastrodiae rhizome
Glechoma hederacea L.	Tender shoot, young leaf	Glechoma herb
Grifola frondosa	Fruiting body	Leaf mushroom
Helianthus annuus L.	Seed, leaf	Sunflower
Hibiscus sabdariffa	Petal	Hibiscus
Hovenia dulcis Thunberg	Fruit	Hovenia seed
Hydranzea serrata Seringe	Leaf	Hydranzea
Hygrophorus russula Quel.	Fruiting body	Brown cherry mushroom
Hyssopus officinalis L.	Flower, leaf	Hyssop
Illicum verum	Fruit, seed	Star anise
Imperata cylindrica Beauvois	Root	Cogongrass
Lagerstroemia speciosa Pers.	Fruit, leaf	Banaba
Lavandula angustifolia	Flower, leaf	Lavender
L. *officinalis* Chaix/L. vera		
Lilium auratum	Root	Mountain lily
Lysimachia vulgaris L. var. davurica Led.	Tender shoot, young leaf	Loosestrife
Malva sylvestris L.	Flower, leaf	Common mallow
Marrubium vulgare	Leaf, flower	Horehound
Melissa officinalis L.	Leaf	Balm leaves, lemon balm
Momordicae grosvenori	Fruit	Momordicae fruit
Morinda citrifolia	Fruit	Noni
Morus alba L.	Fruit, leaf, young stem	Mulberry
Nasturtium officinale	Leaf	Cresson
Nelumbo nucifera Gaertner	Root	Lotus root
Opuntia ficus-indica	Fruit, flesh of stem	Cactus
Paecilomyces japonica	Fruiting body and larva	Paecilomyces fungus in larva
P. tenuipes		
Passiflora incarnata L.	Fruit, leaf	Passon flower
Paullinia cupana H.B.K.	Fruit	Guaiana

TABLE 7.2 (CONTINUED) List of Rare Natural Products Used as Major Ingredients in Foods[16,19]

Scientific name	Edible portion	Common name
Phlomis umbrosa Turcz.	Young leaf, root	Dipsaci root
Pimpinella anisum L.	Fruit	Anise seed
Pinus densiflora Sieb & Zucc. *P. sylvestris* L.	Pollen, shoot, leaf, stem, stalk	Pine tree
Pinus koraiensis S. et Z.	Seed, leaf	Pine tree
Portulaca oleracea L.	Young leaf, tender shoot	Purslane
Prunella vulgaris L.	Tender shoot, tender leaf	Self-heal
Rana catesbeiana	Flesh	Bull frog
Rosa spp.	Fruit, petal, shoot	Rose
Rubus spp.	Fruit, leaf	Raspberry
Rubus spp.	Fruit, leaf	Blackberry
Rubus suavissimus S. Lee	Leaf	Tencha
Rumex acetocella L.	Tender shoot, young leaf	Sheep sorrel
Rumex acetosa L.	Leaf, root	Sorrel
Russula subdepallens	Fruiting body	Russula mushroom
Salvia officinalis L.	Leaf	Sage
Siegesbeckia glabrescens Makino	Tender shoot, young leaf	Siegesbeckia herb
Silybum marianum L.		Milk thistle
Stachys sieboldii Miq.	Root	Chinese artichoke
Stevia rebaudiana	Leaf	Stevia
Suaeda asparagoides Makino	Tender shoot, tender leaf	Sea-blite
Taraxacum mongolicum H. Mazz.	Tender shoot, young leaf, root	Dandellion
Taraxacum officinale Wiggers	Tender shoot, young leaf, root	Common dandellion
Tilia spp.	Flower, leaf	Linden
Torreya nucifera S. et Z.	Fruit	Japanese nutmeg
Trifolium pratense L.	Young leaf	Red clover
Trigonella foenum-graecum	Seed	Fenugreek
Vaccinium macroparpon	Fruit	Cranberry
Viola mandshurica W. Becker	Tender shoot, young leaf	Violet

TABLE 7.3 List of Rare Natural Products Used as Minor Ingredients in Foods[16,19]

Scientific name	Edible portion	Common name
Achillea millefolium L.	Leaf	Yarrow
Achyranthes japonica Nakai	Root	Achyranthes root
Acorus gramineus Soland.	Root(stem)	Sweet flag
Agastache rugosa O. Kuntze	Ground part	Pogostemi herb
Alnus japonica (Thunb.) Steudel	Bark, leaf	Black alder
Amomum xanthioides Wallich	Seed	Amomum fruit
Aralia continentalis Kitagawa	Root	Udo
Aralia cordata		
Atractylodes japonica Koidzumi	Root, peeled stem	Atractylodes rhizome white
Atractylodes lancea	Root, stem	Atractylodes rhizome
Biota orientalis Endlicher	Leaf	Arborvitae leaf
Cervus nippon T./C. elaphus L.	Antler	Overgrown antler
Cervus nippon T./C. elaphus L.	Antler	Young antler
Cnidium officinale Makino	Root	Angelica
Codonopsis pilosula Nannfeldt	Root	Codonopsis root
Commelina communis L	Whole plant	Laphatheri herb
Cornus officinalis S. et Z.	Fruit(flesh)	Cornus fruit
Crataegus pinnatifida Bunge	Fruit	Crataegus fruit
Curcuma domestica (=C. longa)	Root(stem)	Turmeric
Curcuma zedoaria Roscoe	Root(stem)	Curcuma root
Cuscuta chinensis Lamark	Seed	Cuscuta seed
Diospyros kaki Thunberg	Leaf	Persimmon
Gardenia jasminoides Ellis	Seed	Gardenia
Geoclemys reevesii Gray	Carapace	Turtle shell
Hepatica asiatica Nakai	Root	Liverleaf
Houttuynia cordata Thunb.	Whole plant	Houttuynia herb
Inula britannica	Flower	Elecampane
Inula japonica Thunberg		
Juniperus communis	Fruit	Juniperberry
Juniperus rigida		
Liriope platyphylla Wang et Tang	Root(tuber)	Liriope tuber
Lonicera japonica Thunberg	Flower, leaf, stem	honeysuckle
Nelumbo nucifera Gaertner	Flower, leaf, seed	Lotus
Paeonia albiflora Pallas var. trichocarpa Bunge	Root	White peony
Paeonia japonica var. pilosa Nakai		

TABLE 7.3 (CONTINUED) List of Rare Natural Products Used as Minor Ingredients in Foods[16,19]

Scientific name	Edible portion	Common name
Paeonia albiflora var. hortensis Makino	Root	Red peony
Paeonia obovata Maximowicz		
Polygala tenuifolia Wildenow	Root	Polygala root
Poria cocos Wolf	Sclerotia	Hoelen
Pueraria thunbergiana Benth.	Flower	Arrowroot flower
Rehmannia glutinosa Liboschitz	Root	Rehmannia root
Salvia miltiorrhiza Bunge	Root	Sage plant
Saururus chinensis Baill.	Ground part	Saururus herb
Torilis japonica Decandolle	Fruit	Cnidii monnieri fruit
Trifolium pratense	Flower	Red clover
Valerian officinalis L.	Root	Valerian root
Zea mays L.	Style	Corn
Zizyphus jujuba	Seed	Jujube
Zizyphus vulgaris Lamarck		

REGULATION AND MANAGEMENT OF FUNCTIONAL FOODS

There are two major laws regulating foods and medicines in Korea. Foods are defined in the Food Sanitation Law as any commodities provided for dietary intake, excluding medicines. On the other hand, medicines are defined in the Medicinal Affairs Law as any materials used for diagnosis, cure, mitigation, treatment, or prevention of diseases and providing pharmacological effects on structure and function in humans and animals. Sanitary goods, instruments, and machinery that are used for health care are excluded from medicines.

At present, there are two major groups of functional foods (health-aid foods and specified nutritious foods as shown in Table 7.4) according to the Food Sanitation Law.[19] "Health-aid foods" are defined in the Food Standards Codex as manufactured and processed foods from materials containing specified constituents by means of extraction, concentration, purification, and mixing for the purpose of health-aid. There are 24 kinds of such foods including purified fish oil, royal jelly, yeast food,

TABLE 7.4   Classification and Definition of Function-Related Foods According to the Food Standards Codex in Korea[16]

Food group	Definition	Food items
Specified nutritious foods	Manufactured and processed foods by mixing foods or nutrients in order to meet the nutritional requirements of infants, children, diseased, aged, obese persons and pregnant women	1. Formulated milk 2. Formulated foods for infants 3. Foods for growing period 4. Formulated cereals for infants and children 5. Other foods for infants and children 6. Foods for nutritional supplementation 7. Diets for patients 8. Substitute foods for daily diets
Health-aid foods	Manufactured and processed foods from materials containing specified constituents by isolation, extraction, concentration, purification, mixing, etc. or directly from constituents in food materials, for ingestion expecting effectiveness from physiological aspects	1. Purified fish oil products 2. Royal jelly products 3. Yeast products 4. Processed pollen products 5. Squalene products 6. Enzyme products 7. Lactic bacteria-containing foods 8. Algal products 9. -Linolenic acid-cont'g foods 10. Processed germ products 11. Processed lecithin products 12. Octacosanol cont'g foods 13. Alkoxyglycerol cont'g foods 14. Grape seed oil products 15. Fermented plant extractives 16. Mucopolysaccharide/protein cont'g foods 17. Chlorophyll-cont'g foods 18. Processed mushroom products 19. Aloe products 20. Japanese apricot extractives 21. Processed snapping turtle products 22. -Carotene-cont'g foods 23. Processed chitosan products 24. Propolis extractives

TABLE 7.4 (CONTINUED) Classification and Definition of Function-Related Foods According to the Food Standards Codex in Korea[16]

Food group	Definition	Food items
Ginseng products	Manufactured and processed products mainly from ginseng or red ginseng	1. Concentrated ginseng and red ginseng products 2. Powdered ginseng and red ginseng products 3. Ginseng and red ginseng teas 4. Ginseng and red ginseng drinks 5. Bottled and canned ginseng products 6. Ginseng candies and gums 7. Sugared ginseng 8. Capsuled ginseng and red ginseng 9. Other ginseng and red ginseng products
Tea products	Favorite foods manufactured and processed mainly from plant materials for drinking purpose	1. Exudated tea 2. Extracted processed tea 3. Powdered instant tea 4. Processed fruit tea 5. Coffee products

pollen, gamma-linolenic acid, lecithin, etc. and 2,000 commercial items produced by 270 manufacturers, amounting to yearly sales of about 1 trillion Korean won (equivalent to 1 billion U.S. dollars). Specialty food products made from ginseng and tea materials are listed separately in the Codex, but they should be classified under this group.

"Specified nutritious foods" are defined as manufactured and processed foods by adding or deleting nutrients into food materials for the purpose of providing them to infants, children, diseased, and obese persons. There are eight kinds of such foods including formulated milk, formulated foods for infants/children, nutrient-fortified foods, foods for patients, and substitute foods, which are produced by 129 manufacturers, amounting to yearly sales of 870 billion Korean won (equivalent to about 720 million U.S. dollars).

"Functional foods," as an ill-defined term, are controlled by the labeling system for specification of application range and compositional standards as well as by the premarket review system in the advertisements. It is strictly regulated to express the effectiveness of the functional foods as follows:

- Expressions for the general promotion of body functions are allowed whereas expressions for prevention and curing of diseases are prohibited. For example, expressions such as health maintenance, health promotion, improvement of physical constitution, diet therapy, and nutritional supplement are allowed whereas expressions such as preventing or curing of diabetes, constipation, etc. are not allowed.
- Expressions for the publicly recognized facts on the basis of food and nutritional sciences are allowed. For example, nutritional supplement during pregnancy and lactation, nutritional supplement during recovery from disease, nutritional supplement for aged persons, nutritional aid for patients, etc. are accepted.
- Expressions for the food-nutritional and physiological role and action of major components contained in the commodity toward body function are allowed. For example, role and action of vitamins, calcium, iron, amino acids, fatty acids, etc. are accepted.

There are many different terminologies in relation to functional foods in the world. The definitions, concepts, and regulatory schemes are different from country to country. In order to avoid any confusion in their application and trade, classification of function-related foods into the following five categories were proposed.[19]

Functional Foods

Functional foods in a narrow sense should be defined as any food commodities having the similar appearance with regular foods and consumed as a part of daily diets; they should exhibit physiological function as well as nutritional value. Functional constituents may be present in the original foods or they may be added.

For instance, any hamburger to which conjugated linoleic acid was added may be called a functional food. It should be regulated under the specifications of regular hamburger or functional food. If the standards are in duplicate in both food groups, increasingly stricter ones should be applied. The functionality, effectiveness, and safety of functional constituents should be evaluated on the basis of scientific evidence. Presently, some ginseng products such as ginseng tea, ginseng crackers, some tea products such as exudated tea, extracted tea, and some extraction-processed products may be included in this category.

Nutraceuticals

Nutraceuticals are defined as any commodities of appearance different from regular foods such as powder, granules, liquids, tablets, capsules, and other drug forms; they may not exhibit any nutritional role, but should have a definite physiological function. Most of the health-aid foods listed in the current Food Standards Codex may belong to this category, if their functionality, effectiveness, and safety are proved on the basis of scientific evidence.

Dietary Supplements

These are not regarded as a common food or a part of diet or meal, but are commodities prepared in the form of powder, granules, liquid, tablets, capsules, etc. by blending foods or nutrients to supplement deficient nutrients. According to the current regulation, a part of health-aid foods such as those containing EPA, DHA and -linolenic acid, and nutrient-supplementing foods among specified nutritious foods may belong to this category.

Medical Foods

These are the formulated diets used to meet specific nutritional requirements or therapeutic purposes according to the diagnosis of medical doctors. These foods should be provided to persons with nutritional requirements different from common people. The current foods for patients may belong to this category.

Foods for Special Dietary Use

These are formulated foods used to meet the need of special dietary conditions such as to prevent allergy, having the same nutritional requirements with common people. Physical state of raw materials for these may be different from those of common foods, but with the similar nutrients. The current meal-substituting foods and weight control diets may belong to this category.

In August 2002 a new act entitled "The Law on Health-Functional Foods" was passed in Korea and enacted in 2003. Here, the health-functional foods were defined as the manufactured and processed foods in the forms of tablet, capsule, powder, granule, liquid, pill, etc. for the purpose of providing health effects by means of nutrient control or physiological action in the structure and function of the human body. The new law controls health-aid foods, specified nutritious foods, and ginseng products that have been managed under the Food Sanitation Law, on the basis of scientific evaluation for their safety and functionality. Intensive studies to establish the testing and labeling requirements for functionality of such products are under way. The current Food Sanitation Law may regulate any claim for health-related functionality of traditional natural foods.

REFERENCES

1. National Statistical Office. *Korea Statistical Yearbook*. Vol. 46, Seoul, Korea, 1999.

2. Ministry of Health and Welfare. *National Nutrition Survey Reports*. Seoul, Korea, 1970–1995.

3. KoSFoST. *30-Years History of Korean Society of Food Science and Technology*, Seoul, Korea: Korean Society of Food Science and Technology, 1998.

4. SR Lee. Retrospect and prospects on food research in Korea. Paper presented at the 11th World Congress of Food Science and Technology, in Seoul, Korea, on April 22–27, 2001. *Food Sci Ind* (KoSFoST) 34(2): 47–54, 2001.

5. Korean Academy of Science and Technology. Prospects and research strategy for agricultural and fishery science in Korea toward 2030s. *Hallim Symposium Proceedings* Vol. 2, Seoul, Korea, 1998, pp 1–117.

6. CH Lee. Trend analysis for the dietary habits and nutritional status during the last century. *Proceedings on the Food, Nutrition and Economic Evaluation on the Dietary Regime of Koreans by the Korean Forum on the Food, Nutrition and Economics,* Seoul, Korea, 1988, pp 1–54.

7. CH Lee. Health concepts in traditional Korean diet. *Bull Res Inst Food Sci* (Kyoto Univ) 62:1–17, 1999.

8. KY Park. The nutritional evaluation, and antimutagenic and anticancer effects of Kimchi. *J Korean Soc Food Nutr* 24:169–182, 1995.

9. KY Park, KA Baek, SH Rhee, HS Cheigh. Antimutagenic effect of Kimchi. *Foods Biotechnol* 4:141–145, 1995.

10. C Choi (Ed). The physiological activity function of traditional soy fermented products. *Proceedings of 2nd Symposium on Soybean Fermented Foods by Yeungnam Univ.,* Daegu, Korea, 1999, pp 9–181.

11. Korean Society of Food Science and Technology. *Proceedings of 2nd International Symposium on Soybean and Human Health,* Seoul, Korea, 2002, pp 1–142.

12. M Messina, JW Erdman Jr (Ed). *Proceedings of 1st International Symposium on the Role of Soy in Preventing and Treating Chronic Disease. J Nutr* 125:567s–798s, 1995.

13. HW Bae (Ed). *Korean Ginseng,* Seoul, Korea: Ginseng Research Institute, 1978, pp 115–214.

14. JY Shin. *Dongeui-Bogum Based on Body Character: Body Characters.* Seoul, Korea: Hakwonsa, 2001, pp 31–80.

15. JY Shin. *Dongeui-Bogum Based on Body Characters: Chinese Remedy.* Seoul, Korea: Hakwonsa, 2001, pp 298–490.

16. Korea Food and Drug Administration. *Korea Food Standards Codex.* Seoul, Korea: Korea FDA, 2000, pp 21–30.

17. National Rural Living Science Institute. *Food Composition Table.* Suwon, Korea: Rural Development Administration, 5th ed., 1996, pp 7–355.

18. Food and Drug Administration. *Korean Food Composition Table.* Seoul, Korea: Ministry of Health and Welfare, 1996, pp 54–149.

19. Testing Laboratory for Korean Medicinals. *Specifications for Korean Medicinal Herbs.* Seoul, Korea: Trade Association for Medicinals, 2000, pp 1–622.

20. MS Chung. Classification and management of functional foods. *Proceedings of Workshop by Korean Society of Food Science and Technology.* Cheju, Korea, 2001, pp 7–15.

8

Evolution of Korean Dietary Culture and Health Food Concepts

CHERL-HO LEE

Graduate School of Biotechnology, Korea
University, Seoul, Korea

TAI-WAN KWON

Department of Food and Nutrition, Inje
University, Kimhae, Korea

NORTHEAST ASIA IN THE PALEOLITHIC AGE

Western society has commonly amalgamated all Northeast Asian culture into the category of Chinese culture, when it is in fact comprised of many cultures and segments of ethnic groups that have developed their own identities and distinctive cultures throughout history. At present, these cultures are grouped according to nation: China, Mongolia, Korea, Japan, and part of Russia (Siberia). However, 15 centuries ago, the ethnic group (or tribe) was more important than the nation in distinguishing the way of life for a people.[1]

The early existence of human beings in this region is indicated by Early Paleolithic remains (1,800,000 to 300,000 years ago, B.P.) of the Early/Middle Pleistocene period on the Northern Chinese Mainland and Korean Peninsula. Evidences of the existence of *Homo erectus* (1,800,000–650,000 B.P.) were found in the Xihoudu, Lantian, and Zhoukoudian sites on the Northern Chinese mainland; in Jinniushan, in the Manchurian Basin; and in the Sokchangni and Chungbuk Keumkul sites on the Korean Peninsula. Zhoukoudian Cave, Locality 1, near Beijing, has yielded the largest number of *Homo erectus* fossils in the world; 40-odd individuals, together with thousands of animal bones.[2] *Homo sapiens* fossils were found in Yokpo Cave (500,000 B.P.) and in Sangwon Cave (400,000 B.P.), near Pyungyang on the Korean peninsula.[3] Recently, several Middle Paleolithic (350,000 to 40,000 B.P.) remains were found on the Korean Peninsula. The stone tools and animal fauna of the Seungrisan, Jommal Yonggul, Durubong, and Chongongni sites were similar to those of the Dingcun site in China. The fauna and stone tools of the Sokchangni seventh and eighth layers, Chongchongam Cave, Gulpori 1, and Sangmu Yongni were comparable to those of Xujiayao site in Northern China. The earliest Paleolithic remains found in Siberia at the Irkutsk site on the Kamchatka Peninsula, were those of 130,000 to 70,000 years ago, similar to those of the Gulpori site on the Korean Peninsula.[4]

Numerous Late Paleolithic (400,000 to 10,000 B.P.) sites were found on the Korean Peninsula, in South Manchuria, and on Japanese Islands as well as the Chinese mainland. These sites indicate the increase in population and the spreading out of the people in this region during the Paleolithic Age.[5] Throughout the glacial periods (Günz, Mindel, Riss, and Würm) of the Pleistocene, the Yellow Plain and the Seto Plain were exposed by lower sea levels, and the East Sea became merely a large lake, which drained through the present Korea Strait. These increased land areas facilitated the movement of humans and animals among and between parts of East Asia.[6] It is also assumed that the Asian Mongolides moved to the American continent over the Bering Strait in the course of these periods.[2] On the other hand, during the warm interglacial period, the sea levels rose to the present level, and the Korean peninsula became a land bridge, connecting the Japanese Islands to Manchuria and the Maritime Province of Siberia. The sites of Paleolithic remains excavated in Northeast Asia, and the possible migratory routes indicates that mobile hunters chased after large animals moving seasonally from southern Kyushu to northern Manchuria and Siberia through the Korean Peninsula.[6] Animal meat, intestine, and blood were probably the main foodstuff for these people, and the use of vegetable supplements, such as grass seeds, tree nuts, and wild fruits and roots, increased at the later stage of Paleolithic era. The people probably began living in mountain caves, and then gradually moved to the lower plains and riverbanks at the Late Paleolithic Age.

THE IMPORTANCE OF THE PRIMITIVE POTTERY AGE IN KOREAN DIETARY CULTURE

The migratory forager's life of Paleolithic men following the periodical and seasonal movements continued until the use of textured pottery. Textured pottery was probably invented by the people in the Far Eastern region, which includes the southern parts of the Japanese Islands, the Korean Peninsula, and the Bohai (Balhae) Corridor between the years 10,000 and 6000 B.C. The use of *chulmun* (Korean) or *jomon* (Japanese) pottery had spread over the region by 6000 B.C., and it gradually changed the migratory hunter's life into the littoral forager's life along the coastal line. The littoral foragers, using textured pottery as the main tools for food processing and storage, probably existed in the Korea Strait region between 8000 to 3000 B.C., prior to Neolithic agricultural settlement. The authors suggest naming this period "Primitive Pottery Age" in order to distinguish it from the European Mesolithic culture.[7,8] The numerous shell mounds excavated along the coastline and major rivers in the Korean Peninsula indicate that the people were engaged in hunting with bow and arrow, and fishing with carved bone tools and fishing equipment.

Animals provided men with meat, gut, and blood. In addition they may have eaten plants such as acorn, chestnut, wild grape, arrowroot, and other wild roots and vegetables.[6] Gradually they developed the skill of food storage by drying. Knowing they could obtain plenty of food around the dwelling sites they stayed longer in these areas. As long as they habited in one place, they reduced mobile hunting practice, and instead obtained more food collecting seeds of grass and barnyard grass, millet, and wild beans. Step by step they became accustomed to collecting frog and snail in the damp ground, and clams and shellfish in the river or beach. However, these marine foods were difficult to dry, and easily decomposed by autolysis and were rapidly spoiled by microbial growth, so they had to consume them instantly without storage, and therefore did not rely on them.

At this time, crockery was invented, and the event must have changed the primitive people's dietary life greatly. Earthenware enabled them to cook perishable foods easily, handle the wet materials, and store them longer for eating. The earthenware at the initial stage was very weak and clumsy, and water absorbency was too high to be used for proper cooking and long-time storage of liquid foods. Although their use for cooking and storage must be very limited for this reason at the beginning, the people became aware that marine foods could be dried easily after boiling and stored longer, like meat. The initial development of pottery technology must have focused on acquiring this effect, and in this way the Northeast Asian people could have experienced major technical advancements in food processing and culinary cuisine.[8]

The Origin of *Chigae* Culture

Before man knew salty taste by using marine foods, people used to take this mineral ingredient from either animal blood or intestine. They came to crave that salty taste in vegetables and plant foods, which they survived on when game was scanty. The people in Primitive Pottery Age who knew the salty taste and the source, lived near the seashore, so seawater and seafood were used to make food with vegetables, roots, and grains. This must be the origin of *chigae* culture, which is the most characteristic Korean food culture. *Chigae* is a stew made by boiling slices of vegetables, seaweeds, clam/fish/meat in salty bouillon. It is used as a side dish for a rice meal. In fact the tribes of Papua New Guinea living in the coastal region today still use seawater as a salty ingredient for cooking.[9]

Northeast Asian people, who survived mainly on captured animals, changed their staple food to fish, shellfish, and vegetables after they began to use earthenware in the Korea Strait region. It was the recipe of *chigae* with marine products that furnished a clue as to how salt could enhance the palatability of vegetable food during cooking. The existence of salt would be naturally discovered in the process of boiling seafood in pottery. Although we do not know exactly when the making of salt started, we can presume that people knew about salt from the beginning of *chigae* culture by observing the white powder left around *chigae* bowl when seawater or seafood were boiled. According to Ishige,[9] the pottery for making salt discovered in the Kanto Province has been dated to approximately 500 B.C. after the *Jomon* period in Japan. It was claimed as one of the oldest archeological evidences of making salt by boiling seawater in pottery. He concluded that the production and consumption of the edible salt started at a much later date, after full development of the Agricultural Age. However, if primitive pottery was made at around 6000 B.C. all over the coast of Korea Strait, and people used the pottery to cook *chigae*, they must have known about salt and its production much earlier than Ishige's assumption. From this point of view, we suggest that the production of edible salt from seawater began at the early stage of Primitive Pottery Age.

Nuruk and the Origin of Fermentation Technology

In areas with high temperature and high humidity, mold growth is a natural process in a container storing wet starchy materials, such as plant seeds, millet, barnyard millet, nuts, beans, and tubers. Some molds like *Rhizopus* species produce enzymes, which can hydrolyze raw starch and convert it into sugars. When sufficient amount of moisture is provided, the sugar is transformed into alcohol by the yeast existing in nature. An alcoholic food or beverage having an attractive aroma is produced within 3 to 4 days in the summer after adding a small amount of water to cooked starchy material in a crock. This is a natural process, which can be easily observed even by early man. When useful microorganisms are grown primarily on the wet seeds and grains, it is called *Nuruk*, the traditional fermentation starter of cereal alcoholic beverage used in Northeast Asian countries. When *Nuruk* is mixed with cooked rice and water in about a 1:1:4 ratio, alcoholic fermentation takes place and is normally completed within one week in summer season. When it is strained with a sieve, turbid liquid is produced, so-called rice-beer, *Makkolli* or *Takju*, and when filtered with a fine filter cloth into a clear liquid it becomes rice-wine, *Chongju*. It appears that the beginning of cereal alcoholic fermentation started by using uncooked starchy ingredients, thus the use of pottery may imply the start of cereal fermentation.

According to the literature, the history of alcoholic beverages is deep-rooted. Chinese literature credits the daughter of King Woo, a legendary king of China who lived around 2100 B.C., as first making an alcoholic beverage.[10] The term "Yojuchonjong (thousand wines in Yao)" implies that alcoholic beverages were made much earlier than Woo's period, and may date from the Yao Shun period, the earliest legendary nation in China. Alcohol fermentation is considered one of the oldest food processing technologies man has ever had, and some believe that alcoholic food or beverages existed from the time human being appeared on earth. The oldest archeological evidence of alcohol fermentation is the rice-wine crock found in the remains of Shang period around 1600 B.C.[10] However, alcohol has been a common beverage from the Myth Era of Northeast Asia dating to 4000 to 3000 B.C. and numerous myths related with alcohol exist in this region.

As stated above, the grain brewery in Northeast Asia presumably started in the early Primitive Pottery Age with the invention of pottery. Although the full-scale production of grain wine began after the farming culture stage of around 3000 B.C., primitive alcoholic foods must have been known for a long time from the use of primitive pottery. It can be also explained by the fact that alcohol made from grain actually heightened the importance of grain and so may have encouraged the farming culture in this region.

Origin of *Kimchi* Fermentation

It is possible to observe lactic acid fermentation of vegetables yielding sour taste by keeping withered cabbage or turnip slices immersed in 2% brine for 3 to 4 days. This condition resembles that of primitive men putting foraged vegetables into a container holding seawater, and with no exception the result would be lactic acid fermentation. In such condition, *Leuconostoc mesenteroides* will be the suitable candidate dominating the system at the initial stage of fermentation.[11] It is heterofermentative bacteria producing both lactic acid and acetic acid from sugars in vegetable and growing actively until the pH goes down to 4.8. When *L. mesenteroides* cease growth at lower pH, other homofermentative bacteria like *Lactobacillus plantarum*, which produce mainly lactic acid only, start to grow, and the vegetable become very sour like sauerkraut of Germany.

This phenomenon is a natural fermentation, which occurs in any region at any time when the right conditions are provided, and it would be no exception for the people of Primitive Pottery Age. The representative traditional foods are *kimchi* in Korea, sauerkraut in Germany, *dhamuoi* in Vietnam, *dakguadong* in Thailand, and *burong mustala* in the Phillipines.[11] Many of the lactic acid fermented vegetables are made under anaerobic conditions by packing vegetables in sealed containers like ensilage, resulting in very sour products. The vegetable pickles described as "zer" in ancient Chinese literature appears to be this type of product, and are much different from *kimchi,* which is made with brine. *Zer* appears in Shiching, one of the oldest Chinese literatures. In a book on Confucius written in 200 B.C. *zer* was described as follows, "Since King Mun of Zhou enjoyed the taste of *zer,*

Confucius who respected him, tried to eat this pickle with a frown face to follow his every action. Three years later, he finally was able to enjoy the taste like the king." From this story, we can assume that Chinese *zer* had very strong sour taste to the degree that he had to frown his face.[10] The Chinese dictionary written in 100 B.C. also describes *zer* as "sour vegetable pickle."

On the other hand, the vegetable pickles traditionally made in Northeast Asia including the Korean Peninsula are made by salting and subsequent lactic acid fermentation, and have a meeker sour taste. This indicates that the Korean style pickles originated from the natural fermentation of withered vegetables stored in seawater. At the beginning, putrefaction may have occurred due to the low concentration of salt in seawater, and people had to increase the salt concentration in order to keep the vegetable longer and palatable. At around 1000 B.C. salted vegetables with very high salt concentration, 20% or more, were widely made. The most unique factor of Korean *kimchi* is that it has the balance of taste, going through the lactic acid fermentation with relatively low concentration of salt, 3 to 6%, and the addition of other vegetables and spices to help the multiplication of lactic-acid bacteria and to prevent other microbes from growing.

Origin of Fish Fermentation

The Paleolithic men of the Korea Strait came to invent and use earthenware to quickly cook by heating and storing the marine products they had gathered, hence there must have existed and developed some kinds of seafood storage techniques. There would not have been enough salt available at the early stage of this period to be able to make fish sauce and fish paste similar to today's products. Under these conditions, there were not many ways to put seafood in earthenware vessels and to store them for a long time. One possible method is mixing the half-dried seafood with vegetables preserved by the lactic acid fermentation process or with alcoholic foods, as explained above, or else with acidic fruits such as wild berries, grapes, and plum. If one mixes the seafood, which easily putrefies, with lactic acid fermented vegetables and lowers the pH to under 4.5, one can prevent the proliferation of harmful microorganisms, and therefore it can be stored over a long period of time and be consumed. Under this condition, because of the low salt concentration, the fish decompose rapidly by autolysis due to the intestinal enzymes, and a strong flavor or putrid stench is formed. The smell and taste created in this process would be an unacceptably strong putrid stench to modern men, but to the people of primitive era, who relied on rough plant materials like acorns, plant roots, grass seeds etc., it reminded them of the savory taste of animal meats and intestines. In fact, some fermented fish products made in different regions of the world have too strong a flavor to be consumed by other people. Therefore under conditions where harmful microorganisms do not prevail, the putrefaction and fermentation are distinguished only according to the subjective judgment of consumers.

Seen from such perspective, the mixture of low-salt cured seafood with lactic acid fermented vegetables would be an essential condiment for the people at the transitory stage between a meat diet and vegetarian diet, and can be an archetype of lactic acid fermented fish products, like *sikhae* in Korea, which are widely consumed in East Asia nowadays.[12] It seems that the rapid decomposition of whole fish and the emergence of concomitant strong smell or putrid stench would have been the target to improve, and as a result, the salt concentration would have been gradually raised. There are several ways to increase salt content in the fermentation system without using crystal salt. For example, seawater in earthenware is concentrated by heating, and cooled and then half-dried fish is added. By these means, high-salt fermented fish containing 20% salt can be easily prepared. In case of high-salt curing, lactic acid fermentation with cereals and vegetables or addition of acidic fruit is not necessary. The high-salt fermented fish products, *joetkal,* would have been developed in such way in the Korean Peninsula.

At an even later stage, when, having raised the salt concentration, people came to add *nuruk* in order to achieve rapid decomposition of fish as well as to reduce strong putrid stench by the action of the enzymes in *nuruk*. This is the origin of *jang*, which has been used widely in Northeast Asia and China as the major preserved food and condiment. The first description on *jang* appears in Juolii written in 200 B.C. in China. It describes two types of *jang*, *hae* and *hie*, and records the methods of preparation. *Hae* is made from sun-dried meats of foul, beast, and fish, and ground into powder, mixed with rice-wine, salt, and *nuruk* made from millet, and packed in a jar, sealed, and aged for 100 days. *Hie* is made from the same materials as *hae*, but acidic plum juice is added to provide a sour taste. It is apparent that *jang* was originally made from meat, and is a kind of meat sauce, not fermented soybean products, which *jang* is commonly called today.[10] It can be said that *jang* is a high-class condiment developed through thousands of years of experience, and applies the same fermentation principles that might have been developed by the people in Korea Strait region during the Primitive Pottery Age

The *chigae* culture and fermentation technique that developed together with the use of earthenware is deemed to have exerted a huge influence on the nutritional condition and social development of the inhabitants of Northeast Asia, especially in Korea. The stewing method of *chigae*, whereby various ingredients were mixed together and boiled, made it possible to provide a more nutritionally balanced diet and from the hygienic standpoint made it possible to have a higher developed food culture. Once the saltiness of food can be adjusted by means of seawater, the taste of food improved and it became possible to use various ingredients that could not be used before. Furthermore, since the fermentation technique made it possible to store seafood and vegetables that putrefy easily, for a long period of time, a stable food supply and improvement of the food taste became possible. This technical development is considered to have greatly improved the nutritional condition of the people of the Primitive Pottery Age compared to that of Paleolithic Man, and consequently resulted in the extension of life span and increased birthrate, and it probably brought about a sharp population increase. Such social development would have accelerated the development of agriculture and the formation of tribal nations around 4000 B.C. and also would have become the driving force that nurtured the rise of the leading tribe of Northeast Asian megalithic culture named Dong-Yi, the Eastern Tribe, who opened the early monarchical system of the region.

DEVELOPMENT OF KOREAN DIETARY CULTURE

Food habits of a people are primarily decided upon by the availability of food material obtainable in their natural environment. Other influencing factors of food culture include religious belief and thought, influx of foreign culture by war and invasion, knowledge in health and nutrition, and technological developments.[6] Korean dietary culture has evolved from the Primitive Pottery Age culture, which is characterized by the abundant use of marine foods incorporated with fermentation technology.

The Influence of Northern Nomads

When the horse riding people of the north, the *Yemaek* tribe of northeastern *Dong-yi*, came south to the Korean Peninsula to form agrarian communities, they needed to have a stable protein source to replace meat from the animal herds. They invented the use of wild soybean as food by soaking it in water and cooking it properly to be edible and also to eliminate the antinutritional factors in the bean. The *Maek* tribes are considered the first consumers of soybean as food in history.[10] It was cultivated by the nomads who began settlement farming around Mt. Baekdu, South Manchuria, and the Korean Peninsula at the beginning of the Bronze Age (1500 B.C.). In a Bronze Age excavation in Paldang, near Seoul, a smooth earthen vessel having the traces of soybean on the surface was discovered. Botanists believe that the origin of soybean is the line from South Manchuria to the Korean Peninsula where most abundant varieties of wild soybeans are found. The first record on soybean appears in *Shijing*, a Chinese literature written in the seventh century B.C. The story of soybean expansion into China follows that soybean was brought into China from Sanyung (South Manchuria) in the early seventh century B.C. by Hwangong of the Chhi Dynasty as he conquered Sanyung during the Chhun Chhiu Period, and it was therefore called *yungsuk*.[10]

The early cereal grain cultivated and utilized by the people in Northeastern Asia and the Korean Peninsula appears to be millet, which is the native plant in this region. The origin of short grain rice in this region is obscure, but numerous carbonized rice grains dated to be of the Bronze Age or earlier have been excavated. Soybean played an important role not only in supplementing protein but also providing palatability in the form of fermented soybean products to the bland cereal and vegetable diet. *Weyjyh, Dong-yi joen, Kokuryo cho* of *Sanguojyh*, a history book written in the sixth century in China, describes the people of Kokuryo (one of the three Korean Kingdoms) as experts in preparing fermented soybean products.

The production of soy sauce by the *Maek* tribe, who were originally meat-eating nomads, created a typical Korean dish, *Bulgoki* "fire" beef, the grilled meat marinated with soybean sauce. In Chin (B.C. 221 to 206) of China, the marinated grilled meat was called *Maek-chok*, which meant Korean grilled meat. The meat diet of the nomads gradually changed because of their changing settlement patterns, as they adapted to the cereal-based food diet of the natives on the southern plains.

The Influence of Buddhism

The introduction of Buddhism to the Korean Peninsula in A.D. 372 (Koguryo) and in A.D. 528 (Silla), accelerated the reduction of animal food consumption and encouraged the spread of vegetarian food habits. According to *Samguksaki* (1145), the oldest document of Korean history, rice, wine, oil, honey, soy sauce, soybean paste, dried meat and fish sauce were all important food items that were prepared for a wedding in the royal family in Silla in the year of 683 A.D. The people of the Unified Silla and succeeding Koryo dynasty were strong Buddhists. During these thousand years of the period, the nomadic animal food habits disappeared. The extensive use of salted vegetables and soybean, as the major source of protein, resulted from this change. The technologies of soy-sauce fermentation and rice-wine making were well developed and transferred to neighboring countries. The document of *Shoso-in* (752 A.D.) of Japan describes *Miso*, the Japanese name of soybean paste, as a dialect from *Koryo* (Korea) and often called *Koryo Jang*.[10] The ancient Japanese history book, *Kojiki*, mentions that a man from *Baekje* taught them how to make rice-wine. The memorial tablet of a man called *Chin* of *Silla* is kept in a shrine, the Matuo Taisha in Kyoto, as a god of rice-wine. The rice-wine producers in Japan today attend an annual worship ceremony for him, in order to pray for success in their own wine brewing.

The Influence of Mongol (Yuan) Invasion and Confucianism

The Chinese *Yuan* (Mongol) invasion of Koryo in the 13th century (1259 to 1356) and the respect for Confucianism in Chosun Kingdom brought about the suppression of Buddhism and restored the animal food habit of Korea. Another important change in the Korean diet took place when red pepper was introduced, in the 17th century. The route of the propagation of red pepper into Korea is unknown. Korean literature describes how it was introduced from Japan during the Korean-Japanese War in the 1600s, while some Japanese literature records that it was introduced through Korea into Japan. With the introduction of red pepper, the traditional salted-vegetable dish was transformed into today's *kimchi*. *Kochujang*, a typical hot soybean paste of Korea was also developed through the introduction of red pepper.

During the Chosun Kingdom (1382 to 1910), a well-balanced variety of foods, of both animal and vegetable origins, were utilized. *Imwon sibyukchi*, an encyclopedia written in 1827 by *Soe YuGu*, describes 11 kinds of water, 36 kinds of cereal, 72 kinds of vegetables, 13 kinds of poultry, 34 kinds of fish, and 8 kinds of spice, as major food materials that were used in the 19th century of Korea. The ideal diet for Koreans was standardized between the 15th and the 19th centuries. Records of an ideal standard meal for Koreans appear in much of the literature of the Choson Kingdom, for example, in *Shiui Chonso*, written in the 19th century. The literature written between the 17th and the 19th centuries outline a standard meal consisting of a bowl of cooked rice, a bowl of soup, and a dish of kimchi as the basic constituents. To this basic menu, side dishes are added, forming a three-dish meal (*samchop bansang*), a five-dish meal (*ochop bansang*), a seven-dish meal (*chilchop bansang*) (Table 8.4), and so on.[13] A 12-dish meal was an extravagance served only for the king.

The Influence of Western Culture and Korean War

Korea opened her gate to Western countries in the 1870s, much later than Japan and China. The European and Russian diplomats, as well as missionaries from America introduced cakes and coffee. However, it was soon overshadowed by the Japanese invasion of Korea, and she was annexed to Japan in 1910 for 36 years. One of the statistical records of the colonial regime shows that one-third of rice produced in Korea was extorted to Japan every year during this period.[14]

The people suffered greatly with the shortage of food and even defatted soybean flake was rationed as a substitution of rice. Soon after the rehabilitation in 1945, hundreds of thousand people moved from Communist North Korea to South Korea. The total number of refugees from North to the South after the Korean War (1950 to 1953) was estimated at 2 million.

The famine during the Korean War was barely overcome by wheat flour and nonfat dry milk given by the U.S. Aid Program. Milk porridges were rationed to the starved people who had been nonmilk-eating people. After severe lactose intolerance symptoms, people gradually adapted to eating milk porridge. It triggered the explosive consumption of milk products during the economic growth of the 1970s to 1980s, and the rapid Westernization of Korean food habits afterward.[14]

HEALTH CONCEPTS IN TRADITIONAL MEDICINE

The early classics of Chinese literature are the products of a long history of philosophy, religion, culture, and wisdom of the many tribes in this region. The early historians in China described the lives of neighboring countries. The Eastern Tribe inhabited a wide range of Northeast Asia, from the Shandong Peninsula to the Bohai Corridor, the Manchurian Basin, the Liadong Peninsula, and the Korean Peninsula, which was mostly ruled by Koguryo until the fifth century A.D.[2] Taoism, the folk religion that originated from the shamanistic beliefs of this region, forms the basis of the health concepts found in the traditional diet and medicine of the Northeast Asian people.

Taoism

Korean thought on life and health is based on the shamanistic folk philosophy, Taoism, which sets as the ultimate goal a healthy eternal life. The established Taoism, as developed by early Chinese philosophers teaches that this goal can be achieved by discipline, mainly through the control of breath, sex, and food. The principle of control is the harmony of *yin* and *yang*, the negative and positive nature of the universe.[15]

The pictographs on the engraved tortoise shells found in China show that the basic principles of *yin* and *yang* were a part of the Shang dynasty, and that they originated from the legendary saint, Bok-Eui (3000 B.C.), the God of Divination. The Chinese characters, which are used today, were formed in the Jou dynasty (1100 to 220 B.C.). The Theory of Interchange developed through the Jou dynasty for 3,000 years led to Taoism and Confucianism.

YIN AND YANG AND THE FIVE PHASES THEORY

The Book of Changes, *Yijing*, is the basis of the *yin* and *yang* theory and the Principles of *Five Phases*, and it contains the principles that explain changes in the universe and in nature.[15] Examples of *yin* and *yang* that are commonly found in nature are dark/bright, female/male, inside/outside, center/circumference, weak/strong, empty/full, cold/hot, rise/descend, plants/animals, death/life, moisture/dryness, big/small, sparse/dense, and electron/proton. The important principles applied to the *yin-yang* relationship are mutual suppression and repulsion, mutual dependence, mutual compensation for equilibrium, and mutual transformation. The principle implies that there is no absolute *yin* (negative) or *yang* (positive) in nature, and that everything is relative.

Wood, Fire, Earth, Metal, and Water represent the Principle of Five Phases. It implies transition, movement, or passage, rather than the stable, homogeneous chemical constituents such as Earth, Air, Fire, and Water, the four eternal elements of ancient Greek science. The Five Phases is the principle of changes linked by the relationships of generation and destruction (or suppression), as shown in Figure 8.1.[16]

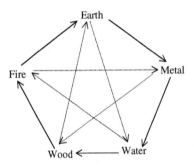

Figure 8.1 The Five Phases. As individual names or labels for the finer ramifications of yin and yang, the Five Phases represent aspects in the cycle of changes. The Five Phases are linked by relationships of generation and destruction. Patterns of destruction may be summarized as follows: water puts out fire; fire melts metal; a metal ax will cut wood; a wooden plow will turn the earth; an earthen dam will stop the flow of water. The cycle of generation proceeds as water produces the wood of trees; wood produces fire; fire creates ash, or earth; earth is the source of metal; and when metals are heated, they flow like water. (From LN Magmer, A History of Medicine, New York; Marcel Dekker, 1992, p. 46.)

TABLE 8.1   Classification of the *Five Phases*

5 Phases	Taste	Organs	Intestines	Senses	Tissues
Wood	Sour	Liver	Gall bladder	Eye	Tendon
Fire	Bitter	Heart	Small intestine	Tongue	Pulse
Earth	Sweet	Spleen	Stomach	Mouth	Meat
Metal	Hot	Lung	Large intestine	Nose	Skin and hair
Water	Salty	Kidney	Bladder	Ear	Bone

According to *yin-yang* and the Five Phases Theory, all food materials are classified by their properties and their different tastes. The properties are cool, as *yin*; neutral; and warm as *yang*. For example, fruits on the tree are considered to have *yang* property, while roots in the soil have *yin* property. *Yin* property also represents material entities such as nutrients; while *yang* property represents functions, like energy. Taste is divided into five groups, representing the Five Phases; sour-Wood, bitter-Fire, sweet-Earth, pungent-Metal, and salty-Water. As shown in Table 8.1, taste can be related to the human body and its organs, senses, and feelings, and even to color, the weather, and the seasons, through classification into the Five Phases. Antagonistic or affinitive relations between tastes and organs/senses are also judged or predicted by the principles of the *Five Phases*. For example, sour (*wood*) generates heart (*fire*) but suppresses spleen (*earth*), and salty is related to kidney and generates liver and suppresses heart. Though simplified unrealistically, it explains the basic notion of *Five Phases* applied to food and health practice.[8]

Eastern Medicine

The oldest Chinese medicinal book, The Yellow Emperor's Classic of Medicine, written in the Chin and Han period of China (220 B.C. to 220 A.D.) contains theories of man–universe unity, *yin* and *yang*, the *Five Phases*, the Ten Calendar Signs (the decimal system), Earth's Twelve Branches (the duodecimal system), and other fundamental principles of medical treatment.[17] This book was first introduced into Korea in the period of Koguryo King Pyungwon, in year 3 (561 A.D.). Since then, Chinese medicinal knowledge has greatly influenced the health concepts and food habits of the Korean people. It has contributed to the development of Eastern Medicine, in combination with traditional folk medicine in Korea, as recorded by Hur Jun in 1611.[18] Eastern Medicine was further developed during the 18th and 19th centuries, and grew into Sasang Medicine: as described by Lee Je-Ma in 1894.[19] Sasang is a unique theory of categorizing people into four body types according to their physical constitutions, Tae Yang, Tae Eum, So Yang, and So Eum. It emphasizes the importance of individual body type in the diagnosis and treatment of diseases and suggests prescribing different medicinal treatments and food supplies for each.

HEALTH CONCEPTS IN TRADITIONAL KOREAN DIET

The basic idea of traditional Korean nutrition is to harmonize properties and tastes in the diet on the basis of *yin* and *yang* and the *Five Phases*. A diet that emphasizes one property or extreme taste is considered to be unhealthy. Korean meals are prepared to harmonize the properties and tastes through selecting the proper ingredients and process. A Korean meal containing a bowl of rice, mugwort soup, cabbage kimchi, shepherd's purse salad, broiled plant root (*Codonopsis lanceolata*) fernbrake salad, stewed yellow corvina, and leek pancakes was analyzed in terms of *yin-yang* and the *Five Phases*, as shown in Table 8.2.[20] It showed that the composition of the meal is well balanced in terms of *yin* and *yang* and the *Five Phases*.

Recently, K. B. Lee, Emeritus Professor of the Medical School of Seoul National University, compiled several food lists categorizing foods as desirable or undesirable for people of different body type according to Sasang Medicine, as shown in Table 8.3. He developed a simple test to identify the body types of individuals, which is known as the "O-Ring Method."[21] Table 8.3 shows that some foods, like rice, Italian millet, and corn, are desirable for all types of people; while glutinous rice

TABLE 8.2 Analysis of a Korean Meal in Terms of *Yin-Yang* and the *Five Phases*

	Wood (Sour)	Fire (Bitter)	Earth (Sweet)	Metal (Pungent)	Water (Salty)
Yang (warm)	Leek	Mugwort	Shepherd's purse, wheat flour	Green onion, garlic, ginger, black pepper, sesame	Salt
Neutral			Water, rice, soybeans, yellow corvina		
Yin (cool)	Vinegar	Plant root, fernbrake	Cabbage	Onion	Soy sauce, soybean paste

TABLE 8.3 Desirable (O) and Undesirable (X) Foods for the People of Different Body Type

Food	TY	SY	TE	SE	Food	TY	SY	TE	SE	Food	TY	SY	TE	SE	Food	TY	SY	TE	SE
Polished rice	O	O	O	O	Sugar	O	O	O	O	Tomato	X	X	X	X	Walnut	X	X	O	O
Brown rice	O	O	O	O	Glucose	O	O	O	O	Mustard	X	O	X	O	Gingko nut	X	X	O	O
Glutinous rice	X	X	O	O	Chocolate	X	X	O	O	Pepper	X	O	X	X	Pinenut	O	O	O	X
Barley	O	O	O	X						Curry	X	O	X	X	Peach	O	X	O	O
Wheat bran	X	O	O	X	Korean cabbage	O	O	O	X	Chinese radish	X	O	X	X	Sea mustard	X	O	O	O
Buck wheat	O	O	O	X	Cabbage	O	O	O	X	Carrot	X	O	X	X	Laver	O	X	O	O
White soybeans	X	X	O	O	Kale	X	X	O	O	Lotus root	O	O	X	X	Sea tangle	O	X	O	O
Black soybeans	O	O	X	X	Lettuce	O	O	X	X	Root of Chinese bellflower	X	O	O	O					
Colored beans	O	O	O	O	Young radish	O	O	O	O	Codonopsis lanceolata	X	O	O	O	Beef	X	O	O	O
Kidney beans	O	O	O	O	Spinach	O	O	O	O	Burdock	O	O	O	O	Pork	X	O	O	X
Peanuts	X	O	O	X	Crown daisy	X	O	O	X	Hemp	O	O	O	O	Chicken	X	X	O	O
Gray redbeans	O	O	O	X	Celery	O	O	O	X	Musk melon	X	O	X	X	Dog meat	X	X	O	O

Food	TY	SY	TE	SE
Red beans	O	X	O	O
Adlay	X	X	O	O
Italian millet	O	O	O	O
Indian millet	X	O	X	X
Sorghum	X	O	X	X
Corn	O	O	O	O
Mung beans	O	X	O	O
White sesame	X	O	X	X
Black sesame	O	X	O	O
Perilla	O	X	O	O
Potato	O	X	X	X
Sweet potato	O	O	X	O
Honey	X	X	X	X
Ginseng	X	O	X	O
Parsley	O	O	X	O
Watercress	X	O	X	X
Green onion	O	O	O	X
Onion	O	O	X	X
Leek	X	O	X	O
Red pepper	O	O	X	X
Ginger	X	X	X	O
Garlic	O	O	X	X
Sesame leaf	X	O	X	O
Pumpkin	X	O	O	X
Eggplant	O	O	X	O
Cucumber	O	O	O	O
Young antler	O	O	X	X
Youngji	O	O	X	X
Watermelon	O	O	X	O
Strawberry	X	X	O	O
Persimmon	O	X	O	O
Pear	O	X	O	O
Apple	O	X	O	O
Citrus fruit	O	X	O	O
Orange	O	X	O	O
Lemon	O	X	O	O
Grape	X	O	O	X
Banana	O	O	O	O
Jujube	O	O	O	O
Chestnut	O	O	X	O
Ogapi	O	X	O	O
Vitamin B	X	X	O	X
Milk	X	O	O	O
Egg	X	O	O	O
Shellfish	O	O	X	X
Shrimp	O	O	X	X
Crab	O	O	X	X
Oyster	O	O	X	X
Squid	O	O	X	X
Hairtail	O	O	X	X
Mackerel	O	O	X	X
Herring	O	O	X	X
Yellow corvenia	O	X	O	O
Vitamin E	X	O	X	X
Vitamin C	O	O	O	O

Note: TY=Tae Yang; SY=So Yang; TE=Tae Eum; SE=So Eum

is desirable only for *yin*-type people and cabbage only for *yang*-type people. The reliability of this categorization is not confirmed, but it provides an example of how to select the foods that are desirable for an individual body. This kind of thinking forms the basis of the therapeutic food concepts of the Korean people.

NUTRITIONAL VALUE OF THE TRADITIONAL KOREAN DIET

On the basis of the philosophical ideas and medical knowledge developed in China and Korea, the Korean people have developed a standardized ideal meal, within a systematic menu program, that is called *Chop Bansang*. Recently, the nutritional value of the Korean traditional diet was analyzed using the seven-dish meal menu of Kim Ho-Jik (1944) and the standard weekly menu of Pang Sing-Young (1957) and was compared to the current Recommended Dietary Allowances (RDA) for Koreans.[20] Table 8.4 shows the nutritional value of a traditional Korean meal in the menu of Kim Ho-Jik as calculated by the current Food Composition Table of Korean Food. The basic meal consisting of a bowl of cooked rice, a bowl of soup, and a dish of kimchi, could supply 40% of the energy and 48.7% of the protein of the RDA. When three dishes were added to the basic meal, the three-dish meal (*samchop bansang*) contained 47.2% of the energy and 94.3% of the protein of the RDA. Sufficient amounts of minerals and vitamins were supplied by the three-dish meal. Carbohydrates contributed 77% and 64.4% of the total energy in the basic meal and the three-dish meal, respectively; while lipid contributed only 8.3% and 11.6%. The energy from lipid did not exceed 12% of the total energy supply until a five-dish meal, which was considered a luxury, was analyzed.

The Korean traditional diet was estimated to be able to supply from 2,000 to 2,500 calories and from 80 to 90 grams of protein per day. The energy constituents were 73 to 77% carbohydrates, 15 to 18% proteins, and 10 to 12% lipids. Animal protein was 20 to 30% of the total protein. The contribution of lipid energy in total calorie intake did not significantly

TABLE 8.4 Evaluation of the Nutritional Value of a Traditional Korean Meal in the Menu of Kim Ho-Jik (1944)

Type of menu	Basic meal	Three-dish meal	Five-dish meal	Seven-dish meal
Composition of menu	Cooked rice, soup, kimchi	Basic meal + spinach, roasted beef, dried fish	Three-dish meal with stew + meat jelly, fermented fish roe	Five-dish meal + panned oysters. radish kimchi
Total energy (kcal)	995 (40.0)	1181 (47.2)	1320 (52.8)	1672 (66.8)
Carbohydrate (%)	77.0	64.4	60.1	53.4
Protein (%)	14.7	24.0	28.0	27.7
Lipid (%)	8.3	11.6	11.9	18.9
Total protein (g)	36.5 (48.7)	70.7 (94.3)	92.5 (123.3)	115.5 (154.0)
Animal protein (g)	28.7	59.5	69.0	72.3
Ca (mg)	161.1 (26.9)	216.3 (36.1)	255 (42.5)	596 (99.3)
Fe (mg)	12.1 (121.9)	23 (230)	26.8 (268)	40.3 (403)
Vitamin A (I.U)	426.2 (17.1)	8,7616.6 (350.5)	9,129 (365.2)	9,965 (398.6)
Vitamin B_1 (mg)	0.62 (47.6)	0.86 (66.2)	1,08 (8.1)	2.16 (166.2)
Vitamin B_2 (mg)	1.92 (127.9)	3.03 (202.2)	3.44 (229.3)	4.35 (290.4)
Niacin (mg)	11.6 (68.3)	28.9 (169.9)	37.1 (218.2)	45.8 (269.4)
Vitamin C (mg)	19.7 (35.9)	83.7 (152.2)	86.4 (157.2)	99.6 (181.2)

Note: ( ) percent of RDA

TABLE 8.5 Estimated Dietary Goals as Shown in the
Traditional Korean Standard Meal

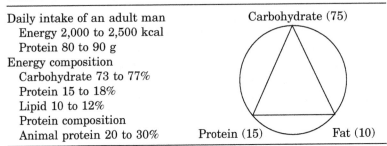

Daily intake of an adult man	Carbohydrate (75)
Energy 2,000 to 2,500 kcal	
Protein 80 to 90 g	
Energy composition	
Carbohydrate 73 to 77%	
Protein 15 to 18%	
Lipid 10 to 12%	
Protein composition	
Animal protein 20 to 30%	Protein (15) Fat (10)

change by increasing the number of side dishes up to five, but
that of protein did increase. It appears that the Korean traditional diet could supply amounts of protein, minerals, and
vitamins sufficient to nourish an adult male whose energy
intake exceeded 2,000 calories per day.

The dietary goal to be achieved in the Korean standard
meal appears to be for an adult man to be supplied daily with
2,000 to 2,500 kcal energy, made up from 75% carbohydrate,
15% protein, and 10% lipid, as shown in Table 8.5. The daily
intake of energy to be achieved in a Korean standard meal is
similar to present-day recommendations, but the composition
of the energy is different. The traditional diet emphasizes the
low intake of lipids, only 10% of total energy, which are one
half of today's recommendation and a quarter of the average
American diet. The large amounts of carbohydrates, which
are mainly supplied from cereals and vegetables, and the
small amounts of animal meat and fat are characteristic of
the traditional Korean diet.

FOOD AS MEDICINE

In the traditional Korean culture, food was considered to be
the fundamental source of health, and it was believed that
all diseases could be cured by the control of food intake.
Without any knowledge of the chemical composition of foods,
nutritional value could be evaluated solely through the medicinal effects on human subjects. While the science of nutrition

in Western society was tested mainly through animal experiments, Korean concepts of food and nutrition developed through long experience with human trials.

On the basis of the health and nutritional concepts of Korea, Hong Seon Pyo proposed dietary guidelines in his Book of Korean Cookery, published in 1940, as follows[22]:

1. Eat only when hungry
2. Eat hard materials, with adequate mastication
3. Stop eating before achieving satisfaction
4. Eat raw food wherever possible

He suggested using certain principles in selecting ingredients for the preparation of healthy food:

1. fresh
2. raw
3. natural
4. long-lived plants and animals
5. dense texture
6. young plants and animals
7. materials produced nearby
8. nonstimulating foods

He also recommended the reduction of salt and fine sugar intake. His dietary guidelines and his principles of selecting food materials are widely accepted today.

Considering food to be medicine, practitioners of traditional medicine studied each food ingredient for its property, taste, and medicinal effects. Their knowledge has been compiled in numerous medicinal books for thousands of years, and has been practiced in everyday life at the household level as a part of Korean dietary custom. Food preparation was likened to prescribing medicine for the individuals in a household. The word *yaknyum,* the general term for seasoning, means "thought of medicine." This mentality refuses to accept processed food made in a mass production system. The enormous size of the health food market today in Korea reflects the country's tradition of "food as medicine."

A recent survey of consumers' attitudes toward health food and their perceptions on health and food habits in Korea revealed that the people considered their food habits as being the most important factor in the maintenance of health, followed by physical exercise. More than 90% of the people believed that food habits were the most important factor determining the health condition of human beings, and those diseases could be prevented and cured by adjusting food habits.[23] One half of the subjects had made use of health foods, and 68 percent of them believed in their effectiveness.[24]

Conclusion

Many reports suggested that low fat intake and high plant food intake of the Koreans might be part of the reason for the lower prevalence of obesity, lower death rates due to coronary heart disease, high blood pressure, and the lowest rate of breast cancer and prostate and colon cancers than in many other Asian and Western countries.[25] Koreans believe the adage of "food as medicine." Therefore, herbs or fruit ingredients such as ginger, cinnamon, adlay, mugwort, pomegranate, citron, mushroom, ginseng etc., were used in cooking, and also used for their therapeutic effects. Some of the well-known food supplementary ingredients today in the United States such as ginger, garlic, dates, chestnuts, gingko, soybeans, and others have been used as spices in traditional Korean dishes for generations. Therefore, today the Western term "functional foods" is the same as the traditional Eastern term of therapeutic foods or health foods that have been commonly used in Korea for many years.

REFERENCES

1. AC Nahm. *Korea, Tradition and Transformation*. Holly International Corp., Elizabeth, N.J., 1988.

2. GL Barnes. *China, Korea and Japan, The Rise of Civilization in East Asia*. London: Thames and Hudson, 1993.

3. PK Sohn. Human race and residence, in *Hankuksaron* (Korean History). *Korean Archeology Ia*. 12:187–211, 1983.

4. BK Choi. Comparison of Paleolithic cultures in Northeast Asia, in *Hankuksaron* (Korean History). *Korean Archeology Ia.* 12:228–306, 1983.

5. YC Lee. A chronicle, in *Hankuksaron* (Korean History). *Korean Archeology Ia.* 12:367–414, 1983.

6. CH Lee. The food ways of Paleolithic man in the Northeast Asia and Korean Peninsula. *Korean Culture Research* 31:415–458, 1998.

7. CH Lee. The Primitive Pottery Age of Northeast Asia and its importance in Korean food history. *Korean Culture Research* 32: 325–457, 1999.

8. CH Lee. *Fermentation Technology in Korea.* Seoul: Korea University Press, 2001.

9. N Ishige. Salt of Kumupa, Research Report of National Ethnic Museum of Japan 1:357–372, 1976.

10. SW Lee. *Hankuk Sikpum Sahoesa* (History of Korean Food and Society), Kyomunsa, Seoul, 1984, p. 168.

11. CH Lee. Importance of lactic acid bacteria in non-dairy food fermentation. In: CH Lee, J Adler-Nissen, G. Baewald, Eds. *Lactic Acid Fermentation of Non-Dairy Food and Beverages.* Seoul: Hanrimwon, 1994, pp 8–25.

12. CH Lee, KH Steinkraus, PJA Reilly, Eds. *Fish Fermentation Technology.* Tokyo: UNU Press, 1993.

13. SS Yoon. *Hankuk Sikpumsa Yongu* (Korean Dietary Culture Studies). Seoul: Shin Kwang Publish Co., 1993, p. 28.

14. CH Lee. Changes in the dietary patterns, health, and nutritional status of Koreans during the last century. *Korean and Korean American Studies Bulletin* 6:32–47, 1995.

15. KB Lee. *The Principles of Eum and Yang, in Korean Dietary Culture.* International Course for Food Fermentation Technology, Korea University, Seoul, 1992, p. 93.

16. LN Magner. *A History of Medicine.* New York: Marcel Dekker, 1992, p. 46.

17. MS Ni. *The Yellow Emperor's Classic of Medicine.* Boston: Shambhala, 1995.

18. J Hur. *Dongeui Bogam* (The Classic of Eastern Medicine), Korea, 1610.

19. JM Lee. *Dongeui Susebowon,* Korea, 1894.

20. MH Kim. *Nature, Man and Traditional Korean Medicine*, Seoul: Yoksa Bipyungsa, 1995, p 187.

21. KB Lee. *Oriental Medicine, in Korean Dietary Culture.* International Course for Food Fermentation Technology, Korea University, Seoul, 1992, p. 123.

22. SP Hong. *Chosun Yorihak* (Book of Korean Cookery), Korea, 1940.

23. EJ Lee, SO Ro, CH Lee. A survey of consumer attitude toward health food in Korea (1) Consumer perceptions of health and food habits. *Korean J Dietary Culture* 11:475–485, 1996.

24. EJ Lee, SO Ro, CH Lee. A survey of consumer attitude toward health food in Korea (2) Consumer perceptions of health food. *Korean J Dietary Culture* 11:487–495, 1996.

25. CY Lee, CH Lee, TW Kwon. Future of Foods: Harmonization of Eastern and Western Food Systems. http://www.Worldfoodscience. org/vol3-2/focus-lee.html.2002

9

Asian Fish Sauce as a Source of Nutrition

CHAUFAH THONGTHAI

Mahidol University, Bangkok, Thailand

ASBJØRN GILDBERG

Norwegian Institute of Fisheries and
Aquaculture Research, Tromsø, Norway

INTRODUCTION

The fish sauce of Asia is a nutritious condiment made from a traditionally fermented fish and salt mixture. Its appearance makes the sauce easily mistaken for a well-aged whisky. The richness in its content of amino acids and oligopeptides, short-chain fatty acids and aldehydes, together with minerals, impart a characteristic cheesy and meaty aroma, apart from its sharp and salty taste.[1-3] Having vitamin B12 as an indigenous constituent makes fish sauce a unique product in the class of condiment.[4] Consumption of fish sauce is almost obligatory among the people of Southeast Asia. The sauce is known locally as *nampla* in Thailand and Laos, *nuocmam* in Vietnam, *tuk-trei* in Cambodia and *patis* in the Philippines. Other local names, of less widely consumed fish sauces, include *yulu* or *yeesui* in China, *shottsuru* in Japan, *aekjeot* in Korea and *ngan-pya-ye* in Myanmar.[5–7] Fish sauce can be used by direct addition to dishes for its saltiness and flavor, or made into a dip together with other spicy ingredients. The daily consumption of 15-30 ml per person is estimated to supply up to 7.5% of dietary protein.[8] This amount is also sufficient to protect against megaloblastic anemia with its vitamin B12 content.[4]

To date, genuine fish sauce manufacturing remains a traditional process. Heavily salted fresh and whole pelagic fishes, mainly of *Stolephorus* species, are tightly packed into clay containers. Fish protein hydrolysis and indigenous fermentation are allowed to proceed naturally at ambient conditions. A clear brown liquid is drawn from a spigot at the base of the fermenting broth and further aged before use. The total storage time is up to 12 months. Salt saturated brine leaching constitutes second-grade fish sauce (Table 9.1).[5,9]

HISTORY

Old Traditions of Southeast Asia

From time immemorial, salting of fish has served as a simple and inexpensive method of food preservation. The practice is associated with the transition from a nomadic to an agricultural economy,[5] particularly in the Southeast Asian region, where the long coastlines provide both plentiful fish raw material and salt. The hot and humid environment naturally accelerates fish spoilage. Salting not only prevents spoilage of highly perishable fish, but also allows fermentation of many fermented food products of Asia. A strong correlation between the use of fermented fish products and the use of cereals, especially rice and vegetable, was observed throughout the world. Mackie et al.[5] has compiled a detailed report on the processing of fermented fish products in various parts of the world. More

TABLE 9.1 Local Names of Fish Sauce and Some Properties

Local name	Country of origin	Colour	Clarity	Fish spp.	Reference
Nampla	Thailand	Amber	Clear	*Stolephorus* spp., *Rastrellinger* spp., *Clupea* spp., *Sardinella* spp., *Cirrhinus* spp.	5, 9, 10
Nuoc mam	Vietnam	Amber	Clear	*Stolephorus* spp., *Engraulis* spp., *Clupeoides* spp. *Dorosoma* spp. *Decapterus* spp.,	5, 16, 17
Patis	Philippines	Amber	Clear	*Stolephorus* spp., *Ostrea* spp., *Clupea* spp.	5, 86, 170
Yee-sui or Yu-lu	Hong Kong	Amber	Clear	*Sardineela* spp. *Engraulis* spp.	11, 30, 171
Shottsuru	Japan	Yellow	Clear	*Astroscopus* spp. Sardines, Anchovies and mollusk especially squids	5, 72, 101
Aekjeot, Jeot-kuk	Korea	Yellow to brown	Clear	*Astroscopus* spp. *Engraulis* spp.	10, 11, 30
Ketjab-ikan	Indonesia	Brown	Clear	*Stolephorus* spp. *Leiagnasthus* spp. *Osteochilus* spp. (freshwater fish)	5, 10, 134

TABLE 9.1 (CONTINUED) Local Names of Fish Sauce and Some Properties

Local name	Country of origin	Colour	Clarity	Fish spp.	Reference
Nga Ngan-pya-ye[a]	Myanmar	Dark brown	Turbid without sediments	Not available	5, 7
Budu	Malaysia, Southern Thailand	Dark brown	Turbid with heavy sediments	*Stolephorus* spp. *Sardinella* spp., *Clupeoides* spp., and Selar boops	5, 134, 167

[a] Produced normally as by-product of Nga-pi, the fish paste.

recent reviews on various East Asian and Southeast Asian fermented fish products were published together with their local names, some available processing descriptions, with biochemical and microbiological data.[6,10]

Most fermented fish products are indigenous and originated locally. They are broadly separated, with respect to their appearance, into three categories which include fish sauce, fish paste, and the fermented fish bits.[5] Other classification systems are based on the processing methods,[8] the extent of salting,[11] and substrates used in the fermentation.[12]

In Thailand, fish paste or the parallel product, shrimp paste, has a smooth and pasty consistency of a partially sundried, heavily salted (i.e., more than 20% and wholly comminuted) fish muscle. The fermented fish bits appear more solid with recognizable pieces of whole or dressed or even macerated fish muscle of larger fish species. They contain a lower salt concentration, i.e., from 6 to 18% and characteristically include a lactic acid type of fermentation. A wide variety of fermented fish bits are influenced by the diverse carbohydrates added during processing. They range from cooked rice, roasted rice, rice bran, red rice to palm sugars.[13] Essentially, the types of carbohydrate added, salt concentration, method of dressing of the fish including gutting, filleting or maceration, as well as fermentation conditions exert a direct influence on the texture, flavor and aroma of the particular products.

Among the indigenous fermented fish products, Asian fish sauce is the most well known and enjoys increasing popularity worldwide. Processing of the fish sauce contrasts with that of fish paste. The heavily salted fresh, normally small pelagic, fish is kept away from sun-drying but allowed to liquefy naturally to its completion under rather anaerobic conditions. Traditionally the salt saturated fish and salt mixture are usually packed tightly in a container, topped heavily with more salt. Any air space is driven off with brine and the system is left to ferment naturally. The first filtrate is drawn after 8 to 12 months, and up to three leachings for lower quality fish sauce are obtainable. For well-fermented sauce, maturation and blending of the sauce are optional rather than a rule.

The time duration involved in the processing of the three categories of indigenous fermented fish products differs significantly. While the fermented fish bits may take from a few days up to several weeks, the fish paste may take from several weeks to a few months. Traditional fermentation of fish sauce is the most time consuming as it may take from several months up to a year and a half to obtain the characteristic flavor and aroma.

Fish sauce is an indispensable item in the Thai kitchen and presumably similar situations are found in Vietnam and the Philippines. Its consumption by the Thai population is estimated to be at 17 to 20 ml per person per day (Department of Industry Promotion, Ministry of Industry, Thailand, personal communication). A similar estimate in Vietnam was 15 to 30 ml.[8] Therefore, it is not unreasonable to believe that Southeast Asians generally consume, both sensibly and insensibly, approximately 20 ml of fish sauce per person per day.

The voluminous fish sauce production in Thailand is mainly consumed locally, and export comprises a minor quantity. Export of fish sauce from Thailand was reported to average about 26,000 metric tons per annum during the period of 1995 to 1999. This represented almost a sixfold increase in metric tons per annum during the period from 1977 to 1982 (Custom Department, Ministry of Finance, Thailand, personal communication). The United States of America alone accounted for 34% of the total export, while Hong Kong, Japan, Australia and France together accounted for 37%. Currently Thai fish sauce is sold in about 100 countries around the world. Nevertheless, the export volume of fish sauce represents an average of 5% of the total annual output. In 1998, a total amount of 426 million liters (about 500,000 tons) of fish sauce was produced, of which 32 million liters were exported. Almost all of the fish processed were anchovies of *Stolephorus* Spp. with only a small amount of Indo-Pacific mackerels and other species. Seventy-five out of the total 88 plants of medium-to-large-scale fish sauce production were located in the east coast of Thailand. Together they account for 98% of the total output.[14] The situations in Vietnam and the Philippines may differ with respect to the type of fish used in the processing (Table 9.1), nevertheless the extent of local consumption of fish sauce presumably is similar. In 1990 the estimate of annual fish sauce production in Southeast Asia was a modest 250,000 tons.[15] Although statistics are not available, the authors suggest that the present annual world production is at least 1 million tons, Thailand being the major producer.

Traditional fish sauce fermentation is an old technology. The process involved was probably first documented by Roe in Vietnam (cited in Reference 5).[16,17] Presumably fish sauce fermentation originated as a household recipe. With increasing demand it gradually developed into cottage industries. Although thousands of small-scale fish sauce producers spreading throughout the Southeast Asian region adhere to the traditional method of processing, medium- and large-scale plants, for instance in Thailand, benefit from modern mechanization in many steps of the production line. These include salting, filtration, pasteurization, bottling and packaging. Despite the modern facilities in processing, the actual fermentation is still generally carried out in large and covered concrete vats which are built into the ground. Since traditional fish sauce fermentation is a time-consuming process, attempts to shorten the fermentation time have been a popular topic of research. Various innovations include largely the use of exogenous sources of enzymes, resulting in varying degrees of success.

Ancient Traditions and Recent Developments in Europe and North America

Although fish sauce generally is acknowledged as a typical Asian product, it was also made and highly valued by the ancient Romans and Greeks. The Roman author, Pliny the Elder, living in the first century, describes fish sauce as an exquisite liquid.[18] *Garos* and *aimeteon* were Greek names for fish sauces. The Roman equivalent to *garos* was *garum*. In addition, *liquamen*, *allex* and *muria* were various kinds of fish sauces used by the Romans. Whereas *garum* and *liquamen* were high-quality fish sauces sold at prices similar to perfume, *allex* and *muria* were second-grade qualities made from residual fractions of the former.[6,18]

Fish viscera, blood, liver and other by-products from mackerel, herring and tunny fish were the most frequently used raw materials,[6,18] and the heavily salted mixtures were fermented for at least 9 months before the sauce was harvested.[6,19] In some cases fish sauce was mixed with wine, vinegar or oil before use. In regions south of Rome fish sauce production became a thriving industry, but fish sauce was made all around the Mediterranean sea including Asia Minor and Northern Africa.[18]

Just as with the present Asian fish sauces, the ancient Greek and Roman sauces were used as condiments for many kinds of dishes, but in most cases they were thicker and had a very strong flavor.[19] Although the Romans were fully aware of the risk of halitosis from consuming too much fish sauce, they also believed that it had curative effects or could be used as a remedy for various other purposes. Garum mixed with oil or vinegar was used as a laxative. On the other hand, garum with lentils was used to treat chronic diarrhea.[18] Fish sauce was also used for external medication as a treatment for healing wounds. According to Pliny cited in Corcoran,[18] burns could be treated with garum, however Pliny warns, it was only efficient if fish sauce was not mentioned by name during treatment. The cheaper sauce, muria, was believed to be especially useful in curing diseases as varying as dysentery and sciatica.

Although the Mediterranean fish sauce tradition gradually vanished, two types of fish sauce, *botargue* and *ootarides*, were still produced in Italy and southern Greece until the 19th century.[6] Recently work has been carried out in Spain to investigate the possibility of making fish sauce similar to the traditional *garum* by modern technology.[19] Eviscerated mackerel and tuna liver were mixed in a ratio of 1 : 1 and preserved by 5 to 25% salt. Several ingredients like oregano, coriander, thyme, citric acid and antioxidants were added before the mixture was hydrolyzed by addition of various proteolytic enzyme preparations. After 2 days of storage at 35 to 37°C a good yield of fish sauce was obtained using the commercially available Neutrase as the proteolytic enzyme. The sauce obtained was supposed to be similar to the ancient *garum*, being a thick sauce with equal amounts of lipid and crude protein. The sauce contained about 0.4 and 1%, of eicosapentaenoic acid (EPA) and docosahexaenoic acid (DHA), respectively. The authors conclude that production of *garum* by this rapid method may be a viable possibility of preparing a nutritious food ingredient from fish viscera and underutilized fatty fish species.

Since the late 1980s the possibility of preparing fish sauce from several cold water fish species has been evaluated in Norway. This includes by-products from Atlantic salmon, Atlantic cod viscera, cod waste from the salt fish industry, whole sprat and male Arctic capelin.[20-25] Laboratory and pilot scale experiments reveal that fish sauce similar to Asian fish sauce can be made from these raw materials. However, the flavor is generally too weak unless specific cultures of halophilic or halotolerant bacteria are added. In most cases supplementation with intestines or other raw materials rich in proteases active at neutral conditions is necessary to obtain acceptable recovery of fish sauce. Canadian researchers achieved good quality fish sauce from capelin when some enzyme-rich squid hepatopancreas was added to the raw material.[26-28]

Recently, work has been carried out in the U.S. to explore the possibility of making fish sauce from Pacific whiting by-products in surimi production.[19] A good fish sauce recovery was achieved after only 60 days of fermentation. However, a very weak color of the filtered sauce indicated that further storage would be necessary to obtain a fully ripened product.

After almost 2,000 years the interest for fish sauce production is currently reviving in the Western Hemisphere. Two reasons for this may be the rapid internationalization of food habits and the reduced availability of suitable raw material for fish sauce production in Southeast Asia. Although a substantial commercial production of fish sauce has not yet been established either in Europe or America, this will most probably soon occur.

FUNCTIONAL PROPERTIES

Nutritional Aspects

Nutritionally the Asian fish sauce is more than a mere palatable condiment. Being made from the whole fish by fermentation under stringent conditions of large quantities of salt, the particulate fish tissues slowly dissolve with the aid of a myriad of indigenous enzymes to a clear liquid at the end of the yearlong storage. This resultant liquid is biochemically a concentrated mixture of various free amino acids, oligopeptides, nucleosides and their respective bases. Short-chain organic acids, aldehydes, and esters together with vitamins and minerals are contained in the sauce.[30] A batch of traditionally fermented fish sauce made from anchovy of *Stolephorus* spp. shows the typical concentration of amino acids as summarized in Table 9.2.[10,31]

Amino Acids and Nonamino Acid
N- and C-Compounds

After 10 months of fermentation the total soluble free and peptidic amino acids present in the fish sauce was approximately 10 g per 100 ml. Up to three quarters of the amino acids were present in the free form rendering them readily absorbable without any further digestion in the gut. A portion of the pepidic amino acids, i.e., di- and tripeptides present in the fish sauce are also transported directly into the mucosal cells, while some larger oligopeptides may also play a role as bioactive peptides. Genuine fish sauce contains larger quantities of essential amino acids. Over 40% of the total amino acids present in fish sauce are indispensable amino acids. In a decreasing order they include lysine, 1,319; valine, 575; leucine, 507; histidine, 442; threonine, 470; isoleucine, 423; phenylalanine, 319; and methionine, 274 mg per 100 ml fish sauce. These eight indispensable amino acids are uniformly present in all Asian fish sauces.[30] In the same study, tryptophan and arginine were less consistently present. This may be due to the indigenous microflora conversion to other products or instability during chemical analysis.

TABLE 9.2 Concentrations of Free and Total Amino Acids in a Single Batch of Nampla Made From Anchovies (*Stolephorus* spp.) After 10 Months of Traditional Fermentation

| | Concentration (mg/100 ml) | | Fish muscle (Ref. 10) |
| | Fish sauce (Ref. 31) | | |
	Free amino acid	Total amino acid	Total amino acid
Indispensable			
Histidine	305	442	576
Isoleucine	394	423	800
Leucine	448	507	1472
Lysine	1,000	1,319	1692
Methionine	275	274	432
Phenylalanine	222	319	752
Threonine	510	470	834
Tryptophan	NA	NA	NA
Valine	507	575	928
Subtotal	3,661	4,329 (~43%)	7,486
Dispensable			
Alanine	757	891	1,136
Aspartic acid	698	1,001	1,800
Glutamic acid	1,318	2,051	2,704
Serine	457	440	924
Subtotal	3,230	4,383 (~44%)	6,564
Conditionally indispensable			
Arginine	±	±	1,120
Cysteine	NA	NA	NA
Glycine	366	674	816
Proline	392	544	NA
Tyrosine	54*	97[a]	656
Subtotal	812	1,315 (~13%)	2,592
Total	7,703 (100%)	10,027 (100%)	16,642 (100%)

Note: Figures are expressed as mg per 100 ml. ±: undetectable; NA: not available.
[a] Low solubility

Fundamentally, amino acids are the primary subunits of proteins which dictate their foldings to their respective ultimate functioning structures. This includes contractile proteins, enzymes, storage proteins, transport proteins, protective proteins, cell receptors, peptide hormones, neurotransmitters, and modulators of various physiological processes. It is well documented that deficiency in any of the indispensable amino acids will eventually result in retarding growth and development as well as dysfunctioning either physiologically or mentally or both.

The three most abundant amino acids present in fish sauce, respectively, are glutamic acid, lysine and aspartic acid. Together they comprise about 40% of total amino acids. The valine, leucine and isoleucine group alone made up to 15%. The indispensable amino acids fraction represents almost half of the total amino acids. Lysine is especially rich in fish sauce, being present at 1,319 mg per 100 ml (Table 9.2) or between 10 to 13% of total amino acids in Asian fish sauces.[30] Methionine, however, is present at a relatively lower concentration, i.e., at about 3% of the total amino acid content. Nonetheless, both lysine and methionine in the fish sauce are effectively compensating for the imbalance of cereal proteins in the Asian traditional diet which is consumed by two-thirds of the world's population. An estimate of daily consumption of 15 to 30 ml of fish sauce per person could supply as much as 7.5% of dietary protein in Vietnam.[8]

The various absorbed dietary amino acids and oligopeptides broadly function in the human body through a myriad of cellular and organic functions. Review by Kamiya[32] includes muscle protein maintenance, immune stimulation and potentiation, signal transmission in the brain, tissue repair acceleration after burn or trauma, liver protection from toxic agents, pain relief effects, lowering blood pressure, modulating cholesterol metabolism, stimulating insulin or growth hormone secretion, and detoxifying blood ammonia, among others.

Glutamic acid is one of the most abundant amino acids present in nature, including fish sauce. It can be synthesized within the cell. Hence it is regarded as a dispensable (nonessential) amino acid. Biochemically, glutamine and glutamic acid with proline, histidine, arginine and ornithine constitute the "glutamate family" of amino acids, which are convertible to glutamate. However, under the extreme conditions of major trauma as in major surgery, sepsis, bone marrow transplantation, intense chemotherapy and radiotherapy, where consumption of glutamate and glutamine exceed its synthesis, it becomes a conditional essential amino acid.[33] In the gut and also the placenta, glutamate, glutamine and aspartate together supply up to 70% of protein energy. They are catabolized mainly for energy to CO_2, alanine and lactate. As a result, the absorbed dietary glucose and liver storage of glycogen was spared for the constant requirement in the brain.[34,35] About 20% of dietary glutamate is used for cellular biosynthesis as protein, glutathione, proline and arginine. Only 4% of food glutamate enters the body directly. As a component of glutathione, glutamate indirectly plays a role in antioxidant defense, through the efficient protection against cellular oxidative stresses of glutathione.[36] More directly glutamate and its interconversion to glutamine function as a detoxifying mechanism of ammonia, stabilizing pH of body fluid as well as carrier of nitrogen molecule between organs in the body. As for its role in palatability, glutamic acid or its sodium compound, monosodium glutamate (MSG) is perceived by human taste receptor as the characteristic *umami* or savory taste of foods.[37,38] Hence, monosodium glutamate is commonly used as a food flavoring agent either alone or synergistically with 5-nucleotides such as 5-guanylate (5-GMP) or 5-inosinate (5-IMP) to further potentiate the taste.[39] Thus, it is evident that fish sauce, besides being nutritious, is responsible for the palatability of the dish to which it is added, and is especially potent in the plain cooked rice and vegetable.

Excessive amounts of MSG has been implicated to associate with "Chinese restaurant syndrome" (CRS) in sensitive individuals. However, experiments using large doses of MSG reported inconsistent symptoms and the allegation remains unconfirmed.[40]

Contrasting to the abundance of glutamic acid in nature, lysine, an important and indispensable amino acid is probably most limited in the food chain. Nonetheless its content is second most abundant in fish sauce, at 10 to 13% of total amino acids (Table 9.2). Lysine, together with arginine, is well documented in playing an important role in normal human growth and development, especially in bone metabolism and growth.[41] It is added in diet as a supplement to compensate for the imbalanced amino acid profile of cereals. Besides, lysine together with glycine is reported to play a role in delaying cataractogenesis in aging and diabetic individuals by some yet unknown mechanisms.[42,43]

About half of the total soluble nitrogen found in the fish sauce after 10 months of traditional fermentation is nonaminacid nitrogen.[44] Some of these nonprotein nitrogenous compounds are identified as nucleosides and their corresponding bases, creatine and creatinine, in addition to ammonia. The concentrations of nucleosides together with purine and pyrimidine bases are present up to about 300 mg per 100 ml with roughly one-third being hypoxanthine. The balance is made up by a mixture of, in a decreasing order, uracil, guanine, cytosine, uridine, adenosine, xanthine, cytidine, thymine, thymidine, inosine, guanosine, and adenine, respectively.[30] Both nucleosides and their bases function as essential anabolites for the cellular syntheses of nucleic acids, and various coenzymes and vitamins which are the prerequisites for cell growth and division.

Creatine and creatinine together are present at concentrations about 300 mg per 100 ml.[30] Both substances were present originally in the fish muscle as phosphocreatinine. They were recovered almost entirely in fish sauce after the fermentation. In the human body, urinary creatinine bears a direct relation to the muscle mass of the individual and excess dietary creatine and creatinine are excreted in the urine. Recently the possible contribution of creatine supplementation in enhancing muscular performance, though unsettled, is a topic of active research interest.[45]

Histamine, a decarboxylated product of the amino acid histidine by putrefactive bacterial enzymes, is found to be present in fish sauces at a range of 2 to 76 mg per 100 ml.[46] It is a potent vasodilator and hence, undesirable in food. A maximal daily consumption of 30 ml of fish sauce per person gives from 0.1 to 28 mg of histamine. Since the threshold toxic dose for histamine in man is not precisely known,[47] histamine levels at 200 mg per kg of sample is taken as a low limit for combroid poisoning.[48] Hence, the traditional consumption of fish sauce is not likely to cause vasodilatory problems. Production of fish sauce commonly generates a certain amount of ammonia and carbon dioxide as by-products of proteolysis, peptidyl and amino acid hydrolysis and oxidation. A certain amount of ammonia is formed mainly through deamination, some of which are recycled for synthesis of other amines, polyamines and amides, while an appreciable amount remains as ammonia and ammonium salts in the sauce. These compounds tend to raise the pH of the fish sauce as well as contribute to the pungent flavor and aroma characteristics to fish sauce.[1,2,49]

Fish sauce contains a mixture of short chain organic acids. Lactate and acetate constitute about half of the total short chain fatty acids in the sauce. Succinate, formate, citrate and malate are present in minute quantities while pyroglutamate is a major constituent in this fraction. Especially in *nuoc mam* pyroglutamate where it alone accounts for almost half of the 1.5 g of total organic acids in 100 ml fish sauce, whereas in *nampla, aekjeot* and *shottsuru* it accounts for only one-third of this fraction.[30] Together with other products from Strekker's degradation of amino acids which include aldehydes, alcohols and esters, these low carbon compounds together with the dissolved carbon dioxide contribute to lowering the pH as well as to the cheesy aroma among other desirable attributes of fish sauce.[1,2,49]

Micronutrients

Asian fish sauce is a good source for many micronutrients. Because fish sauce is made from animal protein, it contains vitamin B12. Plant cells do not contain vitamin B12-dependent enzymes therefore they do not produce cobalamine, a more scientific name for vitamin B12. Fish sauce from Thailand contains an average of 1.91 µg per 100 ml of vitamin B12 ($n = 108$). Fermented soybean sauce contains no ($n = 13$) or low concentration of vitamin B12, i.e., 0.14 ($n = 35$) µg per

100 ml.[50] The small amount of cobalamine present in some soybean sauce samples is attributed to microbial synthesis. Fermented fish and fish sauce are part of the traditional diet of Thailand, and their high content of cobalamine is believed to protect the Thai population from megaloblastic anemia, an anemia caused by vitamin B12 deficiency.[50] The estimated average requirement (EAR) for cobalamine is only 2 µg per day,[51] an amount easily met in the traditional Thai diet.

The High Salt Problem

Other than vitamin B12, there are few studies on vitamin and mineral content in fish sauce. Some small amounts of magnesium and iron, at 50 and 3 mg per 100 ml respectively, are found in fish sauce.[52] Traditionally fermented fish sauce contains a high concentration of salt in the final product and this consequently limits its intake. An average of 20.5 ± 2.8 g of sodium chloride was found for 100 ml of fish sauce from seven Asian countries including Thailand ($n = 10$), Vietnam ($n = 20$), Myanmar ($n = 7$), Laos ($n = 2$), China ($n = 2$), South Korea ($n = 9$), and Japan ($n = 11$).[30] An estimated daily consumption of fish sauce per person in Vietnam is between 15 to 30 ml, which gives about 1.2 to 2.4 g sodium. This coincidentally matches the range of minimal sodium requirement[53] and the upper limit of recommended daily sodium intake,[54] respectively. Studies show great variation among countries in daily sodium intakes, in decreasing order, Japanese men rank highest in daily sodium intake at 5.4 g; the United States, Thailand, and New Zealand at 3.9 g; a Polynesian Island, 1.4 g; Amazon jungle, new Guinea highland and Kalahari desert at only 0.69 g.[55,56] Taking into consideration that consumers of fish sauce already obtain adequate amounts of sodium in the portion of fish sauce added in the cooking and at the table, therefore extra sodium including those already present in the food as raw materials or added as other sodium compounds beside fish sauce will undoubtedly contribute to the sodium load of the consumer. Dietary requirement for sodium has not yet been determined, but normal intakes usually provide for more sodium than is needed. Sodium is essentially absorbed completely in the small intestine. From 90% to 98% of ingested sodium is usually excreted mostly in urine with some small and variable amount of sodium being excreted in feces and perspiration as well as nonperspiration losses from skin.[57] While the minimal sodium allowance for adults is 1.25 g per day,[53] Food and Nutrition Board Committee on Diet and Health recommends that daily intakes of sodium be limited to no more than 2.4 g per day.[54] Generally salt is recognized as a basic flavor and plays a role in maintaining the osmolarity of extracellular fluid compartments of the human body. However, consumption of a large amount of salt might be noxious. Case studies limited to industrialized societies have often failed to find a conclusive relationship between blood pressure and sodium excretion.[58] Most individuals can eat large amounts of sodium without remarkably affecting blood pressure because they excrete sodium adequately.[59-61] About 2 to 4 g of sodium are excreted per day by normal individuals.[62] Exceptions are individuals with kidney diseases who tend to retain sodium due to slow excretion of the sodium load. Sodium retention is believed to be important in the pathogenesis of hypertension.[63,64] Among the many initiating causes of hypertension include genetic defects and decrease in renal mass due to surgical intervention.[65] In the well-known *Intersalt Studies* which involved 52 participating centers, 10,000 people from 23 countries, a positive correlation between blood pressure and sodium intake was shown. However, removal of data from the four centers which reported the lowest sodium intake found the correlation to be less significant.[66] Tobian[67] reasons that persons genetically resistant to hypertension outnumber those genetically susceptible, and a correlation between sodium intake and blood pressure is apparent only in those who are susceptible. However, it remains that total amount of sodium and chloride determine the size of the extracellular space in the human body.[68] Within limits, in normal individuals the excess intake of sodium chloride consumed is excreted in urine. The body sodium concentration is regulated by various factors both extrarenal and intrarenal so that blood pressure can be maintained in a balance.[69] It remains crucial to restrict sodium intake only in individuals with hypertension or with renal insufficiency.

Fish Sauce Fortification

As fish sauce is consumed regularly by the majority of Southeast Asian population, regardless of socioeconomic classes, it can be used effectively as a micronutrient fortification vehicle. In Thailand the prevalence of iron deficiency anemia (IDA) and iodine deficiency disorder (IDD) though presently a reduced problem, the vulnerable groups are charted to be remedied with fortifications.[69] Studies on iron fortification of fish sauce and double fortification of fish sauce with iron and iodine have been successfully tested for stability and sensory properties in home usage.[70,71]

Autolytic Protein Digestion

A vast number of biochemical reactions take place during fish sauce fermentation, but the autolytic protein digestion is of premium importance for both recovery and functional properties of the final product. Partly digested protein, peptides and free amino acids yield the major organic component in fish sauce. The content of crude protein may vary in the range 5 to 17% (w/v).[8,72]

The endogenous fish enzymes, particularly the tryptic enzymes, play a major role in fish tissue solubilization.[73] Also microbial enzymes participate, but a significant autolytic capacity of bacterial proteases only occurs if the number of bacteria approaches 10^7 cfu/g.[72] Although the content of certain proteolytic halophiles like *Halobacterium salinarum* sometimes may exceed this level,[74,75] the cfu of fish sauce is normally much lower.[10,76] In special sauce products, like the Indonesian *bakasang*, carbohydrate is added before fermentation. This provides a rather high bacterial activity which may contribute to protein degradation.[77]

The initial step in fish sauce preparation is mixing whole small fish, minced medium-size fish or fish viscera with salt. The large variety of proteases present in the raw material immediately starts digesting tissue and sarcoplasmic proteins.

As a result of this digestion, called autolysis, most of the tissues are solubilized and most proteins are hydrolyzed to peptides and free amino acids. The speed and extent of this process is greatly dependent on a number of factors like kind of raw material, level of enzyme activity, presence of enzyme inhibitors, concentration and quality of salt, pH and temperature. Normally small pelagic species like anchovy or sardine are more easily autolyzed than larger ground fish.[6,8] This may partly be due to a higher level of proteolytic enzymes in the small pelagic species and partly due to a lower content of poorly digestible connective tissues. Autolysis is also greatly influenced by seasonal variations both in the content of hydrolytic enzymes and connective tissues. Logically the content of digestive proteases has a peak level during heavy feeding seasons[78]; experiments with herring larvae showed that secretion of pancreatic enzymes like trypsin stopped after 6 to 8 days of starvation.[79] Strengthening of collagenous connective tissues during starvation periods[80–82] also inhibits solubilization, since native collagen is poorly digestible by most proteases. Hence, small pelagic species caught in or shortly after heavy feeding periods may be the most suitable raw material for fish sauce production, having a high content of digestive proteases and a relatively low content of connective tissues.

Fish muscle and particularly blood contain enzyme inhibitors.[83,84] These are serum proteins which inhibit pancreatic enzymes such as trypsin, chymotrypsin and elastase. The influence of such inhibitors during fish sauce fermentation is uncertain, but most probably they reduce the activity of tryptic enzymes at least at the initial fermentation stage.[73] When fish viscera is used as a raw material, it may be wise to drain off the blood, since this contains about 40 times higher concentration of the serum inhibitors than muscle tissues.[83] The autolytic protease activity is also significantly inhibited by the high salt concentration in fish sauce,[20,24,73,85] and this is probably the main reason why tissue solubilization during fish sauce fermentation proceeds quite slowly. The autolysis is influenced not only by salt concentration, but also by the purity of the salt employed. Traditionally solar dried sea salt is most frequently used and apparently the best autolysis is obtained with essentially pure NaCl-preparations (cited in Reference 5).[86]

Normally fish sauce fermentation proceeds under weak acid conditions. Actually this coincides with the minimum range for protein digestion by fish digestive enzymes. Although both pepsins and trypsins have the ability of digesting fish proteins at such conditions,[87] it is evident that tryptic digestion is most important since pepsins are very unstable under weak acid conditions and also more efficiently inhibited by salt than the trypsins.[20,88]

In Southeast Asia fish sauce fermentation takes place under ambient tropical temperatures (about 30°C). This is convenient since most endogenous enzymes in tropical fish are very active and quite stable at such temperatures. Enzymes in cold water species are less thermostable and more active at lower temperatures, but still fermentation temperatures as high as about 25°C seem to give the best recovery.[25]

Being the favorite raw material for fish sauce production, particular attention has been paid to anchovies. It has been shown that trypsin and chymotrypsin are major autolytic agents, but also elastase and aminopeptidases are important particularly during late stages of digestion.[89] In addition to the pancreatic trypsins, tryptic enzymes active at neutral pH have been detected in anchovy muscle.[90] Apparently, trypsin is most important during the initial stage, splitting myofibrillar proteins at the carbonyl side of lysine and arginine.[91] Successively, the role of the less specific chymotrypsin, splitting at the carbonyl side of aromatic amino acids as well as leucine, becomes more important.[92] Generally chymotrypsin seems to be more stable than trypsin during long-term storage under high salt concentrations.[22] Peptides with a hydrophobic amino end, occurring from chymotrypsin digestion, often have interesting functional properties.

Although the serine proteases trypsin and chymotrypsin play a key role in protein digestion during fish sauce fermentation, a great number of endogenous proteinases may contribute to various extents. These include muscle cathepsins active at neutral and weak acid conditions,[90,93,94] the calcium-activated calpains,[95] high molecular weight multicatalytic fish muscle proteinases,[96–99] as well as elastases[100] and even collagenolytic enzymes which may be present in the digestive tract of certain carnivorous fish species.[101] Finally the digestive exopeptidases play an important role in releasing considerable amounts of free amino acids.[89,102]

Normally the degree of protein hydrolysis in mature fish sauce is in the range 0.4 to 0.7.[8,25,72,103] Accordingly, the fish sauce protein is highly, although not extensively, hydrolyzed. Ultrafiltration experiments indicated that almost half of the crude protein in Thai fish sauce made from anchovy included small- and medium-size peptides ⟨500 to 3000 d⟩,[104] whereas Park et al.[30,105] reported that only 20% of the crude protein was peptides and almost 70% free amino acids. Bacterial activity during fermentation may cause transformation of amino acids from the native L form to D form, but the rate of such transformation is low in fish sauce with a high salt concentration (>20%).[106] In addition to common amino acids fish sauce also contains significant amounts of taurine which has some interesting functional properties.[30] Furthermore, fish sauce is a treasure of bioactive peptides, in addition to being highly nutritious and a rich source of essential amino acids.

Bioactive Peptides

Although several splendid biological properties have been attributed to fish sauce since ancient times,[18] little of this has been verified through scientific research. However, since fish sauce contains a complex mixture of peptides, it may contain a number of peptides with interesting biological activity.

Bioactive peptides may principally occur in two very different manners. In living organisms bioactive peptides are released as a result of the activation of regulatory enzymes which split off specific peptides from specific proteins. In this case the peptides have predestined biological functions. However, identical peptides may also occur, more or less randomly, as a result of autolytic digestion by various proteolytic enzymes acting on a mixture of protein. Obviously, the probability of obtaining at least small bioactive peptides as a result of autolysis is significant, and in some cases even di- and tripeptides may express biological activities.[107–109]

Research on bioactive peptides in fish sauce is very scarce, but there are several reports on the investigation of various biological activities in other fish protein hydrolyzates which have relevance also to fish sauce. A major part of the fish protein hydrolyzate produced worldwide is used for animal or fish feed. Hence, it has been of particular interest to investigate the content of immunostimulating peptides in such products. *In vitro* experiments have shown that acid peptide fractions from a cod stomach hydrolyzate strongly stimulated respiratory burst reactions in salmon leucocytes,[110] whereas *in vivo* stimulation of salmon leucocytes was achieved with medium-size peptides (500 to 3000 Da) from a cod muscle hydrolyzate.[111] Vinot et al.[112] showed that low concentration (4 µg/ml) of acid peptide fractions from a fish protein hydrolyzate strongly stimulated the proliferation of mouse lymphocytes during *in vitro* cultivation.

Although little is known about bioactive peptides in fish sauce, one *in vitro* experiment with human monocytes revealed some proliferation stimulation by addition of peptides from fish sauce. Different concentrations of medium-size peptides (500-3000 Da) isolated by ultrafiltration of commercial fish sauce made from anchovy were added to the cell cultures. Some proliferation stimulation was achieved when 5 µg/ml was added, whereas no effect was observed either at lower or higher peptide concentrations (Figure 9.1). This result underlines the delicate dose/response relationship existing with many immunostimulatory substances.[113]

Anticarcinogenic activity has been atributed to some small peptides like the tetrapeptide tuftsin (Thr-Lys-Pro-Arg). *In vivo* studies with mice, infected intraperitoneally with leukemia cells, showed improved survival with the mice also receiving intraperitoneal injections with tuftsin.[114] It was indicated that tuftsin stimulated cytotoxicity in both NK cells, macrophages and granulocytes.

Studies on enzymatic fish protein hydrolyzates of various sources showed that whereas the high molecular weight peptide fractions seemed to contain cell growth factor activities,

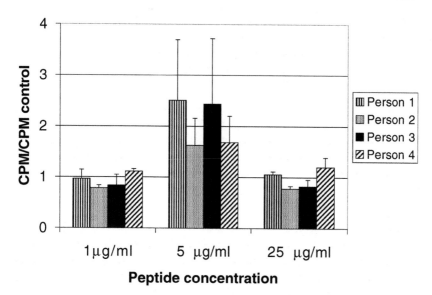

Figure 9.1 Relative values for *in vitro* proliferation of human monocytes after addition of three different concentrations of fish sauce peptides. Medium-size peptides (500 to 3000 Da) separated by ultrafiltration of a commercial Thai fish sauce made from anchovy were added to the cultures. Cell proliferation was determined as [³H]thymidine incorporation, and the results are given as cell proliferation in cultures with fish sauce peptides added compared to cell proliferation in control cultures. (From Steiro and Gildberg, unpublished results.)

the low molecular weight fraction often showed secretagogue activities like gastrin and cholecystokinin.[115] Peptides inhibiting the activity of angiotensin I converting enzyme (ACE) have been detected in a number of fish protein hydrolyzates of various origin including sardine, bonito viscera, shrimp waste and cod head.[109,116,117] Such peptides may reduce hypertension and high blood pressure. Particularly tripeptides with a basic amino acid in the middle, and proline in the carboxyl terminal position, were found to be strong ACE inhibitors.[109] Obviously such peptides may frequently occur by extensive enzymatic digestion of connective tissue proteins like collagen which has a high content of proline as well as arginine and lysine. Also opoid-like activities have been detected in certain fractions of protein hydrolyzates made from shrimp waste, cod head and sardine.[117]

As mentioned earlier, trypsin digestion yields peptides with either lysine or arginine at the carboxyl terminal position, whereas chymotrypsin yields peptides with hydrophobic carboxyl terminal residues (normally tyrosine or phenylalanine). Often such peptides may express biological activities. Studies on milk proteins have shown that chymotrypsin digestion may produce small immunostimulating peptides, whereas trypsin releases low molecular weight ACE inhibitory peptides.[108] Also the anticarcinogenic tetrapeptide tuftsin has a carboxyl terminal arginine and may be released from a protein by trypsin digestion.[114] Since trypsin and chymotrypsin apparently are the most active proteases during fish sauce fermentation, similar or identical peptides may very well be present in fish sauce. Park et al.[105] recently isolated several dipeptides from fish sauce with either phenylalanine or tyrosine at carboxyl terminal position.

Considerable attention has recently been paid to the health benefits derived from the presence of natural antioxidants in foods. This is particularly related to antioxidative protection against coronary heart diseases and cancer.[118,119] Some small- or medium-size peptides may also provide antioxidative capacity. Hatate et al.[120] showed that a pepsin hydrolyzate of sardine myofibrillar protein acted synergistically with commercial antioxidants inhibiting oxidation of linoleic acid. Medium-size peptides from a pollack skin hydrolysate reduced the speed of lipid oxidation significantly and also acted synergistically with α-tocopherol.[121] Two antioxidative peptides (13 and 16 amino acid residues) were isolated, both with the carboxyl terminal sequence Gly-Pro-Hyp.

Obviously fish sauce made from whole fish or fish waste fractions must contain considerable amounts of hydrolyzed collagen/gelatine which may yield antioxidative peptides. It has also been suggested that oral administration of gelatine hydrolyzates may enhance wound healing and have beneficial effects on degenerative diseases of the musculo-skeletal system.[122]

Undoubtedly fish sauces possess many compounds, including peptides, with biological activities. However, the great variety of raw materials and fermentation conditions employed certainly yield unique final products, not only different in nutritional quality, but also inhabiting a variable spectrum of bioactive substances.

Microbial Activity

Halophilic Bacteria

In the traditional fish sauce fermentation process, the initial salting of fresh small fishes has adverse effects on most endogenous microorganisms in the starting materials. The non-halophiles die off rapidly during the first few days, while the halotolerants stay for weeks and the salt-loving halophiles stay for months. The spore-forming bacilli, though low in number remain the longest. Through natural selection a variety of common species have adapted to high salt environment and the conditions in the fermentation vats, i.e., pH 5–6, 29 to 32°C and saturated NaCl concentration, hence, they are found to grow at various points during the fermentation.[74] Among the eubacteria include *Micrococcus, Bacillus, Vibrio, Achromobacter, Flavobacteriurn,*[9,77,123] *Pseudomonas, Coryneform, Sarcina, Lactobacillus, Streptococcus.*[124] *Halobacterium* and *Halococcus* species are identified among the Archaea in fermenting Thai fish sauce.[75,125] Halophilic lactic acid-producing bacteria are of quantitative significance culturable at 10^7 CFU per ml[126] and *Halobacterium salinarium* at >10^8 CFU per ml[74] during the first month of fermentation. Yeast and fungi are not found to be of any quantitative significance, mostly they are found below 10^3 CFU per ml.[74]

Proteolytic Activity

Most eubacteria and archaea isolated from specimens of fermenting fish sauces and residues exhibit gelatinase and protease activities.[125] Of the strong halophiles are *Halobacterium, Halococcus, Bacillus* and *Coryneform groups*, respectively. The protease-producing halotolerants include *Micrococcus, Staphylococcus, Streptococcus, Proteus* and *Pseudomonas*.[127] A moderately halophilic spore-former is identified as *Halobacillus thailandensis* sp. nov. from fish sauce and found to highly express three extracellular proteolytic enzymes, two serine proteases and one metalloprotease.[128] The organism exhibits high similarities to *Halobacillus litoralis* and *Halobacillus halophilus* with respect to their 16 s rRNA sequences.

Since autolytic fish enzyme solubilization is well recognized to be the principal proteolytic activity in traditional fish sauce fermentation process, microbial proteolytic activities as evidenced from various eubacterial and archaeal isolates are believed to play a complementary minor role apart from flavor formation. However, the salt-resistant proteases should be explored to reach their potentials.

A pertinent study on protease activities in a single vat of traditional process of fish sauce fermentation revealed that the bulk of protease activity was present initially in the liquid while only 15% of the protease activity was associated with the fish particulates.[129] With respect to salt stability, while most liquid-associated protease activities were NaCl-stable for up to 25%, the opposite was found for fish residue-associated protease activities. Such proteolytic activities rapidly disappeared after the first week of fermentation and only some residual activities remained detectable.[129] Thus, it appears that the large amount of autolytic fish solubilizing activities are inactivated soon after heavy mixing of salt with fish prior to vat-packing of the mixture. Consequently, the salt-stable proteolytic activities are those contributed by microbial load of the system. Analysis of free amino acids and total amino acid in a time-course release in a fish sauce fermentation process suggests very active exopeptidases in good cooperation with endopeptidases at the early stage, from the beginning to the fifth week, while at later stages only endopeptidases remain active.[44] Together, it is most likely that once the salted fish mixture is immersed in brine, the halophilic microbial proteases dominate the slow and lingering release of oligopeptides and amino acids in the later stage of fermentation. Since high salt concentration is of utmost necessity in fish sauce fermentation in order to maintain safe products[130] with good aroma, then appropriate halophile inocula should be explored and used for industrial production.

Lipolytic Activity

Rapid lipid hydrolysis is well known to occur in the flesh of whole fish, such as herring and sardine kept without refrigeration. The enzymes are present in the flesh itself as well as in the visceral tissues. Canadian studies[131] on lipid hydrolysis in cod flesh during both light and heavy salt curing show complete inhibition of hydrolysis when lipids contain about 50% free fatty acids. The cessation of hydrolysis is due to enzyme inhibition by salt or inhibition by reaction products, or by both mechanisms.

As in proteolytic enzymes, lipolytic enzymes existing in fishes themselves presumably work first in fish sauce fermentation. In an experiment using individual extracts of marine fish stomach, intestine, pyloric caeca and liver, it was found that all fraction gave positive lipase activities. Pyloric caeca exhibited the highest lipolytic activities[132] and coincidentally, the highest protease activities. Furthermore, this lipolytic activity was salt-stable for up to at least 15%.[132] Hydrolysis of fats by lipases yields both fatty acids and glycerol. Since small amounts of short- and medium-chain fatty acids generally give good food flavor good aroma and glycerol is useful for aroma, it is therefore certain that the action of lipase is useful for both flavor and aroma in fish sauce.

Lipase producing microorganisms are less studied. Lipolytic activities are most frequently found in nonhalophilic bacteria of the genera Bacillus, Coryneform and Staphylococcus. Lipases are less prevalently encountered in halotolerant or moderately halophilic (10% NaCl), respectively, *Staphylococcus, Coryneform,* and some *Micrococcus* and *Bacillus.* At 25% NaCl in the isolation conditions, only *Halococcus* and *Halobacterium* were lipase producers.[127] Other studies using microaerobic conditions to isolate lipase producers reported genus *Clostridium* spp. which can grow at high salt concentrations (15% up). In a minimal medium incorporated with anchovy fish fat, volatile fatty acids including acetic acid, propionic acid, butyric acid and isovaleric acid were produced, all of which were found to be present in commercial fish sauces. Interestingly the organisms involved do not produce proteolytic enzyme but can digest starch.[133] This same genus studied by Beddows in Vietnamese fish sauce is associated with volatile acids as well as protease production.[134]

Although the content of amino acids and beneficial oligopeptides carry the weight for the nutritional quality of fish sauces, the flavor and aroma as determined by the complex blend of volatile acids, as well as numerous other character impact components in the source, will ultimately exert a strong influence on consumers' choice of fish sauce. Further research in this area of lipases, especially the origins as well as their role in the generation of fish sauce flavor and aroma, is needed.

Microbial Metabolites

The endogenous microbial varieties and loads are the natural input of eubacteria and archaea in the traditional fish sauce fermentation system. The salt-saturated brine also contributes to the initial populations in the vat fermentation. However, it remains that the physico-chemical conditions encompassing the fishes, salt and brine will ultimately be the selective factors. Temperature, salt concentrations and oxygen availability or its absence bear major influences on the selection of the surviving microbes. As time proceeds, some micronutrients originally available in the vat become exhausted and salts concentrations reach an equilibrium. This situation together with the surplus amount of enzyme end products (including the consequent changes in the pH) makes almost all selected and thriving eubacteria and archaea cease to grow, leaving only a few spore formers. At the end of a well-fermented batch, the microbial populations will have almost entirely sterilized themselves out from the fish sauce.[74] At this stage the increased concentrations of all biochemical constituents in the fish sauce will cause the NaCl to crystallize as characterized by the appreciable drop of NaCl concentration in the fermented liquid. This drop of NaCl is also accompanied by the appearance of salt crystals at the bottom of the vat or hardened salt plates covering the surface. In certain experimental batches of exogenous enzyme-added fermentation, white crystals of tyrosine are observed, due to its low solubility.[135] After filtration through the layers of fish residues the clear liquid filtrate is the nutritious concentrate of a mixture of whole fish-fermented hydrolyzate. A product with unique flavor and aroma which is abundant in free amino acids and oligopeptides and rich in minerals and vitamins is obtained. The amino acid profiles of various fish sauces differ noticeably with respect to species of fishes while the flavor and aroma vary even more with respect to the conditions of fermentation, both before and after brine immersion of the fish and salt mixtures. Lopetcharat and coworkers[6] have compiled a comprehensive review on various Asian fish sauces. Representative amino acid profiles of fish sauces from China, Korea, the Philippines, Thailand and Vietnam are presented.[6]

Fish Sauce and Nitroso Compounds

The global cancer statistics in the year 2000 documents the top five type of most common new cases as carcinomas of lung, breast, colorectal, stomach and liver, at 1.2, 1.0, 0.9, 0.8 and 0.6 million cases, respectively.[136] The profiles in various populations differ greatly. Evidences suggest that variations due to different lifestyles and environmental factors are amenable to preventive interventions. It is generally recognized that a variety of exogenous and endogenous factors inclusive of dietary, chronic microbial infections, chemical, radiological as well as certain genetic deficiencies, to a varying extent are associated with carcinogenicity in man. Diet-associated factors are estimated to account for about 30% of cancers in developed countries. Obesity increases the risk of many types of cancers involving esophagus, colorectum, breast, endometrium and kidney, while alcohol does similarly for cancers involving oral cavity, pharynx, larynx, esophagus, liver and breast.[137]

Special attention on dietary components is emphasized on *in vivo* nitrosation of various food components which may give rise to mutagens, carcinogens and their precursors. A large number of foods and foodstuffs including fish sauce have been analyzed for formation of potential mutagens and carcinogens as well as their related precursors, both as preformed volatile nitroso compounds in the foods or as nitroso and related compounds after nitrite treatment.

In a large case-control study in Chaoshan, the area of high risk in South China for cancers involving narsopharyngeal, thyroid and stomach, it was concluded that consumption of Yulu, a Chinese fish sauce, substantially increased risk associated with esophageal cancer.[138] Esophageal carcinoma is an uncommon malignancy except in China and Japan with a higher incidence.[139] Among other factors, the risk of squamous carcinoma associates with tobacco and alcohol separately and also with synergism between them. Earlier studies using animal experiments show less conclusive results. While administration of homemade yulu[140] proved to be cocarcinogenic in the forestomach of mice and esophageal epithelium of rats, administration of commercial yulu did not render similar carcinogenic effects.[140] After nitrosation treatment, a nitroso compound in fish sauce was characterized as N-nitrosomethyl urea (NNU), however, no quantitation was reported.[141] Low concentrations of the relatively less potent compound, *N*-nitrosodimethylamine (NDMA) were reported in Thai fish sauce, Chinese fish sauce and soy sauce. Thai fish sauce contained an average of 0.1 ppb ($n = 10$, range 0.0 to 1.1) of NDMA while *N*-nitrosopiperidine (NPIP) and *N*-nitrosopyrrolidine (NPYR) were not detectable.[142] Chinese soy sauce contained a relatively higher concentration of NDMA than fish sauce, i.e., at 0.5 ppb ($n = 74$, range 0.0 to 6.7 ppb) and 0.1 ($n = 10$, range 0.0 to 0.2 ppb) respectively.[143]

Generally nitrite concentration in human saliva and gastric juice does not exceed 50 ppm, therefore the amount of nitroso compound generated from salivery nitrite and nitroso compound precursors in the diet taken can not be very large. Hence, it is unlikely that fish sauce and soy sauce, which are normally taken in small quantities, will contribute significantly to mutagenicity *in vivo*.[144] Due to the low concentration of detectable nitroso compounds in fish sauce and the limited quantity of intake, the health risk from fish sauce consumption should be minimal. Beside fish sauce, voluminous literature have accumulated concerning the presence of NDMA and related compounds in other foods, including beers,[145] cured meat products (cited in Reference 143),[146] marine foods (cited in Reference 143),[147] pickled vegetables (cited in References 143 and 149),[148,149] and milk products (cited in Reference 143).[150] Interested readers should consult Appendix I in Shephard et al.[151]

However, Chinese in South China have a high risk of narsopharyngeal, esopharyngeal and gastric carcinomas owing to several predisposing factors such as heavy tobacco smoking and habitual very hot tea and hot soup consumption. These and other factors are recognized as being etiologically significant for the above-mentioned cancers.[152–154]

Two lines of evidence offer protective effect against harmful dietary nitroso compounds formation. Incorporation of 200 to 2,000 ppm of ascorbic acid or vitamin C, which is present abundantly in citrus fruits and fresh vegetables, into the fish sauce and other foods prior to nitrosation inhibits the formation of N-nitrosomethyl urea formation.[155] Beside ascorbic acid, proteins and amino acids when present in high concentrations, as it occurs in the stomach, effectively scavenge nitrite and thus inhibit the formation of mutagens and carcinogens from nitrite.[156] Reactions include conversion of proline and cysteine to nonmutagenic nitrosoproline and S-nitrosocysteine, respectively. About 50% of nitrosodimethylamine formation are inhibited by most amino acids in the reaction between 200 mM dimethylamine and 50 mM nitrite at pH 3 and 37°C. From the above, it is logical that Western diets emphasize consumption of fresh vegetables, as in salads, and protein in main dishes while Eastern, especially Thai traditional diets, emphasize the use of fresh vegetables and lime juice in moderate protein mixes as in "yum" (hot and sour dishes) of all kinds.

ACCELERATING FISH SAUCE FERMENTATION

The traditional recipe for fish sauce fermentation recommends a 6-to-12-month fermentation period to achieve satisfactory yield and sensory properties.[8] The long production time demands considerable capital investment in storage tank capacity. Many strategies for reducing the production time have been evaluated. Most of the methods suggested may improve the recovery rate significantly, but an equivalent acceleration of the flavor development still remains a challenging task.

Since endogenous proteolytic enzymes are essential for tissue solubilization, it is important that the raw material used for fish sauce production has a satisfactory level of such enzymes. Most of the fish sauce is made from whole small pelagic species, and the content of digestive enzymes at the time of capture is of vital importance for autolytic tissue solubilization. Generally fish caught in a feeding period have the highest content of digestive enzymes.[78,157] Fish caught in a nonfeeding season may also be used for fish sauce production, but then some enzyme-rich raw material should be added to accelerate tissue degradation. Raksakulthai and Haard[28] supplemented male Arctic capelin caught in the low feeding season with 2.5% enzyme-rich squid (*Illex illecebrosus*) hepatopancreas and recovered a high-quality fish sauce after 6 months of storage at room temperature. Cathepsin C, which has a specificity similar to chymotrypsin, was the major protease in the squid pancreas. Similar results were obtained by addition of cod pylorus (10%) rich in trypsin and chymotrypsin.[25,158] Even waste from heavily salted cod (bachalao) may be used for fish sauce production if cod intestines and a culture of *Halobacterium salinarum* is added.[23] However, in all these cases a 6-month storage period was necessary to obtain acceptable sensory properties.

As described previously, the autolytic conditions during traditional fish sauce fermentation are far from optimal. The weak acid conditions coincide with the minimal activity level of the combined mixture of digestive proteases,[20,89] and the high salt concentration strongly inhibits most of the proteolytic enzymes.[20,73,90,95] Hence, several reports have proposed solutions to this problem, and most of them imply an initial incubation at low salt concentration. In a method given by Gildberg et al.[159] the pH of lightly salted anchovy was reduced to 4 by addition of hydrochloric acid and the mixture was incubated 5 days before it was neutralized by sodium hydroxide and salt was added to a final concentration of 25%. This provided favorable conditions for pepsin digestion, and an acceptable fish sauce was obtained after only 2 months of storage. By incubation of anchovy at very acid initial conditions (pH 2-3), a good recovery was achieved after 1 week, but this product had very little aroma and taste[160].

Also initial incubation at low salt and alkaline conditions has been recommended. The advantage with this procedure is that it favors both stability and activity of the tryptic enzymes which are key enzymes during traditional fish sauce fermentation.[73] Improved recovery was achieved when salmon filleting waste was stored for 2 days at pH 8.7 and low salt concentration before neutralization and further salt addition.[20] Addition of histidine (0.5 to 2%) accelerated the autolysis during sardine sauce fermentation.[161] This was probably due to the weak initial alkali provided by supplementation of the basic amino acid. Similar results were obtained by initial adjustment to pH 8 with minced capelin and cod pylorus as raw materials.[25] In both latter cases the pH of the final products became similar to untreated controls without further pH adjustment. Yoshinaka et al.[162] described a rapid production method where an extract of sardine viscera was mixed with minced sardine muscle and incubated under optimal conditions (pH 8 and 50°C) for 5 hours. After incubation the mixture was centrifuged and the clear supernatant preserved by addition of 25% salt. The sensory properties of this rapid fish sauce were rated equally good as for commercial Japanese fish sauce (*shottsuru*).

Not only pH and salt concentration, but also the storage temperature is of vital importance for quick autolysis. Within the stability range the activity of an enzyme is approximately doubled if the temperature is raised by 10°C. When fish sauce is made from tropical fish species with fairly thermostable digestive proteases, the tissue solubilization may be accelerated significantly by increasing the storage temperature.[87] In a pilot experiment with round scad *(Decapterus aecrosoma)* a good quality fish sauce was recovered after 2 months, if the fish was stored first for 50 days at ambient tropical temperatures and then aerated and stored further for 10 days at 45 to 50°C.[163]

Several groups have investigated the effect of supplementing commercial proteases of plant or microbial origin to speed up the fermentation process.[19,26,103,160,164] In most cases where external enzymes are used, the product recovery is good, whereas the sensory properties are inferior to the properties of traditional fish sauce. The best results were achieved with bromelain, a protease extracted from pineapple waste. The reason for this may be that the specificity of bromelain is similar to both trypsin and chymotrypsin which are the most important proteases during natural fish sauce fermentation.[73,165] Most likely several of the methods mentioned above have been adopted in commercial fish sauce production, but details about production procedures are normally not disclosed by the industry.

FUTURE POTENTIAL FOR FISH SAUCE

Whether fish sauce was first developed in Asia or Europe remains uncertain. A more interesting question is why the fish sauce tradition vanished in Europe, but became a thriving industry in Southeast Asia. Lee[11] explains that there may be a connection between the extensive use of bland tasting rice as a staple food in East Asia and the need for some salty and tasty protein-rich food supplement. In the Southern region fish was the most convenient raw material for such a product, whereas soy sauce, first developed in Japan, became more important in the North.[166]

During the second half of the 20th century soy sauce has become a popular food item in every corner of the globe,[166] whereas the fish sauce until recently has been looked upon as an exotic product. The main reason for this is probably the ease of soybean cultivation, and hence production of soy sauce in many parts of the world. Fish sauce is produced from a wide variety of fish with less predictable availability and sometimes doubtful microbial quality documented by a too high level of histamine.[6,72] Due to this, fish sauce is still considered as an unsafe product by many people in the industrialized countries. At present, however, fish sauce is also gaining popularity in the Western world. Due to the high number of Asian immigrants and the greatly expanded tourism to Southeast Asia, more people have acknowledged fish sauce as an interesting new food experience which is easily adaptable as an alternative to soy sauce in many dishes.[72]

To achieve a better image for fish sauce worldwide in the future it is necessary to put more emphasis on factors like more predictable availability and better traceability of the raw materials used, improved hygiene and microbial security during production and a better labeling and declaration of content on the final product. Most recommendations mentioned above may be fulfilled by improving raw material logistics, transportation, and production facilities, however, the availability of fish for an expanding market is a great challenge. During the last decades the annual world fish catch has culminated and most probably any increase in future production will come from aquaculture.[167] However, to a great extent aquaculture is also based on feed ingredients of fish origin, and hence, a competitor for its use in other applications.[168]

Generally various species of anchovy are the preferred raw material for fish sauce production. At present the availability of such fish is limited in Southeast Asia, but at the Pacific coast of South America huge quantities of such fish are caught and processed to fish meal and oil. Provided that fish sauce production offers a better feasibility than fish meal production, it should be possible to establish production based on at least a small portion of this raw material. Other small pelagic species available for such production in the Western Hemisphere may be Arctic capelin, blue whiting, and sprat.[24,25,27,28]

Increasing amounts of by-products occur as a result of more extensive industrial fish processing. By-products account for as much as 60% of the fish weight during production of white fish fillets.[167] It has been shown that fish sauce can be produced from various filleting wastes from Atlantic cod and Atlantic salmon and from by-products obtained during surimi production of Pacific whiting.[20,22,29] Waste fractions from salt fish industry are of particular interest, since such material is not suitable for animal feed production. Pedersen and Skjerdal[23] found that fish sauce made from salted cod fish waste had a bland taste, but after addition of a small amount of Asian fish sauce as a microbial inoculum a pleasant flavor developed after a few months of further storage.

Fish sauce made from special by-product fractions are popular in certain population groups. In Korea fish sauce made from cod gills, yielding about 2.5% of the fish weight, is highly valued (Cherl Ho Lee, personal communication), whereas fish milt is an attractive raw material for various hydrolyzed products in Japan.[169]

It is likely that high-quality fish sauce can be made from many nontraditional raw materials if suitable enzymes and bacterial cultures are added. In addition to by-catch and by-products from the fisheries sector, the prosperous fish aquaculture industry may provide predictable quantities of high-quality raw materials for sauce production. Apparently the initial establishment of such production is now taking place both in Europe and North America.[19,25,27–29]

REFERENCES

1. J Dougan, GE Howard. Some flavouring constituents of fermented fish sauce. *J Sci Food Agric* 26:887–894, 1975.

2. V Suvanich, I Maciel-Pedrote, KR Cadwallader, EB Moser, C Thongthai and W Prinyawiwatkul. Aroma Indices of Southeast Asian Fish Sauce. Abstracts of International Food Technology Annual Meeting and Food Expo, New Orleans, 2001b, 73F16.

3. J Worapong. Roles of Microorganism in the Traditional Process of Fish Sauce Fermentation: Flavor and Aroma. MS dissertation, Mahidol University, Bangkok, 1995.

4. S Areekul, R Thearawibul, D Matrakul. Vitamin B$_{12}$ content in fermented fish sauce and soybean sauce. *Southeast Asian J Trop Med Public Health* 5:461, 1974.

5. TM Mackie, R Hardy, G Hobbs. Fermented fish product FAO. *Fisheries Reports,* No. 100, Rome, 1971, pp. 1–54.

6. K Lopetcharat, YJ Choi, JW Park, MA Daeschel. Fish sauce products and manufacturing: a review. *Food Rev Internat* 17:65–88, 2001.

7. MT Tyn. Traditional and modified methods of fish sauce production in Myanmar. In: PJA Reilly, RWH Parry, LE Barile, Eds., *Post-Harvest Technology, Preservation and Quality of Fish in Southeast Asia.* Manila, Echanis Press, 1990, pp. 15–19.

8. K Amano. The influence of fermentation on nutritive value of fish with special reference to fermented fish products of South-East Asia. In: E Heen, R Kreuzer, Eds., *Fish in Nutrition.* London: Fishing News Books, 1962, pp. 180–197.

9. P Saisithi, B Kasemsarn, J Liston, AM Dollar. Microbiology and chemistry of fermented fish. *J Food Sci* 31: 105–110, 1966.

10. P Saisithi. Traditional fermented fish: fish sauce production. In: AM Martin, Ed. *Fisheries Processing: Biotechnological Applications.* London: Chapman & Hall, 1994, pp. 111–131.

11. CH Lee. Fish fermentation technology — review. In: PJA Reilly, RWH Parry, LE Barile, Eds. *Post-Harvest Technology, Preservation and Quality of Fish in Southeast Asia.* Manila: Echanis Press, 1990, pp. 1–13.

12. MR Adams, RD Cook, P Rattagool. Fermented fish products of Southeast Asia. *Trop Sci* 25:61-73, 1985.

13. M Sundahakul, W Daengsubha, P Suyanandana. A brief description of Thailand's traditional fermented food products. *Thai J Agric Sci* 8:205–219, 1975.

14. Department of Fisheries, Ministry of Agriculture and Cooperative, Bangkok, Thailand, Rep No 4, 2001, pp. 1–39.

15. G Stefansson, U Steingrimsdottir. Application of enzymes for fish processing in Iceland — present and future aspects. In: MN Voigt, JR Botta, Eds. *Advances in Fisheries Technology and Biotechnology for Increased Profitability.* Lancaster, Technomic Publ Comp Inc, 1990, pp. 237–250.

16. E Rose. Recherches sur la fabrication et la composition chimique du 'nuoc-mam.' *Bull econ, Indochine* NS, 20 (129):155, 1918a.

17. E Rose. Le 'nuoc-mams' du nord (Nord Centre-Annam et Tonkin) composition chimique et fabrication. *Bull econ Indochine*, 20 (132):955, 1918b.

18. TH Corcoran. Roman fish sauces. *Classical J* 58:204–210, 1963.

19. Y Aquerreta, I Astiasaran, J Bello. Use of exogenous enzymes to elaborate the Roman fish sauce garum. *J Sci Food Agric* 82:107–112, 2002.

20. A Gildberg. Accelerated fish sauce fermentation by initial alkalification at low salt concentration. In: S Miyachi, I Karube, Y Ishida, Eds. *Current Topics in Marine Biotechnology.* Tokyo: Fuji Technology Press, 1989, pp. 101-104.

21. A Gildberg. Recovery of proteinases and protein hydrolysates from fish viscera. *Bioresource Technol* 39:271–276, 1992.

22. A Gildberg, S Xian-Quan. Recovery of tryptic enzymes from fish sauce. *Process Biochem* 29:151–155, 1994.

23. G Pedersen, T Skjerdal. Utilisation of Waste from the Salt Fish Industry in Production of Fermented Fish Sauce and Other Possible Products. Proceedings of 29th WEFTA Meeting, Thessaloniki, 2000, pp. 392–394.

24. A Gildberg, C Thongthai. The effect of reduced salt content and addition of halophilic lactic acid bacteria on quality and composition of fish sauce made from sprat. *J Aquatic Food Product Technol* 10:77–87, 2001.

25. A Gildberg. Utilisation of male arctic capelin and Atlantic cod intestines for fish sauce production – evaluation of fermentation conditions. *Bioresource Technol* 76:119–123, 2001.

26. N Raksakulthai, YZ Lee, NF Haard. Effect of enzyme supplements on the production of fish sauce from male capelin *(Mallotus villosus)*. *Can Inst Food Sci Technol J* 19:28–33, 1986.

27. N Raksakulthai, NF Haard. Correlation between the concentration of peptides and amino acids and the flavour of fish sauce. *ASEAN Food J* 7:86–90, 1992a.

28. N Raksakulthai, NF Haard. Fish sauce from capelin *(Mallotus villosus)*: Contribution of cathepsin C to the fermentation. *ASEAN Food J* 7:147–151, 1992b.

29. K Lopetcharat, JW Park. Characteristics of fish sauce made from Pacific whiting and surimi by-products during fermentation stage. *Food Chem Toxicol* 67:511–516, 2002.

30. JN Park, Y Fukumoto, E Fujita, T Tanaka, T Washio, S Otsuka, T Shimizu, K Watanabe, H Abe. Chemical composition of fish sauces produced in Southeast and East Asian countries. *J Food Comp Analys* 14:113–125, 2001.

31. C Thongthai and the late H Okada. Unpublished results.

32. T Kamiya. Biological functions and health benefits of amino acids. *Food Ferm J* Japan 206:33–44, 2002.

33. H Tapiero, G Mathe, P Couvreur, KD Tew. The metabolic basis of arginine nutrition and pharmacotherapy: II Glutamine and Glutamate. *Biomed Pharmacotherapy* 56:446–457, 2002.

34. PJ Reed, DG Burrin, B Stool, Johoor. Intestinal glutamate metabolism. *J Nutr* 130:978S–982S, 2000.

35. EB Chang. Digestion and absorption of carbohydrate and protein. In: EB Chang, MD Sitrin, DD Black, Eds. *Gastrointestinal, Hepatobiliary and Nutritional Physiology.* Philadelphia NY: Lippincott-Raven, 1996, pp. 121–140.

36. NHP Cnubben, IMCM Rietjens, H Wortelbore, J van Zanden, PJ van Bladeress. The interplay of glutathione-related processes in antioxidant defense. *Environ Toxicol Pharmacol* 10:141–152, 2001.

37. S Yamaguchi, K Ninomiya. What is umami? *Food Rev Int* 14:123–138, 1998.

38. F Bellisle. Glutamate and the UMAMI taste: Sensory, metabolic, nutritional and behavioural considerations. A review of the literature published in the last 10 years. *Neurosci Behavioural Rev* 23:423–438, 1999.

39. TE Furia, N Bellanca. Flavor potentiation. In: TE Furia, N Bellanca, Eds. *Fenaroli's Handbook of Flavor and Ingredients.* 2nd ed., Vol 1, Boca Raton FL: CRC Press, 1975, pp. 239–242.

40. RS Geha, A Beiser, C Ren, R Patterson, PA Greenberger et al. Review of alleged reaction to monosodium glutamate and outcome of a multicenter double-blind placebo-controlled study. *J Nutr* 130:1058S–1062S, 2000.

41. M Fini, P Toricelli, G Giavoresi, A Carpi, A Nicolini, R Giardino. Effect of L-lysine and L-Arginine on primary osteoblast cultures from normal and osteopenic rats. *Biomed Pharmacotherapy* 55:213–220, 2001.

42. S Ramakrishnan, KN Sulochana. Decrease in glycation of lens proteins by lysine and glycine by scavenging of glucose and possible mitigation of cataractogenesis. *Exp Eye Res* 57:623–628, 2002.

43. MAM van Boekel, HJ Hoenders. Glycation of crystalline in lenses from aging and diabetic individuals. *FEBS Lett* 314:1–4, 1992.

44. C Thongthai, H Okada. Changes in nitrogenous compounds during nam pla fermentation. In: H Taguchi, Ed. Microbial Utilisation of Renewable Resource, Vol. 1. International Center of Cooperative Research and Development in Microbial Engineering, Osaka, Japan, 1980, pp. 101–107.

45. MA Saint-Pierre, J Poortmans, L Liger. Creatine supplementation — A review of the literature. *Sci and Sports* 17:55–77, 2002.

46. NG Sanceda, E Suzuki, M Ohashi and T Kurata. Histamine behavior during the fermentation process in the manufacture of fish sauce. *J Agric Food Chem* 47:3596–3600, 1999.

47. A Lahsen. Histamine food poisoning: an update. *Fish Technol News FAO*, 11:3–5, 9, 1991.

48. BA Bartholomew, PR Berry, JC Rodhouse, RJ Gilbert and CK Merray. Scombrotoxic fish poisoning in Britain: features of over 250 suspected incidents from 1976 to 1986. *Epidemiol Infect* 99:775–782, 1987.

49. V Suvanich, I Maciel-Pedrote, KR Cadwallader, EB Moser, C Thongthai, W Prinyawiwatkul. Development of Fish Sauces Quantitative Descriptive Sensory Aroma Profile. Abstract of International Food Technology Annual Meeting and Food Expo, New Orleans, 2001a, 73G-34.

50. S Areekul and Y Chantachum. Vitamin B12 content in some Thai foods. *Siriraj Hosp Gaz* 32:73–78, 1980.

51. Food and Nutrition Board, Institute of Medicine. Dietary Reference Intakes. Washington, D.C., National Academy Press, 1997.

52. P Puwastien, M Raroengwichit, P Sungpuag and K Judprasong. Thai Food Composition Table. 1st ed. Institute of Nutrition, Mahidol University, Thailand, 1999, pp. 54–55.

53. Health and Welfare Canada: Nutrition Recommendations, Ottawa: Canadian Government Publishing Centre. 1990.

54. National Research Council: Diet and Health: Implication for Reducing Chronic Disease Risk. Report of the Committee on Diet and Health, Food and Nutrition Board, Washington, D.C.: National Academy of Science Press, 1989.

55. FO Simpson. Sodium intake, body sodium, and sodium excretion. *Lancet* 2:25–28, 1988.

56. FO Simpson. Blood pressure and sodium intake. In: CJ Bulpitt, Ed. *Handbook of Hypertension,* Vol 6. *Epidemiology of Hypertension.* Amsterdam: Elsevier, 1985, pp. 175–190.

57. AJ Vander. *Renal Physiology,* 5th ed. New York: McGraw-Hill, 1995, pp. 89–115.

58. G Pickering. Salt intake and essential hypertension. *Cardiovasc Rev Rep* 1:13–17, 1980.

59. DA McCarron. The dietary guideline for sodium: should we shake it up? Yes! *Am. J Clin Nutr* 71:1013–1019, 2000.

60. NM Kaplan. The dietary guideline for sodium: should we shake it up? No! *Am J Nutr* 71:1020–1026, 2000.

61. FC Luft. Salt and hypertension at the close of the millennium. *Wien Klin Wochenschr* 110:459–466, 1998.

62. A White, P Handler and EL Smith. Renal function and the composition of urine. In: *Principles of Biochemistry.* 3rd ed., New York: McGraw-Hill, 1964, pp. 724–743.

63. FJ Haddy and MB Pamnani. Role of dietary salt in hypertension. *J Am Coll Nutr* 14:428–438, 1995.

64. FJ Haddy, HW Overbeck. The role of humoral agents in volume-expanded hypertension. *Life Sci* 19:935–948, 1976.

65. HG Pruss. Sodium, chloride and potassium. In: BA Bowman, RM Russel, Eds. *Present Knowledge in Nutrition*, 8th ed., Washington, D.C.: International Life Sciences Institute ILSI Press, 2001, pp. 302–310.

66. Intersalt: An International Study of Electrolyte Excretion and Blood Pressure. Result for 24-hour Urinary Sodium and Potassium Excretion. Intersalt Cooperative Research Group, 1988.

67. L Tobian. Human essential hypertension: implications of animal studies. *Ann Intern Med* 98:729–734, 1983.

68. HE de Wardener and GA Macgregor. The relation of a circulating sodium transport inhibitor (the natriuretic hormone?) to hypertension. *Medicine* 62:310–326, 1983.

69. K Tontisirin, Y Kachondham and P Winichagoon. Trends in the development of Thailand's nutrition and health plans and programs. *Asia Pacific J Clin Nutr* 1:231–238, 1992.

70. L Garby and S Areekul. Iron supplementation in Thai fish sauce. *Annal Trop Med and Parasitol* 68:467–476, 1974.

71. P Nopburabutra. Double Fortification of Fish Sauce, Mixed Fish Sauce and Salt Brine for Cooking: Nutrient Stability and Sensory Acceptability. MS dissertation, Mahidol University, Bangkok, Thailand, 2002.

72. P Virulhakul. The processing of Thai fish sauce. *Infofish Int* 5/2000:49–53, 2000.

73. FM Orejana, J Liston. Agents of proteolysis and its inhibition in patis (fish sauce) fermentation. *J Food Sci* 47:198–203(209), 1981.

74. C Thongthai and M Siriwongpairat. The sequential quantitation of microorganisms in traditionally fermented fish sauce (Nam Pla). In: PJA Reilly, RWH Parry, Eds. *Post-Harvest Technology, Preservation and Quality of Fish in Southeast Asia*, Manila: Echanis Press, 1990, pp. 23–28.

75. C Thongthai, TJ McGenity, P Suntinanalert, WD Grant. Isolation and characterization of an extremely halophilic archaeobacterium from traditionally fermented Thai fish sauce (nam pla). *Ltrs Appl Microbiol* 14:111–114, 1992.

76. C Thongthai, M Siriwongpairat. Changes in the viable bacterial population, pH, and chloride concentration during the first month of nam pla (fish sauce) fermentation. *J Sci Soc* Thailand 4:73–78, 1978.

77. FG Ijong, Y Ohta. Physico-chemical and microbiological changes associated with bakasang processing — a traditional Indonesian fermented fish sauce. *J Sci Food Agric* 71:69–74, 1996.

78. A Gildberg. Proteolytic activity and the frequency of burst bellies in capelin. *J Food Technol* 13:409–416, 1978.

79. BH Pedersen, K Hjelmeland. Fate of trypsin and assimilation efficiency in larval herring *(Clupea harengus)* following digestion of copepods. *Mar Biol* 97:467–476, 1988.

80. JR McBride, RA MacLeod, DR Idler. Seasonal variation in the collagen content of Pacific herring tissues. *J Fish Res Bd Can* 17:913–918, 1060.

81. RB Hughes. Collagen and cohesiveness in heat-processed herring, and observations on a seasonal variation in collagen content. *J Sci Food Agric* 14:432–441, 1963.

82. J Lavety, RM Love. The strengthening of cod connective tissue during starvation. *Comp Biochem Physiol* 41A:39–42, 1972.

83. K Hjelmeland. Proteinase inhibitors in the muscle and serum of cod *(Gadus morhua)*. Isolation and characterization. *Comp Biochem Physiol* 76B:365–372, 1983.

84. H Toyohara, Y Makinodan, S Ikeda. Purification and some properties of a trypsin inhibitor from carp *(Cyprinus carpio)* muscle. *Comp Biochem Physiol* 80B:949–954, 1985.

85. H Toyohara, M Kinoshita, Y Makinodan, Y Shimizu. Effect of NaCl on the heating activation of the heat-stable alkaline proteinases from various animal muscles. *Comp Biochem Physiol* 92B:715–719, 1989a.

86. WS Hamm, JA Clague. Temperature and Salt Purity Effects on the Manufacture of Fish Paste and Sauce. In: Research Report 24, Fish and Wildlife Service, U.S. Dept of the Interior, Washington DC: U.S. Govt Printing Office, 1950, 11p.

87. A Gildberg. Autolysis of Fish Tissue — General Aspects. PhD dissertation, University of Tromsø, Tromsø, Norway, 1982.

88. A Martinez, RL Olsen. Characterization of pepsins from cod. *USB Corp* 16:22–23(20), 1989.

89. A Martinez, JL Serra. Proteolytic activities in the digestive tract of anchovy *(Engraulis encrasicholus)*. *Comp Biochem Physiol* 93B:61–66, 1989.

90. M Ishida, N Sugiyama, M Sato, F Nagayama. Two kinds of neutral serine proteinases in salted muscle of anchovy, *Engraulis japonica*. *Biosci Biotech Biochem* 59:1107–1112, 1995.

91. A Martinez, RL Olsen, JL Serra. Purification and characterization of two trypsin-like enzymes from the digestive tract of anchovy *Engraulis encrasicholus*. *Comp Biochem Physiol* 91B:677–684, 1988.

92. MS Heu, HR Kim, JH Pyeun. Comparison of trypsin and chymotrypsin from the viscera of anchovy, *Engraulis japonica*. *Comp Biochem Physiol* 112B:557–567, 1995.

93. ST Jiang, YT Wang, CS Chen. Lysosomal enzyme effects on the postmortem changes in tilapia *(Tilapia nilotica XT. Aurea)* muscle myofibrils. *J Food Sci* 57:277–279(282), 1992.

94. M Yamashita, S Konagaya. Differentiation and localization of catheptic proteinases responsible for extensive autolysis of mature chum salmon muscle *(Oncorhynchus keta)*. *Comp Biochem Physiol* 103B:999–1003, 1992.

95. H Toyohara, Y Makinodan. Comparison of calpain I and calpain II from carp muscle. *Comp Biochem Physiol* 92B:577–581, 1989b.

96. L Busconi, EJ Folco, C Studdert, JJ Sanchez. Purification and characterization of a latent form of multicatalytic proteinase from fish muscle. *Comp Biochem Physiol* 102B:303–309, 1992.

97. EJ Folco, L Busconi, C Studdert, CA Casalongue, JJ Sanchez. Distribution of multicatalytic proteinase in fish tissues. *Comp Biochem Physiol* 102B:311–313, 1992.

98. I Stoknes, T Rustad. Purification and characterization of a multicatalytic proteinase from Atlantic salmon *(Salmo salar)* muscle. *Comp Biochem Physiol* 111B:587–596, 1995.

99. JJ Sanchez, EJ Folco, L Busconi, CB Martone. Multicatalytic proteinase in fish muscle. *Mol Biol Rep* 21:63–69, 1995.

100. B Asgeirsson, JB Bjarnason. Properties of elastase from Atlantic cod, a cold-adapted proteinase. *Biochim Biophys Acta* 1164:91–100, 1993.

101. MM Kristjansson, S Gudmundsdottir, JW Fox, JB Bjarnason. Characterization of a collagenolytic serine proteinase from the Atlantic cod *(Gadus morhua). Comp Biochem Physiol* 110B:707–717, 1995.

102. J Overnell. Digestive enzymes of the pyloric caeca and of their associated mesentery in the cod *(Gadus morhua). Comp Biochem Physiol* 46B:519–531, 1973.

103. CG Beddows, M Ismail, KH Steinkraus. The use of bromelain in the hydrolysis of mackerel and the investigation of fermented fish aroma. *J Food Technol* 11:379–388, 1976.

104. A Gildberg. Unpublished results.

105. JN Park, K Ishida, T Watanabe, K Endoh, K Watanabe, M Murakami, H Abe. Taste effects of oligopeptides in a Vietnamese fish sauce. *Fish Sci* 68:921–928, 2002.

106. H Abe, JN Park, Y Fukumoto, E Fujita, T Tanaka, T Washio, S Otsuka, T Shimizu, K Watanabe. Occurrence of D-amino acids in fish sauces and other fermented fish products. *Fish Sci* 65:637–641, 1999.

107. H Kayser, H Meisel. Stimulation of human peripheral blood lymphocytes by bioactive peptides derived from bovine milk proteins. *FEBS Lett* 383:18–20, 1996.

108. H Meisel, E Schlimme. Bioactive peptides derived from milk proteins: ingredients for functional foods? *Kieler Milchwirtschaft Forschungsber* 48:343–357, 1996.

109. N Matsumura, M Fujii, Y Takeda, K Sugita, T Shimizu. Angiotensin I-converting enzyme inhibitory peptides derived from bonito bowels autolysate. *Biosci Biotech Biochem* 57:695–697, 1993.

110. A Gildberg, J Bøgwald, A Johansen, E Stenberg. Isolation of acid peptide fractions from a fish protein hydrolysate with strong stimulatory effect on Atlantic salmon *(Salmo salar)* head kidney leucocytes. *Comp Biochem Physiol* 114B:97–101, 1996.

260 Thongthai and Gildberg

111. J Bøgwald, RA Dalmo, RM Leifson, E Stenberg, A Gildberg. The stimulatory effect of a muscle protein hydrolysate from Atlantic cod, *Gadus morhua* L., head kidney leucocytes. *Fish Shellfish Immunol* 6:3–16, 1996.

112. C Vinot, P Bouchez, P Durand. Extraction and purification of peptides from fish protein hydrolysates. In: S Miyachi, I Karube, Y Ishida, Eds. *Current Topics in Marine Biotechnology*. Tokyo: Fuji Technology Press, 1989, pp. 361–364.

113. J Raa. The use of immunostimulatory substances in fish and shellfish farming. *Rev Fish Sci* 4:229–288, 1996.

114. K Nishioka, GF Babcock, JH Phillips, RA Banks, AA Amoscato. *In vivo* and *in vitro* antitumor activities of tuftsin. *Ann NY Acad Sci* 419:234–241, 1983.

115. I Cancre, R Ravallec, A Van Wormhoudt, E Stenberg, A Gildberg, Y Le Gal. Secretagogues and growth factors in fish and crustacean protein hydrolysates. *Mar Biotechnol* 1:489–494, 1999.

116. K Sugiyama, K Takada, M Egawa, I Yamamoto, H Onzuka, K Oba. Hypotensive effect of fish protein hydrolysate. *Nippon Nogeikagaku Kaishi* 65:35–43, 1991.

117. S Bordenave, I Fruitier, I Ballandier, F Sannier, A Gildberg, I Batista, JM Piot. HPLC preparation of fish waste hydrolysate fractions, effect on guinea pig ileum and ACE activity. *Prep Biochem Biotechnol* 32:65–77, 2002.

118. JA Lovegrove, KG Jackson. Coronary heart disease. In: GR Gibson, CM Williams, Eds. *Functional Foods — Concept to Product*. Cambridge: Woodhead Publishing, 2000, pp. 97–139.

119. IT Johnson. Antitumour properties. In: GR Gibson, CM Williams, Eds. *Functional Foods — Concept to Product*. Cambridge: Woodhead Publishing, 2000, pp. 141–166.

120. H Hatate, Y Numata, M Kochi. Synergistic effect of sardine myofibril protein hydrolysates with antioxidants. *Nippon Suisan Gakkaishi* 56:1011, 1990.

121. SK Kim, YT Kim, HG Byun, KS Nam, DS Joo, F Shahidi. Isolation and characterization of antioxidative peptides from gelatin hydrolysate of Alaska pollack skin. *J Agric Food Chem* 49:1984–1989, 2001.

122. S Oesser, M Adam, W Babel, J Seifert. Oral administration of C-14 labeled gelatin hydrolysate leads to an accumulation of radioactivity in cartilage of mice. *J Nutr* 129:1891–1895, 1999.

123. B Zenitani. Studies on fermented fish product. I. On the arobic bacteria in "Shiokara." *Bull Jap Soc Sci Fish* 21:280–283, 1955.

124. S Liptasiri. Studies on Some Properties of Certain Bacteria Isolated from Thai Fish Sauce. MS thesis submitted to the Faculty of Graduate Studies. Kasetsart University, Bangkok, 1975.

125. C Thongthai, P Suntinanalert. Halophiles in Thai fish sauce (Nam Pla). In: F Rodriguez-Valera Ed., *General and Applied Aspects of Halophilic Microorganisms*, Plenum Press, New York, pp. 381–388, 1991.

126. C Thongthai, unpublished results.

127. P Suntinanalert. Roles of Microorganism in the Fermentation of Nampla in Thailand: Relationship of the Bacteria Isolated from Nampla Produced from Different Geographical Localities in Thailand. MS dissertation, Mahidol University, Bangkok, 1979.

128. S Chaiyanan, S Chaiyanan, T Maugel, A Huq, FT Robb, RR Colwell. Polyphasic taxonomy of a novel *Halobacillus, Halobacillus thailandensis* sp. nov. isolated from fish sauce. *System Appl Microbiol* 22:360–365, 1999.

129. C Thongthai, W Panbangred, C Khoprasert, S Dhaveetiyanond. Protease activities in the traditional process of fish sauce fermentation. In: PJA Reilly, RWH Parry, LE Barile, Eds., *Post-Harvest Technology, Preservation and Quality of Fish in Southeast Asia*, Manila: Echanis Press, 1990, pp. 61–65.

130. JD Owens, L S Mendoza. Enzymatically hydrolyzed and bacterially fermented fishery products. *J Food Tech* 20:273–293, 1985.

131. JA Lovern. The lipid of fish and changes occurring in them during processing and storage. In: E Heen, R Kreuzer Eds., *Fish in Nutrition*. Fishing News (Books) Ltd, London, pp. 86–111, 1962.

132. T Yamane, C Thongthai. Role of Lipase in Fish Sauce (Name Pla) Fermentation. Annual Report of International Center for Biotechnology, International Center for Cooperative Research and Development in Microbial Engineering. Osaka, Japan, vol. 9, pp. 277–278, 1986.

133. P Wilaipan. Halophilic Bacteria Producing Lipase in Fish Sauce. MS dissertation, Chulalongkorn University, Bangkok, 1989.

134. CG Beddows. Fermented fish and fish products. In: HT Chan Jr., Ed., *Fermented Fish Products*, New York, Marcel Dekker Inc., 1983, pp. 255–295.

135. C Thongthai, A. Srisutipruti. Occurrence of tyrosine crystals in Kem-sapparods in rapidly processed Nam Pla (fish sauce). In: PJA Reilly, RWH Parry, LE Barile, Eds., *Post-Harvest Technology, Preservation and Quality of Fish in Southeast Asia.* Manila: Echanis Press, 1990, pp. 23–28.

136. DM Parkin. Global cancer statistics in the year 2000. *The Lancet Oncology* pp. 533–543, 2001.

137. TJ Key, NE Allen, EA Spencer, RC Travis. The effect of diet on risk of cancer. *The Lancet* 360:861–868, 2002.

138. L Ke, P Yu, X Zhang. Novel epidemiologic evidence for the association between fermented fish sauce and esophageal cancer in South China. *Int J Cancer* 99:424–426, 2002.

139. MS Levine, RA Halvorsen. Carcinoma of the esophagus. In: RM Gor, MS Levine, Eds., *Textbook of Gastrointestinal Radiology.* Philadelphia: WB Saunders, pp. 403–433, 2000.

140. PZ Lin, JS Zhang, ZW Ding et al. Carcinogenic and promoting effect of fish juice, preserved rice and salted dry fish on the forestomach epithelium of mice and esophageal epithelium of rats (Chinese). *Zhonghua Zhong Liu Za Zhi* 8:332–335, 1986.

141. DJ Deng, SM Yang, T Li et al. Confirmation of N-(nitrosomethyl) urea as a nitroso urea derived by nitrosation of fish sauce. *Biomed Environ Sci* 12: 54-61, 1999.

142. EJ Mitacek, KD Brunnemann, M Suttajit, N Martin, T Limsila, H Ohshima, LS Caplan. Exposure to N-nitroso compounds in a population of high liver cancer regions in Thailand: volatile nitrosamine (VNA) levels in Thai food. *Food and Chem Toxicol* 37:279–305, 1999.

143. PJ Song, JF Hu. N-nitrosamines in Chinese foods. *Food Chem Toxicol* 26:205–208, 1988.

144. A Nakahara, K Ohshita and S Nasuno. Relation of nitrite concentration to mutagen formation in soy sauce. *Food Chem Toxicol* 24:13–15, 1986.

145. D Adrzejewski, DC Havery, T Fazio. Determination and confirmation of N-nitrosodimethylamine in beer. *J Ass Off Analyt Chem* 64:1457, 1981.

146. TA Gough, MF McPhail, KS Webb, BJ Wood, RF Coleman. An examination of some foodstuffs for the presence of volatile nitrosamines. *J Sci Food Agric* 28:345, 1977.

147. DP Huang, JHC Ho. Volatile nitrosamines in salt-preserved fish before and after cooking. *Food Cosmet Toxicol* 19:167, 1981.

148. T Kawabata, J Uibu, H Ohshima, M Matsui, M Hamano, H Tokiwa. Occurrence, formation and precursors of N-nitroso compounds in the Japanese diet. In: EA Walker, L Griciute, M Castegnaro, Börsönyi, Eds., *N-nitroso Compounds: Analysis, Formation and Occurrence.* IARC Scient Publ No. 31. International Agency for Research on Cancer, Lyon. 1980, pp. 418.

149. DJ Seel, T Kawabata, M Nakamura, T Ishibashi, M Hamano, M Mashimo, SH Shin, K Sakamoto, EC Jhee, S Watanabe. Nitroso compounds in two nitrosated food products in Southwest Korea. *Food Chem Toxicol* 32:1117–1123, 1994.

150. LM Libbey, RA Scanlan, JF Barbour. N-nitrosodimethylamine in dried dairy products. *Food Cosmet Toxicol* 18:459, 1980.

151. SE Shephard, CH Schlatter, WK Lutz. Assessment of the risk of formation of carcinogenic N-nitroso compounds from dietary precursors in the stomach. *Food Chem Toxic* 25:91–108, 1987.

152. YS Choi, H Kahyo. Effect of cigarette smoking and alcohol consumption in the ethiology of cancers of the digestive tract. *Int J Cancer* 49:381–386, 1991.

153. TY Gao, W Zheng, NR Gao, F Jin. Tobacco smoking and its effect on health in China. In: IK O'Neill, J Chen, H Bartsch, Ed., *Relevance to Human Cancer of N-nitroso compounds, Tobacco Smoke and Mycotoxins.* IARC Scientific Publication no. 105. International Agency for Research on Cancer, Lyon, 1991, pp. 62–67.

154. AS Chan, KF To, KW Lo, M Ding, XH Li, P Johnson, DP Huang. Frequent chromosome 9p losses in histologically normal nasopharyngeal epithelia from southern Chinese. *Int J Cancer* 102(3):300–303, 2002.

155. NP Sen, SW Seaman, PA Baddoo, C Burgess, D Weber. Formation of N-nitroso-N-methylurea in various samples of smoked/dried fish, fish sauce, sea foods and ethnic fermented/pickled vegetables following incubation with nitrite under acidic conditions. *J Agri Food Chem* 49:2096–2103, 2001.

156. T Kato, K Kikugawa.Proteins and amino acids as scavengers of nitrite: inhibitory effect on the formation of nitrosodimethylamine and diazoquinone. *Food Chem Toxicol* 30:617–626, 1992.

157. ZE Sikorski, E Kolakowski. Endogenous enzyme activity and seafood quality. In: NF Haard, BK Simpson, Eds., *Seafood Enzymes*. New York: Marcel Dekker, 2000, pp. 451–487.

158. AJ Raae, BT Walther. Purification and characterization of chymotrypsin, trypsin and elastase like proteinases from cod *(Gadus morhua* L.). *Comp Biochem Physiol* 93B:317–324, 1989.

159. A Gildberg, JE Hermes, FM Orejana. Acceleration of autolysis during fish sauce fermentation by adding acid and reducing salt content. *J Sci Food Agric* 35:1363–1369, 1984.

160. CG Beddows, AG Ardeshir. The production of soluble fish protein solution for use in fish sauce manufacture. II. The use of acids at ambient temperature. *J Food Technol* 14:613–623, 1979b.

161. NG Sanceda, T Kurata, N Arakawa. Accelerated fermentation process for the manufacture of fish sauce using histidine. *J Food Sci* 61:220–222, 225, 1996.

162. R Yoshinaka, M Sato, N Tsuchiya, S Ikeda. Production of fish sauce from sardine by utilization of its visceral enzymes. *Bull Jap Soc Sci Fish* 49:463–469, 1983.

163. RC Mabesa, EV Carpio, LB Mabesa. An accelerated process for fish sauce (patis) production. In: PJA Reilly, RWH Parry, LE Barile, Eds., *Post-Harvest Technology, Preservation and Quality of Fish in Southeast Asia*. Manila: Echanis Press, 1990, pp. 45–49.

164. CG Beddows, AG Ardeshir. The production of soluble fish protein solution for use in fish sauce manufacture. I. The use of added enzymes. *J Food Technol* 14: 603–612, 1979a.

165. JS Bond. Commercially available proteases. In: RJ Beynon, JS Bond, Eds., *Proteolytic Enzymes a Practical Approach*. Oxford: OIRL Press, 1989, pp. 232–249.

166. Hirotomo, DE Blendford. The soy sauce explosion. *Food Process Ind* Aug/1980, 25,27,29, 1980.

167. A Gildberg. Enhancing returns from greater utilization. In: HA Bremner, Ed., *Safety and Quality Issues in Fish Processing.* Cambridge: Woodhead Publ Ltd, 2002, pp. 425–449.

168. RL Maylor, RJ Goldburg, JH Primavera, N Kautsky, MCM Beveridge, J Clay, C Folke, J Lubchenco, H Mooney, M Troell. Effects of aquaculture on world fish supplies. *Nature* 405:1017–1024, 2000.

169. Y Murata, T Hayashi, E Watanabe, K Toyama. Preparation of skipjack spermary extract by enzymolysis. *Nippon Suisan Gakkaishi* 57:1127–1132, 1991.

170. NG Sanceda, T Kurata, N Arakawa. Formation and possible derivation of volatile fatty acids in the production of fish sauce. *J Home Econ Jpn* 41:939–945, 1990.

171. RC McIver, RI Books, GA Reinneccius. Flavor of fermented fish sauce. *J Agri Food Chem* 30:1017–1020, 1982.

172. H Ito, H Tachi, S Kikuchi. Fish fermentation technology in Japan. In: CH Lee, KH Steinkraus, PJA Reilly, Eds., *Fish Fermentation Technology.* Seoul: YuRim Publ Co., 1989b, pp. 161–170.

10

Nutraceuticals from Seafood and Seafood By-Products

FEREIDOON SHAHIDI

Department of Biochemistry, Memorial
University of Newfoundland,
St. John's, NL, Canada

INTRODUCTION

Marine foods have traditionally been used because of their variety of flavor, color, and texture. More recently, seafoods have been appreciated because of their role in health promotion arising primarily from constituent long-chain omega-3 fatty acids, among others. Nutraceuticals from marine resources and the potential application areas are varied, as listed in Table 10.1. Processing of the catch brings about a considerable amount of by-products accounting for 10 to 80% of the total landing weight. The components of interest include lipids, proteins, flavorants, minerals, carotenoids, enzymes,

TABLE 10.1 Nutraceutical and Bioactive Components from Marine Resources and Their Application Area

Component (source)	Application area
Chitin, chitosan, glucosamine	Nutraceuticals, agriculture, food, water purification, juice clarification, etc.
Carotenoids, carotenoproteins	Nutraceuticals, fish feed
Omega-3 fatty acids	Nutraceuticals, foods, baby formula, etc.
Biopeptides	Nutraceuticals, immune-enhancing agents
Minerals (Calcium, etc.)	Food, nutraceuticals
Algae (Omega-3, minerals, carotenoids)	Nutraceuticals
Chondroitin sulfate	Arthritic pain relief
Squalene	Skin care
Specialty chemicals	Miscellaneous

and chitin, among others. The raw material from such resources may be isolated and used in different applications, including functional foods and as nutraceuticals. The importance of omega-3 fatty acids in reducing the incidence of heart disease, certain types of cancer, diabetes, autoimmune disorders, and arthritis has been well recognized. In addition, the residual protein in seafoods and their by-products may be separated mechanically or via a hydrolysis process. The bioactive peptides so obtained may be used in a variety of food and nonfood applications. The bioactives from marine resources and their application areas are generally diverse. This chapter provides a cursory account of nutraceuticals and bioactives from selected seafood by-products.

MARINE OILS

The long-chain omega-3 polyunsaturated fatty acids (PUFA) are of considerable interest because of their proven or perceived health benefits.[1-3] These fatty acids are found almost exclusively in aquatic resources (algae, fish, marine mammals, etc.) and exist in varying amounts and ratios. While algal sources also provide minerals, such as iodine, as well as carotenoids and xanthophylls, fish body oil contains mainly tri-acylglycerols, and fish liver oils serve as a source of vitamin A, among others. In addition, liver from other aquatic species, such as shark, contain squalene and other bioactives. Another source of long-chain omega-3 fatty acids is the blubber of marine mammals which contains eicosapentaenoic acid (EPA) and docosahexaenoic acid (DHA), similar to fish oils, as well as docosapentaenoic acid (DPA). It is worth noting that myristic acid is present in much smaller levels in the blubber oil from marine mammals than algal or fish oils; this is a definite advantage when considering the atherogenic properties of myristic acid. In humans, DHA accumulates at a relatively high level in organs with electrical activity, such as retinal tissues of the eye and the neural system of the heart. While DHA and other long-chain omega-3 fatty acids may be formed from alpha linolenic acid (ALA) (Figure 10.1), the conversion efficiency for this transformation is very limited in healthy human adults and is approximately 3 to 5%.[4] In adults with certain ailments, the conversion of ALA to DHA is less than 1%.[5] As shown in Figure 10.1, DHA may be retroconverted to DPA and EPA. Human feeding trials have indicated a retro conversion of DHA to EPA of about 10%.[6]

The beneficial health effects of marine oils in reducing the incidences of coronary heart disease (CHD) have been attributed to their omega-3 fatty acid constituents.[1] Omega-3 fatty acids are known to reduce the incidence of CHD by lowering the level of serum triacylglycerols and possibly cholesterol and also to lower the blood pressure in individuals with high blood pressure as well as to decrease the venticular arrhythmias, among others. In addition, omega-3 fatty acids are known to relieve arthritic swelling and possibly pain, relieve type II diabetes, and enhance body immunity. However, omega-3 fatty acids may increase fluidity of the blood, and hence their consumption by patients on blood thinners such as coumadin and aspirin should be carefully considered in order to avoid any unnecessary complication due to vasodilation and possible rupture of capillaries. The omega-3 fatty

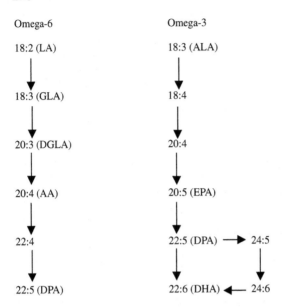

Figure 10.1 Essential fatty acids of the omega-6 and omega-3 families. Symbols are: LA, linoleic acid; GLA, gamma linolenic acid; DGLA, dihomo-gamma linolenic acid; AA, arachidonic acid; DPA, docosapantaenoic acid; ALA, "alpha linolenic acid; EPA, eicosapentaenoic acid; and DHA, docosahexaenoic acid.

acids, especially DHA, are known to dominate the fatty acid spectrum of brain and retina lipids and play an essential role in the development of fetus and infants as well as in the health status and body requirements of pregnant and lactating women.

Consideration of the three-dimensional structures of unsaturated fatty acids demonstrates that bending of the molecules increases with an increase in the number of double bonds in their chemical structures, and this is further influenced by the position of the double bonds (i.e., omega-3 vs. omega-6). These structural features in the triacylglycerol molecules as well as the location of the fatty acids in the glycerol molecule (i.e., sn-1, sn-2, and sn-3) may have a major effect on the bioavailability of fatty acids involved and their potential health benefits.

TABLE 10.2 Major Fatty Acids of Omega-3 Rich Marine and Algal Oils

Fatty acid	Seal blubber	Cod liver	Menhaden	Algal (DHASCO)
14:0	3.73	3.33	8.32	14.9
16:0	5.58	11.01	17.4	9.05
16:1T7	18.0	7.85	11.4	2.20
18:0	0.88	3.89	3.33	0.20
18:1T9+T11	26.0	21.2	12.1	18.9
20:1T9	12.2	10.4	1.44	—
20:5T3	6.41	11.2	13.2	—
22:1T11	2.01	9.07	0.12	—
22:5T3	4.66	1.14	2.40	0.51
22:6T3	7.58	14.8	10.1	47.4

Note: Units are weight percents of total fatty acids. DHASCO = docosahexaenoic acid single cell oil.

Two important sources of omega-3 fatty acids, namely menhaden oil (MO) and seal blubber oil (SBO) were considered in our work. Table 10.2 summarizes the fatty acids of MO, SBO, cod liver oil, and a commercial algal oil known as DHASCO (docosahexaenoic acid cell oil). While omega-3 fatty acids, especially DHA, are primarily located in the sn-2 position in menhaden oil, they are mainly in the sn-1 and sn-3 positions of seal blubber oil (Table 10.3).[7] These differences

TABLE 10.3 Distribution of Long-Chain Omega-3 Fatty Acids in Menhaden and Seal Blubber Oils

Fatty acid	Seal blubber			Menhaden		
	sn-1	sn-2	sn-3	sn-1	sn-2	sn-3
EPA	8.36	1.60	11.2	3.12	17.5	16.3
DPA	3.99	0.79	8.21	1.12	3.11	2.31
DHA	10.5	2.27	17.9	4.11	17.2	6.12

Note: Units are weight percents of total fatty acids. EPA = eicosapentaenoic acid; DPA = docosapentaenoic acid; and DHA = docosahexaenoic acid.

TABLE 10.4   Food Application of Omega-3 Oils

Food	Country
Bread/hard bread	Australia, France, Germany, Ireland, Denmark
Cereals, crackers & noodles	France, Korea, Taiwan
Bars	USA
Pasta and cakes	France, UK
Infant formula	Australia, Brazil, Japan, New Zealand, Taiwan, UK
Milk, fortified	Argentina, Indonesia, Italy, Spain, UK
Juices, fortified	Brazil, Germany, Spain
Mayonnaise & salad dressings	Korea
Margarines & spreads	Ireland, Japan, UK
Eggs	USA, UK
Canned tuna steak & seafood	Japan, USA
Tuna burger	USA

undoubtedly have a definite influence on their assimilation, absorption, and health benefits as well as reactions in which they are involved.

Regardless of the source of long-chain omega-3 fatty acids, such oils must undergo appropriate processing in order to provide a bland-tasting product devoid of contaminants. Therefore, refining, bleaching, deodorization, and addition of appropriate antioxidant stabilizers must be practiced in order to allow their use in food formulations. The type of food in which such omega-3 oils may be incorporated is listed in Table 10.4. These include foods that could be used within a short period of time and in products which do not develop off-flavors during their expected shelf life.

OMEGA-3 CONCENTRATES

For therapeutic purposes the natural sources of omega-3 fatty acids, as such, may not provide the necessary amounts of these fatty acids and hence production and use of concentrates of omega-3 fatty acids may be required.[8] The omega-3 fatty acid concentrates may be produced in the free fatty acid, simple alkyl ester and acylglycerol forms. To achieve this, physical, chemical, and enzymatic processes may be employed for concentrate production. The available methods suitable for large-scale production include low-temperature crystallization, fractional or molecular distillation, chromatography, super-critical fluid extraction, urea complexation, and enzymatic splitting.[9]

Among the simplest methods for concentrate production is fractional crystallization, which takes advantage of the existing differences in the melting points of different fatty acids, as neat compounds or in different solvent systems. The more saturated fatty acids have higher melting points and may crystallize out of the mixtures and hence leave behind, in the liquid form, the more unsaturated fatty acids. Obviously, the free fatty acids and simple alkyl esters are more amenable to provide a higher concentration of omega-3 fatty acids than acylglycerols. This is because the latter mixtures consist of fatty acids with varying chain lengths and degrees of unsaturation in many different combinations in the triacyl-glycerol molecules.

Fractional distillation is another facile process for separation of mixtures of fatty acid esters under reduced pressure (0.1 to 1.0 mm Hg).[10] However, due to sensitivity of more highly unsaturated fatty acids to oxidation, one may use a spinning band column, which does not impose such limitations.[11] While fractional distillation of menhaden oil ethyl esters increased the content of EPA from 15.9 to 28.4%, and DHA from 9.0 to 43.9%, molecular distillation afforded DHA with 90% parity.[12]

Reverse phase chromatography has been used by Nakahara[13] to produce a DHA and DPA concentrate from marine microalgae. Teshima[14] used a silver nitrate-impregnated silica gel column to separate EPA and DHA from squid liver oil fatty acid methyl esters. The yield of the process for these fatty acids was 39 and 48%, respectively, with 85 to 96% EPA and 95 to 98% DHA purity. Similar studies on a variety of other oils have recently appeared in the literature using high-performance liquid chromatography.[15–18] More recently, centrifugal partition chromatography (CPC) has gained attention for production of omega-3 concentrates.[19,20] Wanasundara[21] used a CPC technique to produce highly concentrated fatty acids such as EPA and DHA with a near quantitative yield.

Supercritical fluid extraction (SFE) is a relatively new process, which is desirable for separation of PUFA. Since this method is based on separation of compounds based on their molecular weight and not their degree of unsaturation, a prior concentration step may be required in order to concentrate the omega-3 PUFA. Thus omega-3 fatty acids have been concentrated by SFE from fish oil and seaweed.[22-24] Fish oil esters were fractionated by SFE to obtain an oil with 60 to 65% DHA.[25]

Another possibility for concentration of omega-3 fatty acids is urea complexation. The natural acylglycerols are hydrolyzed to their fatty acid constituents in ethanol and the resultant components are allowed to crystallize in the presence of urea. The highly unsaturated fatty acids, which deviate more and more from a near linear shape, are not included in the urea crystals and remain in the liquid form, referred to as nonurea complexing fraction (NUCF). Meanwhile, saturated fatty acids and, to a lesser extent, mono- and diunsaturated fatty acids may be included in the urea to afford the urea complexing fraction (UCF). In this manner, depending on the variables involved, e.g., the amount of solvent, urea, and time and temperature, optimum conditions may be employed for the preparation of concentrates. If necessary, the urea complexation process may be repeated in order to enhance the concentration of certain fatty acids in the final products. We have used such techniques to prepare concentrates dominated by DHA, EPA, or DPA. The total omega-3 fatty acids in one such preparation from seal blubber oil was 88.2% and this was dominated by DHA (67%).[26]

Finally, enzymatic procedures may be used to produce concentrates of omega-3 fatty acids. Depending on the type of enzyme, reaction time, temperature, and the concentration of the reactants and enzyme, it is possible to produce concentrates in different forms, e.g., as free fatty acids or as acylglycerols. Thus, processes such as transesterification, acidolysis, alcoholysis, and hydrolysis as well as esterification of fatty acids with alcohols or glycerol may be employed.

Wanasundara and Shahidi[27] have shown that enzymes might be used to selectively hydrolyze saturated and less unsaturated lipids from triacylglycerols, hence concentrating the omega-3 fatty acids in seal blubber and menhaden oils in the acylglycerol form. In this manner, the omega-3 PUFA content was nearly doubled. Furthermore, following urea complexation, omega-3 concentrates obtained may be subjected to esterification with glycerol to produce concentrated acylglycerols. Upon glycerolysis of specialty alkyl esters from seal blubber oil, we found that monoacylglycerols (MAG), diacylglycerols (DAG), and triacylglycerols (TAG) were formed simultaneously. The amount of MAGs decreased continuously while that of TAGs increased.[28] Depending on the structural characteristics of final products, the stability of acylglycerols was found to be better than that of their corresponding ethyl esters. Possible loss of natural antioxidants during processing may also affect the stability of products involved. Therefore, it is important to stabilize the modified oils using any of the recommended synthetic antioxidants or preferably natural stabilizers. Thus, TBHQ (tertiary-butylhydroquinone) at 200 ppm was able to inhibit oxidation of menhaden oil at 60°C over a 7-day storage period. Meanwhile, the inhibition effects were 32.5% for mixed tocopherols (500 ppm), 18.0% for alpha-tocopherol (500 ppm), 39.8% for mixed green tea catechins (200 ppm), 45.1% for EC (epicatechin), 48.2% for ECG (epicatechin gallate), 51.3% for EGC (epigallocatechin), and 50% for EGCG (epigallocatechin-3 gallate).[29] For seal blubber oil, the best protection of 56.3% was rendered by TBHQ at 200 ppm and 58.6% by ECG (200 ppm).

STRUCTURED LIPIDS

Structured lipids (SL) are TAGs containing combinations of short-chain fatty acids (SCFA), medium-chain fatty acids (MCFA) and long-chain fatty acids (LCFA) located in the same glycerol molecule, and these may be produced by chemical or enzymatic means.[30,31] SLs are developed to fully optimize the benefits of their fatty acid varieties in order to affect metabolic parameters such as immune function, nitrogen balance and lipid clearance from the bloodstream. These specialty lipids may be produced via direct esterification, acidolysis, and hydrolysis or interestification.

MCFA are those with 6 to 12 carbon atoms and are often used for production of SLs. As mentioned earlier, MCFA are highly susceptible to oxidation.[32] These fatty acids are not stored in the adipose tissues and are often used in the diet of patients with maldigestion and malabsorption.[33] They have also been employed in total parenteral nutrition and formulas for preterm infants. Production of SLs via acidolysis of blubber oil with capric acid was recently reported.[34] Lipozyme-IM from *Mucor miehei* was used as a biocatalyst at an oil to fatty acid mole ratio of 1:3 in hexane, at 45°C for 24 h and 1% (w/w) water.[34,35] Under these conditions, a SL containing 2.3% EPA and 7.6% DHA at 27.1% capric acid (CA) was obtained. In this product, CA molecules were primarily located in the sn-1 and sn-3 positions (see Table 10.5), thus serving as a readily available source of energy to be released upon the action of pancreatic lipase. Incorporation of CA into fish oil TAG using immobilized lipase from *Rhizomucor miehei* (IM-60) was also

TABLE 10.5   Enzymatic Modification of Seal Blubber Oil with Capric Acid

Fatty acid	Unmodified	Modified	Sn-1 & Sn-3
10:0	—	27.1	85.1
14:0	3.4	2.7	48.1
14:1	1.0	0.8	58.3
16:0	5.0	3.7	46.8
16:1T7	15.1	11.9	55.5
18:1T9 and T11	26.4	19.3	56.1
18:2T6	1.3	1.7	66.7
20:T9	15.0	9.1	72.5
20:5T3	5.4	2.3	31.9
22:1T11	3.6	1.9	52.6
22:5T3	4.9	3.0	76.7
22:6T3	7.0	7.6	82.1

Note: Percent of modified fatty acid in sn-1 and Sn-3 positions.
Units are percents of total fatty acids.
The enzyme used was lypozyme-IM from *Mucor miehei*.

reported.[36] After a 24 h incubation in hexane, 43% CA was incorporated into fish oil while the content of EPA and DHA in the product was reduced to 27.8 and 23.5%, respectively. Similar results were obtained upon acidolysis of seal blubber with lauric acid.[35]

In an effort to produce specialty lipids containing both omega-3 PUFA and gamma linolenic acid (GLA), preparation of such products under optimum conditions was reported.[31] GLA is found in relatively large amounts in borage oil (20 to 25%), evening primrose oil (8 to 10%) and blackcurrant oil (15 to 18%). Using borage oil, urea complexation process afforded a concentrate with 91% GLA under optimum reaction conditions.

Lipase-catalyzed acidolysis of seal blubber oil and menhaden oil with GLA concentrate,[37] under optimum conditions of GLA to TAG mole ratio of 3:1, reaction temperature of 40°C over 24 h and 500 units enzyme per gram oil afforded products with 37.1 and 39.6% GLA incorporation, respectively. Of the two enzymes tested, lipase PS-30 from *Pseudomonas sp.* served better in the acidolysis process than *Mucor miehei*.[38] Incorporation of GLA was in all positions and its content in the sn-2 position of both seal blubber oil and menhaden oil was 22.1 and 25.7%, respectively (Table 10.6). Thus, PS-30

TABLE 10.6 Fatty Acids of Seal Blubber Oil (SBO), Menhaden Oil (MO) and Their Acidolysis Products with Gamma linolenic Acid (GLA, 18:3T6)

	SBO			MO		
Fatty acid	Unmodified	Modified	Sn-1 and Sn-3[a]	Unmodified	Modified	Sn-1 and Sn-3[a]
14:0	3.36	2.40	58.3	8.18	4.55	53.3
16:0	5.14	3.04	51.1	19.89	8.78	53.5
18:1T9	22.6	14.1	46.6	9.86	4.24	53.7
18:3T6	0.59	37.1	77.9	0.43	39.6	74.3
20:1T9	17.3	8.30	55.4	1.62	0.83	20.0
20:5T3	5.40	3.80	84.6	12.9	11.0	65.9
22:5T3	5.07	2.99	78.0	2.48	2.07	66.7
22:6T3	7.73	4.36	79.2	10.0	6.56	77.4

served in a nonspecific manner in the acidolysis process. The structured lipids containing GLA, EPA, and DHA so produced may have health benefits above those exerted by use of their physical mixtures.

Production of structured lipids containing GLA, EPA, and DHA may also be achieved using borage and evening primrose oils as sources of GLA and either EPA or DHA or their combinations.[39,40] The products so obtained, while similar to those produced by incorporation of GLA into marine oils, differ in the composition and distribution of fatty acids involved.

BIOACTIVE PEPTIDES FROM MARINE RESOURCES

Protein hydrolysis leads to the formation of peptides with different numbers of amino acids as well as free amino acids. Depending on reaction variables as well as the type of enzyme, the degree of hydrolysis of proteins may differ considerably. The peptides produced from the action of a specific enzyme may be subjected to further hydrolysis by other enzymes. Thus, the use of an enzyme mixture or several enzymes in a sequential manner may be advantageous. The peptides so obtained may be subjected to chromatographic separation and then evaluated for their amino acid sequence as well as their antioxidant and other activities.

In a study on capelin protein hydrolyzates, four peptide fractions were separated using Sephadex G-10. While one fraction exerted a strong antioxidant activity in a beta-carotene-linoleate model system, two fractions possessed a weak antioxidant activity and the fourth one had a prooxidant effect. Two-dimensional HPTLC (high performance thin layer chromatography) separation showed spots with both pro- and antioxidative effects.[41] Meanwhile, protein hydrolyzates prepared from seal meat were found to serve as phosphate alternatives in processed meats and reduced the cooking loss considerably.[42] Furthermore, Alaska pollock skin hydrolyzate was prepared using a multienzyme system in a sequential manner. The enzymes used were in the order of alcalase,

TABLE 10.7 Antioxidative Peptides from Gelatin Hydrolyzate of Alaska Pollock Skin in Comparison with That of Soy Conglycinin

Peptide	Amino acid sequence
	Alaska Pollock Skin
P$_1$	Gly-Glu-Hyp-Gly-Pro-Hyp-Gly-Pro-Hyp-Gly-Pro-Hyp-Gly-Pro-Hyp-Gly
P$_2$	Gly-Pro-Hyp-Gly-Pro-Hyp-Gly-Pro-Hyp-Gly-Pro-Hyp-Gly
	Soy Conglycinin
P$_1$	Val-Asn-Pro-His-Asp-His-Glu-Asn
P$_2$	Leu-Val-Asn-Pro-His-Asp-His-Glu-Asn
P$_3$	Leu-Leu-Pro-His-His
P$_4$	Leu-leu-Pro-His-His-Ala-Asp-Ala-Asp-Tyr
P$_5$	Val-Ile-Pro-Ala-Gly-Tyr-Pro
P$_6$	Leu-Glu-Ser-Gly-Asp-Ala-Leu-Arg-Val-Pro-Ser-Gly-Thr-Tyr-Tyr

pronase E, and collagenase. The fraction from the second step, which was hydrolyzed by pronase E, was composed of peptides ranging from 1.5 to 4.5 kDa and showed a high antioxidant activity. Two peptides were isolated, using a combination of chromatographic procedures, and these were composed of 13 and 16 amino acid residues.[43] The sequence of the peptides involved is given in Table 10.7 and compared with those of soy conglycinin hydrolyzates.[44] These peptides exert their antioxidant activity via both free radical scavenging as well as chelation effects. Recently, proteases from shrimp processing discards were characterized.[45]

CHITIN, CHITOSAN AND RELATED COMPOUNDS

Chitin may be recovered from processing discards of shrimp, crab, lobster, and crayfish following deproteinization and demineralization.[46,47] The chitin so obtained may then be deacetylated to afford chitosan.[46] Depending on the duration of the deacetylation process, the chitosan produced may assume different viscosities and molecular weights. The chitosans produced are soluble in weak acid solutions, thus chitosan ascorbate, chitosan acetate, chitosan lactate, and chitosan malate, among others, may be obtained and these are all soluble in water. Chitosan has a variety of health benefits and may be employed in a number of nutraceutical and health-related applications. Chitosan derivatives may also be produced in order to obtain more effective products for certain applications. However, to have the products solubilized in water without the use of acids, enzymatic processes may be carried out to produce chitosan oligomers. Due to their solubility in water, chitosan oligomers serve best in rendering their benefits under normal physiological conditions and in foods with neutral pH. Furthermore, depending on the type of enzyme employed, chitosan oligomers with specific chain length may be produced for certain applications.[48]

Chitosans with different viscosities were prepared (Table 10.8) and used in an experiment designed to protect both raw and cooked fish against oxidation as well as microbial spoilage.[49–51] The content of propanal, an indicator of oxidation of omega-3 fatty acids, was decreased when chitosan was used as an edible invisible film in herring. Furthermore, the effects were more pronounced as the molecular weight of the chitosan

TABLE 10.8 Characteristics of Three Different Kinds of Chitosans Prepared from Crab Shell Waste[a]

| | Chitosan | | |
Properties	I	II	III
Deacetylation time[b]	4 h	10 h	20 h
Moisture (%)	4.50 ± 0.30	3.95 ± 0.34	3.75 ± 0.21
Nitrogen (%)	7.55 ± 0.10	7.70 ± 0.19	7.63 ± 0.08
Ash (%)	0.30 ± 0.03	0.25 ± 0.02	0.30 ± 0.00
AV[a] (cps)[d]	360	57	14
DA[c] (%)	86.3 ± 2.1	91.3 ± 1.2	94.5 ± 1.3
Mv[c] (dalton)	1,816,732	963,884	695,122

[a] Results are expressed as mean ∀ standard deviation of three determinations.

[b] Deacetylation for chitosan I, II and III was achieved using 50% NaOH at 100°C.

[c] Mv = viscosity molecular weight; AV = apparent viscosity; and DA = degree of deacetylation.

[d] cps = cycles per second.

TABLE 10.9 Content of Propanal mg/kg (dwb) in Headspace of Chitosan-Coated Herring Samples Stored at 4°C[a]

Chitosan	Storage period (days)					
	0	2	4	6	8	10
Uncoated	12.6 ± 3.4[a]	23.7 ± 4.2[b]	29.9 ± 4.2[c]	34.3 ± 1.9[c]	44.1 ± 4.0[c]	46.3 ± 2.4[c]
14 cps	13.8 ± 2.1[a]	18.3 ± 3.0[a]	24.6 ± 1.2[b]	30.9 ± 2.9[bc]	33.0 ± 0.8[b]	39.7 ± 0.9[b]
57 cps	12.6 ± 3.0[a]	15.5 ± 2.1[a]	19.7 ± 2.6[a]	24.9 ± 1.6[a]	22.8 ± 1.9[a]	24.2 ± 1.9[a]
360 cps	14.2 ± 2.4[a]	15.7 ± 2.6[a]	17.6 ± 2.2[a]	20.2 ± 1.4[a]	18.3 ± 2.4[a]	22.7 ± 1.3[a]

[a] Results are expressed as mean standard deviation of three determinations. Values with the same superscripts within each column are not significantly different (P<0.05).

increased (Table 10.9). In addition, inhibitory effects of chitosan coatings in the total microbial counts for cod and herring showed an approximately 1.5 and 2.0 log cycles difference between coated and uncoated samples, respectively, after 10 days of refrigerated storage (results not shown). The monomer of chitin, N-acetylglucosamine, has been shown to possess anti-inflammatory properties. Meanwhile, glucosamine, the monomer of chitosan, prepared via HCl hydrolysis, is marketed as glucosamine sulfate. This formulation is prepared by addition of ferrous sulfate to the preparation. Glucosamine products may also be sold in formulation containing chondroitin 4- and chondroitin 6-sulfates. While glucosamine helps to form proteoglycans that sit within the space in the cartilege, chondroitin sulfate acts like a liquid magnet. Thus, glucosamine and chondroitin work in a complementary manner to improve the health of the joint cartilage.

The by-products in chitin extraction process from shellfish include carotenoids/carotenoproteins and enzymes.[52–54] These components may also be isolated for further utilization in a variety of applications.

REFERENCES

1. Simopoulos, A.P. Omega-3 fatty acids in health and disease and growth and development. *Am. J. Clin. Nutr.* 1991, 54, 438–463.

2. Abeywardena, M.Y., Head, R.J. Long-chain n-3 polyunsaturated fatty acids and blood vessel function. *Cardiovascular Res.* 2001, 52, 361–371.

3. Shahidi, F., Kim, S-K. Marine lipids as affected by processing and their quality preservation by natural antioxidants. In: *Bioactive Compound in Foods: Effects of Processing and Storage*, Lee, T.C., Ho, C-T. Eds., ACS Symposium Series 816, American Chemical Society, Washington, D.C., 2002, pp. 1–13,

4. Emken, E.A., Adlof, R.O., Gulley, R.M. Dietary linoleic acid influences desaturation and acylation of deuterium-labeled linoleic and linolenic acids in young adult males. *Biochem. Biophys. Acta.* 1994, 1213, 277–288.

5. Salem, N. Jr., Wegher B., Mena, P., Uauy, R. Arachidonic and docosahexaenoic acids are biosynthesized from their 18-carbon precursors in human infants. *Proc. Natl. Acad. Sci. USA*, 1996, 93, 49–54.

6. Coquer, J.A., Holub, B.J. Supplementation with an algae source of docosahexaenoic acid increases (n-3) fatty acid status and alters selected risk factors for heart disease in vegetarian subjects. *J. Nutr.* 1996, 126, 3032–3039.

7. Wanasundara, U.N., Shahidi, F. Positional distribution of fatty acids in triacylglycerols of seal blubber oil. *J. Food Lipids* 1997a, 4, 51–64.

8. Wanasundara, U.N., Wanasundara, J., Shahidi, F. Omega-3 fatty acid concentrates: a review of production technologies. In: *Seafoods: Quality, technology and nutraceutical applications.* Alasalvar, C., Taylor, T. Eds., Springer, Berlin and New York, NY, 2002, pp 157–174.

9. Wanasundara, U.N., Shahidi, F. Structural characteristics of marine lipids and preparation of omega-3 concentrates. In: *Flavor and lipid chemistry of seafoods.* Shahidi, F., Cadwallader, K.R., Eds., ACS Symposium Series 674, American Chemical Society, Washington, D.C. 1997b; pp. 240–254.

10. Brown, J.B., Kolb, D.K. Application of low-temperature crystallization in the separation of polyunsaturated fatty acids and their compounds. *Prog. Chem. Fats Other Lipids* 1995, 3, 57–94.

11. Haraldsson, G.G. Separation of saturated/unsaturated fatty acids. *J. Am. Oil Chem. Soc.* 1984, 61, 219–222.

12. Max, Z. High-Concentration Mixture of Polyunsaturated Fatty Acids and Their Esters from Animal and/or Vegetable Oils and Their Prophylactic or Therapeutic Uses, UK Patent GB2, 218, 984, 1989.

13. Nakahara, T., Yokochi, T., Higashihara, T., Tanaka, S., Yaguchi, T., Honda, D. Production of docosaxhaenoic and docosapentaenoic acids by *Schizochytrium* sp. Isolated from Yap Island. *J. Am. Oil Chem. Soc.* 1996, 73, 1421–1426.

14. Teshima, S., Kanazawa, A., Tokiwa, S. Separation of polyunsaturated fatty acids by column chromatography on silver nitrate-impregnated silica gel. *Bull. Jp. Soc. Sci. Fish* 1978, 44, 927–929.

15. Tekiwa, S., Kanazawa, A., Teshima, S. Preparation of eicosapentaenoic and docosahexaenoic acids by reversed phase high-performance liquid chromatography. *Bull. Jp. Soc. Sci. Fish* 1981, 47, 675–678.

16. Adlof, R.O., Emiken, E.A. The isolation of omega-3 polyunsaturated fatty acids and methyl esters of fish oils by silver resin chromatography. *J. Am. Oil Chem. Soc.* 1985, 62, 1592–1595.

17. Hayashi, K., Kishimura, H. Preparation of n-3 PUFA ethyl ester concentrates from fish oil by column chromatography on silicic acid. *Nippon Suisan Gakkaishi* 1993, 59, 1429–1435.

18. Corley, D.G., Zeller, S.G., James, P., Duffin, K. Process for Separating a Triglyceride Comprising a Docosahexaenoic Acid Residue from a Mixture of Triglycerides. International Patent PCT/US so/04166, 2000.

19. Murayama, W., Kosuge, Y., Nakaya, N., Nunogaki, K., Cazes, J., Nunogaki, H. Preparative separation of unsaturated fatty acid esters by centrifugal partition chromatography. *J. Liq. Chromatogr.* 1988, 11, 283–300.

20. Goffic, F.L. Countercurrent and centrifugal partition chromatography as new tools for preparative-scale lipid purification. *Lipid Technol.* 1997, 9, 148–150.

21. Wanasundara, U.N. Isolation and Purification of Bioactive Compounds by Centrifugal Partition Chromatography (CPC) and Other Technologies. Presented at the 2nd International Conference and Exhibition on Nutraceuticals and Functional Foods (Worldnutra 2001). November 26-December 2, Portland, OR, 2001.

22. Choi, K.I., Nakhost, Z., Krukonis, V.I., Karel, M. Supercritical fluid extraction and characterization of lipid from algal *Scenedesmus obliques*. *Food Biotechnol.* 1987, 1, 263–271.

23. Yamagouchi, K., Muraksmi, M., Nakano, H., Konosu, S., Kokura, T., Yamamoto, H., Kosaka, M., Hata, K. Supercritical carbon dioxide extraction of oils from Antarctic krill. *J. Agric. Food Chem.* 1986, 34, 904–907.

24. Mishra, V.K., Temelli, F., Oraikul, B. Extraction and purification of T-3 fatty acids with an emphasis on supercritical fluid extraction: a film demonstration. *Ber Bunsenges Phys. Chem.* 1993, 88, 882–887.

25. Stout, V.F., Spinelli, J. Polyunsaturated Fatty Acids from Fish Oils. US Patent 4: 675, 132, 1987.

26. Shahidi, F., Wanasundara, U.N. Concentration of omega-3 polyunsaturated fatty acids of seal blubber oil by urea complexation: Optimization of reaction conditions. *Food Chem.* 1999, 65, 41–49.

27. Wanasundara, U.N., Shahidi, F. Lipase-assisted concentration of T-3 polyunsaturated fatty acids in acylglycerol forms from marine oils. *J. Am. Oil Chem. Soc.* 1998, 75, 945–951.

28. He, Y., Shahidi, F. Enzymatic esterification of T-3 fatty acid concentrates from seal blubber oil with glycerol. *J. Am. Oil Chem. Soc.* 1997, 74, 1133–1136.

29 Shahidi, F., Wanasundara, U.N. Seal blubber oil and its nutraceutical products. In: *Omega-3 Fatty Acids*, Shahidi, F., Finley, J.W. Eds., ACS Symposium Series 778, American Chemical Society, Washington, D.C., 2001, pp.

30. Lee, K., Akoh, C.C. Structured lipids: Synthesis and applications. *Food Rev. Int.* 1998, 14, 17–34.

31. Senanayake, S.P.J.N., Shahidi, F. Structural lipids containing long-chain omega-3 polyunsaturated fatty acids. In: *Seafoods in health and nutrition: Transformation in fisheries and aquaculture: Global perspectives.* Shahidi, F., Ed., ScienceTech Publishing Co., St. John's, Canada, 2000; pp. 29–44.

32. Odle, J. New insights into the utilization of medium-chain triglycerides by neonate: Observations from a piglet model. *J. Nutr.* 1997, 127, 1061–1067.

33. Willis, W.M., Lencki, R.W., Marangoni, A.G. Lipid modification strategies in the production of nutritionally functional fats and oils. *Crit. Rev. Food Sci. Nutr.* 1998, 38, 639–674.

34. Senanayake, S.P.J.N., Shahidi, F. Enzyme-catalyzed synthesis of structured lipids via acidolysis of seal (*Phoca groenlandica*) blubber oil with capric acid. *Food Res. Int.* 2002, 35, 745–752.

35. Senanayake, S.P.J.N., Shahidi, F. Enzymatic modification of seal (Phoca groenlandica) blubber oil: incorporation of lauric acid, *Food Chem.*, 2005.

36. Jennings, B.H., Akoh, C.C. Enzymatic modification of triacylglycerols of high eicosapentaenoic and docosahexaenoic acids content to produce structured lipids. *J. Am. Oil Chem. Soc.* 1999, 76, 1133–1137.

37. Spurvey, S.A., Shahidi, F. Concentration of gamma linolenic acid (GLA) from borage oil by urea complexation: optimization of reaction conditions. *J. Food Lipids* 2000, 7, 163–174.

38. Spurvey, S.A., Senanayake, S.P.J.N., Shahidi, F. Enzymatic-assisted acidolysis of menhaden and seal blubber oils with gamma linolenic acid. *J. Am. Oil Chem. Soc.* 2001, 78, 1105–1112.

39. Senanayake, S.P.J.N., Shahidi, F. Enzyme-assisted acidolysis of borage (*Borago officinalis* L.) and evening primrose (*Oenothera biennis* L.) oils: Incorporation of omega-3 polyunsaturated fatty acids. *J. Agric. Food Chem.* 1999a, 47, 3105–3112.

40. Senanayake, S.P.J.N., Shahidi, F. Enzymatic incorporation of docosahexaenoic acid into borage oil. *J. Am. Oil Chem. Soc.* 1999b, 76, 1009–1015.

41. Amarowicz, R., Shahidi, F. Antioxidant activity of peptide fractions of capelin protein hydrolysates. *Food Chem.* 1997, 58, 355–359.

42. Shahidi, F., Synowiecki, J. Protein hydrolysates from seal meat as phosphate alternatives in food processing applications. *Food Chem.* 1997, 60, 29–32.

43. Kim, S-K., Kim, Y-T., Byun, H-G., Nam, K-S., Joo, D-S., Shahidi, F. Isolation and characterization of antioxidative peptides from gelatin hydrolyzate of Alaska pollack skin. *J. Agric. Food Chem.* 2001, 49, 1984–1989.

44. Chen, H-M., Muramoto, K., Yamauchi, F. Structural analysis of antioxidative peptides from soybean β-conglycinin. *J. Agric. Food Chem.* 1995, 43, 574–578.

45. Heu, M.S., Kim, J-S., Shahidi, F., Jeong, Y., Yeon, Y-J. 2003. Characteristics of proteases from shrimp processing discards. *J. Food Biochem.* 2003, 27, 221–236.

46. Shahidi, F., Synowiecki, J. Isolation and characterization of nutrients and value-added products from snow crab (*Chionoecetes opilio*) and shrimp (*Pandalus borealis*) processing discards. *J. Agric. Food Chem.* 1991, 39, 1527–1532.

47. Shahidi, F., Arachchi, J.K.V., Jeon, Y-J. Food application of chitin and chitosans. *Trends Food Sci. Technol.* 1999, 10, 37–51.

48. Jeon, Y.-J., Shahidi, F., Kim, S.K. Preparation of chitin and chitosan oligomers and their application in physiological functional foods. *Food Rev. Int.* 2000, 16, 159–176.

49. Jeon, Y.-J., Kamil, J.Y.V.A., Shahidi, F. Chitosan as an edible invisible film for quality preservation of herring and Atlantic cod. *J. Agric. Food Chem.* 2002, 50, 5167–5178.

50. Kamil, J.Y.V.A., Jeon, Y.-J., Shahidi, F. Antioxidative activity of different viscosity chitosans in cooked comminuted flesh of herring (*Clupea harengus*). *Food Chem.* 2002, 79, 69–77.

51. Shahidi, F., Kamil, J., Jeon, Y.-J., Kim, S-K. Antioxidant role of chitosan in a cooked cod (*Gadus morhua*) model systems. *J. Food Lipids* 2002, 9, 57–64.

52. Shahidi, F. Role of chemistry and biotechnology in value-added utilization of shellfish processing discards. *Can. Chem. Neur.* 1995, 10, 25–29.

53. Shahidi, F., Metusalach, Brown, J.A. Carotenoid pigments in seafoods and aquaculture. *Crit. Rev. Food Sci. Nutr.* 1998, 38, 1–67.

54. Shahidi, F., Kamil, Y.V.J. Enzymes from fish and aquatic invertebrates and their application in the food industry. *Trends Food Sci. Technol.* 2001, 12, 435–464.

11

Functional Foods from Meat Animals

YOULING L. XIONG

Department of Animal Sciences,
University of Kentucky, USA

INTRODUCTION

Meat and meat products have been an important human
dietary component in the Asian cultures since antiquity.
Besides providing unique palatability and satiety, this group
of foods offers essential nutrients for growth and sustaining
of human life. The various nutritional benefits of consuming
meat, poultry, and seafood, categorically referred to as "muscle
foods," have been well documented. For instance, muscle foods
provide high-quality proteins having an amino acid composition closely resembling that of the human body. Furthermore,
they are excellent sources of minerals, especially iron, copper,

and zinc, and vitamins, particularly thiamine (pork) and vitamin B-12. In a nutrition study, it was found that pregnant women in Asia who consumed more meat tended to have a greater fetal growth than the other populations that consumed less or no meat.[1]

The health benefits of mammalian, poultry, and aquatic animal meats have also long been recognized. People in Asian countries, especially Chinese, Japanese, Koreans, and Indians, have for thousands of years treated certain particular meat and meat products as special health-healing foods (Table 11.1). Essentially, meat from all the common domestic animal species — beef, pork, lamb (mutton), chicken, duck, and goose — can be beneficial to health if consumed properly. Many of the medicinal effects of meat are described in Chinese medicine books written by ancient doctors, and they are circulated among the people and passed from generation to generation. In addition, a number of nontraditional animal-derived foods, such as sea cucumber, shark cartilage, snake, turtle, and donkey skin, have emerged in recent years in the Asian supermarkets as health-promoting food items. A famous Chinese book, *Huang Di Nei Jing* (The Yellow Emperor's Internal Classic), which was written 2,500 years ago, summarized how to preserve one's health by eating a balanced diet that includes animal food.[21] Thus, in many parts of Asia, a variety of animal-derived foods have been used, with considerable success, either to aid in the cure or alleviation of certain ailments or to enhance one's physical and mental performances. It is worth noting that meats intended for health and remedy are generally cooked with certain herbal medicines or vegetables with known therapeutic effects. Unfortunately, due to a general scarcity of research data, much of the recognized health benefit is understood only at the empirical level. In the wake of tremendous public interest in functional foods, it is expected that fundamental research into the physiological functions and molecular mechanisms of some of the health-promoting muscle food items will be conducted, which undoubtedly will yield scientific evidence to unravel the mystery.

TABLE 11.1   Health-Promoting Effects of Selected Functional Foods
from Meat Animals

Item	Known active component	Purported function	Reference
Pork	Proteins, vitamins, minerals	Benefits the liver	2
Beef	Proteins, vitamins, minerals	Nourishes the blood; benefits the spleen	2
Lamb (mutton)	Proteins, vitamins, minerals	'Yang'; keeps body warm; dispels the cold; benefits the heart	2
Hen	Proteins, vitamins, minerals	'Yang'; nourishes lactating women; stimulates milk production; treats general weakness; benefits the lung	2
Duck	Proteins, vitamins, minerals	'Yin'; balances hot and cold; good for the hot season	
Organ meats	Proteins, vitamins, minerals	Tonics to strengthen the same type of organs; speed up recovery from organ diseases	3, 4
Black-bone chicken	Essential amino acids, B-vitamins, minerals, ergosterol	Nourishes vital energy ('Qi') and blood; tonic for spleen, liver and kidney; treats hepatic fever; alleviates general debility and stress; strengthens muscle and joints	2, 5, 6
Sea cucumber	Polysaccharides, chondroitin sulfate, saponins, essential fatty acids	Treats arthritis and other musculo-skeletal inflam-matory diseases; anticoagulation	7–10
Shark collagen	Peptides, mucopolysaccharides, proteoglycans	Suppresses tumors by inhibiting angio-genesis; alleviates arthritis and inflammation of the joints, bowel lining and muscle tissue	11–18

Table 11.1 (continued)   Health-Promoting Effects of Selected
Functional Foods from Meat Animals

Item	Known active component	Purported function	Reference
Fermented meats	Lactic acid bacteria; peptides	Improves gastro-intestinal functions; suppresses enteric pathogens; modulates immune functions	19,20

GENERAL MUSCLE FOODS

Fresh Meat and Poultry

The notion of *Yin-Yang* has been used to explain some of the
health benefits and often plays a role in one's selection for a
particular meat or poultry product. *Yin* and *Yang* represent
two different energies existing in the human body, and they
are regulated by the type of food consumed, which can be
either *Yin* or *Yang*. A careful balance by eating more of one
type over the other can affect one's health. For example, meat
from a layer hen is considered "hot" or rich in *Yang* while that
from a duck is generally regarded to be "cold" or concentrated
with *Yin*. For this reason, it is a common practice in China
that lactating women, who are generally weak due to loss of
blood, are recommended to include hen meat in their diet.
The meat is believed to contain the type of ingredients that
facilitate speedy recovery of the mother from labor and delivery as well as to stimulate milk production. According to *USA
Today*, the January 6 to 8, 1995, weekend issue, research
conducted by some Western scientists confirmed that chicken
soup can be a good remedy for colds, and the therapeutic effect
is attributed to cysteine, an amino acid released from chicken
in cooking that chemically resembles the drug acetylcysteine
prescribed for respiratory problems. Usually, a whole hen is
slowly cooked either by stewing or steaming to "extract" the
Yang essence (components) as a concentrated form of medicine

or remedy food. Often, tonic herbs or herbal medicines are mixed with the chicken before cooking.

Similarly, rooster meat is thought to enhance one's vital energy and virility. Therefore, it is consumed by people engaged in hard physical work. Lamb, a popular meat item in the northern part of China and in Mongolia, is consumed in large quantities in the winter time because of its "body-warming" or *Yang* effect. According to the Chinese culture, the following common animal meats are beneficial to human health: beef for the spleen, lamb (mutton) for the heart, pork for the liver, and chicken for the lung.[2]

Variety Meats

Many of the organ meats, also known as "by-product meats" or "variety meats" by Westerners, are thought to have particular healing effects for people who suffer chronic illness or physiological disorders. For example, in China, pork kidney and stomach are consumed by people who have ailments related to the kidney or stomach functions. This dietary intervention stems from the belief that "you are what you eat" or belief that damaged human body parts can be repaired by ingested counterparts from the animals. Because of the remarkable resemblance between human and the hog in body composition, muscle as well as organ meats from hogs are generally preferred to those from other species in the diets of the Chinese and many other Asian countries. In fact, studies in the 1920s had confirmed that animal organ meats (liver, kidney, heart, etc.) are particularly effective for treating anemia.[3,4] It is likely that the curing effect is due not only to the high concentrations of hemoglobin in these organ meats but also to the presence of other compounds that may help reestablish the patient's normal organ functions. Furthermore, coagulated and cooked blood, collected from the duck, chicken, or hog, is consumed by many individuals because blood is believed to have a cleansing effect for the lungs. Blood is a rich source of high-quality proteins, amino acids, and antioxidant compounds. When transported into the human body, these components may exert certain physiological functions yet to be elucidated.

Processed Meats

For processed meat products, it is generally believed that they contain special health-promoting components, including those from exogenous sources. For example, fermented sausages contain probiotic bacteria, particularly those that belong to the family of lactic acid-producing organisms. People in the West Asia region, including countries in the Middle East, have consumed fermented meats for thousands of years. The Japanese have also long used fermentation techniques to produce fermented fish products as health-promoting foods. Because fermented meats are usually cooked before consumption, one should not expect ingestion of live probiotic bacteria grown in the sausage. However, metabolites and end products produced by some strains of *Lactobacillus*, *Streptococcus*, and other homo- and hetero-fermentative cultures in fermented meats (e.g., lactic acid, enterocins, and other bioactive peptides) may have prebiotic effects in maintaining an optimum gastrointestinal condition by suppressing pathogenic microorganisms, or they can be absorbed and subsequently participate in the regulation of internal physiological processes.[19,20]

SPECIAL MUSCLE FOODS

Black-Bone Chicken

Several other particular meat and poultry items have also long been regarded to be functional or medicinal foods. Among them is the black-bone chicken, *Gallus Domesticus*. A genetically unique type of poultry originating from China during the Tang Dynasty in A.D. 618, black-bone chicken is a small bird having black-colored bones but with different feather and meat colors — white or black (Figure 11.1). Because of its rarity, black-bone chicken was once prepared only for the royal family, hence, acquiring the phrase "Food for the Kings" in ancient China. Today, black-bone chickens are grown and consumed in most Asian countries as well as in some parts of

Figure 11.1 Black-bone chicken, *Gallus Domesticus*. Photograph courtesy of John Mok (www.e2121.com).[22]

Europe. In the United States, a counterpart, called "Silkie Bantam," is also grown, but the birds are raised primarily as pets. Black-bone chicken can also be found as frozen items or capsulated extracts at Asian supermarkets in North America. There are many purported health benefits of black-bone chicken. For example, it is consumed to improve the function of the liver and the kidney, to nourish the blood for those who suffer irregular menstruation, leucorrhea, and dysmenorrheal, to alleviate general debility and stress, and to help build muscle and strengthen the joints.[2,5] Because of these observed physiological functions, black-bone chicken is a favored meat item for Chinese women during the course of pregnancy and subsequent lactation. While the exact chemical and biochemical mechanism remains unknown, nutritional analyses show that black-bone chicken meat is rich in essential amino acids, minerals, and B-vitamins.[5]

Sea Cucumber

A variety of marine species are also consumed in Asian countries as functional foods. Sea cucumber, *Stichopus chloronotus*, an animal related to starfish and urchins, has been used as a treatment for arthritis. There are over 1,000 cucumber species; most of them have a cucumber-shaped, elongated muscular flexible body (Figure 11.2). The popular Chinese name for sea cucumber is "Haishen," meaning ginseng of the

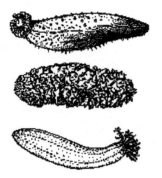

Figure 11.2 Examples of sea cucumber, *Stichopus chloronotus.*
Source: http://www.starfish.ch/reef/echinoderms.html#brittle.[23]

sea. Fresh harvested sea cucumber is cleaned and boiled in
salty water and then air dried for a long shelf life. When used
as a delicacy or medicinal food, it is softened in water and
then cooked.

Sea cucumber has been used to help treat musculo-skeletal
inflammatory diseases, including rheumatoid arthritis,
osteoarthritis, and ankylosing spondylitis, a rheumatic ail-
ment that affects the spine. Although few clinical studies have
been conducted to substantiate the observed medicinal effects,
analyses of the chemical composition revealed high contents
of mucopolysaccharides (sulfated fucans) and chondroitins in
this marine species, which are often deficient in people with
arthritis and connective tissue disorders.[9] The glycosamino-
glycan from sea cucumber has the same structure as mammalian chondroitin sulfate, and thus, may function similarly
to inhibit inflammatory arthritis. Furthermore, sea cucumber
fucosylated chondroitin sulfate has been shown to inhibit
smooth muscle cell proliferation as heparin, and it has a
potent enhancing effect on endothelial cell proliferation.[10] Its
ability to modulate *in vitro* fibroblast growth factor 2-depen-
dent angiogenesis indicates potential inhibitory effects on
tumors. Sulfated fucans also exhibit strong anticoagulant
activities.[7]

An excellent source of essential fatty acids, sea cucumber
may have the ability to regulate the synthesis of prostaglandins

that are involved in the inflammatory and wound-healing processes.[8] Saponin glycosides are also present in sea cucumber meat. These compounds have a structure resembling the active constituents found in ginseng, ganoderma, and other famous tonic herbs that are believed to nourish the blood and vital essence. Thus, the Chinese utilize sea cucumber to treat weakness, debility, and frequent urination resulting from disorders of blood and kidney functions. For modern applications, dried or extracted sea cucumber is used as nutritional supplement, manufactured in the form of capsule or tablet.

Shark Cartilage

Another functional seafood item is shark cartilage, which is widely marketed in Asia primarily as an anticancer substance. Shark cartilage contains a number of bioactive components, some of which are proteins and peptides that act as angiogenesis inhibitors, i.e., suppressing the development of new blood vessels.[14,17] This biochemical function lends shark cartilage itself to potentially fighting against the development and growth of tumor cells, which is normally accompanied by the development of new networks of blood vessels for nutrient supply. The therapeutic effect of shark cartilage on human patients with advanced cancer has been controversial. While some clinical studies demonstrated a high efficacy of shark cartilage for treating advanced cancer patients,[24] other studies showed that shark cartilage as a single agent was inactive in patients with advanced cancer.[15] Analgesic and anti-inflammatory effects of shark cartilage have also been reported.[12,13] Inflammatory conditions for which shark cartilage may be useful include arthritis, regional enteritis (inflammation of the lining of the bowel), and psoriasis — a common inflammatory skin disease with dilation of capillaries as an early histological change.[11,16] Response of some of the above conditions to shark cartilage may result from the presence of mucopolysaccharides and proteoglycans present in the cartilage tissue that act to stimulate the immune system.[13,18] To prepare shark cartilage feasible for ingestion, the tough elastic skeleton of the shark is dried and pulverized.

EMERGING NEW MUSCLE FOODS

Lean Meat Products

Due to the difficulty in clearly defining physiological functions of muscle foods, a new strategy for functional muscle foods has been developed in recent years. The concept is to produce "healthier" meat and meat products either by removing or by reducing some of the undesirable meat components (e.g., fat) that negatively impact human health. Thus, as is in the Western countries, new animal breeding techniques and biotechnologies as well as nutritional interventions have been incorporated into animal production strategies in Asia to yield leaner meat animals. Furthermore, a variety of muscle foods with reduced (removed) fat are being produced by the Asian meat industry. A particular product of this type, called "ham sausage," which contains as little as 1% animal fat but as much as 10% starch and 30% or more added water, has been developed and currently dominates in the Chinese convenience food market. Ham sausage is processed by curing fresh meat with sodium nitrite to elicit a characteristic cured meat aroma and pinkish color. Cured meat is stuffed in heat-resistant plastic casing and subsequently cooked in a sterilizing retort. The product requires no refrigeration for storage and has a shelf life of at least 6 months. Some processed meat items also contain added nutraceuticals, e.g., antioxidants, dietary fibers, probiotics, and prebiotics to boost the product health benefits.

Muscle Foods from Nutrition-Modified Animals

In addition to the manipulation of meat-raw materials and product formulations, there are production strategies aimed at modifying and optimizing the muscle composition for specific nutritional purposes. Based on the recent success in the Western countries in producing "functional" meat through feeding livestock antioxidant vitamins, selenium yeast protein, fish oil diets with a high content of health-promoting fatty acids (conjugated linoleic acid or CLA, docosahexaenoic acid or DHA, eicosapentaenoic acid or EPA, etc.), meat producers in Asia have started some preliminary trials. Pork, beef, poultry, and fish raised on these special diets have been shown to have elevated levels in these nutrients.[25-27] These emerging nutritional and medical interventions are still in their infancy, and, unlike in the Western countries, they are being carried out on a very limited scale by the Asian meat researchers.

CONCLUSIONS

The health-promoting and remedy effect of certain meat, poultry, and seafood products has not been recognized by the vast majority of the world's population when compared to other functional food categories, notably those derived from plant sources. On the contrary, with an increasing awareness of the possible side effect associated with meat overconsumption, an increasing number of Asian consumers are restricting meat in their regular diet. The relative high contents of saturated fatty acids and cholesterol present in meat, which are shown by medical professionals to cause coronary cardiovascular diseases, have somewhat overshadowed the nutritional benefits from meat consumption, let alone the possible functional or medical benefits. Despite its empirical nature, the health-promoting or medicinal effect of certain yet-to-be-identified meat constituents or meat products seems to be undeniable. As evidenced, the Chinese have long used chicken and pork as remedy foods and they seemed to work. Thus, the true challenge for muscle food researchers, nutritionists, and medical professionals is not one that attempts to establish whether some muscle foods are of both nutritional and medicinal values, but one that determines the molecular mechanism underlying the health benefits. Some progress has been made in the past few years in this emerging research field, especially with aquatic meat species in which a number of therapeutic and bioactive compounds have been identified and their linkage to the prevention and remedy of illness established. However, more fundamental and clinical studies that involve all the other meat animal species are required in order to wisely utilize muscle foods to improve the welfare of the general public.

REFERENCES

1. PA Wharton, PM Eaton, BA Wharton. Subethnic variation in the diets of Moslem, Sikh and Hindu pregnant women at Sorrento Maternity Hospital, Birmingham. *Br J Nutr* 52:469–479, 1984.

2. L Li, L Yin, J Zhang, X Zhang, Z Lin. Functionalities of Traditional Foods in China. Proceedings of 9th JIRCAS International Symposium, Epochal Tsukuba, Japan, 2002, pp 140–144.

3. FS Robscheit-Robbins, GH Whipple. Blood regeneration in severe anemia. II. Favorable influence of liver, heart and skeletal muscle in diet. *Am J Physiol* 72:408, 1925.

4. FS Robscheit-Robbins, GH Whipple. Blood regeneration in severe anemia. VI. Influence of kidney, chicken and fish livers, and whole fish. *Am J Physiol* 79:271, 1927.

5. E Zhang. White Phoenix Bolus of Black-Bone Chicken. In: E Zhang, Ed. Highly Efficacious Chinese Patent Medicines. Shanghai, China: Publishing House of Shanghai University of Traditional Chinese Medicine, 1990.

6. YH Li, XL Li. Determination of ergosterol in cordyceps sinensis and black-bone chicken capsules by HPLC. *Yao Xue Xue Bao* (in Chinese) 26:768–771, 1991.

7. O Berteau, B Mulloy. Sulfated fucans, fresh perspectives: structures, functions, and biological properties of sulfated fucans and an overview of enzymes active toward this class of polysaccharides. *Glycobiology* 13R:29R–40R, 2003.

8. BD Fredalina, BH Ridzwan, AA Abidin, MA Kaswandi, H Zaiton, I Zali, P Kittakoop, AM Jais. Fatty acid compositions in local sea cucumber, *Stichopus chloronotus*, for wound healing. *Gen Pharmacol* 33:337–340, 1999.

9. PA Mourao, IG Bastos. High acidic glycans from sea cucumbers. Isolation and fractionation of fucose-rich sulfated polysaccharides from the body wall of Ludwigothurea grisea. *Eur J Biochem* 166:639–645, 1987.

10. J Tapon-Bretaudiere, D Chabut, M Zierer, D Helley, A Bros, PAS Pourao, AM Helley. A fucosylated chondroitin sulfate from Echinoderm modulates *in vitro* fibroblast growth factor 2-dependent angiogenesis. *Mol Cancer Res* 192:96–102, 2002.

11. E Dupont, PE Savard, C Journain, C Juneau, A Thibodeau, N Ross, K Marenus, DH Maes, G Pelletier, DN Sauder. Antiangiogenic properties of a novel shark cartilage extract: potential role in the treatment of psoriasis, *J Cutan Med Surg* 293:146–152, 1998.

12. JB Fontenele, GB Araujo, JW de Alencar, GS Viana. The analgesic and anti-inflammatory effects of shark cartilage are due to a peptide molecule and are nitric oxide (NO) system dependent. *Biol Pharmaceut Bull* 20:1151–1154, 1997.

13. JA Kralovec, Y Guan, K Metera, RI Carr. Immunomodulating principles from shark cartilage. Part 1. Isolation and biological assessment *in vitro*. *Int Immunopharmacol* 3:657–669, 2003.

14. A Lee, R Langer. Shark cartilage contains inhibitors of tumor angiogenesis. *Science* 221:1185–1187, 1983.

15. DR Miller, GT Anderson, JJ Stark, JL Granick, D Richardson. Phase I/II trial of the safety and efficacy of shark cartilage in the treatment of advanced cancer. *J Clinic Oncol* 16:3649–1655, 1998.

16. M Milner. A guide to the use of shark cartilage in the treatment of arthritis and other inflammatory joint diseases. *Am Chiropractor* 21(4):40–42, 1998.

17. JR Sheu, CC Fu, MI Tsai, WJ Chung. Effect of U-995, a potent shark cartilage-derived angiogenesis inhibitor, on antiangiogenesis and antitumor activities. *Anticancer Res* 18:4435–4441, 1998.

18. DX Li, SQ Zhang, P Liu. Study on the purification of SCAMP and the preparation of its sulfated polysaccharides. *Guang Pu Xue Yu/Guang Pu Fen Xi* 22(1):59–62, 2002.

19. MR Foulquie Moreno, R Callewaert, B Devreese, J Van Beeumen, L de Vuyst. Isolation and biochemical characterization of enterocins produced by enterococci from different sources. *J Appl Microb* 94:214–229, 2003.

20. RSD Read. Macronutrient innovations and their educational implications: proteins, peptides and amino acids. *Asia Pacific J Clinic Nutr* 11(S6):S174–S183, 2002.

21. S Li. The scientific way to eat and keep good health in ancient books. *Chinese Foods* (Chinese) 3:217, 1996.

22. http://www.e2121.com.

23. http://www.starfish.ch/reef/echinoderms.html#brittle.

24. IW Lane, E Contreras. High rate of bioactivity (reduction in gross tumor size) observed in advanced cancer patients treated with shark cartilage material. *J Naturopathic Med* 31: 86–88, 1992.

25. GF Combs, Jr, JM Regenstein. Influence of selenium, vitamin E and ethoxyquin on lipid peroxidation in muscle tissues from fowl during low-temperature storage. *Poult Sci* 59:347–351, 1980.

26. DC Mahan, TR Cline, B Richert. Effects of dietary levels of selenium-enriched yeast and sodium selenite as selenium sources fed to growing-finishing pigs on performance, tissue selenium, serum glutathione peroxidase activity, carcass characteristics and loin quality. *J Anim Sci* 77:2172–2179, 1999.

27. BD Choi, SJ Kang, YL Ha, RG Ackman. Accumulation of conjugated linoleic acid (CLA) in tissues of fish fed diets containing various levels of CLA. In: YL Xiong, CT Ho, F Shahidi, Eds. *Quality Attributes of Muscle Foods*. New York: Kluwer Academic/Plenum Publishers, 1999, pp 61–71.

12

Functional Foods from Fruit and Fruit Products

JOHN SHI

Guelph Food Research Center, Agriculture
and Agri-Food Canada, Canada

JAMES H. MOY

Department of Molecular Biosciences and
Biosystems Engineering, University of Hawaii
at Manoa, USA

INTRODUCTION

Most Asian countries are located in the tropical, subtropical, and temperate zones. In these regions, agriculture dates back several thousand years, with fruits as one of the major sources of food and nutrition. There are many varieties of fruits that grow abundantly in different regions. Many of them are eaten fresh or processed into products. Different end products, such as dried fruits, jam, marmalade, juice, preserves, pickles, puree, canned products, and many other forms such as whole, slices, pieces, bars, powders, flakes, or leathers have been available in the market for a long time. In traditional diet and medicine, fruits are used not only as foods, but also for medicinal purposes. Some fruits and their products have been used historically as natural materials to maintain health and to prevent or cure diseases for humans.

Health authorities, medical practitioners, and nutritionists in recent years have repeatedly urged people to consume a generous portion of fruits and vegetables in their daily diet. Many components in fruits such as carotenoids, flavonoids, polyphenols, isothiocyanates, and fiber have been demonstrated to confer health benefits such as antimutagenicity, antioxidation, and inhibition of tumor promotion. Increased consumption of fruits improves protection against oxidative damage that may play a role in carcinogenesis and some chronic diseases, namely oxidative DNA damage and lipid peroxidation. Modern epidemiological studies have demonstrated that fruit consumption reduces risks of certain forms of cancer, especially cancers of the gastrointestinal tract.[1,2] Because fruits are rich in substances with antioxidant activity, it has been proposed that the antioxidant properties of fruits are responsible for maintaining human health.

Presented below are 38 major and 8 minor fruits, and several forms of preserved fruits and fruit products that have various functional components believed to be helpful to the human body. Some of the statements are derived from tradition and observations, and some are based on scientific studies. Not all the benefits of these functional components have been scientifically or clinically proven, but many of the results have been observed and carried from generation to generation, some over a period of a thousand years or more. The fact remains that consuming fruits is healthy, and causes no harm. Its implied safety is in line with the concept of GRAS (generally regarded as safe), long adopted by the U.S. Food and Drug Administration.

FRUIT PRODUCTS

Annatto Fruits

The annatto (*Bixa orellana* L.) is cultivated in many tropical countries in Asia. The fruits inside are generally found in cavities containing 10 to 50 small seeds, about the size of grape seeds. The seeds are covered with a thin, highly colored, orange to red resinous layer from which the natural color is obtained. The main application of annatto fruits is for coloring cheese and other dairy products such as ice cream, butter mixes, yogurt, meat (sausages), fish margarine, snacks, dressings, sauces, and confectionery. Among the naturally occurring colorants, annatto colorant ranks second in economic importance after caramel.

Around 80% of the carotenoids in annatto consist of bixin, a diapocarotenoid that contains 24 carbon atoms in the skeleton and a (Z)-double bond. Also norbixin, the corresponding dicarboxylic acid with the (all-*E*)-configuration has been isolated. It is well established that the carotenoids possess a wide range of biological activities, with potential health benefits. Some minor carotenoids in annatto and some 15 minor compounds have been identified recently.[3]

Apples

Apples (*Malus sylvestris*) are grown in China, Korea, Japan, and India. Most of apples are eaten fresh. Fresh apples can be made into puree, then dried and grounded into powder. Apple powder is a special product used mainly in the treatment of infant diarrhea. Apple pectin in apple powder has a strong bacteriostatic action on *Staphylococcus aureus*, *Streptococcus faecalis*, *Pseudomonas aeruginosa*, and *Escherichia coli*.[4] Eberhardt et al.[5] suggested that the strong inhibition of tumor-cell proliferation *in vitro* could be due to apples' combination of phytochemicals (phenolic acids and flavonoids), as these are natural antioxidants. Apple pectin inhibits azoxymethane (AOM)-induced colon carcinogenesis. Therefore, apple pectin may be expected to have a strong influence on the intestinal microflora and bacterial enzyme activities. The same intestinal bacteria may reportedly play a significant role in the pathogenesis of colon cancer because their enzymes are important in the metabolism of procarcinogens, and the production of tumor promoters in the colon.[6]

Apple pectin has a stronger bacteriostatic action against pathogenic bacteria than citrus pectin. The induction of colon neoplasia by AOM is dose-dependently inhibited by apple pectin. Fecal tryptophanase activity tends to decrease in the apple pectin group compared with that in the control group. The reduced level of tryptophan metabolites in the colon might be related to the inhibitory effect of apple pectin on colon carcinogenesis.[4] Apple pectin exerts an antitumor effect and prevents cancer metastasis and carcinogenesis by modifying host immune function, and altering the intestinal flora. The inhibition of hepatic metastasis by oral administration of apple pectin (apple powder) suggests it may be effective for the prevention of hepatic metastasis and residual cancer cells remaining after surgery.

Aronia Fruits

Aronia (*Aronia melanocarpa* Michx.)trees are organically grown as a hardy plant, and aronia berries are not attacked by insects. Aronia trees grow as high as 6 feet, and the dark-colored berries are picked in late August or early September. Aronia berries contain one of the highest concentrations of anthocyanin pigments of any cultivated plant. The pigment is located throughout the berry. The majority of the anthocyanin pigment remains with the pomace when the juice is squeezed. Thus, this waste product is an excellent source of anthocyanin pigments.

Aronia berries contain four kinds of anthocyanins. All of the anthocyanins identified are 3-substitued monoglycosides and they are present in the following percentages, galactose (68%), arabinose (28%), glucose (1.5%), and xylose (2.5%).[7] Only two of these compounds (galactose and arabinose) account for more than 95% of the material. The anthocyanins obtained from aronia provide one of the least complex mixtures found in the plant kingdom.

Aronia fruits (chokeberry) provide a healthy fruit beverage and are a natural colorant. Aronia berry juice has an astringent taste, very similar to that of cranberry and black currant. Aronia berry is rich in vitamins, minerals, antioxidants, and other health-beneficial materials. Aronia juice concentrate is attractive to consumers. The aronia beverage is high in flavonoids, specifically condensed tannins and anthocyanins, and is considered beneficial to human health.

Avocados

Avocados (*Persea Americana* Mill.) are commercially grown in the Philippines, Hawaii, and Israel. People usually prefer the avocado fruits sweetened with sugar, or combined with other fruits such as pineapples, oranges, grapefruits, dates, and bananas. Avocados have a high lipid content of 5 to 25%. Among the saturated fatty acids the myristic level may be 0.1%, palmitic 14 to 21%, and stearic 0.6 to 1.7%. The oil in the flesh is rich in Vitamins A, B, G, E. The fruit peel is considered as an antibiotic for vermifuge and a remedy for dysentery.

Bananas

Bananas (*Misa acuminata* Colla.) are grown in the humid tropical regions and constitute one of the largest fruit crops of the world. India is the leading banana producer in Asia. Other producers are Taiwan, Indonesia, and the Philippines. Most of bananas are eaten fresh. Some are dried in various forms such as banana powder and some are made into puree. Diced banana products are used as raisin substitutes in food ingredients. They can be eaten as a snack food or used in making fruit cake and bakery products. Banana puree is by far the most important processed product from the pulp of ripe fruits. The puree has a creamy white to golden yellow color, free from musty or off-flavors. Banana puree is an important infant food. Puree canned in drums by the aseptic canning process is a new product for the baking and ice cream industry.

All parts of banana fruits, peel and flesh, have medicinal applications. Banana pulp soup is taken to control dysentery and diarrhea and also used for treating malignant ulcers. Antifungal and antibiotic principles are found in the peel and pulp of ripe bananas. Norepinephrine, dopamine, and serotonin are also present in the ripe bananas, which give bananas a functionality in elevated blood pressure and inhibiting gastric secretion and stimulating the smooth muscle of intestines.[5]

Bilberries

The bilberry (*Vaccinium myrtillus* L.) is a low-growing shrub native to Asia and northern Europe. The small dark blue fruit is eaten fresh or made into juice and preserves. The dried berries can be used as snacks. Bilberries can be used as a colorant for several food products. Bilberry juice is one of the most anti-mutagenic fruit products and is effective in reducing mutagenicity caused by the polycyclic aromatic hydrocarbons.[9]

Gallic acid, an astringent, and an unusual phenolic acid, melilotic acid are identified in bilberry fruits and leaves. Other phenolic compounds are also found in the plant. Bilberry fruits contain flavonoids (quercitrin, isoquercitrin, hyperoside, avicularin, meratine, and astragaline), catechin tannis, oligomeric proanthocyanidins, iridoid monoterpenes (asperuloside and monotropein), phenolic acids (chlorogenic, salicylic, and gentisic), quinolizidine alkaloids (myrtine, epimyrtine and occasionally arbutine), and some pectin.[10] Bilberries have a wide range of fiber values on a fresh weight basis: insoluble dietary fiber 1.9 to 3.2%, soluble dietary fiber 0.4 to 1.1%, total dietary fiber 2.3 to 3.9%.[11] Bilberry fruit and leaves are used for a variety of medical conditions. The functional components of bilberry products appear to be the phenolic compounds, particularly the anthocyanins. Several anthocyanins are found in the fruits. The pigments are located primarily in the skin of the berries.

Morazzoni and Bombardelli[12] reviewed the history of medical uses for bilberries from the Middle Ages to the present. Surveys of 1,994 cases found 183 products containing bilberry extract as an ingredient.[13] Bilberry anthocyanins also reduce platelet aggregation *in vitro*.[14] Anthocyanin extracts inhibit porcine elastase *in vitro*. The fruits of bilberry are used to treat many conditions resulting from diabetes. The high levels of glucose in the blood of diabetics trigger many deteriorative events in the body. Bilberry extract is believed to help improve eyesight, particularly night vision. This health benefit is the primary reason for the product's popularity in Japan and Korea.[15] Anthocyanins are important for regeneration of visual purple.[16] Bilberry extracts appear to benefit vision in several ways: improving night vision by enhanced regeneration of retinal pigments, increasing circulation within the capillaries of the retina, inhibiting of Maillard reactions in the lens to reduce cataract formation, and protection from ultraviolet light.[16]

The antioxidant properties of bilberry extracts may be responsible for these health benefits. Antioxidants have been suggested to retard oxidation in the lens and slow retinal angiopathy that occurs in age-related macular degeneration and diabetic retinopathy.[17] Tannins in bilberry are considered to be responsible for their ability to treat acute diarrhea and mild inflammations of the mouth and throat.[18]

Black Prunes

The black prune (*Prunus armeniace,* Thunb.) is also known as the Japanese apricot. The fruit is sour and tart to taste. Major chemical components are glucoside prudomenin, malic acid, and succinic acid.[19] Some of its medicinal functions are to act as an astringent, antipyretic, and vermicidal, to stimulate contraction of the muscles of intestinal parasites and gallbladder, and to cause relaxation of the bile duct.[20] It is also an antimicrobial agent. People usually use it for the treatment of chronic diarrhea and dysentery, feverish thirst, achlorhydria, no appetite, residue coughing, chronic malaria, biliary ascariasis, hookworms, abdominal pain, cholecystitis, and gallstones. The fruits are commonly preserved as snack foods, or made into a beverage or wine.

Carambolas

The Carambola (*Averrhoa carambola* L.) is originated in Sri
Lanka. The common name is "star fruit" due to its shape when
cut in cross section. The major production areas are in East
Asia, including Indonesia, Malaysia, Sri Lanka, Taiwan, the
southern part of China, and Vietnam. There are two distinct
cultivars: sweet and sour cultivars. The sour cultivar is rich
in flavor, with more oxalic acid.[8] The sweet cultivar is mild
flavored, rather bland, with less oxalic acid. The sweet cultivar
is used for fresh consumption and juice processing, while the
sour cultivar is processed into jam, jelly, canned fruits, sweetened nectar, or other preserves. Juice products are by far the
most important processed commodities of carambola fruits.

Fermented carambola juice is a traditional health drink
in China and India. It is served as a cooling beverage, and
good for smoothing some uncomfortable body conditions, such
as to quench thirst, increase the salivary secretion, and allay
fever. The fruit pulp is considered to allay biliousness and
diarrhea, and relieve a hangover from excessive indulgence in
alcohol. In India, the ripe fruit is used to halt hemorrhages
and to relieve bleeding hemorrhoids. Carambola is recommended as a diuretic in kidney and bladder complaints, and is
believed to have a beneficial effect in the treatment of eczema.[8]

Cherimoyas

Cherimoya (*Annona cherimola* Mill.) is growing in the Philippines, India, and Sri Lanka. The flesh of ripe cherimoya
fruits is most commonly eaten out of hand. Fruits also can be
made into juice and salad, or fermented into alcoholic beverage, or dried into powder. The cherimoya fruit's powder is
used to kill lice and is applied on parasitic skin disorders, and
also to relieve pneumonia.[8]

Chinese Dates

Chinese dates (*Zizyphus vulgaris var. spinosa*) are grown in
the Hunan, Shandong, Zhejiang, and Shanxi provinces of
China. Chinese dates are dried in the sun, or by dryers;

depending on the technique used for drying, the final dried dates products have different names: "red dates" and "black dates." The final products have moisture levels of 18 to 20%.

1. **Red Dates:** Fully ripe Chinese dates are blanched and dried as whole fruits by the sun. The product has a dark-red color, golden-yellow meat, elastic texture, and sweet taste.
2. **Black Dates:** Fully ripe Chinese dates are selected, blanched, then dried and fumed at 60 to 70°C for 20 to 24 h.[21] The product has a dark-violet color, wrinkled surface, sweet taste, and elastic texture.

Chinese date products have a special function in invigorating blood circulation according to Chinese traditional medicine. Some major medicinal functions are to strengthen the spleen and stomach, moisturize the heart and lungs, and regulate various medications. Chinese dates products are used for treatment of weak stomach and spleen, anemia, inadequate energy (fatigue), and salivation. Dates and rice cooked into gruel is nutritious.

Citrus Fruits

Citrus fruits are widely known to contain various types of chemopreventers such as D-limonene, limonoids and their glucosides, flavonoids, and carotenoids. Levels of auraptene vary from high (408 to 585 mg/kg fresh wt.) in the peels of natsumikan and hassaku oranges, moderate (101 to 120 mg/kg fresh wt.), and absent (<1 mg/kg fresh wt.) in the peels of the Satsuma mandarin (tangerine), Valencia orange, navel orange, lemon, and lime. The auraptene content in the sarcocarps of the above fruits is similar to that in the peels.[22] Commercial juices from natsumikan and hassaku oranges showed higher contents of auraptene (0.87 to 1.80 mg/L). Auraptene, a citrus coumarin, is an effective cancer chemopreventer. These characteristics together with high chemopreventive potency make it an appropriate source substance for the creation of physiologically functional foods. The citrus juices have hypocholesterolemic effects in heart disease.[23]

Oranges

The juice of oranges (*Citrus sinensis*) grown in China, India, and Japan has a deep orange color. Orange juice concentrate is prepared from either freshly extracted and pasteurized single-strength juice or from a storage and pasteurized single-strength juice. Spraying and drum drying produce dehydrated orange juice. The final powder has less than 0.6% moisture and maintains its quality when stored at room temperature. Orange products are traditionally taken to allay fever and catarrh. The roasted pulp is prepared as a poultice for skin diseases. The immature fruits are also made into infusion (or tea) to relieve stomach and intestinal complaints.[8]

Red Tangerine

Red tangerine (*Citrus reliculata* Blanco.) peels are the red-colored external layer of the pericarp. The major chemical components are citral, geraniol, linalool, methylanthranilate, stachydrine, putrescine, apyrocatechol, and glucosides (naringin, poncirin, hesperidin, neohespiridin, and nobiletin).[8] Some of the medicinal functions are to correct energy circulation, strengthen the lungs, and resolve phlegm. Red tangerine peels are used for treatment of fullness in chest and indigestion, eliminating sputum and coughing.[8] People use it as a tea, or prepare a red tangerine peel gruel made by decocting and cooking, in which the ingredients include red tangerine peel and bitter apricot kernel. The red tangerine peels can also be steamed with chicken and wine.[28]

Mandarin Oranges

Mandarin orange (*Citrus reticulata*) is considered a native of southeastern Asia and the Philippines. It is now abundantly grown in China, Japan, and India. Mandarin oranges are eaten fresh, or used in fruit salads, gelatins, puddings, or cakes. The essential oil from the peel is produced commercially as a flavoring ingredient. Mandarin orange peels are dried peels that are bitter and acrid to taste. Major chemical components are bitter-tasting flavone glycosides (neohesperidin and naringin, neohesperidose), nonbitter flavonoids (hesperidin, rutoside, sinensetin, nobiletin, and tangeratin), 1 to 25 essential oils (limonene), and pectin.[8] Some medicinal functions are to correct energy circulation, strengthen the spleen, counteract excessive moisture in the body, and resolve phlegm. People often use it for easing of fullness in chest and abdomen, regurgitation and vomiting, chest and abdominal pains, poor appetite, productive coughing, indigestion, and diarrhea. It also can be used as a popular dish called mandarin orange peels beef, or as preserved fruit products to enhance the digestive system and blood circulation.

Kumquats

Kumquats (*Fortunella margarita*), also called "Golden Orange" in China, originated in Northern China, and are produced primarily in China and the Philippines, with limited production in other areas of the world such as in Southeast Asia and Japan. Kumquats are eaten fresh or are candied or cured whole, and are unique in that the entire fruit, including the peel, is generally eaten. The cured products have a golden color, translucent texture, dry surface without sugar particles, and strong fresh flavor. The cured kumquat products are usually used as confections to improve appetite.[8]

Pummelos

Pummelos (*Citrus maxima* Merr.) are the largest citrus fruit, native to southeastern Asia, and are grown in China, India, Indonesia, Japan, Malaysia, Thailand, Taiwan, and the Philippines. People like to eat the juicy pulp, which is used for salads, desserts, or made into preserves. Pummelo juice makes an excellent beverage. The pulp and peel have a sedative effect in cases of epilepsy, chorea, and convulsive coughing. Pummelo juice is taken as a febrifuge in the Philippines and Southeast Asia.[8]

Cranberries

Cranberry (*Vaccinium macrocarpon* A.) products as dietary supplements are widely available in a variety of food and beverage forms. Cranberry fruits contain phytochemicals, which include flavonoids and phenolic acids with antioxidant and other physiologically beneficial activities. Classes of cranberry flavonoids include anthocyanins, flavonols, flavan-3-ols, and proanthocyanidins. Each of these classes of compounds has interesting physiological activities. Anthocyanins are the pigments that give cranberries their rich, red color. Cranberry fruits also contain ellagic acid, which has been shown to have a broad range of anticarcinogenic activities.[24]

Cranberries and cranberry products have long been associated with a variety of health benefits. Cranberries appear to have a relatively unique menu of components that have interesting value in human nutrition, particularly in maintaining health and wellness. Cranberries, and particularly cranberry juice products have long enjoyed a folk reputation as a treatment for urinary tract infections. Although the low pH of the fruit is considered as the antimicrobial agent, fructose and high molecular weight phenolic compounds have been found to prevent the adhesion of *Escherichia coli* cells *in vitro*.[24] Fructose and polyphenols prevent mamnnose-resistant adhesions on certain P-fimbriated *E. coli* isolated from attaching epithelial tissues in the urinary tract.[25] Purified cranberry proanthocyanidins are reported to possess antiadherence properties in an *in vitro* assay.[26] It is said that cranberries can dress wounds and prevent inflammation, and were used aboard ships to help prevent scurvy, although their level of vitamin C is well below that of most citrus fruits. Cranberries are thought to help relieve the symptoms of urinary tract infections, even prevent their occurrence. Much anecdotal information is responsible for the medical myth that surrounds the fruits and their products.

Durians

Durians (*Durio zibethinus* L.) are large fruits covered with hard, hexagonal, stubby spines. It is a heavy fruit reaching the size of a honey melon. Skin of the ripened fruit turns from brown to bright yellow. Durian is a delicious tropical fruit and well known throughout Southeast Asia, Thailand, Malaysia,

and South Vietnam; the southern Philippines are important
producers of durians. Durian flesh is mostly eaten fresh but
is also canned in syrup, or dried, or made into paste. Durian
is a good source of iron, B vitamins, and ascorbic acid. The
thick, pudding-like texture of the aril is due to gums, pectin,
and hemicellulose.[27] The flesh of durian is said to serve as a
vermifuge. Durian flesh is also widely considered an aphro-
disiac in Thailand. In India, durian products are marketed to
provide energy, to keep the body vigorous and tireless, the
mind alert with faculties undimmed and spirit youthful.

Embalics

Embalics (*Phyllanthus emblica* L.), native to Southeastern
Asia, are grown in tropical Asia such as Bangladesh, Cambodia, India, Malaysia, Pakistan, Sri Lanka, Southern China,
Thailand, Vietnam, and the Philippines. Both ripe and half-
ripe fruits can be canned or made into jam and juice. Embalics
can be combined with other fruits in making chutney and
pickles. The embalics are important in Asian traditional medicine as an antiscorbutic and in the treatment of diverse
ailments, especially with the digestive organs. Embalic's are
often considered diuretic and laxative. The embalics pulp and
juice, especially after fermentation, are helpful for indigestion, anemia, jaundice, dyspepsia, coughs, nasal congestion,
retention of urine, and some cardiac problems. The embalic
powder is an effective expectorant as it stimulates the bronchial glands.[8]

Figs

Figs (*Ficus carica* L.) are believed to be indigenous to Western
Asia, and grown in mild temperate climates, and have been
commercially produced in most of the countries bordering the
Mediterranean Sea. Fig flesh is usually eaten out of hand,
but the fruits are also cooked in pies, cakes, bread, cookies,
or ice cream. Fruits also can be prepared into jam, marmalade,
and paste. They are usually sun-dried, but dehydration is also
practiced to produce low-moisture fig products. Turkey is one

of the most important fig-producing countries. Dried figs are eaten as a snack, or used as cake fillings.

Dried figs are a good nutrient and energy source because of their carbohydrate content. Figs are an especially good source of fiber, which aids in the anticonstipation process. Dried figs contain 5.6% fiber. In addition, the potassium salts of organic acids in figs help maintain acid-alkaline balance in the body by neutralizing the excess acids present. Dried figs exert a positive effect on the alkaline reserves in the body. Figs and fig extracts have been used for medicinal purposes such as in the treatment of Ehrlich sarcoma. Dried figs have long been appreciated for their laxative action. The latex is widely used for treating warts, skin ulcers and sores, and taken as a purgative and vermifuge. A decoction of the fruits is gargled to relieve sore throat. The fig fruits are used as poultices on tumors and other abnormal growths.[8]

Grapes

Grapes (*Vitis vinifera*) are processed primarily into wine, juice, raisins, and brandy. Other products include grape-seed oil, grape pomace, hydrocolloids, and anthocyanins. The components of grapes and grape products play a significant role in preventing or delaying the onset of diseases including cancer and cardiovascular diseases.[28–30] Phenolic compounds and other health-promoting compounds are secondary plant metabolites that significantly contribute to the flavor and color characteristics of grapes, grape juices, and wines. The phenolic compounds of grapes include phenolic acids, anthocyanins, flavonols, flavan-3-ols, and tannins. The flavonoids (C_6-C_3-C_6), which include the amthocyanins, flavonols, and flavan-3-ols, are powerful antioxidants, and are found in high concentration in grapes and grape products.[31] These compounds exhibit a wide range of biochemical and pharmacological effects, including anti-inflammatory and antiallergic effects. Other grape flavonoids such as quercetin, kaempferol, and myricetin also inhibit carcinogen-induced tumors.[32–34]

Grape is rich in anthocyanins, which have known pharmacological properties and are used by humans for therapeutic purposes. Applied orally or by intravenal or intramuscular injection, pharmaceutical preparations of anthocyanins reduce capillary permeability and fragility.[35] This anti-inflammatory activity of anthocyanins accounts for their significant antiedema properties and their action on diabetic microangiopathy.[36] It has also been reported that anthocyanins possess antiulcer activity, and provide protection against UV radiation.[37]

Ellagic acid and resveratrol are two important components to reduce the risk of cancer and coronary heart diseases.[38,39] Ellagic acid ($C_{14}H_6O_8$) is an acid hydrolytic product of ellagitannin found in grape juice. Resveratrol (3,4,5-trihydroxystilbene), a naturally occurring phytoalexin produced in response to injury, has drawn much attention as a functional component. It is found in large quantities in grapes, and its presence in wine is thought to be responsible for the low mortality from coronary heart disease in wine-drinking populations. Resveratrol is reported to be a cancer chemopreventive agent having shown activity in assays representing three stages of carcinogenesis.[40] It is also shown to be an antioxidant, inhibiting lipopolysaccharide or phorbol ester-induced superoxide radical and hydrogen peroxide production by macrophages. In muscadine grapes, the skins have the highest concentration of resveratrol.[41]

Guavas

Guavas (*Psidium guajava* L.) are grown in Malaysia, Indonesia, India, Vietnam, Thailand, South China, and the Philippines. Guava flesh is often eaten fresh as dessert and salads. Many commercial products use guava flesh in pies, cakes, puddings, sauce, ice cream, juice, nectar, jam, jelly, marmalade, chutney, relish, and other products, which may frequently be seen on the markets in India, Pakistan, and Indonesia. The products made from immature fruits are commonly used to halt gastroenteritis, diarrhea, and dysentery throughout the tropical area. It contains several glucosides including avicularin, guaijavarin, and amritoside, and their hydrolyzed genin, quercetin. Fruits of *Fan Shi Liu* exhibit antidiarrheal and antibacterial effects, which are spasmolytic, chiefly from the effect of the glucosides and their genin and quercetin.[42] The fruit has a slight antihyperglycemic effect.[43] The water-based extract also exerts an antimutagenic activity and can counteract the mutagenicity of the direct action of mutagens.[44] The fruit is used to treat dysentery and acute gastrointestinal inflammation.

Hawthorn Fruits

Hawthorn fruits (*Crataegus pinnatifida* Bge.) are grown in China. Hawthorn has long been used to make candies in China. It can be consumed as a snack food such as hawthorn cookies and hawthorn cake. Hawthorn fruits have a sweet-sour taste, and a fresh flavor. The dish of sweet and sour pork with hawthorn is considered a medicinal food, in which hawthorn and licorice are first cooked. It is also used as a sauce for deep-fried pork.[45] The fruit contains chlorogenic acid, caffeic acid, phlobaphene, L-epicatechol, choline, choline acetate, (β)-sitosterol, sorbitol, vitamin C, crategolic acid, hyperin, tartaric acid, citric acid, and certain chromones. Hawthorn is a rich source of the flavan-3-ol (-)-epicatechin and proanthocyanidins related to (-)-epicatechin, e.g., epicatechin- (4β→8)-epicatechin (procyanidin B_2).[20]

This fruit has also long been used in Chinese herbal medicine to provide one of the best tonic remedies for the heart and circulatory system, and for treating swelling. Hawthorn fruits are said to control blood stasis, relieve pains associated with swelling, promote digestive function, and mitigate other conditions, especially in reducing blood pressure. Some of the pharmacological activities, e.g., the hypotensive effects, have been attributed to the chromones.[19] They act in a normalizing way upon the heart, depending on the need, stimulating or depressing its activity. The major medicinal functions are to help digestion, stimulate blood circulation, stop diarrhea, lower blood cholesterol, smooth the surface of the atherosclerotic area, increase blood flow in heart, increase the myocardial contractibility, and lower blood pressure.[20]

Hawthorn products are usually used for treatment of indigestion, infantile marasmus, menstrual cramps, diarrhea, dysentery, hernia, hyperchole sterolemia, angina pectoris, and hypertension.[20] It is often used as a cardiac tonic, and the blossoms are also effective.[46]

Recently a study on 104 hypercholesterolemic patients demonstrated that a daily dose equivalent to 46 g of the fruit for 45 days caused normalization of cholesterol value in 75% of the patients, with an additional 15% of the patients experiencing a 20 milligrams/deciliter (mg/dl) reduction. Daily supplementation of an extract of the fruit (equivalent to 15 g fresh fruit daily) for 12 weeks in 16 coronary artery patients with angina led to outstanding improvement in the conditions of most of the patients, including normalization of exertional electrocardiogram and resting electrocardiogram. There were also substantial reductions in serum triglycerides and cholesterol.[47] Oral supplementation of extracts of the fruit showed effectiveness in lowering blood pressure in hypertensive patients.[48,49] In addition to these human studies, many animal studies also demonstrated that the hawthorn fruits and their extracts can reduce heart muscle fatigue, strengthen the heart muscles, contraction amplitude and pumping power, dilate the coronary artery, and enhance blood supply to the heart muscles.[20] Extracts of hawthorn fruits are now sold in world markets as a health food or cardiac tonic.

Indian Jujubes

Indian jujubes (*Zizyphus mauritiana* Mill.), with other names such as Indian plum, Indian cherry, and Malay jujube, are grown in India, Indonesia, Malaysia, Southern China, Thailand, and the Philippines. The ripe fruits are usually eaten raw, or stewed. Some canned products, juice, and dried powder are also available in the markets. The fruits are traditionally used for cuts and ulcers, for pulmonary ailments and fevers. The dried fruit powder is a mild laxative. Sometimes the fruit pulps are blended with salt and chili for indigestion and biliousness.[8]

Jackfruits

Jackfruits (*Artocarpus heterophyllus* L.) are produced in the Philippines, Malaysia, Thailand, Cambodia, Laos, and Vietnam. Fruit flesh can be made into ice cream, chutney, jam, jelly, paste. Firm types of jackfruits are preferred for canning. Products more attractive than the fresh pulp are called 'vegetable meat.' The fruits also can be dried. The Chinese consider jackfruit pulp a nutritious tonic, cooling and nutritious, and effective in overcoming the influence of alcohol in humans.[8]

Kiwifruits

Kiwifruits (*Actinidia chienesis* Planch.), with a Chinese name "*gooseberry*" or "*Yang Tao*," are grown in the Yangtze River valley. Kiwifruits are rich in Vitamin C and usually eaten fresh, or used as appetizers, in salads, pies, pudding, and cakefilling. Quinic acid predominates in young fruits, then disappears with the formation of ascorbic acid. Kiwifruits contain the proteolytic enzyme actinidin that is said to aid digestion.[8] Kiwifruit flesh is also rich in folic acid, potassium, chromium, and Vitamin E. Kiwifruit juice of optimal flavor is produced from ripe fruits of sound quality. With other fruit juice, a sparkling kiwifruit juice can be made by carbonation.[21,50] In China, the fruit juice is valued for promoting expulsion of the kidney or gallstone.[19]

Loquats

Loquats (*Eriobotrya japonica* Lindl.), also called Japanese plums, probably originated in China, and are adapted for a subtropical to mild-temperate climate. Today China, India, Israel, and Japan are the leading producers of loquats. Loquats are usually eaten fresh. Japan, Taiwan, and Israel have exported canned loquats in syrup to the world markets. Canned loquats are consumed largely as dessert fruits. Canned products retain a golden color and fresh flavor. The fruits are also used in gelatin desserts, as pie-filling, or chopped and cooked as a sauce. The loquat products are traditionally considered to act as a sedative and are taken to halt vomiting, quench thirst, or relieve coughing.[8]

Longans

Longan (*Euphoria longan* Lour.) fruits are produced in Southern China, Taiwan, India, Thailand, Cambodia, Laos, Vietnam, Malaysia, and the Philippines. Longans are mostly eaten fresh. The dried products are black, leathery, and smoky in flavor. They are mainly used in the making of infusion beverages. The main chemical components in Longans are vitamin B, glucose, sucrose, and tartaric acid. The fruit products can be administered as a stomachic, febrifuge, and vermifuge, and are regarded as an antidote for poison. A decoction of the dried flesh is traditionally taken as a tonic and treatment for insomnia and neurasthenic neurosis.[5] Some major medicinal functions are to nourish the spleen, cultivate the heart, and supplement the intellect. Traditionally the fruit products are used for anemia, hyperactive mental activity, and forgetfulness.[19]

Litchi

Litchi or lychee (*Litchi chinensis* Sonn.) originates in the Guangdong province of China and has been grown in China for more than 4,000 years. The Guangdong and Fujian provinces in southern China remain the largest producers of litchi, followed by Vietnam, Thailand, India, Burma, Japan, the Philippines, Taiwan, Pakistan, and Bangladesh. Dried lichi fruits, frequently referred to as "lychee nuts," or "lichi nut," offer interesting opportunities in domestic and foreign markets. During drying, the pericarp or outer skin gradually loses its original color and becomes cinnamon-brown and brittle, while retaining its shape. The pulp turns dark-brown to nearly black as it shrivels around the seed and becomes very pleasant in flavor and raisin-like in texture. Lichi are most relished fresh. Pureed lichi are added to ice cream and hot milk. Canned lichi in sugar syrup has been exported from China and India for many years. Ingested in moderate amounts, lichi are traditionally taken to relieve coughing and have beneficial effects on gastralgia, tumors and enlargements of glands. Fermented lichi are also used in the Chinese medicine.[8,19,20]

Mangoes

Mangoes (*Mangifera indica* L.) are originally from the Indo-Malaysian region. The earliest growing area was Northeastern India and Burma eastward to Indochina. The production of mangoes later extended into many Asian countries and regions such as Southern India, the Philippines, Indonesia, China, Thailand, Malaysia, Sri Lanka, and Israel. India, the Philippines, Pakistan, and Thailand are the leading exporters of processed mango products.

Most people enjoy eating mango flesh as appetizers or dessert. The ripe flesh may be spiced and preserved in jars or canned in syrup, or made into jam, marmalade, jelly, or nectar. Dried mangoes are utilized commercially as a substitute for the mangoes used in chutney manufacture. Dried slices are prepared from ripe fruits. The peeled or unpeeled slices of raw mango are dried in the sun or in a cabinet dryer, then turned into powder used as a souring agent in Indian cuisine. Mango juice has a red-yellow color, and high in fresh-like flavor. Mango juice powder is used in infant and invalid foods. Mango products have the medicinal properties of a laxative, diuretic and a fattening agent according to traditional medicine. Mango juice has a cooling effect and is used during hot weather in the North Indian region. It is also alleged to help cure cholera and plague. Dried mango peel and flowers, containing up to 15% tannins, can be used as astringents in cases of diarrhea, chronic dysentery, catarrh of the bladder, and chronic urethritis resulting from gonorrhea.[8]

Mangosteens

Mangosteens (*Garcinia mangostana* L.) are grown in Burma, India, Malaysia, Sri Lanka, Thailand, the Philippines, and Vietnam. The mangosteen flesh is often eaten fresh as dessert. The flesh amounts to 31% of the fruits. The fruit flesh contains phytin up to 0.68% on a dry basis. The flesh is canned, or made into jam in Malaysia and the Philippines. The dried fruit powder is used to overcome dysentery in traditional medicine, and is also applied on eczema and skin disorders, to relieve chronic diarrhea, cystitis, gonorrhea, and gleet, it is sometimes used for astringent lotion.[8]

Mulberry

Mulberry (*Morus alba* L.) is grown in subtropical areas. Mulberry is sour and tart, yet has a pleasant taste. Some chemical components such as morin, dihydromorin, dihydromorin, dihydrokaempterol, 2,4,4′,6-tetrahydroxybenzophenone, maclurin, mulberrin, mulberrochromene, and cyclomulberrochromene have been isolated from mulberry.[19,20] Major medicinal functions are to strengthen kidneys, aid vision, and nourish blood. People use it for treatment of agitation and insomnia, deafness and blurred vision, white patches in hair and beard, hot intestines and constipation, pain in back and knees, and stiffness of muscles and joints. Famous mulberry gruel is made with mulberry fruits, rice, chicken, and other ingredients, including red *jujubes,* lotus seeds, and pine seeds. The congee is very effective for bronchitis, sinusitis, and asthma. It is said to strengthen the lungs and is used as an antitussive.[51] Mulberry is also processed into fruit beverage.

Papayas

Papayas (*Carica papaya* L.) are grown in Hawaii, India, Malaysia, Sri Lanka, Thailand, and the Philippines. Ripe papayas are most eaten fresh. The ripe flesh is usually made into sauce, or pickled, or preserved as marmalade and jam. Papaya flesh is rich in carotenoids. The major carotenoid is cryptoxanthin.[8] Papaya flesh is also prepared into juice, puree, and nectar. Papaya juice is extracted, then prepared into nectar, a ready-to-drink beverage. Papaya juice has a deep, rich orange color, and contains papain. It is also high in vitamin A and C, and is considered a "health food." Papaya juice concentrate is commonly sold to hospitals and health food stores in the Philippines.

Passion Fruits

Passion fruits (*Passiflora edulis* Deg.) are grown in Southern Asia. Passion fruit juice, due to its unique intense flavor, high acidity, and yellow/orange pulp, has been described as a natural concentrate. Passion fruit juice makes a highly palatable beverage when sweetened and diluted. India, Sri Lanka, Indonesia, Thailand, Malaysia, Taiwan, and the Philippines are important sources of passion fruit products in the world market. The yellow flesh has less ascorbic acid than that in the purple flesh, but is richer in total acid (mainly citric acid) and carotene content. Carotenoids in the flesh are 0.6 to 1.16%.[8] The flesh is a good source of niacin and riboflavin.

The juice can be sweetened, and then diluted with water or other fruit juices, to make cold drinks. Passion fruit juice can be concentrated, then used in the making of sauces, gelatin desserts, candy, ice cream, sherbet, cake filling, meringue or chiffon pie, cold fruit soup, and cocktails. The frozen juice can be kept for 1 year, and is a very appealing product. The juice can also be dehydrated using a freeze-dryer or vacuum-dryer process. According to Chinese traditional medicine, passion fruits (or dried powder) can be prescribed for insomnia, convulsions, nervous breakdown, menopause, fevers, tension, and high blood pressure. It is rich in the nutrient complexes, especially calcium and magnesium. The juice is taken as a digestive stimulant, and used in treatment for gastric cancer. There is currently a revival of interest in the pharmaceutical industry in the use of glycosides as sedatives or tranquilizers.[8]

Persimmons

Persimmons (*Diospyros kaki* L.) are grown all over Asia. Japan is the largest producer and *Kaki* is its popular name in Japan. Other persimmon-producing countries are China, Israel, the Philippines, Indonesia, India, Burma, Vietnam, and Korea. The fully ripe persimmons are usually eaten fresh. The flesh may be added to salads, blended with ice cream, yogurt, cakes, cookies, desserts, puddings, jam, or marmalade. The Japanese dry large quantities of persimmons, which are used as confection or food. Dried persimmon products have white "persimmon sugar" on the surface, with a soft texture, and a sweet taste. Large quantities of persimmons are preserved by drying in the sun. The dried products are flattened into form by pressing, sugar crystals then appear on the surface. In Indonesia, ripe fruits are stewed until soft, then pressed flat and dried in the sun. In Israel, the intestinal compaction from consumption of persimmons has been eliminated by drying the fruits before marketing, and some dried fruits are now being exported to Europe. A decoction of the calyx and fruit products is traditionally taken to relieve hiccups, coughs, and labored respiration in Asian countries.[8]

Pineapples

Over the past 100 years, pineapple (*Ananas comosus* Merr.) has become one of the leading commercial tropical fruits in the world. Major producing areas are Malaysia, Hawaii, Taiwan, the Philippines, and Thailand. Field ripe fruits are best eaten fresh. The flesh of pineapples is cut in pieces and eaten fresh as dessert, in salads, or cooked in pies, cakes, puddings, or made into sauces or preserves. In Malaysia, pineapples are used in curries and meat dishes. In the Philippines, the fermented pineapple pulp is made into a popular sweetmeat called *nata de pina*. Much of the Asian-grown pineapples are canned and are an important value-added product in world markets. The chief sources of the world's canned pineapple and pineapple juice are Bangladesh, India, Malaysia, Taiwan, Thailand, and the Philippines. Thailand is the leading producer and exporter in the world canned pineapple product market. There is a growing demand for pineapple juice. Pineapple juice, nectar, and concentrate are now commercially prepared. Pineapple juice as syrup is used in confections and beverages, or made into powder. Pineapple juice is traditionally taken as a diuretic and to expedite labor, also as a gargle in cases of sore throat and as an antidote for seasickness.

Pomegranates

The pomegranate (*Punica granatum* L.) is a subtropical fruit native to the Middle East. It has long been cultivated in the

Middle East, the Mediterranean region, and other areas in Asia. The most important pomegranate growing regions are China, Afghanistan, Pakistan, Bangladesh, Iran, Iraq, India, Burma, and Saudi Arabia. There are some commercial orchards in Israel on the coastal plain and in the Jordan Valley. People like sucking the fruit sacs from the fresh pulp of pomegranates. In some countries, pomegranate juice is a very popular beverage. An attractive colored juice (purplish red), large juicy grains, mild acid-sweet taste, and tannin content of not more than 0.25% are the qualities desired in the fruits used for the juice processing.[52] For beverage purposes, the juice is usually sweetened. In Saudi Arabia, the juice sacs may be frozen intact or the extracted juice may be concentrated and frozen for future use. Pomegranate juice is widely made into grenadine syrup for use in mixed drinks. It is also made into thick syrup for use as a sauce.

Pomegranate is a source for antioxidants considered to be antiatherogenic. The juice is rich in citric acid and sodium citrate, which can be used for pharmaceutical purposes. Pomegranate juice has been used for treating dyspepsia, and is considered beneficial in leprosy. Recent *in vitro* studies demonstrated a significant dose-dependent antioxidant capability of pomegranate juice against lipid peroxidation in plasma (by up to 33%), in low-density lipoprotein (by up to 43%), and in high-density lipoprotein (by up to 22%).[53] Pomegranate juice not only inhibited low-density lipoprotein oxidation, but also reduced two other related modifications of the lipoprotein, i.e., its retention to proteoglycan and its susceptibility to aggregation.[54] The antioxidative effects of pomegranate juice against lipid peroxidation in whole plasma and in isolated lipoproteins have been also shown *in vivo* in humans. Pomegranate juice consumption by humans increases the activity of their serum paraoxonase, which is high-density lipoprotein-associated esterase that acts as a potent protector against lipid peroxidation.

Sea Buckthorn Fruits

Sea buckthorn (*Hippophae rhamnoides* L.) is distributed widely throughout the Himalayan regions in Asia, and usually

TABLE 12.1 Some Functional Components of Sea
Buckthorn Fruits

Component(s)	Content
Carotene and carotenoids	16-28 mg/100 g fruit
Flavonoid (fruit)	120-2100 mg/100 g fruit
Volatile oil	3.6 mg/100 g fruit
Saturated fatty acid (fruit)	47.0%
Unsaturated fatty acid (fruit)	53.0%

Source: From WC Chen. *Chinese Herb Cooking for Health.* Chin-
Chin Publishing, Taipei, Taiwan. 1997. SE Kudritskaya, LM
Zagorodskaya, EE Shishkina. Carotenoids of the sea buckthorn,
variety obil'naya. *Chem. Natural Compounds* 25:724–725. 1989.
W Franke, H Muller. A contribution to the biology of useful
plants. 2. Quantity and composition of fatty acids in the fat of
the fruit flesh and seed of sea buckthorn. *Angewandre Botanik*
57:77–83. 1983.

on river banks and coastal dunes along the Baltic Coast and
on the Western coast along the Gulf of Bothnia. Sea buckthorn
is a unique and valuable plant species currently being domes-
ticated in various parts of the world. Sea buckthorn fruits are
yellow or orange berries, rich in carbohydrates, protein,
organic acids, amino acids, and vitamins. The contents of
these components vary with fruit maturity, fruit size, species,
and geographic locations (Table 12.1).

Medicinal uses of sea buckthorn are well documented in
Asia. The most important pharmacological functions attributed to sea buckthorn oil include: anti-inflammatory, antimicrobial, pain relief, and the promotion of tissue regeneration.
Sea buckthorn oil is also recommended as a treatment for
oral mucositis, rectum mucositis, vaginal mucositis, cervical
erosion, radiation damage, burns, scalds, duodenal ulcers,
gastric ulcers, chilblains, skin ulcers caused by malnutrition,
and other skin damage.[19,20] Sea buckthorn oil extracted from
seed is popular in cosmetic preparations, such as facial cream.
According to the recent report from China, in a study with
350 patients, beauty cream made with sea buckthorn oil had
positive therapeutic effects on melanosis, senile skin wrinkles,
and freckles.[57] More than 10 different functional foods have

been developed from sea buckthorn fruits in Asia such as liquids, powders, plasters, films, pastes, pills, liniments, suppositories, and aerosols. Other products made from sea buckthorn include beverages and jam from fruits and fermented pulp products.

Santol Fruits

Santol fruits (*Sandoricum koetjape* Merr.) are grown in Cambodia, India, Indonesia, Laos, Malaysia, Thailand, Vietnam, and The Philippines. The fruits are abundant in the local markets. The fruits are usually eaten fresh, sometimes with spices in India. The fruits are also made into jam, jelly, marmalade, or canned, after removing the seed and peeling. The preserved pulp is used for medicinal purposes as an astringent.[8]

Soursop Fruits

Soursop (*Annona muricata* L.) fruits are the largest tropic fruits, and are very common in the markets of Malaysia, Thailand, the Philippines, and Southeast Vietnam. Soursop fruits are eaten fresh, or in refreshing juices throughout the tropical area. The pulp is made into tarts, jelly, syrup, and nectar. The strained and frozen pulp and canned vacuum-concentrated juice are commercial products in the Philippines. The juice of the ripe soursop fruits has a diuretic function and is considered a remedy for hematuria and urethritis. It is also believed that juice can relieve liver ailments and leprosy.[8]

Tamarinds

Tamarind (*Tamarindus indica* L.) fruits are grown in Cambodia, India, Laos, Malaysia, the Philippines, and Vietnam. The pulp is rich in calcium, phosphorous, iron, thiamine, riboflavin, and niacin. The fully ripe fresh fruits are relished and eaten fresh. The tender, immature, and sour pulp is cooked as seasoning with rice, fish, and meats in India. The acid-sweet pulp is also blended with sugar to make into confection, sauce, jam, or nectar. In Southeast Asia, some people use the tamarinds to counteract the ill effects of an overdose of false chaulmoogra. Tamarind pulp is considered useful in the restoration of sensation in cases of paralysis.[8]

Wolfberry

Wolfberry (*Lycium Chinense* Miller.) is grown in subtropical areas. Wolfberry is pleasant to taste. Major chemical components are betaine, zeaxanthin, physalein, and vitamins (carotene, nicotinic acid, and vitamin C).[19,20] Major medicinal functions are to strengthen the kidneys, restore semen, nourish the liver, and clear vision. People usually use it for treatment of nutritional deficiency, eye diseases, diabetes, inadequate liver and kidney function, and seminal emission. A dish called pork kidney with wolfberry (other ingredients include squid and lycium bark) can energize the body and supplement the blood. It can be a mild treatment for diabetes and vision defects.[58] Wolfberry can also be cooked with chicken or rice.[51] It can be decocted as a tea for drinking.

Miscellaneous Fruits

Fruits of *Prunus mume* Sieb. et Zucc. (*Wu Mei*)

Fruits of *Prunus mume* Sieb. et Zucc. (with Chinese name *Wu Mei*) are harvested just before ripening while it is still green, then baked dry at a low temperature. The finished product is black colored and extremely sour. Fruits are very rich in the glucoside prudomenin, malic acid, and succinic acid. The dried fruit is used to impart tartness and flavor in preparing beverage drinks in China. The fruit infusion is commonly used to treat biliary ascariasis and hookworm. The fruit can stimulate contractions of the muscles of intestinal parasites and of the gall bladder, but causes relaxation and beneficial in purging ascaris from the bile duct and intestine.[20] The fruits are also used in the treatment of cholecystitis and gallstone disease. Concoctions of the fruits have been used to treat neoplasia-like conditions.[51]

Fruits of *Trichosanthes Kirilowii maximi*

Fruits of *Trichosanthes kirilowii maximi*, quashlike fruits, are important material for treating neoplasia and many other conditions, particularly at the early stages of cancer. An extract of the fruit skin has been shown to be a potent cytocide against cultured cancer cells.[59] One of the active principles was identified as a small protein, trichosanthin. The fruits also contain many phenolic compounds and alkaloids.

Fruits of *Lycium Chinense* Mill. (*Gou Qi Zi*)

Fruits of *Lycium Chinense* Mill. (with Chinese name as *Gou Qi Zi*) are bright red-colored fruits, and have long been a tonic herb in China. The fruit contains betaine (0.1% of dry basis), and zeaxanthin, physalein, vitamins such as carotene, nicotinic acid, and vitamin C.[19,20] The effects of the fruits include an increase in leukocyte count and nonspecific immunity, and stimulation of tissue development.[20] The infusion of fruits can increase the plasma level of tumor necrosis factor and interleukin-1 in human blood, and can also lower blood pressure and stimulate the heart. *Lycium Chinense* Mill. fruit, as a cool-type herbal medicine, is used to treat certain types of inflammatory conditions and hypertension. It is also used in tonic soups and alcoholic drinks.[19,20] The extract of *Lycium Chinense* Mill. fruits enhance phagocytotic activity of immune cells and promote white cell formation. The fruit infusion can be a mild treatment for diabetes and vision defects. The fruits can be cooked with meat, rice, or used as tea.

Fruits of *Schisandra chinensis*

In China, the dried, purplish-black berry of *Schisandra chinensis*, about a few millimeters in diameter, is used both as a flavoring agent in food and beverage, and also as a medicine. One of the most important modern uses of this berry is in the treatment of bacterial and viral chronic hepatitis and xenobiotic-induced hepatitis. *Schisandra* berries are used in combination with other herbs to treat excessive sweating, certain types of bronchitis and asthma, and other conditions. Modern studies suggest that it has CNS-stimulation effects (77AA). The berry contains many active compounds such as schizandrin, deoxyschizandrin, γ-schizandrin, pseudo-γ-schizandrin, schizandrol, α-chamigrene, chamigrenal, phytosterols, citral, and vitamins C and E.[60] The berry has been used in China for promoting health according to traditional Chinese medicine. Now tablets and powders of *Schisandra* berry have been marketed widely as nutritional supplements for promoting health.

Fruits of *Emblica officinalis* Gaertn

Fruits of *Emblica officinalis* Gaertn have been used extensively as the main ingredient to prevent colds, coughs, and enhance immunity.[61] The fruit is known as one of the best sources of natural vitamin C, which has been found to be more readily assimilated than the synthetic vitamin C. The fruit is also known to contain a significant amount of pectin, a complex polysaccharide containing galactoside residues, which are known to possess anticancer properties. The antioxidant, and strong reducing properties of vitamin C are known to be a free radical scavenger, suggesting its chemotherapeutic potential. The anti-inflammatory property of this fruit is also mentioned in Indian traditional medicine. The *in vitro* antitumor property against human leukemic cell lines, preliminary *in vivo* antitumor property against ehrlich ascites carcinoma, and the anti-inflammatory property of fruit extract, are being further evaluated in clinical studies.[62]

White Chinese Olive

White Chinese olives (*Canarium album* Raeusch) (with Chinese name *Qing Guo*) have a pleasant and acrid taste. The fruits are usually made into preserved products. The major medicinal functions are to remove fever, purify the lungs, eliminate apprehension, stimulate appetite, promote salivation, and detoxify the body.[20] People usually use it for treatment of sore throat, thirst, restlessness, globefish poisoning, and alcohol intoxication.

Fruits of *Poncirus trifoliatea* L. (*Gou Gi*)

In the ripe fruit of *Poncirus trifoliatea* L. (with Chinese name *Gou Gi)*, several substances have been isolated from this fruit, including poncirin, lemonin, imperatorin, bergapten, neohesperidin, citrifoliol and myrcene, camphene, and τ-terpinene.[20] Limonin has a chemopreventive effect against carcinogenesis, and can shorten sleep time. *Gou Gi* is used to treat gastric pain and constipation. It has been used with success to treat prolapse of the uterus or rectum.

Fruits of *Myrica rubra* Lour. Sieb. et Zucc. (*Gou Mei*)

Fruits of *Myrica rubra* Lour.Sieb.et Zucc. (Chinese name of *Gou Mei)* contain myricetin, a genin hydrolyzed from the glucoside myricitrin.[20] The fruit inhibits melanin biosynthesis, attributed to its inhibition of tyrosinase. It can be used as a whitening agent of the skin.[63] The fruit and myricitrin have antifungal and antibacterial activity. They can inhibit the growth of *Cladosporium cucumcimum, Bacillus subtillus,* and *Escherichia coli.*[64] This fruit is used to treat gastric pain, diarrhea, and dysentery.

Some Special Snack Foods — Preserved Fruits

Traditional methods of preserving fruits are by adding sugar, honey, salt, spices, and herb ingredients. These products are commonly called "preserved fruits," "cured fruits," or "candied fruits." Fruits for preserving should be in the firm–ripe stage. Cane sugar, beet sugar, corn syrup, honey, salt, and some herb or spice flavoring ingredients are commonly used. Preserved fruits are the most popular products among Asian people. Preserved fruits are excellently served with a variety of entrees, confections, and snacks to promote appetites.

There are several kinds of cured Chinese olives with different tastes and flavors: aroma preserved olive, multi-taste olive, sweet-preserved olive, sweet-preserved olive (*Soo Larm*), sweet-preserved olive (*Wo Sang Larm*), sweet-preserved olive (*Lar chow larm*), sweet-preserved olive (*Wong Cho Larm*), and salted dry olive.

Preserved prune and plum (*Prumus salicina*) products are produced mainly in Southern China and Malaysia. The product has a sweet and sour taste, and can stimulate the appetite. There are several kinds of preserved prune and plum products with distinctive taste and flavor. They are sweet-preserved prune, *chen-pee* (*Mei Prune, Chen-Pee Mei*), seasoned prune, dried prune, salted preserved prune, half-dried prune, preserved prune, brine-preserved prune, sweet prune cake, sweet-preserved plum (dried plum), sweet-preserved plum (*Poo Tow Lee*), seedless preserved plum, salted- preserved plum, salted dried plum, half-dried plums, and brine-preserved plum.

Red bayberries, also called *Yangmei* in China, are produced in Southern China. Preserved red bayberries have a dark-red color, and round shape.

Cured sweet orange or lemon peels are well-known products, mixing with a sweet herbal powder. The product has a multitaste, yellow-brown color, and high flavor.

Fruit leathers, known commercially as fruit rolls, are manufactured with fruit purees into leathery sheets. The leathers are eaten as a confection or used as a sauce. The dried products have a bright translucent appearance, chewy texture, and distinct fruit flavor. They can be prepared from a wide variety of fruits, including apple, apricot, banana, blackberry, cherry, grape, guava, hawthorn, papaya, peach, pear, pineapple, plum, raspberry, strawberry, and so on. Overripe fruits with high sugar and flavor but low fiber content are suitable for making fruit leathers.

SUMMARY

Fruits constitute an important part of the human diet. They are one of the main food resources that humans need to ingest daily. Most fruits are consumed fresh with little preparation. Approximately half is processed for year-round consumption. Some fruit products are consumed directly as foods, while some are used as ingredients in confectionery, bakery, and diet foods. Some are also used in pharmaceuticals products.

Today, the food trend for the consumer is toward convenience and quality. This trend is best described as the "health-conscious" food preference. High-quality fruit products usually imply freshness with appealing flavor, color, texture, and appearance. Fruit products with additional nutritive value and microbiological quality will offer consumers a healthy product containing natural sources of vitamins, minerals, and many health-promoting components.

Some of the health benefits from various functional components of more than 50 different fruits described here are based on time-honored tradition and observations, and some based on scientific studies and discoveries as elucidated in the references cited here. The key point is that all fruits and their products are a good source of vitamins (Vitamin C and carotenoids), antioxidants (flavonoids, polyphenols), and fiber.

REFERENCES

1. GB Block, B Patterson and A Subar. Fruit, vegetables, and cancer prevention: a review of the epidemiological evidence, *Nutr. Cancer.* 18:1–29. 1992.

2. World Cancer Research Fund and American Institute for Cancer Research. Food, Nutrition and the Prevention of Cancer: A Global Perspective. AICR, Washington, D.C., 1997.

3. AZ Mercadant, A Steck, H Pfander. *J. Agri. Food Chem.*, 45:1050–1055, 1997.

4. K Tazawa, H Yamashita, I Ohnishi, Y Saito, T Okamoto, M Masuyama, K Yamazaki, K Takemori, S Saito, and M Arai. Anticarcinogenic and/or antimetastatic action of apple pectin in experimental rat colon carcinogenesis and on hepatic metastasis rat model. In: *Functional Foods for Disease Prevention, Vol 1, Fruits, Vegetables, and Teas. Metastasis Rat Model, Functional Foods for Disease Prevention, Vol 1, Vegetables, and Teas.* Chapter 9, 96–103, 1997.

5. MV Eberhardt, CY Lee, RH Liu. Antioxidant activity of fresh apples. *Nature*, 405:903–904, 2000.

6. H Ohkami, K Tazawa, I Yamashita, Y Ohnishi, K Kobashi, M Fujimaki. *Jpn. J. Cancer Res.* 86:523, 1995.

7. GC Schloemer, G Kaczmarowicz, J Niedworok. Anthocyanin products from aronia melanocarpa. In: *Natural colorants for food, nutraceuticals, beverages, confectionery and cosmetics: proceedings of the third international symposium, Princeton,* SIC Publishing Company, 1998, pp 309–313.

8. JF Morton. *Fruits of Warm Climates.* Media, Incorporated, Winterville, NC 1987.

9. ME Camire. Bilberries and blueberries as functional foods and pharmaceuticals. In: *Functional Foods: Herbs, Botanicals and Teas,* G. Mazza, B.D. Oomah, Eds. Technomic Press, Lancaster, PA, 2000.

10. J Gruenwald, T Brendler, C Jaenicke. Vaccinium myrtillus, *Physicians Desk Reference for Herbal Medicines,* 1st ed., Medical Economics, Montvale, NJ, pp 1201–1202, 1998.

11. SP Plami, JT Kumpulainen, RL Tahvonen. Total dietary fiber contents in vegetables, fruits and berries consumed in Finland, *J. Sci. Food Agric.* 59(4):545–549, 1992.

12. P Morazzoni, E Bombardelli. Vaccinium myrtillus L., *Fitoterapia.* 67(1):3–29, 1996.

13. W Kalt, S MacEwan, G Miner. Vaccinium Extract in Pharmaceutical Products. Report to the Wild Blueberry Association of North America and the Wild Blueberry Producers Association of Nova Scotia, 1994.

14. F Zaragoza, I Iglesian, J Benedi. Estudio comparativo de los efectos antiagregantes de los antocianosidos y otros agents, *Archiv. Farmacol. Toxicol.* 11(3):183–188, 1985.

15. W Kalt, D Dufour. Health functionality of blueberries, *HortTechnol.* 7(3):216–222, 1997.

16. R Alfieri, P Sole. Influence des anthocyanosides administers par voie oro-perlinguale sur l'adapto-electroretinogramme(AERG) en lumiere rouge chez l'Homme, *Comptes Rendus des Seances de la Societe de Biologie et des Ses Filiales.* 160(8):1590–1593, 1996.

17. JR Trevithick, KP Mitton. Antioxidants and diseases of the eye. In:*Antioxidant Status. Diet, Nutrition, and Health,* AM Pappas, Ed. CRC Press, Boca Raton, FL, pp 545–565, 1999.

18. VE Tyler. *Herbs of Choice,* Haworth Press, Binghampton, NY, pp 51–54, 1994.

19. YW Huang, CY Hung. Traditional Chinese functional foods. In: *Asian Foods, Science and Technology.* CYM Ang, K Lui, YW Huang. Eds. Technomic Publishing Co. Inc., 1999.

20. KC Huang. *The Pharmacology of Chinese Herbs.* CRC Press, Inc., Boca Raton, FL, 1993.

21. ST Chow. *Fruit Processing and Storage.* 2nd ed., Beijing Scientific Press, China, 1991.

22. A Murakami, Y Wataru, H Takahashi, T Yonei, H Tanaka, K Makita, N Wada, M Ueda, Y Haga, Y Nakamura, Y Ohto, OK Kim, H Ohigashi, K Koshimizu. Auraptene, an Alkyloxylated coumarin from citrus natsudaidai HAYATA, inhibits mouse skin tumor promotion and rat colonic aberrant crypt foci formation. *Functional Foods for Disease Prevention, Fruits, Vegetables, and Teas,* 8:86–95, 1997.

23. A Chiralt, X Martinez-Monzo, M Chafer, P Fito. Limonene from citrus. In: *Functional Foods: Processing and Biochemistry Aspects,* J. Shi, G. Mazza, M. Le Maguer, Eds. CRC Press, New York, 2002.

24. I Ofek, J Goldhar, D Zafriri, H Lis, R Adr, N Sharon. Anti-Escherichia coli adhesin activity of cranberry and blueberry juices, *New Engl. J. Med.* 324(22):1599, 1991.

25. I Ofek, J Goldhar, N Sharon. Anti-Escherichia coli adhesin activity of cranberry and blueberry juices. In: *Toward Anti-Adhesion Therapy for Microbial Diseases,* Itzhak Kahane and Itzhak Ofek, Eds. Plenum Press, New York, 179-183, 1996.

26. AB Howell, A Der Marderosian, LY Foo. Inhibition of the adherence of P-fimbriated Escherichia coli to uroepithelial-cell surfaces by proanthocyanidin extracts from cranberries, *New Engl. J. Med.* 339(15):1085–1086, 1996.

27. PW Martin. Durian and mangosteen, tropical and subtropical fruits, S Nagy, PE Shaw, Eds. AVI, Westport, CT, 1980.

28. CV DeWhalley, SM Rankin, RS Hoult, W Jessup, DS Leake. Flavonoids inhibit the oxidative modification of low-density lipoproteins by macrophages, *Biochem. Pharmacol.,* 39:1743–1748, 1980.

29. EN Frankel, J Kanner, JE Kinsella. Inhibition *in vitro* of oxidation of human low-density lipoproteins by phenolic substances in wine, *Lancet,* 341:454–457, 1993.

30. MGL Hertog, PCH Hollman, MB Katan, D Kromhout. Intake of potentially anticarcinogenic flavonoids and their determinants in adults in the Netherlands, *Nutr. Cancer*, 20:21–29, 1993.

31. J Kanner, E Frankel, R Granit, B German, JE Kinsella. Natural antioxidants in grapes and wines, *J. Agric. Food Chem.*, 42:64–69, 1994.

32. H Wei, L Tye, E Bresnick, DF Birt. Inhibitory effect of apigenin, a plant flavonoid, on epidermal ornithine decarboxylase and skin tumor promotion in mice. *Cancer Res.*, 50:499–502, 1990.

33. EE Deschner, J Ruperto, G Wong and HL Newmark. Quercetin and rutin as inhibitors of azoxymethanol-induced colonic neoplasia, *Carcinogenesis*, 7:1193–1196, 1991.

34. E Middleton, C Kandaswami. Effects of flavonoids on immune and inflammatory cell functions, *Biochem. Pharmacol.*, 43:1167–1179, 1992.

35. H Wagner. New plant phenolics of pharmaceutical interest, *Ann. Proc. Phytochem. Soc. Eur.*, Vol. 15. cf. Van Sumere and PJ Lea, Oxford UK:Clarendon Press. pp 409–425, 1985.

36. R Boniface, M Miskulin, L Robert, AM Robert. Pharmacological properties of Myrtillus anthocyanosides: correlation with results of treatment of diabetic microangiophathy, In: *Flavonoids and Bioflavonoids*. L Farkas, M Gabor, F Kally, Eds. Amsterdam: Elsevier. pp 193–201, 1986.

37. E Kano, J Miyakoshi. UV protection effect of keracyanin an anthocyanin derivative on cultured mouse fibroblast L. cells, *J. Radiat. Res.*, 17:55–65, 1976.

38. JL Maas, GI Galletta, GD Stoner. Ellagic acid, an anticarcinogen in fruits, especially strawberries: a review, *HortScience*, 26:10–14, 1991.

39. M Jang, L Cai, GO Udeani, KV Slowing, CF Thomas, CWW Beecher, HSnFong, NR Farnsworth, AD Kinghorn, RG Mehta, RC Moon, JM Pezzuto. Cancer chemopreventive activity of resveratrol, a natural product derived from grapes, *Science*, 275:218–220, 1997.

40. T Okeda, K Yokotsuka. Trans-resveratrol concentrations in berry skins and wines from grapes grown in Japan. *Am. J. Enol. Vitic.*, 47:93–99, 1996.

41. AM Rimando, W Kalt, JB Magee. Resveratrol in vaccinium species and muscadine grapes, Proceedings of Pre-Congress Internet Conference. pp 186–188, 2001.

42. MA Morales, J Tortoriello, M Meckes, D Paz, X Lozoya. Calcium-antagonist effect of quercetin and its relation with the spasmolytic properties of *Psidium guajava* L. *Arch. Med. Res.,* 25 (1):17–21, 1994.

43. R Roman-Ramos, JL Flores-Saenz, FJ Alarcon-Aguilar. Antihyperglycemic effect of some edible plants. *J. Ethnopharmacology,* 48:25–32, 1995.

44. IS Grover, S Bala. Studies on antimutagenic effects of guava (*Psidium guajava*) in *Salmonella typhimurium. Mutation Res.,* 300:1–3, 1993.

45. YH Lee. *Home-Cook Nutritional Recipes.* Haibin Publishing, Hong Kong, 1990.

46. D Potterton, Ed. *Culpepers Color Herbal.* Sterling Publishing, New York, 1983.

47. Shen Yang Army Hospital. Medical Data Base. Vols. 1 and 2 (in Chinese), 1975.

48. Liao Ling See Fung Disease Prevention Center. *Tieh Ling Medicine* 2:33–36, 1972.

49. AN Chau An, County Health Survey. County Health Department. Database on Canton Medicines. 11, 44–49, 1977.

50. DB Zhang. *Fruit and Vegetable Processing Technology.* China Light Industry Press, Beijing, China, 1994.

51. WC Chen. *Chinese Herb Cooking for Health.* Chin-Chin Publishing, Taipei, Taiwan, 1997.

52. RN Adsule, NB Patil. Pomegranate. In: *Handbook of Fruit Science and Technology,* Salunkhe DK, Kadam SS, Eds., Marcel Dekker, New York, 1995.

53. M Aviram, L Dornfeld, M Rosenblat, N Volkova, M Kaplan, T Hayek, D Presser, B Fuhrman. Pomegranate Juice Consumption Decreases Oxidative Stress, Low-Density Lipoprotein Modifications and Atherosclerosis: Studies in the Atherosclerotic Apolipoprotein E-Deficient Mice and in Humans, Proceedings of Pre-Congress Internet Conference. 151–152, 2001.

54. M Aviram, L Dornfeld, M Rosenblat, N Volkova, M Kaplan, T Hayek, D Presser, B Fuhrman. Pomegranate juice consumption reduces oxidative stress, low-density lipoprotein modifications and platelet aggregation: studies in the atherosclerotic apolipoprotein in E deficient mice and in humans. *Am. J, Clin. Nutr.*, 71:1062–1076, 2000.

55. SE Kudritskaya, LM Zagorodskaya, EE Shishkina. Carotenoids of the sea buckthorn, variety obil'naya. *Chem. Natural Compounds* 25:724–725, 1989.

56. W Franke, H Muller. A contribution to the biology of useful plants. 2. Quantity and composition of fatty acids in the fat of the fruit flesh and seed of sea buckthorn. *Angewandre Botanik* 57:77–83, 1983.

57. J Bernath, D Foldesi. See buckthorn (Hippophae rhamnoides L.): a promising new medicinal and food crop. *J. Herbas, Spices, Medicinal Plants*, I:27–35, 1992.

58. WL Su. Oriental Herbal Cook Book for Good Health (1). Shun An Tong Corp., Flushing, NY, 1993.

59. Beijing Medical College Chinese Herbal Medicine Research Group. *J. Beijing Medical College* 1:104. 1959, (in Chinese).

60. R I S Lin. Phytochemicals and antioxidants. In: *Functional Foods, Designer Foods, Pharmafoods, Nutraceuticals*. I Goldberg Ed. Chapman & Hall, New York, 1994.

61. *The Wealth of India. A Dictionary of Indian Raw Materials and Industrial Products; Raw Materials*; C. S. I. R.: New Delhi; vol. 3 pp 168–170, 1952.

62. PG Sur, DK Ganguly, Y Hara, Y Matsuo. Antitumor activity of *Emblica officinalis* Gaertn fruit extract, In: *Functional Foods for Disease Prevention 1. Fruits, Vegetables, and Teas.* Chapter 10, p. 104, 1997.

63. H Matsuda, M Higashono, W Chen, H Tosa, M Iinuma, M Kubol. Studies of cuticle drugs from national sources. III. Inhibitory effect of *Myrica rubra* on melanin biosynthesis. *Biol. Pharm. Bull.*, 18:1148–1150, 1997.

64. S Gatner JL Wolfender, S Mavi, K Hostettmann. Antifungal and antibacterial chalcones from *Myrica serrata. Planta Med.*, 62:67–69, 1996.

13

Functional Foods from Fermented Vegetable Products: Kimchi (Korean Fermented Vegetables) and Functionality

KUN-YOUNG PARK AND SOOK-HEE RHEE

Pusan National University, Korea

INTRODUCTION

Kimchi is a group of traditional fermented vegetable foods consumed as the main side dish in the Korean diet. The word kimchi originated from *chimchae*, the Chinese character, meaning pickled vegetables with salt. *Chimchae* was called *dimchae, dimchi,* and then kimchi in sequence by word of mouth transfer.

There are 161[1] to 187[2] different kimchi preparations, depending on various vegetables or ingredients and processing methods. However, the most popular kimchi is *baechu* kimchi, which is made with Korean *baechu* cabbage (known to Westerners as Chinese cabbage). *Baechu* kimchi is called kimchi in general. Usually fermented kimchi contains high levels of lactic acid bacteria (10^{7-9} CFU/mL), organic acids, and other nutrients such as vitamins, minerals, dietary fibers, and functional components from ingredients or formed during fermentation.[3] Thus, it might be called a lactic acid bacteria fermented food and a functional food.

Kimchi has been known to help increase appetite, reduce constipation, maintain proper intestinal flora, and has been reported to have anticarcinogenic and antiaging effects and other health benefits.[4]

The main ingredients of kimchi are *baechu* cabbage and radish that are brined in salt, with which spices, fermented fishes, and other vegetables are mixed and then fermented. Both the solid and the watery portions of kimchi are consumed for their texture and taste. Koreans generally prefer optimally ripened kimchi, but fresh kimchi and overripened kimchi are also consumed in various ways. Traditionally for Koreans, kimchi is served as a stored food during cold winters. In the 1950s Koreans consumed about 300g/day/person of kimchi, but consumption decreased to 124g/day/person in 1998.[5] Kimchi has usually been prepared at home, but recently a large portion of kimchi in Korea has been produced commercially. Kimchi is also packaged for export to foreign markets.

HISTORY

Salted vegetables as preserved foods were consumed about 2,000 years ago in Korea.[6] Old Chinese books such as *Jeijeon* by Yangseu and *Dongijeon* by Namsa indicated that Kokurye people (37 BC to 668 AD, the Koreans' ancestors) developed and preferred various fermented foods such as fermented soybeans, vegetables, fishes, alcoholic beverages, and others.[6]

Samkuksaki (published in 1145 AD) indicated that fermented vegetables were prepared using stone pickle jars in the Bupju temple at Mt. Sokri during the Shinla dynasty (720 AD). In the Koryo dynasty (918 to 1392 AD) Buddhism prohibited meat consumption, and vegetables were preferred. In *Dongkukisangkukjip* written by Lee Kyubo (AD 1168 to 1241), a turnip kimchi pickled with salt was described, and it was explained that it could last over the winter season as a tasty side dish. These writings also suggested that garnished kimchi appeared at this time, using garnishes of garlic, Chinese pepper, ginger, and tangerine peels. Watery varieties of kimchi such as Nabak kimchi and Dongchimi were the popular kimchi types in the Shinla and Koryo dynasties.[7] The main vegetables used in the Koryo dynasty to make kimchi seemed to be radishes rather than cabbages.

At the time of the Chosun dynasty founded in 1392 AD, various vegetables were introduced from foreign countries to Korea and as a consequence the ingredients of kimchi became more diverse. Methods of making kimchi also became more elaborate in the early Chosun dynasty (1392 to 1660 AD). *Baechu* (Chinese cabbage) and white radishes became the main ingredients of kimchi after 1600 AD, the mid-Chosun dynasty.

Jibongyusol (1613 AD) showed the first records of red peppers, and their use in kimchi as recorded in *Sallimkyongje* (1715). Confucianism was adopted in the Chosun dynasty. Koreans believed that red color could protect or remove the bad luck or evil spirit from the Korean folkway. Red pepper was also a good substitute for salt when the salt stocks were in short supply in the 17th and 18th centuries. Red pepper inclusion in kimchi provided a harmonious taste, good color, and antimicrobial activity. Other ingredients were also employed at this time, such as fermented fishes, meats and other condiments resulting in different and desirable flavors and nutritive values of kimchi.

Approximately 41 different kinds of kimchi are described in the *Jungbosallimkyongje* (1776 AD). This book introduced 50 kinds of vegetables used in kimchi including the main ingredients such as *baechu*, radish, mustard leaf, cucumber, leek, and others, and various minor components as condiments such as garlic, red pepper, ginger, and green onion. This book introduced the cultivation method of *baechu* in detail, this is the most important document to describe the history of kimchi; it described the most popular kimchi of today, *baechu* kimchi, as having red pepper, meat, and fish as ingredients.

In *Imwonshibyukji* (1835), 97 varieties of kimchi were mentioned emphasizing the use of red pepper for kimchi. *Baechu* cabbage became a major ingredient to make kimchi due to the breeding of the cabbage. The characteristics and cultivation method of *baechu* cabbage were reported in *Nong-gawalryongga* (AD 1816). The baechu cabbage leaves were stacked well bred at 20°C to make a better quality kimchi.

PROCESSING OF KIMCHI

The principal processes in the preparation of kimchi are the pretreatment of raw materials, brining, the mixing of ingredients, packaging, and fermentation. The first step in making kimchi is the selection of raw materials. This is the first and an important step. The raw materials for preparation of kimchi are very diverse as shown in Table 13.1. The quality of the raw materials depending on the cultivation method and the kinds or types of materials affects the fermentation behavior, taste, and functional properties of kimchi.

In the preparation of *baechu* kimchi, for example (Figure 13.1), the cabbage is trimmed, washed, brined, and then rinsed, and water is drained from the salted cabbage.[8] The pretreatment of the main raw materials includes grading, washing, and cutting. Ingredients are graded, washed, cut, sliced, or chopped for the proper mixing. The main material of kimchi, *baechu* cabbage or radish, is brined at proper salt concentrations.

The premixture of chopped or sliced subingredients such as garlic, red pepper powder, salt-pickled fishes, and other vegetables are mixed or stuffed between the leaves of the cabbage. In the standardized method of processing *baechu* kimchi, the brined *baechu* cabbage (100%) is mixed with 13% sliced radish, 2% green onion, 3.5% red pepper powder, 1.4% garlic, 0.6% ginger, 2.2% fermented anchovy juice, and 1.0% sugar, and the final salt level is 2.5%. Table 13.2 shows the standardized recipe of *baechu* kimchi[9–11] and the nutrients and functional

TABLE 13.1　Raw Materials Used in the Preparation of Kimchi

Groups	Raw materials
Main raw vegetables	*Baechu* (Korean Chinese cabbage), radish, ponytail (*Chonggak*) radish, young Oriental radish, cucumber, green onion, lettuce western cabbage, leek, green pepper, etc
Subingredients	
Spices	Red pepper, green onion, garlic, ginger, mustard, black pepper, onion, etc
Seasoning	
Salt	Dry salt or brine solution
Salt-pickled	Anchovy, shrimp, calm, hair tail, yellow corvenia, etc.
Other seasoning	Sesame seed, soybean sauce, monosodium glutamate, corn syrup, etc.
Other materials	
Vegetables	Watercress, carrot, crown daisy, parsley, mustard leaves, etc.
Fruits and nuts	Pear, apple, jujube, melon, ginko nut, pine nut, etc.
Cereals	Rice, barley, wheat flour, starch, etc.
Fish and meats	Shrimp, Alaska pollack, squid, yellow corvenia, oyster, beef, pork, etc.
Miscellaneous	Mushroom, etc.

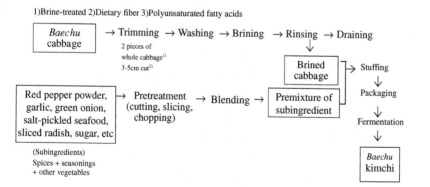

1)Brine-treated 2)Dietary fiber 3)Polyunsaturated fatty acids

Figure 13.1　Flow chart for processing of baechu kimchi: (1) become whole cabbage (Tongbaechu in korean) kimchi; (2) become cut-cabbage (Matbaechu in korean) kimchi

TABLE 13.2   Composition of Kimchi and Nutritional Value

Composition	Ratio	(%) Nutrients/nutraceuticals
Baechu cabbage[1]	100	Sugar,VC, K, DF[2], β-sitosterol, indoles Benzyl isothiocyanate
Red pepper powder	3.5	PUFA[3], sugar, VA, VC, Ca, P, K, DF, capsaicinoid
Crushed garlic	1.4	Sugar, Allyl compounds, Alliin Allicin, diallysulfide
Crushed ginger	0.6	Niacin, K, gingerol
Fermented anchovy juice	2.2	Protein (amino acids), Ca, P, Fe, Na
Sugar	1.0	Sugar
Radish (Sliced)	13.0	Sugar, Niacin, Ca, isothiocyanate
Green onion	2.0	VA, VC, Chlorophyll, sulfur compound
Final salt conc.	2.5	

components of the ingredients. Vitamins, minerals, and various phytochemicals are available in the kimchi ingredients.

However, various other materials can also be added to the premixture of the subingredients depending on family tradition, economic situation, and seasonal and regional availability of the materials. Watercress, mustard leaves, pear, apple, pine nut, chestnut, gingko nut, cereals, fishes, crabs, meats, and others, can be incorporated into kimchi (Table 13.1).

The premixture mixed or stuffed cabbage is packaged and then fermented at different temperatures, but a low temperature (5°C) is better for preparing a good tasting product. The fermentation occurs in the containers. The fermentation conditions such as temperature, anaerobic condition, container type, etc., are important factors that affect taste and functionality.

Traditionally, a large quantity of *baechu* cabbage heads were used for making kimchi for the cold winter season in late November, which was stored in big potteries (earthen jars) underground for the winter, called kimjang, an annual event for the Korean family.[7] The kimchi is traditionally consumed until the following spring. However, a kimchi refrigerator is used for the storage of kimchi for a large proportion of Korean families these days.

FERMENTATION CHARACTERISTICS

Various endogenous microorganisms in the ingredients that can survive in high concentrations of salt and acidity and in anaerobic conditions are involved in the fermentation of kimchi. Lactic acid bacteria (LAB) are the main microorganisms and are supplied from the vegetables naturally. They are facultative anaerobes, microaerobes, or anaerobes and can also survive in acidic and salty conditions.

Salt concentration, temperature, pH, air exposure, and type or population of microorganism control kimchi fermentation. In the typical microbial changes during kimchi fermentation, LAB increases whereas aerobes and pathogens decrease due to the absence of air, salt content, and acid formed during fermentation.[8] Kimchi fermentation is initiated by *Leu. mesenteroides (Leuconostoc* sp.), a heterofermentative LAB and a facultative anaerobe, and it produces lactic acid, acetic acid, CO_2, and ethanol as major end products. As the pH drops between 4.6 to 4.9, due to accumulation of organic acids, *Leu. mesenteroides* is relatively inhibited. *Streptococcus (St. faecalis)* follows almost the same pattern as *Leuconostoc* sp., but with lower numbers. The fermentation continues with other LAB that can endure more acidic conditions. The LAB are *Pediococcus cerevisiae, Lactobacillus brevis, LactobacillusFermentum,* and *Lactobacillus plantarum* (Figure 13.2). The homofermentative LAB, *Lac. plantarum*, is present in the greatest number following the initial fermentation and produces the maximum acidity at the later stages especially at higher temperatures.[12]

Leuconostoc mesenteroides and *Lac. plantarum* are the main LAB for kimchi fermentation. *Leuconostoc* sp. predominate (65.2%) at a lower temperature (5°C), but *Lac. plantarum* predominate (59.7%) at a higher temperature of 25°C, while the levels drop to 28% at 5°C.[13] Thus a lower temperature produces a good tasting kimchi because of the predominance of *Leuconostoc* sp., which is a heterofermentative LAB. The process of kimchi fermentation has distinct phases based on the changes in pH, acid production, CO_2 levels, and reducing sugar contents, all of which are temperature dependent.

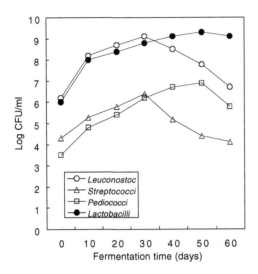

Figure 13.2   Microfloral change of lactic acid bacteria during kimchi fermentation at 5°C.[12]

The first stage has a rapid decrease of pH accompanied by a decrease of reducing sugars. The next stage is a gradual drop in pH, an increase in acidity and CO_2 levels, and a rapid disappearance of reducing sugars. The final stage has no or only slight changes in pH, acidity, CO_2, and reducing sugars.[8] The pH and acidity of optimally fermented kimchi are 4.2 to 4.5 and 0.4 to 0.8%, respectively.

Though *Lac. plantarum* is believed to be the main acidifying or deteriorating microorganism in kimchi fermentation,[14] this LAB can endure at lower pH during digestion and colonize in the colon, and so plays a role as a probiotic microorganism in the health of the colon.[15]

FUNCTIONALITIES OF KIMCHI

Kimchi (*baechu* kimchi) is a low caloric food (18 kcal/g) and contains higher levels of vitamins (vitamin C, β-carotene, vitamin B complex, etc.), minerals (Na, Ca, K, Fe, and P) dietary fibers (24% on a dry basis; 7.8% soluble dietary fiber; 16.2% insoluble dietary fiber), and other functional components.[16]

TABLE 13.3   Functional Characteristics of Kimchi

Increases appetite
Decreases body fats
Prevents constipation and colon cancer
Good source of probiotics (lactic acid bacteria)
Decreases serum cholesterol; increases fibrinolytic activity
Antioxidative effects (antiaging, prevents aging of skin)
Anticancer effects (antimutagenic and antitumor effects)
Increases immune function

The protein and lipid contents can be increased by the addition of fish, clam, oyster, and meat depending on family preference (Table 13.1). Triacylglycerols, polar lipids, free fatty acids, monoacylglycerols, hydrocarbons, sterols, and various free fatty acids are found in the lipids (5.4% on dry basis) of kimchi. Linoleic acid and linolenic acid (44 to 60%) are the main free fatty acids present.

Phytochemicals such as benzyl isothiocyanate, indole compounds, thiocyanate, and sitosterol are the active compounds found in kimchi, which have shown antimicrobial, anticancer, and antiatherosclerotic functions.[17]

Kimchi is mainly prepared with yellow-green vegetables that have been claimed to prevent cancer, increase immune function, retard the aging process, prevent constipation, etc.[3] When kimchi ferments, its taste is enhanced and it becomes a good probiotic (LAB) food. Table 13.3 shows the functionality of kimchi, which has so far been researched.[4] Kimchi increases the appetite by its taste, flavor, and color. Kimchi might have preventive effects in constipation and colon cancer due to the higher content of dietary fibers, organic acids, functional phytochemicals, and LAB. Dietary fibers have been demonstrated to prevent hypertension, diabetes, constipation, and cancers.[16]

Kimchi intake reduces serum cholesterol and increases fibrinolytic activity[18] and thus it has an antiatherosclerotic function. Kimchi might retard aging processes and delay skin aging[19] owing to the antioxidative activities of vitamin C, β-carotene, phenolic compounds, and chlorophylls, etc.

Kimchi also contains β-sitosterol, PUFA (polyunsaturated fatty acids), and their derivatives,[9] glucosinolates, isothiocyanates, indoles, and allyl compounds. These compounds play various roles in the prevention of cancer and enhancement of immune function.[3,17]

As discussed already, different kinds of subingredients can be added when kimchi is prepared, thus specifically they can strengthen the functionality of kimchi; for example, vitamin C-enhanced kimchi, antiatherosclerotic, and anticancer kimchi, and others, can be prepared. Kimchi is a protective food since it contains high levels of vitamins, especially vitamin C, and minerals.

Control of Body Weight

Kimchi has been shown to control the body weight in rats.[20] Capsaicin in red pepper can remove fat in the body by stimulating spinal nerves and thus activating the release of catecholamine in the adrenal glands. This compound increases metabolism and the expenditure of energy.[21] Red pepper powder in kimchi might be the ingredient that produces kimchi's dietary effect, but kimchi showed a better dietary effect than red pepper powder itself.[20] As shown in Table 13.4, when rats were fed a diet containing red pepper powder plus a high fat content, the final body weight of the rats significantly ($p < .05$) decreased compared to rats fed only a high fat diet. However, rats that were fed kimchi containing the same level of red pepper powder plus a high fat content showed an even further decrease ($p < .05$) in their body weight compared to the red pepper diet group after 4 weeks.[20]

Regarding individual organ weights, kimchi plus the high-fat diet significantly reduced liver weight compared to the high-fat diet itself. The liver weight was even lower than that of normal rats. However, the weights of the spleen and kidney were not significantly different among treatment groups. However, epididymal and perirenal fat pad was reduced by red pepper powder and kimchi-added high-fat diets compared to the high-fat diet group control (Table 13.4). The kimchi-added high-fat diet group showed a significantly decreased weight, especially for the perirenal fat pad compared to the red pepper powder-added high-fat diet group and

TABLE 13.4 Changes in Body and Organ Weights of Sprague Dawley (SD) Rats Fed Experimental Diets after 4 Weeks

Weight (g)	Normal	High-fat diet (HFD)	HFD + 5% red pepper powder	HFD + 10% kimchi
Body weight				
Initial weight	171.4 ± 11.9	170.3 ± 10.0	170.7 ± 6.3	171.4 ± 4.2
Final weight	302.5 ± 11.9	338.7 ± 13.3[a]	311.0 ± 9.5[b]	302.5 ± 11.1
Organ weight				
Liver	3.79 ± 0.32[c]	4.39 ± 0.42[a]	4.01 ± 0.39[b]	3.69 ± 0.60[c]
Spleen	0.20 ± 0.03	0.21 ± 0.04	0.21 ± 0.05	0.22 ± 0.05
Kidney	0.91 ± 0.07	0.93 ± 0.11	0.90 ± 0.06	0.90 ± 0.06
Epididymal fat pad	0.96 ± 0.21[c]	1.59 ± 0.42[a]	1.35 ± 0.32[b]	1.32 ± 0.12[b]
Perirenal fat pad	0.89 ± 0.22[d]	1.35 ± 0.25[a]	1.21 ± 0.33[b]	1.14 ± 0.09[c]

[a–d] Different letters in the same row indicate means that are significantly different ($p < 0.05$) by Duncan's multiple range test.

the high-fat diet group. The decrease in fats by kimchi might be due to red pepper powder, garlic, dietary fibers, and other ingredients.

Fermented kimchi stimulates the proliferation of B cells and lowers the lipid accumulation in the epididymal fat pad, and kimchi fermented for 6 weeks at 4°C (optimally fermented kimchi) especially lowers the adipose cell numbers.[22]

Baek et al.[23] investigated the effects of aerobic exercise and/or supplementation of kimchi on changes of the body composition and plasma lipids of obese middle school girls. The exercise group (EG, 8 obese girls) practiced jogging and rope-jumping for 60 minutes 4 times a week and the kimchi group (KG, 12 obese girls) took 3 g of freeze-dried kimchi packed in a 500-mg capsule daily, which is equivalent to 30 g of fresh kimchi. KG + EG showed a greater effect than EG, or KG in reducing body fat. The body mass index (BMI), fat mass, abdominal fat, and triacylglycerol concentration decreased and high-density lipid (HDL)-cholesterol increased in the serum. KG had a greater effect on lowering plasma cholesterol and low-density lipid (LDL)-cholesterol than EG,

Kimchi supplementation while exercising might reduce the obese state by reducing body fat content as well as reducing plasma lipids.

Antiaging Activities

Chlorophyll, phenol compounds, vitamin C, carotenoids, dietary fibers, lactic acid bacteria, and other phytochemicals from raw materials and the fermentation process show possible antiaging activities including antioxidative and antiatherosclerotic activity, prevention of skin aging, and lowering cholesterol levels in the blood.

Antioxidative Activity

The antioxidative compounds in kimchi may remove free radicals formed in the body acting as hydrogen donors.[24] Kimchi inhibited Cu^{2+}-induced LDL oxidation. The dichloromethane fraction of the kimchi showed the highest antioxidant effect against LDL oxidation by inhibiting thiobarbituric acid (TBARS) production or prolonged lag-phase duration by two fold compared to the control.[24] The kimchi dichloromethane fraction-to-1% cholesterol diet considerably decreased plasma and LDL cholesterol, but increased HDL cholesterol.

The liver homogenates of the experimental group containing the dichloromethane fraction of kimchi inhibited LDL oxidation in the presence of Cu^{2+} by 46%.[25] The levels of the activities of catalase, glutathione peroxidase (GSH-Px), Cu, Zn-superoxide (Cu, Zn-SOD), and Mn-superoxide (Mn-SOD) of the kimchi solvent fractions added to the diet group of rats were lower than those of the control group.[25] Low enzyme activities observed from the kimchi solvent fractions added to the diet groups might be due to the fact that the rate of lipid oxidation progressed less in these groups.

Baechu cabbage, red pepper powder, and garlic exerted an antioxidative effect in rabbits that were fed 1% cholesterol for 3 months.[26] Plasma TBARS and peroxide value (POV) level were markedly lowered in both red pepper powder- and garlic-fed rabbits ($p < .05$) compared to the control.[26] Hepatic POV and protein carbonyl values were also lowered in the rabbits fed kimchi ingredients compared to the control. Plasma vitamin E concentration increased in the rabbits that were fed red pepper powder and garlic. In the hepatic antioxidative enzyme activities, catalase activity was significantly increased in the red pepper powder- and garlic-fed rabbits compared to the control. Thus, kimchi ingredients such as red pepper powder and garlic play an important role in rendering antioxidative effects.[26]

The effect of kimchi intake on antiaging characteristics in the brains of senescence-accelerated mice (SAM) in terms of free radical production and antioxidative enzymes was evaluated.[27] Kimchi feeding decreased the increase in free radical production due to aging. Among the kimchi-fed groups, 30% mustard leaf added to *baechu* kimchi and the mustard leaf kimchi groups showed greater inhibiting effects against free radical production than standard *baechu* kimchi.

Cholesterol-Lowering Activity

Kwon et al.[28] reported that plasma cholesterol and triacylglycerol (TG) levels were lowered in rats fed a kimchi diet ($p < .05$). Kimchi intake decreased very low-density lipid (VLDL)-cholesterol, whereas it increased the HDL-cholesterol level in the serum significantly ($p < .05$, Figure 13.3). Especially, the concentration of HDL cholesterol in the 10% kimchi-fed group was the highest. The intake of 5% and 10% kimchi diets also lowered the levels of hepatic cholesterol, TG, total lipids and apolipoprotein B, whereas the levels of fecal total fat, cholesterol, TG, and apolipoprotein A-1 were significantly increased ($p < .05$).

Kim and Lee[22] studied the effect of kimchi on lipid metabolism and immune function in experiments using 63 male Sprague-Dawley (SD) rats fed 6 kinds of *baechu* kimchi during 4 weeks. Three kinds of freeze dried kimchi differing in fermentation period (not fermented, 3-, 6-week-fermented at 4°C) were added at 5% and 10% levels to a diet containing 15% lard. The levels of serum total lipids and triacylglycerol and the content of liver total lipids and triacylglycerol of all kimchi groups were lower than those of the control group. Fermented

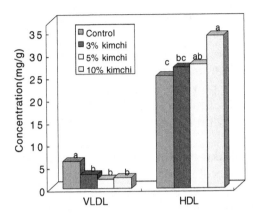

Figure 13.3 The VLDL and HDL concentration in serum of the rats fed, 3%, 5%, and 10% kimchi-added diets. Bars with different superscripts are significantly different at $p < 0.05$.[28]

kimchi had more suppressive effects on the total lipids, cholesterol, and triacylglycerol levels of the epididymal fat pad than the kimchi without fermentation. Thus the fermentation process in kimchi can induce more functional components than the raw materials. The triacylglycerol concentration of feces from the kimchi groups was higher than that of the control group, suggesting that kimchi stimulates lipid mobilization from the epididymal fat pad and lipid excretion via feces. High levels of dietary fibers in kimchi may contribute to capturing cholesterol and TG and removing them via feces.

It is interesting to note that LAB might be involved in reducing serum cholesterol. Ray[29] indicated that LAB metabolize dietary cholesterol and deconjugate bile salts in the colon and prevent their reabsorption in the liver and thus reduce cholesterol levels in the serum.

Kwon et al.[30] studied daily kimchi consumption and its hypolipidemic effect on 102 healthy Korean adult men aged from 40 to 64 years that visited a hospital for physical examination. The physical and biochemical parameters of blood were examined as well as food record, preference for taste, and personal life habit. The intakes of dietary fiber and Ca were found to be increased as kimchi intake increased ($p < .05$).

TABLE 13.5   Correlation Coefficients Between Biochemical Parameters and Kimchi Consumption, BMI, and Preferences of Taste of Korean Adult Men

	Kimchi	BMI	Salt	Hot	Degree of fermentation[)]
TG	0.03	0.28*	0.04	−0.05	−0.02
TCH	−0.09	0.27*	0.10	−0.02	−0.13
HDL-C	0.23*	−0.25*	−0.07	0.07	0.12
LDL-C	−0.17*	0.30*	−0.06	0.01	−0.07
Thyroxine	−0.03	−0.11	−0.02	0.05	−0.09
T3	0.03	0.15	−0.01	−0.01	0.03
SBP	0.17	0.24*	0.02	−0.15*	0.07
DBP	0.11	0.37*	−0.03	−0.12	0.01
BMI	0.03	1.00	−0.12	−0.03	−0.03
AI	0.06	0.27*	0.02	0.08	−0.02

Note: TCH: Total cholesterol; T3: Triiodothytonine; SBP: Systolic blood pressure; DBP: diastolic blood pressure; BMI: Body mass index (kg/m^2); AI: Atherogenic index; Kimchi: Daily kimchi consumption; Salty taste: The preference of salty taste (not salty: 1, moderate: 2, salty: 3); Hot: The preference of hot taste (dislike: 1, like: 2); Degree of fermentation: The preference of fermented kimchi (prefer to fresh kimchi: 1, prefer to fermented kimchi: 2, prefer to over ripened kimchi: 3) significantly different ($p < 0.05$).

(From MJ Kwon, JH Chun, YS Song, YO Song. Daily kimchi consumption and its hypolipidemic effect in middle-aged men. *J Korean Soc Food Sci Nutr* 28:1144–1150, 1999. With permission.)

When correlation coefficient between kimchi consumption and other parameters were analyzed, kimchi consumption (0 to approximately 453 g) was positively correlated with HDL-cholesterol and negatively correlated with LDL-cholesterol ($p < .05$, Table 13.5). The preference for hot taste was negatively correlated with systolic blood pressure. They suggested that kimchi consumption is beneficial in elevating HDL cholesterol and lowering LDL cholesterol.

Prevention of Skin Aging

Human skin cells, keratinocytes (A431, epidermoid carcinoma, human) and fibroblast (CCD-986SK, normal control, human)

were cultured in an oxidative stress condition provoked by paraquat, a superoxide anion generator, and hydrogen peroxide in the absence and presence of kimchi extract. The survival rate of the keratinocyte by the treatment of H_2O_2 was significantly reduced by kimchi extracts on the cells. Especially, a 2-week-fermented kimchi (optimally ripened kimchi) remarkably decreased the cytotoxicity induced by H_2O_2 in the keratinocyte cells.[31] Over 1 mM of paraquat concentration exhibited a strong cell toxicity on the keratinocyte cells, but the extracts from kimchi fermented for 1, 2, and 3 weeks at 8°C showed protective effects in order. Fibroblast cells were significantly affected by H_2O_2 as were keratinocyte cells. Although almost all extracts of kimchi of different fermentation periods showed a protective effect against cell killing at a 0.5 mM concentration of H_2O_2, 2 week-fermented kimchi extract showed the strongest protective effect on fibroblast cells treated with 1 mM H_2O_2 for either 1 day or 4 days. However, most of kimchi extracts showed a weak preventive effect or no effect on oxidative stress produced by paraquat.

In an *in vivo* study, 10% of freeze-dried *baechu* kimchi, leek kimchi, or mustard leaf kimchi was added to the AIN-76 diet, and fed to hairless mice for 20 weeks. At the 16th week, morphological changes in the epidermis and dermis were observed. At the 20th week, the antioxidant effect against UV-induced photoaging and the free radical scavenging effect were investigated.[32] The epidermal thickness of hairless mice was found to be thicker (22 to approximately 37% increase) in the kimchi-fed groups while the stratum corneum was thinner (62 to approximately 58% decrease) compared to those of the control group, thus kimchi seems to keep the skin healthy. Collagen synthesis at the dermis increased with kimchi treatment as the activity of the rough endoplasmic reticulum of the fibroblast observed was greater than those of the control group.[32] Type IV collagen, supporting matrices at the basement membrane, existed in greater amounts in the kimchi-fed groups, especially, the mustard leaf kimchi-fed group (52% increase), than in the control group. Lipid oxidation expressed in the TBARS content of the liver of hairless mice fed a kimchi diet was retarded compared to the control, and the contents of superoxide anion, hydroxy radical, and hydrogen peroxide were also less than in the control group, showing a greater capacity of scavenging free radicals. The free radical scavenging activity of kimchis seems to be due to high levels of chlorophyll, vitamin C, carotenoids, and phenolics in the kimchis.

Antimutagenic/Anticancer Effects

Antimutagenic/Anticarcinogenic Effect

Optimally fermented kimchi exhibited antimutagenicity against aflatoxin B_1 and MNNG (N-methyl-Nnitro-N-nitrosoguanidine) in the Ames test and the SOS chromotest *in vitro*.[33-35] Optimally ripened kimchi showed the highest antimutagenicity compared to freshly prepared kimchi and overripened kimchi.[35] The kimchi extract also showed antimutagenic activity in the *Drosophila* wing test *in vivo*.[36] Kimchi also exhibited anticlastogenic activity in MMC-induced mice using the *in vivo* supravital staining micronucleus assay.[37]

C3H/10T1/2 cells are mouse embryo cells that form foci in culture media when exposed to a carcinogen. The foci that developed as type II and type III correlated well with tumor formations of 50% and 85% in C3H mice, respectively.[38] The transformation of C3H/10T1/2 cells decreased markedly when kimchi extract (methanol soluble fraction, MSF) was added to the test system. When 200 of MSF from 3-week-fermented (at 5°C) kimchi were added along with MCA (3-methylcholanthrene) to the cells, then the numbers of type II and III foci formed was significantly decreased ($p < .05$, 89%).[39]

Anticancer Effect

Kimchi extracts inhibited the survival or growth of human cancer cells (AGS gastric cancer, HT-29 colon cancer, MG-63 osteocarcinoma, HL-60 leukemia and Hep 3B liver cancer) in the SRB (Sulforhodamine B) assay, the MTT (3-(4,5-dimethyl-thiazol-2-yl)-2,5-diphenyl tetrazolium bromide) assay and the growth inhibition test.[9,33] The kimchi dichloromethane fraction inhibited ^3H thymidine incorporation in the cancer cells.[40]

β-sitosterol and a linoleic acid derivative were the main active compounds that showed anticancer activity in kimchi. Kimchi dichloromethane fraction arrested the G2/M phase in the cell cycle and induced apoptosis of HL-60 human leukemia cells.[9] When HIR_C-B cells were treated with dichloromethane fraction from kimchi and β-sitosterol that was identified as an active kimchi compound, followed by microinjection of onco-genic H-ras^{v12}, the DNA synthesis of the cells was decreased, indicating that active compounds of kimchi affected the signal transduction pathway via ras to the nucleus.[9]

In the present studies, sarcoma 180 cells were trans-planted to Balb/c mice and then kimchi-extracted samples were treated. The MSF of the kimchi-treated group resulted in the smallest tumor weight of 1.98 ± 1.8g compared to the control group of 4.32 ± 1.5g. MSF from the 3-week-fermented at 5°C kimchi also reduced malondialdehyde formation com-pared to the control.[41] MSF also reduced the hepatic cytosolic xanthine oxidase activity in sarcoma-180 treated Balb/c mice. On the other hand, MSF increased the hepatic cytosolic gluta-thione content and the activities of glutathione S-transferase and glutathione reductase, indicating that kimchi might be involved in the detoxification of xenotoxic materials in the liver. Kimchi extracts also enhanced the immune function of NK (natural killer) cells and macrophages.[33]

The concentration and kinds of salt used in kimchi prep-aration were important in the chemoprevention of cancer. High levels of salt (8.5%) in kimchi showed comutagenic activity,[42] however, 2.5% salt in kimchi exhibited an antimutagenic/anticancer effect. Also, heat-processed Guwoon salt was very effective in preparing a cancer-preventive kimchi.[43] In exper-imental metastasis with colon 26-M3.1 cells, subcutaneous administration of kimchi extract (0.05 to approximately 1.25 mg/mouse) 1 day after tumor cell inoculation inhibited lung metastasis significantly ($p < .05$, Table 13.6). Functional kimchi significantly inhibited the metastasis of colon 26-M3.1 cells in the lungs of the mice. We prepared functional kimchi by using organically cultivated *baechu* cabbage and added more condiments, mustard leaf, Chinese pepper, etc., and a fermentation method.[44] Thus, the functionality of kimchi can

TABLE 13.6 Inhibitory Effect of Methanol Extracts from Various Kinds of Baechu Cabbage Kimchi on Tumor Metastasis Produced by Colon 26-M3.1 Cells

Treatment	Dose (mg/mouse)	Route	No. of lung metastasis (inhibition,%) Mean ± SD	Range
Control		sc	162 ± 7[a]	153–172
Kimchi	0.05	sc	157 ± 13[a](3)	142–172
	0.25	sc	147 ± 8[ab] (9)	138–157
	1.25	sc	139 ± 5[bc] (14)	131–144
Functional[1] kimchi	0.05	sc	119 ± 4[d] (27)	114–123
	0.25	sc	99 ± 8[e] (39)	89–110
	1.25	sc	83 ± 6[f] (49)	73–91

[a–f]Different letters indicate means that are significantly different ($p < 0.05$) by Duncan's multiple range test.
[1] Organically cultivated *baechu* cabbage used functional kimchi.
(From KY Park, KA Baek, SH Rhee, HS Cheigh. Antimutagenic effect of kimchi, *Foods Biotech* 4:141–145, 1995. With permission.)

be further increased by adjusting the ingredients and the preparation method.[44]

Kimchi LAB

Lactic acid bacteria (LAB) are commonly found in the gastrointestinal tract of humans and animals, in dairy products, and in plants. LAB from yogurt and other dairy products are usually believed to be probiotics, and many studies have reported the functionality of dairy LAB in the form of antimutagenic, antitumor, immunomodulatory properties, etc.[45] There are a few studies on kimchi LAB and their functionality related to chemopreventive effects such as antimutagenic and antitumor effects, and other functionalities.

Antimutagenic Effect of Kimchi LAB

The antimutagenic activities of the cell bodies of several LAB isolated from kimchi were studied using the Ames test and the SOS chromotest.[46,47] The mutagenicities mediated by

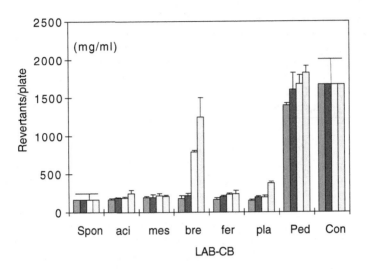

Figure 13.4   Antimutagenic activity of lactic acid bacteria-cell body (LAB-CB) against 4-NQO (0.15.g/plate) on *Salmonella typhimurium* TA100. Spon: spontaneous; aci: *Lactobacillus acidophilus*; mes: *Leuconostoc mesenteroides*; bre: *Lactobacillus brevis*; fer: *Lactobacillus fermentum*; pla: *Lactobacillus planterum*; Ped: *Pediococcus acidilactic*; Con: control.

4-NQO (4-nitroquanoline-1-oxide), MeIQ (2-amino-3,4-dimethylimidazo[4,5-*f*]quinoline), and Trp-P2 (3-amino-1-methyl-5H-pyrido[4,3-b]indole) were effectively suppressed by the kimchi LAB in the tests. The cell body of *Leu. mesenteroides* exhibited a higher antimutagenic activity on 4-NQO, MeIQ and MNNG than any other LABs tested. As shown in Figure 13.4, *Leu. mesenteroides* showed the highest antimutagenic activity among the LAB. However, *Ped. acidilactici* did not show any antimutagenic effect against 4-NQO. The dairy LAB, *Lac. acidophilus*, also showed almost the same antimutagenic activity as kimchi LAB. The antimutagenic activity of LAB was found in the cell wall fraction, rather than in the cytosol fraction.[48]

It was reported that glycopeptide cell wall fragments are responsible for the antimutagenic/antitumor activity.[45] The peptidoglycan is a compositional compound in the cell wall, and this is combined with muramyl peptide. Thus, whether the bacteria are alive or dead, the antimutagenic/antitumor activity is still effective.[48]

Antitumor and Immunostimulant Effects of Kimchi LAB

The administration of lyophilized LAB reduced tumor formation in mice. The kimchi LAB of *Lac. plantarum* and *Leu. mesenteroides* significantly reduced the tumor formation rate in ICR mice treated with sarcoma-180 cells.[49] The inhibition rate was 57% and 39%, respectively, while *Lac. casei* was the most effective, showing an 88% inhibition rate in this study. The tumor formation also decreased when the kimchi LAB, *Lac. plantarum* and *Leu. mesenteroides* were administered using Lewis lung carcinoma in C57BL/6 mice; the inhibition rates were 42% and 44%, respectively. However, *Lac. acidophilus*, a dairy LAB, inhibited the tumor formation rate by 28%, and *Lac. casei* by 78%. It seems that the antitumor activity was somewhat different depending on the LAB strains. However, the activity was not different regardless of the source of LAB (from kimchi or dairy foods).

Shin et al.[50] reported the antitumor effects of kimchi LAB, using mice fed with cell lysate of *Lac. plantarum* from kimchi. The ascites tumor induced by sarcoma-180 was markedly inhibited and the expected life span was extended to 60% in the Balb/c mice fed with *Lac. plantarum* cell lysate for 2 weeks. As the lung was the metastasis site of SOS, the weight of the lung was measured to determine the degree of metastasis inhibition by *Lac. plantarum* of the cell lysate feeding. The rats fed with cell lysate for 1 week showed a remarkable inhibition of lung metastasis by 63% (before) and 46% (after), respectively. These results suggested that the feeding of *Lac. plantarum* cell lysate can induce stimulation of the immune system and these effects result in antitumor activity.

Chae et al.[51] reported the immunostimulation effects on the mice fed with cell lysate of *Lac. plantarum* isolated from kimchi. They observed a general enhancement in the enteric and systemic immune response with a simple oral administration of cell lysate of *Lac. plantarum*. Park[52] also reported that administration to mice of a culture broth of *Lac. plantarum* isolated from kimchi increased phagocytosis of the *Staph. aureus.*

Muramyl dipeptide and its derivatives are involved in stimulating the cell-mediated immune function. When this compound was administered to a macrophage, it stimulated superoxide anion and H_2O_2 production by the marcrophage and killed the tumor cells.[53]

Decreases in the Fecal pH and Activities of Microflora Enzymes Related to Colon Cancer by Kimchi LAB

It was questionable whether kimchi intake changes the composition of human fecal bacteria. Lee et al.[54] examined viable cells of *Lactobacillus* and *Leuconostoc* delivered to the colon. The kimchi LAB counts increased significantly ($p < .05$) during the administration of kimchi, however, other intestinal microflora such as *Bacteroides, Bifidobacterium, Escherischia coli, Streptococcus, Staphylococcus,* and *Clostridium perfringens* did not change significantly, indicating that a portion of LAB present in kimchi can pass through the human stomach and reside in the large intestinal tract. Especially the enzyme levels of β-glucosidase and β-glucuronidase during kimchi intake in humans significantly decreased ($p < .05$). As shown in Figure 13.5, the enzymes that mediate the conversion of procarcinogens to proximal carcinogens involved in colon cancer decreased when kimchi was administered.[54] Oh et al.[17] also reported that β-glucuronidase and nitroreductase activities in the colon were significantly reduced for Koreans and Germans during the kimchi intake. Especially in their experiment, a significant decline in fecal pH was found for both groups during the kimchi intake. Bengmark[15] reported that *Lac. plantarum* can colonize in the intestinal tract for long periods and *Lac. plantarum* can make short-chain fatty acids from dietary fibers, which can induce the apoptosis of colon cancer cells. These results partly confirmed by epidemiological hypothesis that kimchi consumption correlates with a low incidence of colon cancer in Koreans.

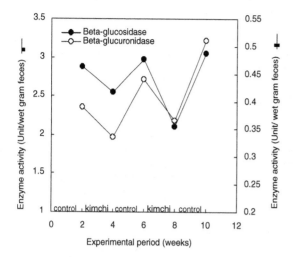

Figure 13.5 Effect of kimchi intake on fecal beta-glucosidase and beta-glucuronidase during experimental period.[54]

Future Research Emphasis on Kimchi LAB

LAB are regarded as probiotics, and much health-related research has been reported on LAB from yogurt and dairy products.[55] Kimchi LAB are also regarded as good probiotics since kimchi has been eaten for centuries. From the previous studies, the functions of kimchi LAB seem almost the same as those of the dairy LAB. Kimchi can be a valuable anticancer food as it carries LAB along with other functional phytochemicals from vegetables and those produced during the fermentation process.

Helicobacter pylori could not colonize in the stomach of *Lac. salivarius*-infected gnotobiotic Balb/c mice, but colonized in large numbers and subsequently caused active gastritis in germ-free mice.[56] In addition, *Lac. salivarius* administered after *Helicobacter pylori* implantation could eliminate colonization by *H. pylori*. These results suggested the possibility of *Lactobacilli* being used as probiotic agents against *H. pylori*. Koreans and Japanese have higher incidence rates of stomach cancer. Kimchi LAB can possibly show this kind of activity. It has also been reported that garlic and red pepper, which are the main subingredients of kimchi, reduced *H. pylori* infections.[57-59] If the kimchi LAB reduce *H. pylori* proliferation, LAB might be used as a preventive agent for stomach cancer.

A number of species within the LAB genera have been shown to exhibit antitumor properties. *Lac. acidophilus* and *Lac. casei* have been most commonly reported to inhibit tumorigenesis. However, kimchi LAB along with other phytochemicals in kimchi are also believed to have antitumor activity. Further research on the probiotic activities, possible mechanisms, and identification of the active compounds of kimchi LAB is necessary.

FUNCTIONALITY OF LEEK (*BUCHU* IN KOREAN) KIMCHI

Leek kimchi is a major traditional special kimchi in Kyungsang province, southern region of South Korea. *Buchu* (Leek, *Allium tuberosum* L.) kimchi is prepared with large quantities of red pepper powder and pickled anchovy, therefore, it is known as a good side dish because of the unique flavor of leeks and its hot taste.[60-62] Leeks, the major ingredient in leek kimchi, have been used as a food or drug for treatment of abdominal pain, diarrhea, hematemesis, snakebite, and asthma in folk remedies from ancient times.[63] Leeks are rich in vitamins A, B_1, and C,[64] and belong to the *Allium* genus that contain large amounts of thiosulfinates and organosulfur compounds, which are responsible for the characteristic odor and flavor of allium.[65-67] The allyl sulfur compounds are known to inhibit chemically induced tumors.[68-72] Leeks also contain high levels of flavonoids.[73,74] Food-derived flavonoids such as flavonols, quercetin, kaempferol, and myricetin have antimutagenic and anticancer effects *in vitro* and *in vivo*.[75,76] In addition, several studies have indicated that high consumption of leek was associated with a reduced risk for colorectal cancer.[77-79]

In the preparation of leek kimchi, the leek is cut into 2 pieces and soaked in 20% salt solution for 20 min at room temperature, then rinsed with tap water twice. The brined leeks are mixed with premixtures of the subingredients. The standardized ingredient ratio of leek kimchi is 11.0 fermented anchovy juice, 7.0 red pepper powder, 5.0 garlic, 2.0 ginger, and 13.0 glutinous rice flour paste in proportion to 100 of salted leeks.

Antimutagenic/Anticancer Effect

The antimutagenic effects of leek kimchi and *baechu* kimchi and their cytotoxic effects against the human cancer cell lines (AGS human gastric cancer cells and HT-29 human colon cancer cells) were investigated in the *Salmonella typhimurium* assay system and MTT assay, respectively.[80] Leek kimchi (optimally ripened) samples showed higher antimutagenic effects against aflatoxin B_1 (AFB$_1$) than optimally ripened *baechu* kimchi against the *Salmonella typhimurium* TA100 strain. Leek exerted stronger antimutagenicity against AFB$_1$ than *baechu* cabbage in the Ames test. In the MTT assay, 6-day-fermented (at 15°C) leek kimchi revealed the highest cytotoxicity against AGS human gastric adenocarcinoma cells, in which 62% and 82% inhibition was observed with the addition of 100g and 400 µg/well, respectively. Leek kimchi samples caused 60 to 70% inhibition of the proliferation of HT-29 human colon adenocarcinoma cells even at 100 µg/well while *baechu* kimchi exhibited 60% inhibition at 400 µg/well. Leek showed higher antiproliferative effect against both AGS cells and HT-29 cells than *baechu* cabbage in the MTT assay. From these results, it is considered that leek and leek kimchi have stronger antimutagenic and anticancer effects than *baechu* cabbage and *baechu* kimchi, and thus the higher inhibition rate of leek kimchi probably results from leek, the major ingredient present in the formulation.[80]

The anticarcinogenic effects of the methanol extracts from leek kimchi and *baechu* kimchi were evaluated using cytotoxicity and transformation tests in C3H/10T1/2 cells.[81] The inhibitory effect of the leek kimchi (6-day fermented at 15°C, pH 4.29) was higher than that of the *baechu* kimchi (4-day fermented at 15°C, pH 4.21) on the cytotoxicity induced by 3-methylcholanthrane (MCA) in a C3H/10T1/2 cell system. While the MCA-treated culture (control) formed 21.0 foci of type II plus type III in C3H/10T1/2 cells, 100 µg/ml of the

TABLE 13.7   Inhibitory Effect of Methanol Extract and
Dichloromethane Fraction (CH_2Cl_2 fr.) from 6-Day-Fermented
Leek Kimchi (pH 4.3) at 15°C on the Transformation of
C3H/10T1/2 Cells Treated with 3-Methylcholanthracene (MCA,
10 μg/mL)[a]

	Total number			
Treatment (50 μg/ml)	Type I foci	Type II foci	Type III foci	Type II + III foci
MCA (control)	15.3	15.0	10.0	25.0
MCA + MeOH ext.	8.3	8.7	4.7	13.0
MCA + CH_2Cl_2 fraction	4.3	3.3	1.7	5.0

[a] 2000 cells were seeded in 60mm/dishes, 10 dishes/group and incubated for
24 h, and then MCA and kimchi fractions were treated for 48 h. Following
treatment, the medium was changed. Subsequently, the medium was changed
at weekly intervals and, at 6 weeks, the dishes were fixed and stained, and
then type I, II and III foci were counted.

methanol extract of the leek kimchi- and of the 4-day fermented *baechu* kimchi-treated cultures reduced the formation
of type II plus type III foci to 7.4 and 11.3, respectively. Among
the fractions of leek kimchi, the dichloromethane fraction
showed the highest inhibitory effect on MCA-induced cytotoxicity in C3H/10T1/2 cells. In the transformation test using
MCA, the dichloromethane fraction considerably reduced the
formation of type II plus type III foci, especially type III foci
(Table 13.7). When 50 μg/ml of dichloromethane fraction from
the leek kimchi was used, the numbers of type III foci mediated by MCA were decreased to 1.7 compared to 10 of the
control.[81] These results indicate that leek kimchi has a stronger anticarcinogenic effect than *baechu* kimchi and that the
dichloromethane fraction of the leek kimchi may contain the
major compound(s) that suppress carcinogenesis in eukaryotic
cells.

In an effort to identify the active anticancer compounds
in leek kimchi, optimally ripened leek kimchi at 15°C for 6 days
was fractionated into 7 groups, such as methanol extract,
hexane extract, methanol soluble fraction, dichloromethane

fraction, ethyl acetate fraction, butanol fraction, and water fraction. Dichloromethane (DCM) fraction from leek kimchi contained major active compound(s) that decrease the growth of the AGS and the HT-29 cancer cells. We identified 2-hydroxy-1-(hydroxy-methyl) ethyl hexadecanoic acid as one of the major antiproliferative active compounds present in DCM fraction by GC-MS.[44] It was also reported that leek contains sulfides,[82] linalool,[83] and flavonoid glycosides.[74,75] Thiosulfinates containing both methyl and 1-propyl groups were identified from dichloromethane extract of leeks by HPLC.[65] β-Sitosterol and β-sitosterol-3-β-D-glucopyranoside were tentatively isolated from the $CHCl_3$ fraction of the leek.[84] These results suggested that the antiproliferative effect of leek kimchi might be due to flavonoids, fatty acids, thiosulfinates, terpenoides and sterol, and others in the DCM fraction.

To elucidate possible mechanisms of the DCM fr.-induced growth inhibition, we further investigated whether the DCM fraction of the leek kimchi affected the cell cycle progression of human leukemia HL-60 cells. Flow-cytometric data of the DCM fraction-treated cells revealed a cell cycle block at the G2/M transition phase. A concentration-dependent decrease in the percentage of cells in G1 phase was observed in the DCM fraction-treated cells. The decrease in G1-phase cells resulting from the DCM fraction treatment was complemented by accumulation of cells in the G2/M phase of the cell cycle. In spite of only 12 h treatment with the DCM fraction (250 µg/ml), the percentage of the cells in the G2/M phase was increased by 1.7 fold compared to the control.[85] These results correlated well with the growth inhibitory effect of the DCM fraction, suggesting that one of the antiproliferative mechanisms of the DCM fraction was due to the block during the G2/M phase and that such cells could not enter the G1 phase.

The DCM fraction of the leek kimchi induced apoptosis, which was demonstrated by direct visualization of morphological nuclear changes.[85] These results suggested that the anticancer effects of the DCM fraction were also related to the induction of the apoptosis, along with the cell cycle arrest at the G2/M transition phase.

FUNCTIONALITY OF MUSTARD LEAF
(ML, *GAT* IN KOREAN) KIMCHI

This is also a popular kimchi in southern regions of South Korea. The standardized ingredient ratio of mustard leaf (ML) kimchi is 100 mustard leaf, 10.5 red pepper powder, 3.0 crushed garlic, 2.0 crushed ginger, 10.5 fermented anchovy juice, 12.0 glutinous rice paste (rice:water = 2:8), 1.4 sugar, 1.5 sesame seed, 15.0 green onion, and the final salt concentration is 2.5%[86]. In the preparation of mustard leaf kimchi, mustard leaf is first washed, then brined in 10% salt water for 2 h and then washed and removed from the water for 2 h and then mixed with a premixture of spices and condiments.[86]

Antimutagenic and Anticancer Effect

Isothiocyanate is a well-known antimutagenic/anticancer active compound. Allyl isothiocyanate and 3-butenyl isothiocyanate are the most volatile substances that are found in ML[87]. Kim et al.[88] identified 4-decanol as one of the active compounds in ML that showed a strong antimutagenic effect on aflatoxin B_1 and MNNG in *Sal. typhimurium* TA100. ML kimchi showed more antimutagenic and anticancer effects than *baechu* kimchi.[86] The fermentation process increased its antimutagenicity (Ames test using aflatoxin B_1 as carcinogen) and anticancer effects in human cancer cells (AGS gastric carcinoma cells and HT-29 colon carcinoma cells) compared to the raw material(s).

Figure 13.6 shows the anticancer effect of raw ML, brined ML, fermented ML, freshly prepared ML kimchi, and 30-day-fermented ML kimchi at 5°C on the growth of HT-29 human colon carcinoma cells in an MTT assay. The fermented ML kimchi was the most effective, followed by fermented ML. Though ML showed an anticancer effect, when ML was fermented the effect was even greater; however, the fermented ML kimchi that contains ML and condiments or other ingredients increased the anticancer activity even more.[86]

Antioxidative Effect

ML contains high levels of chlorophyll and β-carotene, which show anticancer and antioxidative activities. The contents of

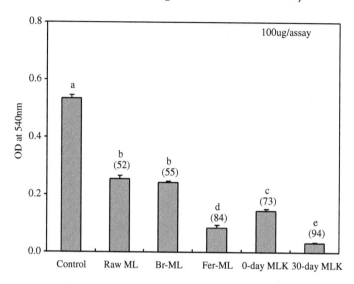

Figure 13.6 Inhibitory effect of methanol extracts from differently prepared mustard leaf (ML) and ML kimchi (MLK) on the growth of HT-29 human colon carcinoma cells in an MTT assay. Br-ML: brined ML. Salted in 10% brine for 2 h. Fer-ML: fermented ML. Fermented the brined ML at 15°C for 7 days (pH 4.3); 30 day MLK: fermented MLK. Fermented the MLK at 5°C for 30 days (pH 4.3); Parenthesis on the bars indicate inhibition rate (%). a–e: Different letters surmounted on the bars are significantly different at $p < 0.05$ by Duncan's multiple range test.

chlorophyll and carotenoids were 523 µg/g and 4.75 mg%, respectively.[89] The active compounds, chlorophylls a and b, are degraded during ML kimchi fermentation for 25 days at 15°C, however, the contents of pheophytins a and b greatly increased.[90] The conversion of chlorophyll to pheophytin is due partly to acids formed during fermentation. Chlorophylls and pheophytins showed almost the same degree of antioxidative activities during autooxidation/enzymatic oxidation of linoleic acid.[90] Song et al.[91] reported that the chlorophyll content remained the same, but 44% of the carotenoids were destroyed during ML kimchi fermentation. Though the fermentation period could not affect the antioxidative effects of the crude chlorophyll and carotenoids mixture from ML kimchi, their antioxidative activities against autooxidation of linoleic acid were significant ($p < .05$) and much higher than those of α-tocopherol.

Hwang et al.[92] reported that the total phenolics content increased during ML kimchi fermentation at 15°C. The TBA value of fermented kimchi (optimally fermented) showed the lowest level in the model system. The antioxidative activity reflected in the inhibition of formation of peroxide levels during auto-oxidation of linoleic acid mixtures in aqueous model systems increased in ripened kimchi compared to freshly prepared kimchi.

Isorhamnetin diglucoside is one of the active phenolic compounds found in the butanol fraction of ML. This compound could convert to isorhamnetin during fermentation in the colon. Intraperitoneal injection of isorhamnetin reduced glucose, TBARS and glycosylated protein levels in STZ-induced diabetic rats, and isorhamnetin showed a higher antioxidative activity.[93,94]

FUNCTIONALITY OF *KAKTUGI* (DICED ORIENTAL RADISH KIMCHI)

Kaktugi is a typical radish kimchi, which is consumed second most frequently among kimchi types in Korea. Oriental radish is the major raw material. The standardized ingredient ratio of *kaktugi* is 3.9 red pepper powder, 2.3 garlic, 1.0 ginger, 5.1 green onions, 1.4 sugar, and 4.1 fermented shrimp sauce in proportion to 100 diced radish. The final salt concentration is 2.5%.[95] In the preparation of *kaktugi* radishes cut into cubes (about 2 × 2 × 2 cm in size) are salted in 7% brine at 5°C for 1 h. The water is drained for 1h and mixed with red pepper powder and then mixed with a premixture of other subingredients.

Antimutagenic/Anticancer Effect

The kinds or amounts of ingredients used and the preparation method affect the antimutagenic/anticancer activities of kimchi. Different varieties or kinds of radish, salt, and red pepper powder used in the recipe affect the antimutagenicity in *Salmonella typhimurium* TA100. Especially, *Saengkum* or *Guwoon* salts (heat processed salts) used in *kaktugi* showed the highest antimutagenicity among salt varieties used.[95]

Red pepper powder showed antimutagenicity (38% inhibition) against aflatoxin B_1 in *Sal. typhimurium* TA100. Seeds of red pepper showed higher antimutagenicity than pericarp. Anticancer activities of the red pepper in AGS human gastric cancer cells and HT-29 human colon carcinoma cells using an MTT assay also showed a similar pattern. The inhibition rate was 82 to 87% for seeds, 10 to 15% for pericarp, and whole red pepper powder exhibited 56 to 64% inhibition for treatment of 100 µg/assay of the methanol extracts. The *kaktugi* prepared with the white part of the green onion (5.1%) significantly increased (84 to 93% inhibition rate) the antimutagenic effect against aflatoxin B_1, but *kaktugi* prepared with green parts or the whole onion decreased antimutagenicity (22 to 27% inhibition rate).[96]

Kaktugi and radish root showed a higher natural killer cell activity (20 to 80% increase)[95] than *baechu* kimchi in mice. *Kaktugi* (methanol extract) showed a higher antimutagenic activity (52 to 59%)[96] than *baechu* kimchi (36 to 41%) and *chongkak* kimchi (26 to 46%) against aflatoxin B_1 in *Sal. typhimurium* TA 100 and MNNG in *E. coli* PQ 37.

Upon treatment with *kaktugi* extract, a concentration-dependent inhibition of the cell viability was observed (the addition of 1 or 1.5% *kaktugi* extract caused about 43 or 74% reduction in the cell growth, respectively).[96] The inhibitory effects were accompanied by distinct morphological changes such as membrane shrinking and cell rounding.[96] Analyses of the cell cycle distribution of A549 human lung cancer cells after exposure to *kaktugi* extract showed that these cells had a marked accumulation of cells in the G2/M phase of the cell cycle, which was accompanied by a significant decrease in their G1 phase as compared with the untreated control cells. G2/M arrest by *kaktugi* extract was associated with the induction of either tumor suppressor p53 or cyclin-dependent kinase inhibitor p21 (*WAF1/CIP1*) in a concentration-dependent manner. In addition, the transcriptional activity of nuclear factor-B (NF-*k*B) using a luciferase reporter assay was dramatically increased by the addition of *kaktugi* extract.[96] Thus, a combined mechanism involving the induction of p21 (*WAF1/CIP1*) and NF-*k*B targets for *kaktugi* extract was noticed, and this may explain some of its anticancer effects.

REFERENCES

1. KH Son. Variety and use of kimchi. *Korean J Diet Cult* 6:503–520, 1991.

2. WS Park, YJ Koo, BH Ahn, SY Choi. Standardization of kimchi-manufacturing process. Report of Korean Food Research Institute, Korea, 1994.

3. KY Park. The nutritional evaluation, and antimutagenic and anticancer effects of kimchi. *J Korean Food Sci Nutr* 24:169–182, 1995.

4. KY Park. Nutrition, Functionality and Anticancer Effect of Kimchi. Proceedings of 2000 Japan·Korea kimchi seminar, Tokyo, 2000, pp 27-50.

5. Report on 1998 National Health and Nutrition Survey (Dietary Intake Survey). Ministry Health and Welfare, Korea Government, 1999, pp 53-68.

6. SB Kim. *Culture; History of Korea Food Life*. Seoul: Kwang moon kak, 1997, pp 146–260.

7. HS Cheigh. *Kimchi Culture and Dietary Life in Korea*. Seoul: Hyoil Publishing Co, 2002, pp 167–193.

8. HS Cheigh, KY Park. Biochemical, microbiological, and nutritional aspects of kimchi (Korean fermented vegetable products). *Crit Rev Sci Nutr* 34:175–203, 1994.

9. EJ Cho. Standardization and Cancer Chemopreventive Activities of Chinese Cabbage Kimchi. PhD dissertation, Pusan National University, Busan, 1999.

10. EJ Cho, KY Park, SH Rhee. Standardization of ingredient ratios of Chinese cabbage kimchi. *Korean Food Sci Technol* 29:1228–1235, 1997.

11. EJ Cho, SM Lee, SH Rhee, KY Park. Studies on the standardization of Chinese cabbage kimchi. *Korean Food Sci Technol* 30:324–332, 1998.

12. CW Lee, CY Ko, DM Ha. Microfloral change of the lactic acid bacteria during kimchi fermentation and identification of the isolates. *Kor J Appl Microbiol Biotechnol* 20:102–109, 1992.

13. CR Lim, HK Park, HU Han. Re-evaluation of isolation and identification of gram-positive bacteria in kimchi. *Korean J Microbiol* 27:404–414, 1998.

14. TI Mheen, TW Kwon. Effect of temperature and salt concentration on kimchi fermentation. *Korea J Food Sci Technol* 16:443–450, 1984.

15. S Bengmark. Use of Prebiotics, Probiotic and Synbiotics in Clinical Immunonutrition. Proceedings of International Symposium on Food, Nutrition and Health for 21st Century, Seoul, 2001, pp.187-213.

16. KY Park, JO Ha, SH Rhee. A study on the contents of dietary fibers and crude fiber in kimchi ingredients and kimchi. *J Korean Soc Food Nutr* 25:69–75, 1996.

17. YJ Oh, IJ Hwang, C Leitzmann. Regular intake of kimchi prevents colon cancer. *Kimchi Sci Ind* 2:9–22, 1993.

18. YS Song, YO Song. Antiatherogenic Effects of Kimchi. Processings of 8th Asian Congress of Nutrition, Seoul, 1999, pp 153-155.

19. SH Ryu, YS Jeon, MJ Kwon, JW Moon, YS Lee, GS Moon. Effect of kimchi extracts to reactive oxygen species in skin cell cytotoxicity. *J Korean Soc Food Sci Nutr* 26:814–821, 1997.

20. SM Choi. Antiobesity and Anticancer Effects of Red Pepper Powder and Kimchi, PhD dissertation, Pusan National Unversity, Busan, 2001.

21. KM Kim. Increases in Swimming Endurance Capacity of Mice by Capsaicin. PhD dissertation, Kyoto University, Kyoto, 1998.

22. JY Kim, YS Lee. The effects of kimchi intake on lipid contents of body and mitogen response of spleen lymphocytes in rats. *J Korean Soc Food Sci Nutr* 26:1200–1207, 1997.

23. YH Baek, JR Kwak, SJ Kim, SS Han, YO Song. Effects of kimchi supplementation and/or exercise training on body composition and plasma lipids in obese middle school girls. *J Korean Soc Food Sci Nutr* 30:906–912, 2001.

24. JW Hwang, YO Song. The effects of solvent fractions of kimchi on plasma lipid concentration of rabbit-fed high-cholesterol diet. *J Korean Soc Food Sci Nutr* 29: 204–210, 2000.

25. HJ Kim, MJ Kwon, YO Song. Effects of solvent fractions of Korean cabbage kimchi on antioxidative enzyme activities and fatty acid composition of phospholipid of rabbit-fed 1% cholesterol diet. *Korean Soc Food Sci Nutr* 29: 900–907, 2000.

26. MJ Kwon, YS Song, YO Song. Antioxidative effect of kimchi ingredients on rabbits fed cholesterol diet. *J Korean Soc Food Sci Nutr* 27:1189–1196, 1998.

27. JH Kim, JD Ryu, HG Lee, JH Park. The effect of kimchi on production of free radicals and antioxidative enzyme activities in the brain of SAM. *J Korean Soc Food Sci Nutr* 31:117–123, 2002.

28. MJ Kwon, YO Song, YS Song. Effects of kimchi on tissue and fecal lipid composition and apoprotein and thyroxine levels in rats. *Korean Soc Food Sci Nutr* 26:507–513, 1997.

29. B Ray. *Fundamental Food Microbiology.* 2nd ed. Boca Raton: CRC Press, 2001, pp. 215–217.

30. MJ Kwon, JH Chun, YS Song, YO Song. Daily kimchi consumption and its hypolipidemic effect in middle-aged men. *J Korean Soc Food Sci Nutr* 28:1144–1150, 1999.

31. SH Ryu, YS Jeon, MJ Kwon, JW Moon, YS Lee, GS Moon. Effect of kimchi extracts to reactive oxygen species in skin cell cytotoxicity. *J Korean Soc Food Sci Nutr* 26:814–821, 1997.

32. BM Ryu. Effect of Kimchi on Inhibition of Skin Aging of Hairless Mouse. PhD dissertation, Pusan National University, Busan, 2000.

33. KY Park, SH Rhee. Nutritional Evaluation and Anticancer Effect of Kimchi. Proceedings of 8th Asian Congress of Nutrition, Seoul, 1999, pp 149–152.

34. KY Park. Antimutagenic and Anticancer Functions of Kimchi. Proceedings of IUFoST'96 Regional Symposium on Non-nutritive Health Factors for Future Foods. Seoul, 1996, pp 139-166.

35. KY Park, KA Baek, SH Rhee, HS Cheigh. Antimutagenic effect of kimchi, *Foods Biotech* 4:141–145, 1995.

36. SY Hwang, YM Hur, YH Choi, SH Rhee, KY Park, WH Lee. Inhibitory effect of kimchi extracts on mutagenesis of aflatoxin B$_1$. *Environ Mut Carcino* 17:133–137, 1997.

37. JC Ryu, KY Park. Anticlastogenic effect of *baechu*(Chinese cabbage) kimchi and *buchu*(leek) kimchi in mitomycin C-induced micronucleus formations by supravital staining of mouse peripheral reticulocytes. *Environ Mut Carcino* 17:133–137, 1997.

38. CA Raznikoff, JS Bertram, DW Brankow, CL Heidelberger. Quantitative and qualititative studies of chemical transformation of cloned C3H mouse embryo cells sensitive to postconfluence inhibition of cell division. *Cancer Res* 33:3239–3429, 1973.

39. MW Choi, KH Kim, SH Kim, KY Park. Inhibitory effects of kimchi extracts on carcinogen-induced cytotoxicity and transformation in C3H10T1/2 cells. *J Food Sci Nutr* 2:241–245, 1997.

40. YM Hur, SH Kim, KY Park. Inhibitory effects of kimchi extracts on the growth and DNA synthesis of human cancer cells. *J Food Sci Nutr* 4:107–112, 1999.

41. YM Hur, SH Kim, JW Choi, KY Park. Inhibition of tumor formation and changes in hepatic enzyme activities by kimchi extracts in sarcoma-180 cell transplanted mice. *J Food Sci Nutr* 5: 48–53, 2000.

42. KY Park, SH Kim, MJ Suh. Comutagenicity of high-salted kimchi in the Salmonella assay system. *J Colle Home Econ Pusan Nat'l Univ* 16:45–50, 1990.

43. JO Ha, KY Park. Comparison of auto-oxidation rate and comutagenic effect of different kinds of salt. *J Korean Assoc Cancer Prev* 4:44–51, 1999.

44. KO Jung. Studies on Enhancing Cancer Chemopreventive (Anticancer) Effects of Kimchi and Safety of Salts and Fermented Anchovy. PhD dissertation, Pusan National University, Busan, 2000.

45. BA Friend, KM Shahani. Antitumor properties of Lactobacilli and dairy products fermented by Lactobacilli. *J Food Prot* 47:717–723, 1984.

46. TJ Son, SH Kim, KY Park. Antimutagenic activities of lactic acid bacteria isolated from kimchi. *J Korean Assoc Cancer Prev* 3:65–74, 1998.

47. KY Park, SH Kim, TJ Son. Antimutagenic activities of cell wall and cytosol fractions of lactic acid bacteria isolated from kimchi. *J Food Sci Nutr* 3:329–333, 1998.

48. TJ Son. Antimutagenic Activities of Lactic Acid Bacteria Isolated from Kimchi. MS thesis, Pusan National University, Busan, 1992.

49. HY Kim, HS Bae, YJ Baek. *In vivo* antitumor effects of lactic acid bacteria on sarcoma-180 and mouse Lewis lung carcinoma. *J Korean Cancer Assoc* 23:188–196, 1991.

50. KS Shin, OW Chae, IC Park, SK Hong, TB Choe. Antitumor effects of mice fed with cell lysate of *Lactobacillus plantarum* isolated from kimchi. *Korean J Biotechnol Bioeng* 13:357–363, 1998.

51. OW Chae, KS Shin, HW Chung, TB Choe. Immunostimulation effects of mice fed with cell lysate of *Lactobacillus plantarum* isolated from kimchi. *Korean J Biotechnol Bioeng* 13:424–430, 1998.

52. IS Park. Function and Physiological Characteristics of Lactic Acid Bacteria Isolated from Kimchi. PhD dissertation, Chung-Ang University, Seoul, 1992.

53. IG Bogdanov, PG Dalev, AI Gurevich, MN Kolosov, VP Malekova, LA Plemyannikova, IB Sorokina. Antitumor glycopeptides from *Lactobacillus bulgaricus* cell wall. *FEBS Lett* 57:259–261, 1975.

54. KE Lee, UH Choi, GE Ji. Effect of kimchi intake on the composition of human large intestinal bacteria. *Korean J Food Sci Technol* 28:981–986, 1996.

55. S Salminen, MA Deighton, Y Benno, SL Gorbach. Lactic acid bacteria in health and disease. In: *Lactic Acid Bacteria*, S Salminen, A Wright, Eds. pp 211–253, Marcel Dekker Inc., New York, 1998.

56. AMA Kabir, Y Aiba, A Takagi, S Kamiya, T Miwa, Y Koga. Prevention of Helicobacter pylori infection by lactobacilli in a gnotobiotic murine model. *Gut* 41:49–55, 1997.

57. L Cellini, ED Campli, M Masulli, SD Bartolomeo, N Allocati. Inhibition of Helicobacter pylori by garlic extract (Allium sativum). *FEMS Immunol Med Microbiol* 13:273–277, 1996.

58. DY Graham, SY Anderson, T Lang. Garlic or jalapeno peppers for treatment of Helicobacter pylori injection. *Am J Gastroenterol* 94:1200–1202, 1999.

59. NL Jones, S Shabib, PM Sherman. Capsaicin as an inhibitor of the growth of the gastric pathogen *Helicobacter pylori*. *FEMS Microbiol Lett* 146:223–227, 1997.

60. KI Lee, SH Rhee, HS Han, KY Park. Kinds and characteristics of traditional special kimchi in Pusan and Kyungnam province. *J Korean Soc Food Nutr* 24:734–743, 1995.

61. JS Han, SH Rhee, KI Lee, KY Park. Standardization of traditional special kimchi in Kyungnam province. *J East Asian Soc Dietary Life* 5:27–32, 1995.

62. KI Lee, KO Jung, SH Rhee, MJ Suh, KY Park. A study on *buchu* (Leek, *Allium odorum*) kimchi–Changes in chemical, microbial and sensory properties, and antimutagenicity of *buchu* kimchi during fermentation. *J Food Sci Nutr* 1:23–29, 1996.

63. YDCD Zhong. *The Dictionary of Chinese Drugs*. Tokyo: Shanghai Science & Technological Publisher, pp 44, 1985.

64. JH Jang. Effect of kimchi as the healthy food. *Kimchi Sci Indust* 2:5–8, 1993.

65. E Block, S Naganathan, D Putman, S Zhao. *Allium* chemistry: HPLC analysis of thiosulfinates from onion, garlic, wild garlic(ramsoms), leek, scallion, shallot, elephant(great-headed) garlic, chive, and chinese chive. Uniquely high allyl to methyl ratios in some garlic samples. *J Agric Food Chem* 40:2418–2430, 1992.

66. E Block, D Putman, S Zhao. *Allium* chemistry: GC-MS analysis of thiosulfinates and related compounds from onion, leek, scallion, shallot, chive, and Chinese chive. *J Agric Food Chem* 40:2431–2438, 1992.

67. GR Fenwick, AB Hanley. The genus *Allium*. *CRC Crit Rev Food Sci Nutr* 22:199–271, 1985.

68. BS Reddy, CV Rao, A Rivenson, G Kelloff. Chemoprevention of colon carcinogenesis by organosulfur compounds. *Cancer Res* 53:3493–3498, 1993.

69. H Sumiyoshi, MJ Wargovich. Chemoprevention of 1,2-dimethylhydrazine-induced colon cancer in mice by naturally occurring organosulfur compounds. *Cancer Res* 50:5084–5087, 1990.

70. EM Schaffer, J Liu, JA Milner. Garlic powder and allyl sulfur compounds enhance the ability of dietary selenite to inhibit 7,12-dimethylbenz[a]anthracene-induced mammary DNA adducts. *Nutr Cancer* 27:162–168, 1997.

71. LW Wattenberg, VL Sparnins, G Barany. Inhibition of N-nitros-odiethylamine carcinogenesis in mice by naturally occurring organosulfur compounds and monoterpenes. *Cancer Res* 49:2689–2692, 1989.

72. MJ Wargovich, C Wood, VWS Eng, LC Stephens, K Gray. Chemoprevention of N-nitrosomethylbenzylamine-induced esophageal cancer in rats by the naturally occurring thioether, diallyl sulfide. *Cancer Res* 48:6872–6875, 1998.

73. MGL Hertog, CH Hollman, MB Katan. Content of potentially anticarcinogenic flavonoids of 28 vegetables and 9 fruits commonly consumed in the Netherlands. *J Agric Food Chem* 40:2379–2383, 1992.

74. A Bilyk, GM Sapers. Distribution of quercetin and kampferol in lettuce, kale, chive, garlic chive, leek, horse radish, red radish and red cabbage tissues. *J Agric Food Chem* 33:226–228, 1985.

75. AR Francis, TK Shetty, RK Bhattacharya. Modifying role of dietary factors on the mutagenicity of aflatoxin B_1: *in vitro* effect of plant flavonoids. *Mutat Res* 222:393–401, 1989.

76. AK Verma, JA Johnson, MN Gould, MA Tanner. Inhibition of 7,12-dimethylbenz[a]anthracene- and N-nitrosomethylurea-induced rat mammary cancer by dietary flavonol quercetin. *Cancer Res* 48:5754–5758, 1988.

77. W Haenszel, FB Locke, M Segi: A case-control study of large bowel cancer in Japan. *J Natl Cancer Inst* 64:17–22, 1980.

78. W Haenszel, JW Berg, M Segi, M Kurihara, F Locke. Large bowel cancer Hawaiian Japanese. *J Natl Cancer Inst* 51:1765–1779, 1973.

79. AJ Tuyns, R Kaaks, M Haelterman. Colorectal cancer and the consumption of foods: a case-control study in Belgium. *Nutr Cancer* 11:189–204, 1988.

80. KO Jung, KI Lee, MJ Suh, KY Park. Antimutagenic and anti-cancer effect of *buchu* kimchi. *J Food Sci Nutr* 4:33–37, 1999.

81. KO Jung, KY Park. The inhibitory effect of leek(*buchu*) kimchi extracts on MCA-induced cytotoxicity and transformation in C3H10T1/2 cells. *J Food Sci Nutr* 4:255–259, 1999.

82. H Kameoka, A Miyake. The constituents of the steam volatile oil from *Allium tuberosum* Rotter. *Nippon Nigeikagaku Kaishi* 48:385–387, 1974.

83. IA Mackenzie, DA Ferns. The composition volatiles from different parts of *Allium tuberosum* plants. *Phytochemistry* 16:763–766, 1977.

84. SB Roh. Studies on the Constituents of Leave of *Allium tuberosum* Rotter. MS thesis, Pusan National University, Busan, 1985.

85. KO Jung, KY Park, LB Bullerman. Anticancer effects of leek kimchi on human cancer cells. *Nutraceu Food* 7: 250–254, 2002.

86. YT Kim. Standardization of Preparation of Cancer Preventive Mustard Leaf Kimchi and its Anticancer Effects. PhD dissertation, Pusan National University, Busan, Korea, 2003.

87. YS Cho, SK Park, SS Chun, JR Park. Analysis of isothiocyanates in Dolsan leaf mustard(*Brassica juncea*). *Korean J Diet Cult* 8:147–151, 1993.

88. JO Kim, MN Kim, KY Park, SH Moon, YL Ha, SH Rhee. Antimutagenic effects of 4-decanol identified from mustard leaf. *J Korean Agric Chem Soc* 36:424–427, 1993.

89. YS Cho, BS Ha, SK Park, SS Chun. Contents of carotenoids and chlorophylls in Dolsan leaf mustard (*Brassica juncea*). *Korean J Diet Cult* 8:153–157, 1993.

90. HS Cheigh, ES Song, YS Jeon. Changes of chemical and antioxidative characteristics of chlorophylls in the model system of mustard leaf kimchi during fermentation. *J Korean Soc Food Sci Nutr* 28:520–525, 1999.

91. ES Song, YS Jeon, HS Cheigh. Changes in chlorophylls and carotenoids of mustard leaf kimchi during fermentation and their oxidative activities on the lipid oxidation. *J Korean Soc Food Sci Nutr* 29:563–568, 1997.

92. JH Hwang, YO Song, HS Cheigh. Fermentation characteristics and antioxidative effect of red mustard leaf kimchi. *J Korean Soc Food Sci Nutr* 30:199–203, 2001.

93. HY Kim. Antioxidative Effects of Mustard Leaves *in vitro* and *in vivo*. PhD dissertation, Pusan National University, Busan, 2002.

94. YS Jo, JR Park, SK Park. Effect of mustard leaf(*Brassica juncea*) on cholesterol metabolism in rats. *Korean J Nutr* 26:13–20, 1993.

95. KM Hwang. Studies on Standardization of Preparation of Fermentation Method and Cancer Prevention Effect of *kaktugi*. MS thesis, Pusan National University, Busan, 1999.

96. YA Kim. Cancer Chemopreventive Effects of *kaktugi* and Conditions of Distribution. MS thesis, Pusan National University, Busan, 2001.

14

Antioxidative Function of Seeds and Nuts and Their Traditional Oils in the Orient

YASUKO FUKUDA

Department of Nutrition and Food Science
Nagoya Women's University, Nagoya, Japan

MAYUMI NAGASHIMA

Department of Culture and Domestic Science,
Nagoya Keizai University Junior College,
Aichi, Japan

INTRODUCTION

Types of Seeds and Nuts Utilized in the Orient

In the Orient many kinds of seeds and nuts have been gathered or cultivated since prehistoric times. Seeds and nuts in the Orient can be categorized into two types based on their ingredients. One type are the oil seeds and nuts, which have a high content of fat and protein, for example, the sweet almond nut (Prunus *amygdalus* BATSCH), Japanese torreya nut (*Torreya nucifera* SIEB et Zucc.), walnut (*Juglans regia* L. and *Juglans subcordiformis* DODE), hazelnut (*Corylus avellana* L.), pine nut (*Pinus koraiensis* SIEB.et ZUCC), peanut (*Arachis hypogaea* L.), coconut (*Cocos nucifera* L.), hemp seed (*Cannabis sativa* L.), sunflower seed (*Helianthus annuns* L.), sesame seed (*Sesamun indicum* L.), rapeseed (*Brassica napus* L.), perilla seed (*Perilla frutescens* BRITON var. japonica HARA.), pumpkin seed (*Cucubita moschate* DUCH), poppy seed (*Papaver somniferum* L.), etc. The other type includes seeds and nuts with a high content of carbohydrates, for example, the ginkgo nut (*Ginkgo biloba* L.), chestnut (*Castanea mollisima* BLUME, *Castanea crenata* SIEB. et ZUCC), Japanese horse chestnut (*Aesculus turbinate* BLUME), lotus seed (*Nelumbo nucifera* GAERTN), water chestnut (*Trapanutans* L. var. *bispinosa* MAKINO), pistachio nut (*Pistacia vera* L.), etc.

The former oil-rich seeds and nuts are used as edible oils or snacks, while the latter carbohydrate-rich seeds and nuts are generally used as a glacé or cakelike rice cake.

Traditional Oils in the Orient and Their Uses

Oil seeds and nuts are valuable as sources of protein and calories. Most of the oils from oil seeds and nuts consist of 60 to 80% unsaturated fatty acids and 20 to 40% saturated fatty acids, and are categorized as liquid oils, except for palm oil, with a high level of saturated and monounsaturated fatty acid. These oils have been used for cooking, frying, lamp, body massage, medicinal purposes, and lubrication. Particularly, in East Asia, the traditional edible oils have mainly been produced from roasted seeds or nuts, for example, groundnut, rapeseed, sesame, perilla, and torreya, since ancient times.

HISTORICAL OVERVIEW

Sesame Seed

Origin of the Sesame Seed in Food Culture

Sesame (*Sesamum indicum* L) seed is an important crop from ancient times, as a popular food and also a representative

TABLE 14.1 Types of Sesame Utilizations in the World

Area	Middle and Near East North Africa Australia	East Asia	North America
Seed color	White (dehulled)	White and black (white, gold and black)	White (dehulled)
Roasting	Slight roast	Well roast	Unroast
Type of use	Paste	Whole, grind, paste	Whole
Oil	Salad (unroasted seed)	Roasted seed oil Salad(unroasted seed)	Salad (unroasted seed)

health food in the Orient. As most of wild *Sesamum* species were found in Africa, the origin of sesame was supported to be in the Savanna regions near the Benne River in North Africa by Nakao et al.[1] based on their field work of cultivation in the Savanna and by Kobayashi et al.[2] based on their genetic research. Sesame spread to Egypt, the Middle East, the Near East, and then to Europe by the way of Silk Road to China, Korea, and Japan, then continued to Egypt and over the ocean to India and Australia. Especially in the regions of North Africa, the Middle, and Near East, China, Korea, and Japan, sesame has long been cultivated and recorded as a highly nutritive, healthy food and the symbol of immortality in early Hindu legend.[3] In Japan, judging from archaeological surveys, it has been thought to have been introduced from Korea during the *Jomon* Period (B.C. 1200).

By a review of the literatures in the world, and the field survey of food intake in Central Asia and Outer Mongolia,[4] we found that sesame is very popular seed all over the world and eaten daily, as a food material or edible oil, particularly in the Orient, as summarized in Table 14.1.

Middle and Near East

The characteristic usage of sesame in the Middle and Near Eastern countries is as a paste (named thahina, tahin, or tehineh) made by milling slightly roasted or raw dehulled white seed. The paste is consumed as a bread condiment, sweetmeat, candy ingredient (halva, helva, or halaweh-sweetened tehineh), many kinds of sauces, garnish on baked goods, creams, and snack foods.

Far East (East Asia)

Compared with the other regions, in East Asia, various sesame products have been developed. Sesame seed is often used after roasting and grinding as a seasoning or dressing. Either white, brown, gold, or black seed is used. Especially in China and Korea, black seed has been well known as an important nutritive and medicinal material for maintaining health and antiaging action from ancient times,[5] for example, roasted and ground black sesame has been eaten as a typical medicinal rice porridge. In Japan, ground sesame has been consumed widely as a kind of salad dressing, named "Aemono" and other daily applications.[6] On the other hand, sesame oil, mainly from roasted sesame seed in Japan, Korea, and China, has been used for cooking, frying, and seasoning.

Other Regions

In India, sesame oil is mainly used for the medicinal purposes, namely, Ayureveda therapy, which dates back to prehistoric times. Medicinal sesame oils have been prepared as follows: unroasted sesame oil added to several dried and half-dried medicinal herbs, which is then heated and then the residues of the herbs are removed. The fat soluble medicinal components in several herbs were dissolved in the sesame oil. Many kinds of medicinal sesame oils were used as body massage oils for many therapies or preventive purposes.[7] Other uses are for confectionaries, for example, at festivals. In the U.S. dehulled white sesame seeds are widely used as the topping of hamburger buns.

Rapeseed

As the rapeseed (*Brassica napus* L) has a cytotoxic goitrogenic compound, named goitorin formed from glucosinolate, the seed has not been eaten as food material, but the oil has been widely used in Japan from the Edo era as medicinal or machinery oil, lamp, and edible frying oil after roasting the seed. Especially dark brown rapeseed oil, called "Akamizu" has been used for flavor and color to prepare fried tofu, traditional, and important health foods. Rapeseed oil has a high level of erucic acid (C22: 1,*cis*-13-docosenoic acid), which causes heart disease. But, nowadays erucic acid-free canola oil has been developed and cultivated in Canada. The canola oil is used widely all over the world as dressing, cooking, and frying oil.

Perilla Seed

Perilla (*Perilla frutescens* BRITON var. japonica HARA.) seed is 45% fat, 30% carbohydrate, 20% protein. The leaf or seed has been consumed in Korea and in Japan, since the *Jomon* Period, around 5000 B.C. as a food material. Its oil has been used for cooking, lamps, paint, coating of paper, and umbrella made with paper, because it is a drying oil with a high content of linolenic acid. The origins were thought to be Asia, but this is still undetermined. In Japan, "E," "Aburae," or "Zyunen" are popular names for perilla seed. The seed has been eaten as a dressing or scattering after roasting and grinding similar to roasted sesame seed. Although the demand for perilla seed has gradually decreased, compared with sesame seed or rapeseed, recently, perilla oil has been attracting much attention for its large content of *n*–3 polyunsaturated fatty acid, α-linolenic acid (approximately 60 to 75% in oil). The *n*–3 series fatty acids were well known to have an inhibitory effect on blood platelet aggregation, anticancer action, etc. Also perilla oil prevents lipid peroxidation *in vivo*[8] and has a more potent serum cholesterol-lowering ability than safflower oil.

Other Seeds and Nuts

Peanut (*Arachis hypogaea. L.*) or ground nut originated in South America. It has been used as an important food material and oil. The composition of the peanut is 50% oil, which consists of 50 to 80% oleic acid and 10 to 25% linoleic acid,

20% protein, and 20% carbohydrates. Traditionally, roasted peanut oil (groundnut oil) is used mainly in India or China. Pratt and Miller[9] reported the flavonoid antioxidant in Spanish peanuts.

Almond nut (*Prunus communis* Fritsch.) and walnut (*Juglans regia* L.) originated in West Asia, but now almond nut is extensively cultivated in California. Both of them are rich in oil (over 50%) and protein. They are used as typical confectionary materials and oils.

Torreyanut (*Torreya nucifera* Sieb. et Zucc.) has been cultivated in the region of Japan, South Korea and South-East China from old times. The nut with high content of protein and fat has also been eaten after roasting. Its oil served as typical frying oil with desirable flavor and high quality in Japan.

Camellia (*Camellia japonica* L.) seed has 40% oil and the oil has a high level of oleic acid, but it is used mainly as a cosmetic oil because of high level of saponin.

Sasanqua tea (*Camellia sasanqua* Thunb.) seed and tea (*Camellia theifera* Dyer.) seed have been used from ancient times in China as medicinal and food materials. They contain 30 to 40% oil and the composition of their fatty acid is oleic acid rich (80%)

CHEMICAL AND PHYSIOLOGICAL PROPERTIES OF SESAME/SESAME LIGNAN

The characteristic properties of sesame as a food material are summarized in Table 14.2. From a nutritional viewpoint, the major components of the seed are fat (51%), protein (20%), carbohydrate (18%), fiber (3%), ash (5%, high level of Ca, Fe, and Se), and vitamins (B_1, 0.95mg; B_2, 0.25mg; E, 25mg/100g seed). However, the marked properties of sesame are dependent upon the physiologically active sesame lignans (1% in both free and bound types), the sesame flavor, pleasant taste, and high resistance to oxidation. The chemical structures of sesame lignans and their physiological functions are summarized in Figure 14.1 and Table 14.3. Sesamin and sesamolin

TABLE 14.2 Food Functions of Sesame Seed

Function	Function of sesame (% fresh weight)
The First (Nutritional)	Nutrients (Fat 51%, Carbohydrate 18%, Protein 20%, Ash 5%(high level of Fe, Ca and Se)) Vitamins (B_1 0.95mg, B_2 0.25mg, E 25mg)Dietary Fiber 3%, Lignans 1%
The Second (Sensory)	Sesame flavor (pyrazines, thiols), brown color, sweat taste and cracking texture developed after roasting
The Third (Physiological)	Lignans: suppressive activity to lipid oxidation *in vivo* (sesamin, sesamolin) enhancement of vitamin E activity (sesamin) suppressive LDL level in serum (sesamin) Lignanphenols:antioxidant, radical scavenger(sesaminol, pinoresinol, lariciresinol, matairesinol, hydroxy matairesinol) suppressive LDL oxidation (sesaminol)

have been known as the major sesame lignans in *Sesamun indicumn* L.[10]; their concentrations are 0.3 to 0.5% and 0.2 to 0.3%, respectively, in the seeds. The four kinds of lignan phenols (sesamolinol, sesaminol, pinoresinol, and P1) with antioxidative activity have been isolated from defatted flour by treatment with β-glucosidase and identified by Fukuda et al.,[11,12] Osawa et al.,[13] and Nagata et al.[14] On the other hand, the chemical structures of sesaminol and pinoresinol glycosides in defatted sesame flour were elucidated by Katsuzaki et al.[15] and their physiological activities were studied by Kang et al.[16] The amounts of larisiresinol, hydroxymatairesinol, and its isomer were identified by Nagashima et al.,[17] and were higher in the water soluble fraction of black seeds than that of white seeds.

Kamal-Eldin and Yousif[18] identified a new lignan in *Sesamum alatus*. Though sesame is one of the superior foodstuffs, it has not been used as the major food material of recipes. Because the flavor or taste is very favorable but rather than oily texture in paste as a peanut butter. The physiological functional components preventing many diseases or aging

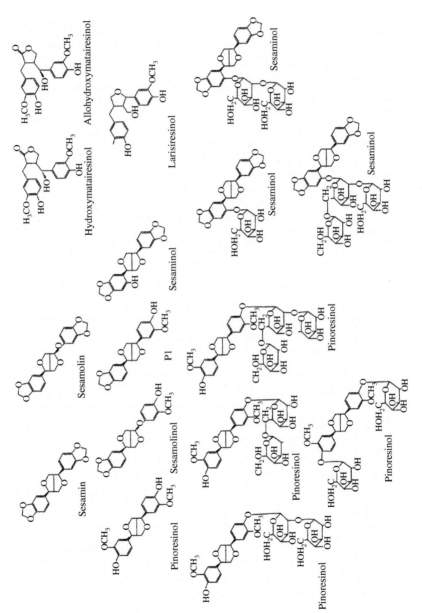

Figure 14.1 Chemical structures of sesame lignans.

TABLE 14.3 Functions of Sesame Lignans

	Contents (%)	Functions	Literatures
Sesamin	0.3–0.5	Heat stable	Fukuda (1986)
		Enhancement of V.E. activity	Yamashita (1995, 1995,2000)
		Enhancing liver activity	Sugano (1990)
		Suppressive activity to lipid oxidation	Akimoto (1995), Nakabayashi (1995)
		Suppressive δ-5'desaturase	Umeda-Sawano (1995)
		Decrease the plasma PGE2	Hirose (1991, 1992)
Sesamolin	0.2–0.3	Decomposition(to sesamol)	Fukuda (1986c)
		Intermolecular transformation (to sesaminol)	Fukuda (1986b)
		Suppression of lipid proxidation	Kang (1998)
Sesangolin	1.5	*S. angolense*	Jone (1962)
2-Episesalatin	0.6–0.8	*S. alatum*	Kamel-Eldin (1992)
		Lignan phenols (glycosides)	
Sesamolinol	Trace (?)	Antioxidant and radical scavenger	Osawa (1985)
Sesaminol -(G),-(2G),-(3G)	0.1	Antioxidant and radical scavenger	Fukuda (1985) Katuzaki (1993) Kang (1999)
Pinoresinol -(G),-(2G),-(3G)	Trace (?)	Antioxidant and radical scavenger	Fukuda (1985) (Katuzaki 1992)
Piperitol	Trace (?)	Antioxidant and radical scavenger	Fukuda (1985)
Lariciresinol	Trace	Antioxidant and radical scavenger	Nagashima (2001)
Hydroxymatairesinol	Trace	Antioxidant and radical scavenger	Nagashima (2001)
allohydroxymatairesinol	Trace	Antioxidant and radical scavenger	Nagashima (2001)

Note: G:Glucose

have been investigated and elucidated. Many studies on the physiological functions of sesame lignans have been carried out and published since the identification of lignan-phenols in sesame seed. Research has revealed that sesamin enhances vitamin E activity against lipid peroxidation in rat,[19–21] has hypocholesterolemic activity, suppresses activity of chemically induced cancer, enhances effect on various liver activities in rats,[22–27] suppresses activity of Δ5desaturase in rat[28] and sesaminol or sesamolin has an inhibitory effect on the oxidation of LDL.[29] Recently Ide et al.[30] reported that sesamin decreases fatty acid synthesis in rat liver accompanying the down-regulation of sterol regulatory element binding protein-1.

The black sesame seed, for instance used as Chinese medicine, has traditionally been accepted to be a more healthy food than white sesame seed in Japan, China, and Korea, but there was no scientific evidence for this assumption. Fukuda et al.[31] reported that the antioxidative activity determined by the thiocyanate method revealed no differences in the 80% ethanol extracts between crushed black sesame seeds and white sesame seeds. However, the water extract of black sesame seed coat showed stronger activity than that of the white sesame seed coat. The water-soluble fraction of black sesame seed coat was black in color and the black pigment was suggested to be anthocyanin. The true black compound or black pigment, however, has not yet been identified. Nagashima et al.[17] reported that recently, four lignan-phenols were isolated from the water extract of black sesame seed coat, namely, pinoresinol, larisiresinol, hydroxymatairesinol and its epimer. They were purified by column chromatography using Amberlite XAD-7, SiO$_2$ and Sephadex LH-20 and by preparative HPLC (Figure 14.2). On the basis of spectroscopic evidence, these lignans belong to bisepoxy-lignan (sesamin type), monoepoxy-lignan (dihydrosesamin type), and hydroxyl-dibenzylbutyrolactone type (Figure 14.1). The larisiresinol, hydroxymatairesinol, and its epimer are already known lignans, but were isolated from sesame seed for the first time. In the past, hydroxymatairesinol and its epimer were isolated from wood, *Picea excelsa,*[32] *Tsuga heterophylla,*[33] and so forth. It was investigated that hydroxymatairesinol has antitumor

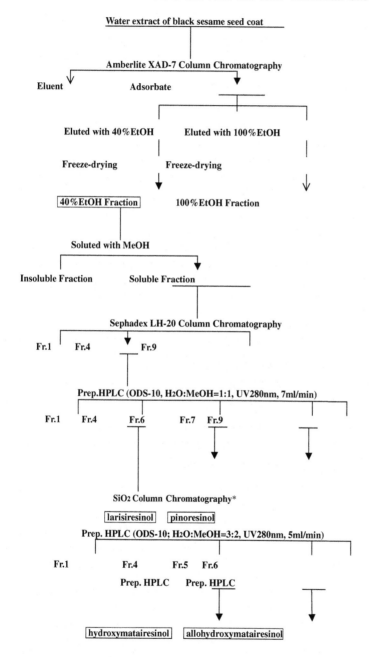

Figure 14.2 The scheme of isolation of antioxidative lignans from water extract of black sesame seed coat.

activity in rat dimethylbenzanthracene (DMBA)-induced
mammary tumor model and antioxidant capacity *in vitro*.[34]
The antioxidative activity of these lignans determined by the
thiocyanate method with 2, 2'-azobis (2-amidinopropane)-
dihydrochloride (AAPH), used as an oxidation accelerator,
were weaker than that of butylated hydroxytoluene (BHT).
On the 1, 1'-diphenyl-2-picrylhydrazyl (DPPH) radical scavenging activity, hydroxymatairesinol was as effective as α-
tocopherol, and allohydroxymatairesinol showed stronger
activity than α-tocopherol (Figure 14.3). Trolox showed a 20%
superoxide radical scavenging activity, obtained by ESR, and
larisiresinol showed a similar activity. The radical scavenging
activity of hydroxymatairesinol was 70%, and was stronger
than that of Trolox (Figure 14.4). The black sesame seed coat
as well as the white sesame seed coat contained these four
lignans. The content of them in black seed was higher than
that in white seed.

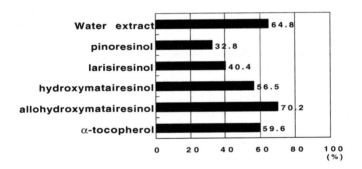

Figure 14.3 DPPH radical scavenging activities by colorimetric
method of water extract of black sesame seed coat, pinoresinol, larisiresinol, hydroxymatairesinol, allohydroxymatairesinol and antioxidant (α-tocophenol). Each result was the average of three assays.
Scavenging activity 100-[(O.D. at 517nm after 30 min of sample)/
(O.D. at 517 nm after 30 min of blanc) × 100] DPPH radical scavenging
activity of α-tocopherol was about 60%. Activities of pinoresinol and
larisiresinol were weaker than that of α-tocopherol. Hydroxymatairesinol showed activity like that of α-tocopherol. Allohydroxymatairesinol showed stronger activity than that of α-tocopherol.

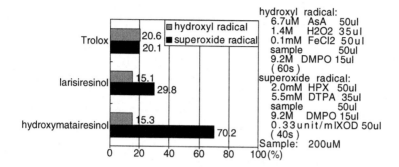

Figure 14.4   Radical scavenging activities by ESR of larisiresinol, hydroxymatairesinol and Trolox used as water soluble tocopherol (antioxidant). Radical scavenging activity 100-(spin adduct of sample/spin adduct of control)100 (%). Hydroxy radical scavenging activity of Trolox was 20%. Activities of larisiresinol and hydroxymatairesinol were weaker than that of Trolox. Superoxide radical scavenging activity of Trolox was 20% and larisiresinol showed activity like that of Trolox. Activity of hydroxymatairesinol was 70%, showed stronger than that of Trolox.

FUNCTIONAL PROPERTIES (AFTER AND DURING ROASTING)

As these Asian regions are in the monsoon zone, many kinds of cereals, oil seeds, and nuts have been grown or cultivated. Most of them are eaten after roasting because during roasting, some reactions among food components or decompositions occur as in the Maillard reaction, characteristically favorable flavor, color, taste, and also high resistance to oxidation may be developed. After roasting and grinding, they have been used as various food materials, for example sauces, dressing, pastes, and sesame tofu in Japan. The traditional edible oils in East Asia have mainly been produced from roasted seeds and nuts, for example, groundnut, rapeseed, sesame, perilla, and Japanese torreya nut.

Properties of Roasted Sesame Seed

Takeda et al.[35,36] investigated the effect of roasting and grinding conditions on the food quality of sesame seeds and elucidated that the most desirable color tone, taste, and flavor were obtained at 200°C for 5 to 10 min roasting in a dry oven. With lower roasting temperatures, free sugars (planteose and sucrose were obtained as major sugars) and free amino acids were relatively retained in the seeds, however, with higher roasting temperatures and longer roasting time, the content of free sugars and free amino acids decreased. These results suggest that the Maillard reaction between reducing sugars and amino acids or caramelization of the sugars occurred inside the skin such as in the residual endosperm tissue or in cells of the cotyledon. As shown in Figure 14.5, it was observed that main free sugars remarkably decreased at 200°C over a 10-min roasting. There was a high correlation ($r = 0.98$) between the total amount of free sugars and the L (lightness) value for surface of samples measured with a color meter.

Concerned with the flavor components formed during roasting, Namiki [37] reviewed a total of 141 flavor compounds

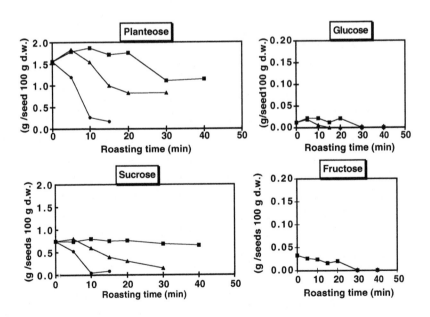

Figure 14.5 Changes in the contents of some saccharides of sesama seeds during roasting; ⊠ 170°C, ▲ 200°C, ● 230°C.

identified by many researchers,[38–45] but most of them had already been identified as components of roasted nut flavor and specifically the key compounds of sesame flavor were not elucidated. Recently, Takei[46] reviewed these investigations by GC-MS,[47–49] 101 flavor compounds were newly identified; the total flavor compounds from roasted seed and oil reached 242. However, little information about key components of sesame flavor and the mechanism of flavor formation was obtained. It is thought that most of these flavor compounds were developed by thermochemical reaction of sugars and amino compounds in sesame seeds by roasting above 180°C.

Comparative studies on the effects of roasting temperatures on antioxidative and radical scavenging activities between roasted sesame seed and almond with equal levels of fat and protein have been carried out.[50] The sesame samples were roasted at 170, 185, and 200°C for 15min and sliced almond at 155, 170, 185°C for 15min. Then, the samples were ground and extracted with *n*-hexane, ethyl acetate-methanol (7:3 v/v, E-M) and methanol (M). Each concentrated fraction of E-M and M tinged with brown, except for oily hexane fraction, was determined both of radical scavenging and antioxidative activities. The browning levels (O.D. at 420 nm) in both sesame and almond fractions were increased with increasing roasting temperature. The highest level of DPPH radical scavenging effects (Figure 14.6) and the strongest inhibition of 4-hydroxynonenal (HNE) by ELISA method[51] formation from linoleic acid (in Figure 14.7) was obtained at the highest roasting temperatures in the E-M fractions rather than in M fractions from both sesame and almond, especially sesame. These results indicated that the browning E-M soluble components (relatively polarity components) formed during roasting seed and nuts have radical scavenging and antioxidative effects, and also in these oil seed and nuts, some important reactions that generate antioxidative functions may take place at relatively high roasting temperatures. These reactions are very important not only in the formation of antioxidative and radical scavenging function, but also, flavors and color.

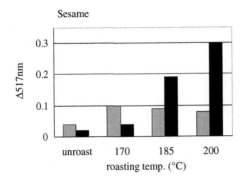

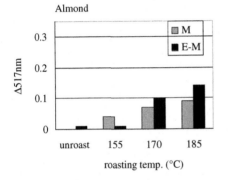

Figure 14.6 DPPH radicals scavenging effect of ethyl acetate-methanol (E-M) and methanol (M) extracts from sesame seeds and roasted sliced almonds. Reaction medium0.1M acetate buffer pH 5.5 + 0.1% sample/EtOH 2ml .5mM DPPH/EtOH 1ml, the value of D517nm=(the value in initiation of the reaction)-(the value after 30 mins). Each value is mean of duplicates.

Antioxidative Stability of Roasted Seed and Nut Oil

The commonly used salad oils from soybean, rapeseed, sunflower, safflower, and cottonseed are produced from unroasted seeds. The processes include steaming, pressing, or extraction with a solvent, degumming, alkali treatment, decolorization, deodorization, and winterizing (Figure 14.8). On the other hand, the traditional refining processes of the oils from roasted seeds and nuts, are steaming, pressing, and decanting

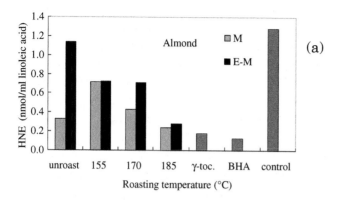

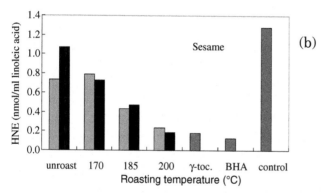

Figure 14.7 Inhibitory effect of toxic HNE on (M) and (E-M) extracts from roasted sesame and almond with ELISA. HNE: 4-hydroxy nonenal, control: linoleic acid. Each value is the mean of duplicates.

of scums. Therefore, many functional components, for example desirable flavor, brown color, and fat soluble antioxidative substances, are retained in the oils in comparison with the salad oils. These characteristic properties depend upon fat soluble compounds generated from the roasting process of seeds or nuts. The processes correspond to the preparation of virgin olive oil. Especially in East Asia, roasted sesame seed, rapeseed, perilla seed, torreya oils, and ground nut oil have been widely used from ancient times. The oxidative stabilities of these roasted oil, seeds and nuts were shown in Table 14.4.[50]

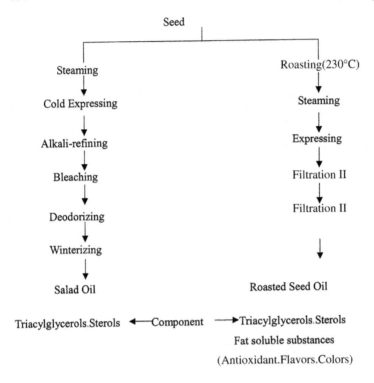

Figure 14.8   Two refining processes of oil in Orient.

The order of oxidative stability was sesame > rapeseed > groundnut > torreya > perilla. These oils were more stable than unroasted nut oils with little correlation to content of tocopherols. As for roasted sesame oil, Yen[53] reported that the activity depends primarily on temperature and superior anti oxidative activity of sesame oil roasted at 200°C rather than at 180°C, which was shown to measure volatiles from oil exposed to fluorescent light. Yoshida et al.[54] reported that the browning level of oil from roasted seed by microwave oven increased in higher temperature and longer time but oxidative stability decreased at higher temperature and longer time. Koizumi et al.[55] and Fukuda et al.[56] investigated the effects of roasting conditions on oxidative stability from a study on 15 kinds of roasted sesame oil obtained from various

TABLE 14.4   Oxidative Stability of Nuts/Seeds-Oils
by Weighting Method

	I.P. (days) [a]	Total toc. (mg/100g oil)[b]
Unroasted seed oil		
Salad[c]	14.07	67.96
Almondnut	15.93	46.79
Hazelnut	11.11	42.03
Walnut	11.11	54.21
Roasted seed oil		
Perilla seed	5.00	57.89
Rapeseed	90.00	88.70
Sesame seed	90.00<	52.70
Groundnut	30.00	30.67
Torreyanut	12.22	32.46

[a] I.P. = Induction period (the days of 5% weight gain) under accelerated oxidation at 60°C
[b] Total toc. (α,β,δ,σ-tocopherol)
[c] Salad: soybeen/rapeseed 7:3

roasting temperatures and durations, and elucidated that the browning level (absorbance at 420 nm) was highly correlated to I.P. (the number of days to reach 5% weight gain of oil under accelerated oxidation at 60°C) as shown at $r = 0.84$ in Figure 14.9. Also, the activity of oil roasted at 200°C for 5 min was superior to that roasted at 180°C for 30 min. The antioxidants of roasted sesame oil have been elucidated by Fukuda et al.[56] Two kinds of antioxidants, tocopherol (γ-tocopherol 0.05% in oil) and sesamol (0.01% in oil) generated from sesamolin (one of the sesame lignans) during roasting, contribute to oxidative stability, but the content of sesamol formed by roasting conditions was not enough to explain the superior antioxidative activity. Therefore the presence of some unknown compounds or synergistic action of some browning compounds (0.01% in oil) during roasting of seed is suggested. Recently, Kumazawa et al.[58] identified small amounts of sesaminol, in roasted sesame oil, which was generated with high roasting temperature. Abon-Gharbia et al.[59] reported that sesame oil from seeds with coat were more stable than those

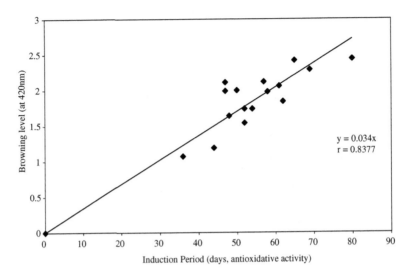

Figure 14.9 Correlation of induction period and browning level.

extracted from dehulled seeds. However, the reason has not yet been elucidated.

Antioxidative Activities of Roasted Rapeseed Oil — Akamizu

Roasted rapeseed oil, called Akamizu in Japanese, has been used for flavor and color to prepare fried tofu, as well as other traditional and important health foods in Japan since 1700, Edo era. It has also been used to lubricate primitive machinery. Hino et al.[61,] and Namiki et al.[62] studied the effect of roasting of rapeseed on flavor, color, and antioxidative properties. It was shown that roasting at 140°C for 10 to 30 min was most favorable for developing a good flavor with minimal bitterness, but in the antioxidative activity determined by weighing method (in Table 14.5), development of the activity was observed even in oils roasted at 140 or 150°C for 30 min, but it increased markedly at above 180°C and gave a very stable oil by roasting at 200°C even for as short as 10 min. As is well known, tocopherol homologs play an important

TABLE 14.5   Effect of Roasting Condition on Antioxidative Activity of Sesame Seed and Rapeseed Oil (I.P., days)

Roasting(/min)	None	150/30	170/10	180/10	190/10	200/10
Sesame seed			28	38	62	70
Rapeseed	3	15		42		70

Note: I.P.: induction period
I.P.: days of 5% weight gain of oil under accelerated oxidation at 60

antioxidative role in conventional vegetable oils, but no significant changes in their contents were observed in rapeseed oil subjected to roasting. These results suggest that the strong antioxidative activity of roasted oils was due to the formation of some products during roasting, which acted as antioxidants or strong synergists with tocopherols and other components similar to roasted sesame oil.

Both roasted sesame and rapeseed oils with high browning level possessed extremely longer I.P., 90 days or beyond 90 days under the accelerated oxidative condition at 60°C than a roasted groundnut oil (I.P. 30 days), in comparison to unroasted safflower salad oil (I.P. 10 days) in Table 14.4. On the other hand, in thermally oxidative condition at 180°C for 6 h, roasted sesame oil was more resistant to oxidation than roasted rapeseed oil. To elucidate oxidative stability of roasted seed oil, the separation of roasted seed oils into fats and fat soluble fractions was carried out with hot methanol and then, the methanol fraction was frozen at −20°C for 20 h and decanted to remove the fat (triacylglycerols). Five percent of each oil's concentrated hot methanol soluble fraction (dark brown) was added to high linoleic safflower oil and heated for 6 h at 180°C. The deterioration of each oil as examined by Anisizine value[63] was 95.8, 31.7, 45.5, and 43.1 for safflower, sesame, rapeseed, and groundnut, respectively. The hot methanol extract from sesame oil had a much higher resistance to thermal oxidation than those of the other two roasted oils. This reason behind this resistance to oxidation in sesame oil may be dependent upon an antioxidant, sesamol generated from sesamolin during frying conditions[64,65] as shown in Figure 14.10.

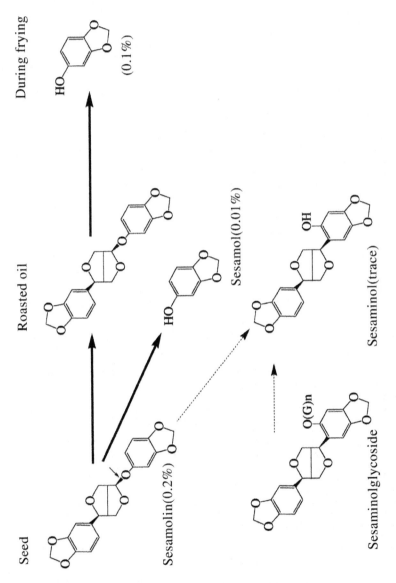

Figure 14.10 Changes of lignans in roasted sesame oil during roasting and frying.

TABLE 14.6   Antioxidative Factors in Roasted Seed Oil

Oil	Antioxidative Activity[a]	Antioxidative Factors
Salada (safflower)	+	Tocopherols
Roasted (rapeseed, groundnut)	+++	Tocopherols Maillard-type components
Roasted sesame	++++	Tocopherols Sesamol Trace of sesaminol Maillard-type components

[a] +; I.P.10, +++; 30I.P.90, ++++;I.P. 90
I.P.: the day of 5% weight gain of oil under accelerated oxidation at 60

The antioxidative factors of roasted or oils are summarized in Table 14.6. During roasting seed nut, the Maillard reaction occurred and fat soluble melanoidins might be exracted together with oil. Hayase et al.[66,67] reported antioxidative and radical scavenging function of high molecular melanoidins from the Maillard reaction, but, the chemical components formed from the Maillard reaction have not yet been elucidated.

The experience by technicians and experts of Tempura restaurant in frying foods has revealed that the roasted seed oils have excellent oxidative stability and can be used for frying many times. Especially in the case of roasted sesame oil, it has been said that tuberculosis is rare among the workers, who were engaged in frying many food materials almost all day long.[68]

The traditional oils with extract and purified processes from roasted seeds have many useful components, flavor, color, taste, antioxidative, and radical scavenging, but a trace of nonuseful components, too. Recently, Namiki et al[69] reported that sesame lignans, antioxidative factors, and also characteristic flavor components could be extracted by supercritical CO_2 procedures from sesame seed or oil. The supercritical CO_2 extraction may be a very available technique in purification of high quality roasted seed's oils.

REFERENCES

1. S Nakao. *Cultivated Crops and Origin of Farm*. Tokyo: Iwanami Shoten, 1966, pp 102.

2. T Kobayashi. Type Diversity and Evolution in Sesamum. Proc. 4th SABRAO Cong. 1, 1982, pp 361–366.

3. T Kobayashi. *The Route of Sesame*. Tokyo Iwanami Shoten, 1986.

4. Y Fukuda. Studies on utilization of sesame by survey. *Vesta* 22:60–65,1995.

5. Rizitin. Honsokohmoku. Chaina: Zinmineiseisyuppansya, 1596.

6. T Takeda, Y Fukuda. Cultural Studies on the usage of sesame as a dietary material in the world (Part 1) A characteristic method of the sesame seed for cooking in each area. *Chori agaku* 29:281–291, 1996.

7. D Bhagwan. *Therapy in Ayurveda*. New Delhi: Concept Publishing Company, 1992.

8. M Okuno, K Kajiwara, S Imai, T Kobayashi, N Honma, T Maki, K Suruga, T Goda, S Takase, Y Muto, H Moriwaki. Perilla oil prevents the excessive growth of visceral adipose tissue in rats by down-regulating adipocyte differentiation. *J. Nutr.* 127:1752–1757, 1997.

9. D.E.Pratt, E.E. Miller. A flavonoid antioxidant in Spanish peanuts. *J. Am. Oil Chem. Soc.* 61:1064–1067, 1984.

10. P Budowski. Recent research on sesamin, sesamolin and related compounds. *J. Am. Oil Chem. Soc.* 41:280–285, 1964.

11. Y Fukuda, T Osawa, M Namiki. Studies on antioxidative substances in sesame seed. *Agric. Biol. Chem.* 49:301–306, 1985.

12. Y Fukuda, M Nagata, T Osawa, M Namiki. Contribution of lignan analogues to antioxidative activity of refined unroasted sesame seed oil. *J. Am. Oil Chem. Soc.* 63:1027–1031, 1986.

13. T Osawa, M Nagata, M Namiki, Y Fukuda. Sesamolino1, a novel antioxidant isolated from sesame seeds. *Agric. Biol. Chem.* 49:3351–3352, 1985.

14. M Nagata, T Osawa, M Namiki, Y Fukuda, T Ozaki. Stereochemical structures of antioxidative biepoxylignans, sesaminol and its isomers, transformed from sesamolin. *Agric. Biol. Chem.* 51:1285–1289, 1987.

15. H Katsuzaki, S Kawakishi, T Osawa. Structure of novel antioxidative lignan glucosides isolated from sesame seed. *Phytochemistry* 35:773–776,1993.

16. MH Kang, M Naito, N Tsujihara, T Osawa. Sesamolin inhibits lipid peroxidation in rat liver and kidney. *J. Nutr.* 128:1018–1023, 1998.

17. M Nagashima, Y Fukuda. Antioxidative lignans from industrial wastewater in cleaning of black sesame seed. *Nippon Shokuhin Kagaku Kogaku Kaishi* 46:382–388, 1999.

18. A Kamal-Eldin, G Yousif. A furofuran lignan from *Sesamum alatus, Phytochemistry* 31:2911–2912, 1992.

19. K Yamashita, Y Nohara, K Katayama, M Namiki. Sesame seed-lignans and -tocopherol act synergistically to produce vitamin E activity in rats. *J. Nutr.* 122:2440–2446, 1992.

20. K Yamashita, Y Iizuka, T Imai, M Namiki. Sesame seed and its lignans produce marked enhancement of vitamin E activity in rats fed a low-tocopherol diet. *Lipids* 30:1019–1028, 1995.

21. K Yamashita, M Kagaya, N Higuti, Y Kiso. Sesamin and -tocopherol synergistically suppress lipidperoxide in rat fed a high docosahexaenoic acid diet. *Biofactors* 11:11–30, 2000.

22. S Ikeda, T Toyama, K Yamashita. Dietary sesame seed and its lignans inhibit 2,7,8-trimethy-2 (2-carboxyethyl)-6-hydroxychroman excretion into urine of rats fed[gamma] -tocopherol. *J. Nutr.* 132:961–966, 2002.

23. M Sugano, T Inoue, K Koda, K Yoshida, N Hirose, Y Shinmen, K Akimoto, T Amachi. Influence of sesame lignans on various lipid parameters in rats. *Agric. Biol. Chem.* 54:2669–2673, 1990.

24. N Hirose, T Inoue, K Nishihara, M Sugano, K Akimoto, S Shimizu, H Yamada. Inhibition of cholesterol absorption and synthesis in rats by sesamin. *J. Lipid Res.* 32:629–638, 1991.

25. N Hirose, F Doi, T Ueki, K Akazawa, K Chijiiwa, M Sugano, K Akimoto, S Shimizu, H Yamada. Suppressive effect of sesamin against 712-dimethylbenza-anthracene induced rat mammary carcinogenesis. *Anticancer Res.* 12:1259–1265, 1992.

26. K Akimoto, Y Kitagawa, T Akamatsu, N Hirose, M Sugano, S Shimizu, H Yamada. Protective effects of sesamin against liver damage caused by alcohol or carbon tetrachloride in rodents. *Ann. Nutr. Metab.* 37:218–224, 1993.

27. A Nakabayashi, Y Kitagawa, Y Suwa, K Akimoto, S Shimizu, N Hirose, M Sugano, H Yamada. alpha-Tocopherol enhances the hypocholesterolemic action of sesamin in rats. *J. Vit. Nutr. Res.* 65:162–168, 1995.

28. S Umeda-Sawada, N Takagashi, O Igarashi. Interaction of sesamin and eicosapentaenoic acid against.5 desaturation and n-6/n-3 ratio of essential fatty acids in rat. *Biosci. Biotech. Biochem.* 59:2268–2273, 1995.

29. MH Kang, Y Kawai, M Naito, T Osawa. Dietary defatted sesame flour decreases susceptibility to oxidative stress in hypercheresterolemic rabbits. *J. Nutr.* 129:1885–1890,1999.

30. T Ide, L Ashakumary, Y Takahashi, M Kushiro, N Fukuda, M Sugano. Sesamin, a sesame lignan, decreases fatty acid synthesis in rat liver accompanying the down-regulation of sterol regulatory element binding protein-1. *Biochim Biophys Acta*, 1534:1–13, 2001.

31. Y Fukuda, T Osawa, S Kawakishi, M Namiki. Antioxidative activities of fractions of components of black sesame seeds. *Nippon Shokuhin Kogyo Gakkaishi* 38:915–919,1991.

32. K Freudenberg and Knof, L., The lignans of spruce wood. *Chem. Ber.* 90:2857–2869, 1957.

33. F Kawamura, H Ohashi, S Kawai, F Teratani, the late Y Kai. Photodiscoloration of Western Hemlok (Tsuga heterophylla) Sapwood. *Mokuzai Gakkaishi* 42:301–307, 1996.

34. N M Saarinen, A Warri, S I Makela, C Eckerman, M Reunanen, M Ahotupa, S Salmi, A A Franke, L Kangas, and R Santti, Hydroxymatairesinol, a novel enterolactone precursor with antitumor properties from coniferous tree (Picea abies). *Nutr. Cancer* 36:207–216, 2000.

35. T Takeda,Y Fukuda. Cooking properties of sesame seeds (part 1) roasting conditions and component changes. *J. Home Ecom. Jpn.* 48:137–143, 1997.

36. T Takeda, H Aono, Y Fukuda, K Hata. A shimada.effect of roasting conditions on the food quality of sesame seeds. *J. Home Ecom. Jpn.* 51:1115–1125, 2000.

37. M Namiki. The chemistry and physiological functions of sesame. *Food Rev. Int.* 11:281–329, 1995.

38. T Yamanishi, S Tokuda, E Okada. *Nosankako Gijutsukenkyuushi.* 7:61, 1960.

39. S Kinoshita, T Yamanishi. Identification of basic aroma components of roasted sesame seeds. *Nippon Nogeikagaku Kaishi.* 47:737–739, 1973.

40. M Soliman, S Kinoshita, T Yamanisi. Aroma of roasted sesame seeds. *Agric Biol.Chem.* 39:973–977,1975.

41. Y Takei, A Naatani, A Kobayashi, T Yamanishi. Studies on the aroma of sesame oil: intermediate and high boiling compounds. *J. Agric. Chem. Soc.* 43:667–674, 1969.

42. Y Takei. Aroma components of roasted sesame seed and roasted huskless sesame seed. *J. Home Econ. Jpn.* 39:803–815, 1988.

43. Y Takei. Volatile components formed by roasting sesame seed fractions. *J.Home Econ. Jpn.* 40:23–34, 1989.

44. Y Takei, Y Fukuda. Effect of roasting temperature of sesame seed on quality of sesame oil. *Choriagaku* 24:10–15,1991.

45. S Nakamura, O Nishimura, H Masuda, S Mihara. Identification of volatile flavor components of the oil from roasted sesame seeds. *Agric. Biol.Chem.* 3:1891–1899, 1989.

46. Y Takei. Flavor of roasted sesame. In: Namiki, Ed. *Goma-Science and Function.* Tokyo: Maruzen planetto, 1998, pp 124–133.

47. P Schicherle, Studies on the Flavour of Roasted White Sesame Seeds, In Progress in Flavor Precursor, Schereier & Alured Publ. Corp., Carol Stroom IL pp 341-360, 1993.

48. P Schicherle. Odour-active compounds in moderately roasted sesame. *Food Chem.* 55:145–152,1996.

49. Y Nakada, M Shimoda, Y Osajima.Flavor descriptive terms for sesame seed oil and principal components analysis of seasory data.*Nippon Shokuhin Kagaku Kogaku Kaishi* 44:848–859, 1997.

50. Y Fukuda, S Nakata. Effects of roasting temperature in sliced almonds and sesame seeds on the antioxidative activities. *Nippon Shokuhin Kagaku Kogaku Kaishi* 46:786–791, 1999.

51. K Uchida, S Toyokuni, K Nishikawa, S Kawakishi, H Oda, H Hiai, ER Stadtman. Michael addition-type 4-hydroxy-2-nonenal adducts in modified low density lipoproteins: markers for atherosclerosis. *Biochemistry* 33:12487–12494, 1994.

52. Y Fukuda, N Kumazaki, Y Shibata. Oxidative stability of edible oils (seeds' and nuts' oils) by tg and sensory evaluation of nuts' oil. *J. of Nagoya Women's University (Home Economics. Natural Science)* 49:117–123, 2003.

53. G C Yen, S L Shyu. Oxidative stability of sesame oil prepared from sesame seed with different roasting temperature. *Food Chem.* 31:215–224,1989.

54. H Yoshida, J Shigezaki, S Takagi, G Kajimoto. Variations in the composition of various acyl lipids, tocopherols and lignans in sesame seed oils roasted in a microwave oven. *J. Sci. Food Agri.* 68:407–415, 1995.

55. Y Koizumi, Y Fukuda, M Namiki. Marked antioxidative activity of seed oils developed by roasting of oil sesame seeds, *Nippon Shokuhin Kagaku Kogaku Kaishi* 43:689–694,1996.

56. Y Fukuda, M Koizumi, R Ito, M Namiki. Synerygistic action of the antioxidative components in roasted sesame seed oil, *Nippon Shokuhin Kagaku Kogaku Kaishi* 43:1272–1277, 1996.

57. Y Fukuda, M Nagata, T Osawa, M Namiki. Chemical aspects of the antioxidative activity of roasted sesame oil, and the effect of using the oil for frying. *Agric. Biol. Chem.* 50:857–862, 1986.

58. S Kumazawa, M Koike, Y Usui, T Nakayama, Y Fukuda. Isolation of sesaminols as antioxidative components from roasted sesame seed oil, *J. O. S.* 52:303–307, 2003.

59. HA Abon-Gharbia, F Shahidi, AAY Shehata, MM Youssef. Effects of processing on oxidative stability of sesame oil extracted from intact and dehulled seeds. *J. Am. Oil Chem. Soc.* 76:203–221,1997.

60. T Hino, A Nishioka, S Yokozeki. Akamizu: The Traditional Oil Processing in Japan, GCIRC Congress Abstracts, 1991, C-26.

61. T Hino, Y Koizumi, G Yabuta, M Namiki. Antioxidant in Rapeseed Oil (Akamizu). Nippon Syokuhin Kagaku Kougakkai Taikai Abstract, Fukuoka, 1999.

62. M Namiki, T Hino, Y Fukuda, Y Koizumi, F Yanagida. AOCS-JOCS Joint Meeting Abstracts, Anaheim, 1993, p6.

63. JOCS (1996) Standard Method for the Analysis of Fats, Oils and Related Materials 2.5.3-1996.

64. Y Fukuda. Change of the amount of antioxidants in two types of sesame oil during deep-fat frying. *J. Home Econ. Jpn.* 38:793–798, 1987.

65. Y Fukuda, T Osawa, S Kawakishi, M Namiki. Oxidative stability of foods fried with sesame oil. *Nippon Shokuhin Kogyo Gakkaishi* 35:28–32,1988.

66. F Hayase, S Hirashima, G Okamoto. Scavenging of active oxygens by melanoidins. *Agric. Biol. Chem.* 53:3383–3385,1989.

67. F Hayase. Glycation and Maillard reaction. *J. Act. Oxyg. Free Rad.* 4:271–284,1993.

68. Y Fukuda. Studies on the kinds of frying oils in tempura restaurants with questionnaires surveys. *Annu. Rep. Stud. Food Life Cult. Jpn.*7:14–25, 1990.

69. M Namiki, Y Fukuda, Y Takei, K Namiki, and Y Koizumi. Changes in Functional Factors of Sesame Seed and Oil During Various Types of Processing. In: TC Lee and CT Ho. Eds. ACS Symposium Series 816; Bioactive Compounds in Foods, Effect of Processing and Strorage. Washington, D.C.: American Chemical Society, 2002, pp 85–104.

15

Antioxidants and Other Functional Extracts from Sugarcane

KENJI KOGE

Planning and Research Section, Chigasaki
Laboratory, Shin Mitsui Sugar Co., Ltd.,
Kanagawa-ken, Japan

MICHAEL SASKA

Audubon Sugar Institute, Louisiana State
University Agricultural Center,
Baton Rouge, LA, USA

CHUNG CHI CHOU

Principal Scientist/Engineer, Cti,
Huntington Station, NY, USA

FUNCTIONS OF PLANT EXTRACTS

Introduction

Many plant extracts have been found to have physiological functions and are used in functional and health foods. Green tea, ginkgo biloba, and grape seed extracts are examples of functional plant products. Green tea extract is reported to prevent cancer, lower blood pressure and cholesterol concentration, and exhibit antibacterial effects.[1-4] Its active compound is a polyphenol (catechin, etc.) and it is used as an ingredient in foods and animal feed. Ginkgo biloba extract improves angiopathy.[5] It is sold as a health food ingredient and is also approved as a medicine in Europe. Grape seed extract, whose active component is a pigment (proanthocyanin), is sold as a functional food ingredient and there are indications that it prevents arteriosclerosis and suppresses the development of gastric ulcer and colon carcinoma.[6] A purified extract of bilberry is rich in anthocyanins and was found effective in human subjects for reducing the clinical symptoms of lowered capillary resistance and increased retinal sensitivity. Extracts of strawberry and spinach were found to enhance the age-related functions of brain in rats, while blueberry extracts reduced the lung damage in rats subjected to pure oxygen. All of these extracts are also known to have antioxidant activities, and a relationship between antioxidant activities and other physiological functions has been noted by many researchers.[7-13]

Although antioxidant activities and related functions, such as anticancer effects and regulation of blood pressure, have been topics of conversation for about 10 years, the effect of improving the immune system was noted more recently. Pale colored vegetables and fruit extracts have been proven to exhibit these activities. The relationship between antioxidant activities and immune reactions has not been clarified.

Sugarcane

Sugarcane is a tropical grass belonging to the same spices as sorghum. The objective of sugarcane harvest is to produce sugarcane stalks with the highest possible sucrose content, ranging from 10 to 15% of the weight of stalks. Most sucrose is stored in the inner portion of the stalks while the majority of valuable sugarcane extracts, including antioxidants, concentrate in the outer component (rind fraction) of the stalks.

Both sucrose and sugarcane extracts are recovered in the form of cane juice via "milling" of sugarcane stalks. The cane juice is further processed to produce white/refined sugar, sugarcane extracts and other products.

Okinawa is a sugarcane cultivating area in Japan, famous for the longevity of its residents. The average life span according to the 1995 data of the Health and Welfare Ministry of Japan is about 85.1 years in women and 77.2 years in men. The elderly people of Okinawa are healthy and continue to work as long as they live. The elderly people eat kokutou, a noncentrifuged sugar, with green tea at teatime. It is a unique diet habit that other Japanese do not follow.

Sugarcane has been reported to contain various effective components.[14–17] The components of kokutou have antioxidant activity[18] and the ability to improve hyperlipemia. Octacosanol from cane wax can enhance physical endurance.[19] This substance is used in health foods. Blackstrap molasses, a by-product of processing of sugarcane, has long been claimed to have therapeutic values albeit with little or no verifiable evidence. It is available in the health food industry and is also reported to have a whitening effect on human skin[20] and is known to possess antimutagenicity. In Japan, it has been used in facial soaps since ancient times.

Recently, increasing concerns among consumers over the use of synthetic chemicals and medicines, such as food additives, antibiotics or hormones used in the domestic animal feed, led to studies of natural materials with physiological functions. The plant extracts mentioned above are some of the examples and sugarcane was thought to possess such effective components. Hence, study of sugarcane extracts was initiated.[21–24]

Physiological Functions of Sugarcane Extracts

Preparation

Four types of sugarcane extracts were produced. Extracts 1 and 4 were prepared from cane juice. Extract 1 was prepared using synthetic adsorbent chromatography whereby the adsorbed substances are concentrated in the extract. Extract

4 was obtained by chromatographic separation with ion exchange resin. Extract 3 was prepared by hot water extraction from bagasse, the fibrous residue of sugarcane. Extract 2 consisted of volatile substances that had been adsorbed on and stripped of a synthetic adsorbent resin. The number of the extract refers to the chronology of the discovery of its effect. First were found the deodorant effects of extracts 1 and 2.[21,22] Most of the physiological functions of extract 1, 3 and 4 were discovered in collaboration with Eisai Co., Ltd. [JAPAN], a producer of pharmaceuticals, food additives and animal feed materials.[21,22,25,26]

Phylactic Effects

Phylactic effects in this case refer to promotion of resistance to viral and bacterial infections. These effects can be exploited to reduce or in some cases eliminate use of antibiotics.

Ten mice (Slc:ICR, male, 5 weeks of age) were used in each experimental group.[21,22] A minimum lethal dose of a virus (*Pseudorabies* virus, originally a swine pathogenic virus) or a bacterium (pathogenic *Escherichia coli*, a strain derived from human) was inoculated subcutaneously into the mice. Each sugarcane extract was orally administered once a day for 3 days after the date of the viral challenge, and only once in the bacterial experiment (the day before the bacterial challenge). The dosage was 500 mg/kg per day. In the control groups, distilled water was administered instead of extracts. Survival rates were determined 7 days after the inoculation for the viral infection, and 4 days after inoculation for the bacterial infection. In both the viral and bacterial experimental groups, all mice in control groups died. In all groups that were administered the extract, at least 7 of 10 mice survived (Figure 15.1 A and B). These results indicate that the extracts have a marked phylactic effects because they did not substantively prevent pathogens from multiplying (*in vitro* test). At present, sugarcane extracts have been developed to be feed materials for chickens, swine, etc. to reduce or in some cases to eliminate use of antibiotics.

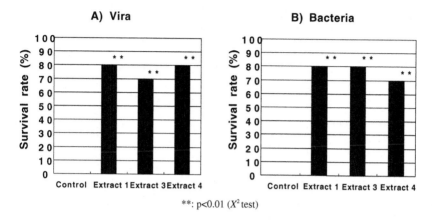

Figure 15.1   Resistance to viral and bacterial infections.

Vaccine Adjuvant Effect

Given at the same time as the vaccination, vaccine adjuvant stimulates the immune response and increases the effectiveness of the vaccine. Domestic animals, especially, are given many vaccines throughout their lives. To elevate the antibody titer level of all animals, vaccines are given many times, but repeated vaccinations stress the animals and adversely affect their growth. Thus, all over the world, the direction has changed toward reducing the number of vaccination by using adjuvants.

Ten mice (Slc:ICR, male, 5 weeks of age) were used in each group of experiment.[21] Each sugarcane extract was administered orally once a day for 6 days from the day of Pseudorabies virus vaccine inoculation. Extract dosage was 500 mg/kg per day. *Pseudorabies* virus challenge occurred 2 weeks after vaccination, and the survival rate was counted on the 7th day of the virus challenge.

All mice died in the group that were not vaccinated and did not receive any extract. Only 20% of the mice survived in the vaccinated group that did not receive the extract. However, the survival rate was 80% in all extract-administered groups (Table 15.1). These results show that the extracts enhanced the effect of the vaccine significantly.

TABLE 15.1   Vaccine Adjuvant Effect

	Dose of adjuvant (mg/kg)	Adjuvant administration route	Vaccination	Survival rate (%)	XX^2 test
Not vaccinated, no extract administered group	None	—	Saline 0.2ml i.m.		
Vaccinated, no extract administered group	None	Oral	Vaccine 0.2ml i.m.	20	
Extract 1	500	Oral	Vaccine 0.2ml i.m.	80	
Extract 3	500	Oral	Vaccine 0.2ml i.m.	80	
Extract 4	500	Oral	Vaccine 0.2ml i.m.	80	

Note: $p < .05$

Protective Effects on Liver Injuries

The number of liver disorders such as hepatitis, fatty liver, and cirrhosis has been increasing recently. Liver disorders are caused by various factors, including foods, alcohol, chemicals, pathogens, etc. The protective effects of sugarcane extract on liver injury models were estimated.[25,26] Five mice (Slc:ICR, male, 5 to 6 weeks of age) were used in each experimental group. Carbon tetrachloride (CCl_4), CCl_4 with phenobarbital (orally administrated 4 days before evocation), ANIT (alfa-naphtyl-isothiocyanate), and D-galactosamine (GalN) were used to induce liver injuries. All models are acute liver injury models, but the mechanisms by which liver injuries are induced differ. Extract 1 was administrated orally once a day for 5 consecutive days, and injury evocation was induced by giving administrations of chemicals on the final day of extract administration. Serum GOT (glutamic oxaloacetic transaminase) and GPT (glutamic pyruvic transaminase) activities (IU/l; JSCC method) were measured the day following induction of liver injury. When liver injury occurs, liver cells are damaged and release these enzymes into blood.

The negative control column shows values for animals not administered any extract and with no induced liver injury (Table 15.2). In the groups given chemical treatment without extract administration, both GOT and GPT activities were higher than in those that had previously been administered extract 1.

The same additional experiments were also conducted using extracts 3 and 4. The results showed the same activities as in extract 1.

Protective Effects on Involution of Lymphoid Organs Exposed to Cold Stress

Two groups of 10 mice each (Slc:ICR, male, 5 weeks of age) were exposed to cold stress in a low temperature room maintained at 5°C for 4, 7, 24 and 24 h on the first, second, third, and fourth days, respectively.[22] Extract 4 was orally administered at a dose of 500mg/kg/day once daily after each exposure. In the negative control group (no exposure to stress) and the positive control group (exposure to stress), distilled water instead of extract 4 was orally administered at a dose of 0.5 ml/mouse/day for 4 consecutive days. Increases in body weight and organ weights were individually measured 1 day after the administration of the last dose of extract 4. In the mice exposed to cold stress, increases in body weight were suppressed and spleen and thymus weights were decreased in the positive control group. However, the oral administration of extract 4 resulted in a body weight increase. The spleen and thymus weights of the extract 4-administered mice were also protected to the same degree as those of the negative control group. Extract 4 is thought to maintain normal immune function and regulation in the mice under cold stress.

Antioxidant Activity

There are many kinds of free radicals and active oxygen species in our bodies (Table 15.3). Some of them are derived from nitrous oxide (NO) that is released by leucocytes. They

TABLE 15.2 Effects of Sugarcane Extract 1 on Liver Injury Models

	Negative control	Chemical treatments			
		CCl4	CCl4 with phenobarbital	ANIT	GalN
GOT No extract	4.3 ± 4.1[b]	2,846 ± 802	4,460 ± 2,130	1,804 ± 616	5,061 ± 3,484
Extract 1	—	1,083 ± 477[b]	403 ± 219[b]	136 ± 117[b]	177 ± 50.7[a]
GPT No extract	18.4 ± 4.1[b]	4,177 ± 1312	9,255 ± 2,272	903 ± 372	7,193 ± 4,064
Extract 1	—	1,059 ± 679[b]	1,382 ± 1,278[b]	57.2 ± 55.7[b]	195 ± 93.3[a]

[a] $p < 0.05$
[b] $p < 0.01$ (t test)

TABLE 15.3   Radicals and Nonradicals

Radical	Nonradical
Oxygen-centered radical	$^1O_2$ Singlet oxygen
HO. Hydroxyl radical	H_2O_2 Hydrogen peroxide
HO_2. Hydroperoxyl radical	HOCl Hypochlorous acid
LO_2. Peroxyl radical	O_3 Ozone
	O_2-. Superoxide anion

Note: O_2- is one of the active oxygen. It is a radical, but its reactivity as a radical is not high and its function as an anion is more important. O_2- scavenging activity was measured using a method that measures SOD activity.

have the important function of attacking cancerous and virus-infected cells. However, they simultaneously damage cells of various organs and may cause many kinds of diseases and aging. At the same time, some enzymes such as superoxide dismutase (SOD) scavenge them[27] and protect the cells from the damage. If this balance is upset, diseases occur and aging progresses.

Currently, some plant extracts get attention because of their antioxidant activity and are used as dietary supplements, functional food, and medicines. Extracts from sugarcane were also evaluated for these activities.[23,28]

DPPH Radical Scavenging Activity

DPPH (1,1-diphenyl-2-picrylhydrazyl) radical is a stable free radical[29] that displays a maximum absorbance at 517 nm. As DPPH-H does not exhibit this maximum, the absorbance is lowered in the presence of DPPH-scavenging antioxidants.

The DPPH radical scavenging activity of sugarcane extracts 1, 3, and 4, catechin, apple extract, and cocoa powder was evaluated. Figure 15.2 shows the antioxidant concentration that can scavenge (reduce the concentration by) 50% of DPPH radical; lower values indicate higher antioxidant activity. Catechin reagent and apple extract, which are polyphenols and representative antioxidants, showed a high level of activity. Extract 1 especially showed a high level of activity.[28]

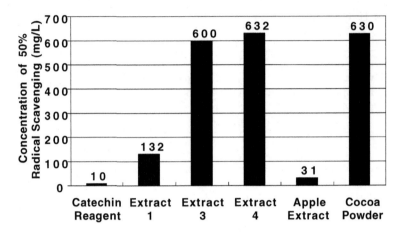

Figure 15.2   DPPH Radical-Scavenging Activity. Concentration of a sample required achieving a 50% reduction of DPPH radical activity.

Extracts 3 and 4 had the same level of activity as cocoa powder, which is known to contain an abundant amount of cacao polyphenols. These results indicate that sugarcane extracts had a relatively high DPPH scavenging activity.

Superoxide Anion Scavenging Activity

Superoxide anion is one of the active oxygen species, and the scavenging activity is measured by determination of superoxide dismutase (SOD) activity. Antioxidants are not enzymes, but some show the same activity as SOD. Table 15.4 shows the scavenging activity converted to enzymatic activity.[28]

TABLE 15.4   Scavenging Activity of the Superoxide Anion ($O_2\cdot$)

Catechin (Reagent)	130,000 U/g
Extract 1	49,000 U/g
Extract 3	6,700 U/g
Extract 4	30,000 U/g
Apple Extract	36,000 U/g
Cocoa Powder	11,000 U/g
Red Wine	270 U/g

Catechin showed the highest activity, and apple extract and cocoa powder activities were relatively high. Extracts 1 and 4 showed the same levels of activity as apple extract, which is sold as plant polyphenols. Extract 3 has a value of 6,700 U/g, which is not high although it does have scavenging activity.

The relationship between antioxidant activity and other physiological functions is not clear, and neither is the mechanism of such effects. It is known, though, that plant extracts having such activities usually have other physiological functions, so the attention to these activities is growing.

Oxygen Radical Absorbance Capacity (ORAC)

ORAC,[30–32] a quantitative method of measuring the antioxidant activity of plasma, foods, and natural extracts, among other has become a standard method, and ORAC values, in μmole TE (Trolox — a soluble analogue of Vitamin E — equivalents) per 100 g are available in the literature (Table 15.5) for a number of common fruits, vegetables and other antioxidant-rich food supplements. In addition, a more recent refinement has been the differentiation between "fast," "slow" and total or "whole" antioxidant capacity, referred to in the following, respectively, as "95% ORAC," "50% ORAC" and "whole ORAC," respectively.[31]

TABLE 15.5 Antioxidant Properties (ORAC Values in mmole TE/100 g) of Various High-Antioxidant Fruits and Vegetables (Weller, 1999)

Prunes	5,800
Raisins	2,800
Blueberries	2,400
Oranges	750
Red grapes	700
Kale	1,800
Spinach	1,300

Five common edible molasses products (Table 15.6) available in the American market were selected and characterized (Tables 15.6 and 15.7). Products A to D were sugarcane-based products, while E was probably a corn-based product with a minor amount of sugarcane liquor blended in.

Of the sugarcane products A to D, only B, based on its high color and sugar composition, corresponded to "blackstrap" molasses. The others were lower color products with higher levels of sugars and lower ash.

The antioxidant capacity of the five products (Table 15.8) correlated very well with their color (Figure 15.3) indicating that high antioxidant polyphenols formed a large part of the sugarcane colorants. With some variations, the "95%" and "50%" ORAC values were much lower than the "whole" ORAC. This suggests that a substantial part of the antioxidant

TABLE 15.6   Five Edible Molasses
Products Available in the American
Retail Market

Code	Product
A	Steen's Home Style Molasses
B	Wholesome Foods Organic Blackstrap
C	Mott's Grandma's Molasses
D	B&G Foods Brer Rabbit
E	Karo's Dark Corn with Refiners' Syrup

TABLE 15.7   Composition of the Five Edible
Molasses Products

Sample	RDS	Sucrose	Glucose	Fructose	Ash	Color
A	80	33	18	17	3.4	38,300
B	79	35	8	10	5.8	186,800
C	78	30	18	17	3.1	69,000
D	79	30	16	18	4.6	89,400
E	76	2	14	1	0.68	4,000

Note: RDS = refractometric dry solids, color in ICUMSA units, all others in g/100 g.

TABLE 15.8 Antioxidant Capacity of the Five
Commercial Edible Molasses Products

Sample	95% ORAC	50% ORAC	Whole ORAC
A	1,170	1,840	4,440
B	6,430	8,860	11,370
C	1,700	2,660	5,340
D	2,640	3,740	6,180
E	160	260	2,830

Note: ORAC in mmole TE/100 g dry solids.

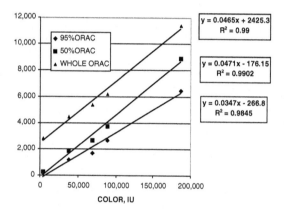

Figure 15.3 Antioxidant capacity of the five edible molasses products correlates well with their color.

capacity originated from components with very slow-acting functionality.

Blackstrap molasses is a final product of sugarcane processing that has been subjected to a number of unit operations, and a possibility exists that some of the antioxidant activity has been lost in the process. Samples of Louisiana sugarcane juice and syrups, i.e., sugarcane juice clarified with two different procedures and concentrated under vacuum were analyzed (Table 15.9). These products have only been subjected to juice extraction, vacuum concentration and, in

TABLE 15.9   Antioxidant Capacity of Louisiana Sugarcane Juice and Syrup

Sample	95% ORAC	50% ORAC	whole ORAC
Conc. cane raw juice	6,100	10,200	26,400
Cane syrup — hot liming	5,700	9,200	27,600
Cane syrup — soda ash	5,400	10,000	26,000

Note: ORAC units per 100 g dry solids.

the case of syrups, to a pH adjustment and settling, and are products with about 80% sucrose on dry solids and a color of about 15,000 ICUMSA units. The ORAC values found were substantially higher than those of the edible molasses and were identical for the concentrated juice and syrups, indicating that neither the lime nor soda ash clarification measurably reduced the antioxidant capacity.

As even prolonged heating of another sample of Louisiana syrup (Table 15.10) did not result in any reduction of its antioxidant capacity, the high antioxidant capacity of the syrup samples that does not conform to the pattern observed in Figure 15.3 for various edible molasses is yet unexplained. Geographical or varietal differences of sugarcane composition, ash components or process chemicals in the industrial process or other factors may be responsible.

Application of granulated activated carbon, bone char, ion exchange resins, crystallization and chromatographic method for separation of colorants, including polyphenols and flavonoids from sugarcane liquors, are well-established industrial processes. Therefore, some of these processes were

TABLE 15.10   Antioxidant Capacity (ORAC units per 100 g dry solids) of a Louisiana Sugarcane Syrup Before (F) and After (G) Heating for 5 Hours at 98°C in a Glass Container

Sample	95% ORAC	50% ORAC	whole ORAC
F	7,800	11,400	35,500
G	8,500	12,400	35,000

TABLE 15.11   Antioxidant Properties (ORAC units per 100 g dry solids) of a Louisiana Sugarcane Syrup and Two Extracts Prepared from the Syrup

Sample	95% ORAC	50% ORAC	whole ORAC
Sugarcane syrup	4,140	6,724	48,930
Concentrate 1	35,220	45,520	56,870
Concentrate 2	826,000	1,021,000	1,232,000

explored to concentrate the antioxidant-rich compounds contained in the sugarcane juices. An example of such an application is in Table 15.11, where the antioxidant capacity is given of one syrup and two kinds of extracts or concentrates. While the concentrate 1 exhibits only a minor improvement over the source syrup, the concentrate 2 is a very antioxidant-rich product. The very high proportion of the "fast" antioxidant capacity is remarkable, and augurs well for its therapeutic potential. Concentrate 1 and 2 are sugarcane extracts produced by different separation processes.

The whole ORAC capacity of the concentrate 2 is comparable to such well-known antioxidants as caffeic and gallic acids, and exceeds that of many existing commercial antioxidant supplements,[33] and, by a factor of 100 or more of such health food favorites (Table 5) as prunes. While its physiological functions still need to be established, it is believed that this natural extract could be produced, as a new natural or even organic product from sugarcane, at a sufficiently low cost and high volume to aid significantly the antioxidant intake of the population. A 250 mg capsule of this product would satisfy the daily recommended intake of 3,000 ORAC units[33] considered as minimum to sufficiently increase the serum antioxidant levels.

Other Functions

Deodorizing Effect

Extracts 1, 2, and 3 have a deodorizing effect.[21] Figure 15.4 shows some of the effects of extract 2. A home steam humidifier was filled with a 0.02% solution of extract 2 in tap water.

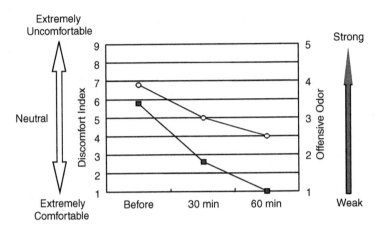

Figure 15.4   Deodorant effect of sugarcane extract 2.

Five people checked the intensity of the offensive odors and discomfort index at the starting point, and after 30 and 60 min of humidifier operation. Figure 15.4 shows both the offensive odor and discomfort index strength were decreased remarkably. If extract 2 has a strong specific smell ("a masking effect"), the offensive odor should decrease and the discomfort index should increase with the passage of time as the concentration of extract 2 in the air increase. Extract 2 is a mixture of volatile component of sugarcane and is useful as a deodorant for room air, clothes, furniture, fabrics, livestock barns, etc., in addition to its application in the food processing industry. Extracts 1 and 3 are useful as deodorizers of food products, such as fish and meat.

Taste and Texture Improvement

Both extracts 1 and 2 have taste improvement effects.[21,34] Figure 15.5 shows that 10-ppm final concentration of extract 1 added to liquid yogurt improved factors in the index. Organoleptic quality, such as off-taste, aftertaste, and stickiness in particular were improved. Considering its concentration was extremely low, the strength of this effect is conspicuous.

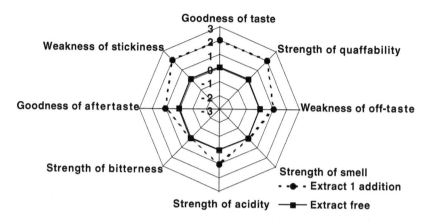

Figure 15.5 Influences of extract 1 on taste of liquid yogurt.

DISCUSSION

Sugarcane extracts have various functions. These functions are very useful and make these extracts effective as ingredients in functional foods, health foods, and functional animal feed. The National Institute of Animal Health [JAPAN] has also investigated immunological effects of extract 4 in chickens. Growth promotion effect in commercially bred chickens (Dekalb) and an immunopotentiation effect and an anti-coccidial infection effect in inbred laboratory chickens were studied.[24,35] These effects of extract 4 are also expected to be an animal feed material.

It is surprising and interesting that sugarcane components have various beneficial physiological functions. Furthermore, they are safe natural products. Sugarcane is mass-cultivated in large areas of the world for sugar production, so that raw materials from sugarcane for extracts are readily available for industrial exploitation and relatively inexpensive as compared with other extraction substrates.

ACKNOWLEDGMENTS

The authors gratefully acknowledge the considerable cooperation, interest and assistance of Ms. Yukie Nagai, Mr. Tadashi

Ebashi, Mr. Hiroshi Iwabe (Shin Mitsui Sugar Co., Ltd.), Dr. Seiichi Araki, Mr. Mamoru Suzuki (Eisai Co., Ltd) and Dr. Kameo Shimura and Dr. Yoshikazu Hirota (National Institute of Animal Health, Japan). Assistance with the ORAC analysis of Dr. Rama Ratham of the Genox Corporation is gratefully acknowledged.

REFERENCES

1. K Sugiyama, P He, S Wada, S Saeki. Liver injury-preventive effect of green tea theanine. *Hissuuaminosan Kenkyu* (in Japanese) 155:67–73, 1999.

2. K Sugiyama, P He, S Wada, S Saeki. Tea and other beverages suppress D-galactosamine-induced liver injury in rats. *J. Nutr.* 129:1361–1367, 1999.

3. K Nakagawa, S Okuda, T Miyazawa. Dose-dependent incorporation of tea catechins, (-)-epigallocatechin-3-gallate and (-)-epigallocatechin, into human plasma. *Biosci. Biotech. Biochem.* 61:1981–1985, 1997.

4. S Wada, P He, N Watanabe, K Sakata, K Sugiyama. Suppression of D-galactosamione-induced rat liver injury by glycosidic flavonoids-rich fraction from green tea. *Biosci. Biotech. Biochem.* 63:570–572, 1999.

5. R Takagaki. Ityoubaekisu no Tokusei to Riyou. *New Food Industry* (in Japanese) 40:1–7, 1998.

6. M Arii. The attractive powers of grape seed polyphenol (proanthocyanidine) on human life. *Syokuhin to Kaihatu* (in Japanese) 35:11–14, 2001.

7. K Kosuna. Tennenbutsuyurai Kousankabussitu no Kaihatsu. *New Food Industry* (in Japanese) 39:17–23, 1997.

8. T Washino. Kousankazai no Seizaika to sono Riyou. *Gekkan Food Chemical* (in Japanese) 8:33–41, 1996.

9. T Tsuda, F Horio, J Kitoh, T Osawa. Protective effects of dietary cyanidin 3-0--D-glucoside on liver ishemia-reperfusion injury in rats. *Arch. Biochem. Biophys.* 368:361–366, 1999.

10. I Suda, S Furuta, Y Nishiba, O Yamakawa, K Matsugano, K Sugita. Reduction of liver injury induced by carbon tetrachloride in rats administered purple-colored sweet potato juice. *Nippon Shokuhin Kagaku Kogaku Kaishi* (in Japanese) 44:315–318, 1997.

11. K Hoffmann-Bohm, H Lotter, O Seligmann, H Wagner. Antihepatotoxic C-glycosylflavones from the leaves of Allophyllus edulis var. edulis and gracilis. *Planta Med.* 58:544–548, 1992.

12. I Suda, O Yamakawa K Matsugano, K Sugita, Y Takaguma, K Irisa, F Tokumaru. Changes of -GTP, GOT, GPT levels in hepatic function-weakling subjects by ingestion of high anthocyanin sweet-potato juice. *Nippon Shokuhin Kagaku Kogaku Kaishi* (in Japanese) 45:611–617, 1998.

13. K Weller. Can foods forestall aging? *Agri. Res.* February 1999.

14. Y Kimura, H Ohminami, H Arichi, H Okuda, S Arichi, T Hayashi. Effects of color componds in raw brown sugar on carbohydrate and lipid metabolism of rats. *Yakugaku-Zasshi* (in Japanese) 102:666–669, 1982.

15. Y Matsuura, Y Kimura, H Okuda. Effect of aromatic glucosides isolated from brown sugar on intestinal absorption of glucose. *J. Med. Pharmaceut. Soc. WAKAN-YAKU* 7:168–172, 1990.

16. H Okuda. Tennenbutsu, toku ni mijikana Syokuhin ni hukumareru Kinouseibussitu ni tuite. *New Food Industry* (in Japanese) 33:77–83, 1991.

17. P Smith, NH Paton. Sugarcane flavonoids. *Sugar Technology Reviews* 12:117–142, 1985.

18. Y Nakasone, K Takara, K Wada, J Tanaka, S Yogi, N Nakatani. Antioxidative compounds isolated from Kokuto, noncentrifuged cane sugar. *Biosci. Biotech. Biochem.* 60:1714–1716, 1996.

19. N Koga. Development and application of new natural functional waxes. *Fragrance Journal* (in Japanese) 8:13–21, 1990.

20. F Yamashita, T Suzuki, I Kesyou. Active components in cane molasses having effect on human skin. *Proc. Res. Soc. Japan Sugar Refineries' Technol.* (in Japanese) 41:43–48, 1993.

21. Y Nagai, T Mizutani, H Iwabe, S Araki, M Suzuki. Physiological Functions of Sugarcane Extracts. Proceedings of the 60th annual meeting of Sugar Industry Technologists, Inc., Taipei, Taiwan, May 2001.

22. K Koge, Y Nagai, M Suzuki, S Araki. Physiological effects of sugarcane extracts. *Up to Date Food Processing.* (in Japanese) 35:39–41, 2000.

23. M Saska, CC Chou. Antioxidant Properties of Sugarcane Extracts, Proc. 1st Biannual World Conference on Recent Developments in Sugar Technologies, Delray Beach, FL, May 16–17, 2002.

24. K Koge, Y Nagai, T Ebashi, H Iwabe, M El-Abasy, M Motobu, K Shimura, Y Hirota. Physiological Functions of Sugarcane Extracts. I. Growth Promotion, Immunopotentiation and Anticoccidial Infection Effects in Chickens. Proc. 61st annual meeting of Sugar Industry Technologists, Inc., Delray Beach, FL, May, 12–15, 2002.

25. K Koge, Y Nagai, M Suzuki, S Araki. Satoukibityuusyutubutu no Kousankakouka to Kansyougai Yokuseikouka. *Gekkan Food Chemical.* (in Japanese) 9:85–88, 2000.

26. K Koge, Y Nagai, M Suzuki, S Araki. Inhibitory effects of sugar cane extracts on liver injuries in mice. *Nippon Shokuhin Kagaku Kaishi.* (in Japanese) 38:231–237, 2001.

27. JM McCord, I Fridovich. A mechanism for the production of ethylene from methanol. *J. Biol. Chem.* 244:6049–6055, 1969.

28. K Koge, Shinsyokuhinsozai: Satoukibityuusyutubutu no Tokusei to Kousankanou. *Gekkan Food Chemical.* (in Japanese) 5:62–65, 2000.

29. T Yamaguchi, H Takamura, T Matoba, and J Terao. HPLC method for evaluation of the free radical-scavenging activity of foods by using 1,1-diphenyl-2-picrylhydrazyl. *Biosci. Biotechnol. Biochem.* 62:1201–1208, 1998.

30. GH Cao, M Alession, RG Cutler. Oxygen-radical absorbance capacity assay for antioxidants, *Free Radical Biology and Medicine*, Vol. 14, 303–311, 1993.

31. G Cao, CP Verdon, AHB Wu, H Wang, RL Prior. Automated assay of oxygen radical absorbance capacity with the COBAS FARA II. *Clin. Chem.*, 41/12, 1738–1744, 1995.

32. Oxygen Radical Absorption Capacity Assay for Measuring Antioxidant Activity, GENOX Corporation, Baltimore, MD, October 2001.

33. RL Prior, G Cao. Variability in dietary antioxidant-related natural product supplements: The need for methods standardization. *J. Am. Nutraceut. Assoc.* Vol. 2, No. 2, 46–56, 1999.

34. Y Nagai, K Koge. [Satoukibityuusyutubutu] no Fuumikaizenkouka. *Gekkan Food Chemical.* (in Japanese) 7:84–88, 2000.

35. M El-Abasy, M Motobu, K Shimura, K-J Na, C-B Kang, K Koge, T Onodera, Y Hirota. Immunostimulating and growth-promoting effects of sugarcane extract (SCE) in chickens. *J. Vet. Med. Sci.* 64:1061–1063, 2002.

16

Functional Foods from Garlic and Onion

TOYOHIKO ARIGA and TAIICHIRO SEKI
Laboratory of Nutrition and Physiology,
Department of Biological and Agricultural
Chemistry, Nihon University College of
Bioresource Sciences,
Fujisawa, Japan

The National Cancer Institute (NCI) initiated the "Designer Food Program" about a decade ago, setting garlic on the top of vegetable-pyramid representing potency in cancer prevention. Many investigators have recognized the effectiveness of garlic and related Allium plants against cancer. Again, it is a common sense among the platelet researchers that blood to be tested should be obtained from the donor who has not taken Chinese dishes involving garlic for at least a week. It is believed that the suppression of platelet aggregation is the most prominent effect of garlic intake, since the effect is measurable by an aggregometer with ease.

Historically, *Allium* plants would undoubtedly be indispensable to protect foods, raw and cooked meats, or boiled beans, from bacterial spoilage. Although, the preservative function of garlic or onion was replaced by refrigeration several decades ago, more of their novel food functions have been revealed scientifically during the past quarter century. The most valuable finding about garlic and onion is that both quantity and quality of their active components change considerably upon processing. Some products contain sulfides, and thus effective for preventing thrombosis, but others have no sulfide, and thus are ineffective. Therefore, utilization of garlic and onion for functional foods is highly promising, if taking *Allium* characters into account.

HISTORY AND BOTANICAL CHARACTERS OF GARLIC AND ONION

Garlic (*Allium sativum* L.) is one of the oldest plants used as a medicinal plant, spice, and food as well as an antidemoniac charm plant. Although the place where garlic grew initially is not known, its cultivation is said to date back about 4,000 years ago. Generally accepted history is that garlic originated from Central Asia, then spread either to the west, the Tigris-Euphrates area, and Egypt, or to the east, China and then to Korea and later to Japan.[1] Modern phytochemical analysis and gene technologies performed on garlic by several researchers also support its Asian origin. According to Tsuneyoshi et al.,[2] and Maas et al.,[3] *Allium longicuspis*, an old garlic species having fertility (flowering and seed forming abilities), could be found only in the west area of the Tien Shan Mountains of China. Therefore, such a fertile clone of garlic plant would have been transported to the east and west, wherein the seed-forming character would be atrophied in a longtime culture history utilizing garlic cloves as seeds (vegetative propagation) in place of true seeds (genetic propagation, Tsuneyoshi et al.[2] classified A. *sativum* into five groups by their restriction fragment length polymorphism (RFLP):

Asian type, Russian type, Yugoslavian type, and European I and II types.

Onion (*Allium cepa* L.) has been cultivated in the Middle and Far East for at least 5,000 years, and prized as a foodstuff by the Asian people in these areas. Onion was widely used for cooking in Sumeria, 4,000 years ago. In Ancient Egypt, onion was one of the staple vegetables of the laborers who worked at building the Great Pyramid at Ghiza (3200 to 2800 B.C.). At present, onions are a staple food, and people will prize them more in the future as the desire for good health increases. This may be assumed true due to the fact that the world annual consumption of onions has increased from 27 million tons to 42 million tons between 1990 and 2000.[4]

On the botanical side, onions have many varieties; *Allium cepa* L. var. cepa (common onion), var. aggregatum (potato onion), var. ascalonicum (shallot), var. proliferum (tree or Egyptian onion), and var. viviparum.[5]

PRODUCTION OF GARLIC AND ONION IN ASIAN NATIONS

Production of Garlic in Asia

Global production of garlic in 2002 was 12,234,220 tons, with Asia accounting for 90%; other contributors were in Americas (5.9%), Africa (2.7%), and Europe (1.4%). In Asia, China contributed 8,694,040 tons (83% of global production), predominating India (4.8%), the Republic of Korea (3.9%), Thailand (1.2%), and Turkey (1.1%). Present global production of garlic (in 2002) shows about a 3.5-fold increase from that in 1962 (the world population doubled during this period), and a 70% increase between 1992 and 2002 (the population increased only 17% in this period). The large production was realized by the increases of both harvesting area (1,135,143 ha) and yield (10.8 tons/ha): those were a 40% and 17% increase in the past decade, respectively. These increases are naturally due to Asian production, especially Chinese production, which expanded the area from 345,718 ha to 630,273 ha (1.82 times) within only 10 years.[4] Under the enormous current production in China, there are nations that began to decrease their own production; in Korea, Pakistan, Armenia, and other nations with small production.

The annual production of garlic in Japan has surprisingly decreased in 2000 (18,228 tons) from that of 1990 (35,381 tons). Therefore, imports from China were required to cover the consumption of garlic of more than 35,000 tons in that year in Japan.[6]

Production of Onion in Asia

The world production of onion in 2001 was about 49.4 million tons, and 63% of it (31.3 million tons) was produced in Asia.[4] These amounts are 4.3 times higher than those of garlic; i.e., twice in harvested area, and twice in yield, as compared to garlic. The onion production represented a 60% increase during the last decade, and personal consumption is estimated to have increased about 40% in these periods. China produced around half of the total amount of onions in Asia, representing thrice the increase in these past 10 years.

GENERAL COMPOSITION AND SULFUR COMPOUNDS OF GARLIC AND ONION

As with other vegetables, both garlic and onion have nutrients, carbohydrates, proteins, lipids as well as vitamins (Tables 16.1 and 16.2). The contents of major nutrients in onion are quite low as compared with those in garlic, and the only component comparable to garlic is vitamin C. However, taking the higher onion consumption of about 5 to 10-fold compared to garlic into account, these lesser amounts of general components in onion should not be neglected. The general composition is, of course, important to the nourishment of both garlic or onion eaters, however, in respect to the food function, the extraordinary high content of sulfur compounds in these vegetables should be much more important.[7] These compounds are present as a group of sulfur-containing amino acids in their intact tissues, especially in the cloves of a garlic bulb[9] or in an onion bulb.

TABLE 16.1 General Composition of Fresh
Garlic and Onion Bulbs

Component	Garlic[a]	Onion[a] White	Onion[a] Red
Energy (kcal)	134	3	38
Water (g)	65.1	89.7	89.6
Protein (g)	6.0	1.0	0.9
Lipids (g)	1	0.1	0.1
Carbohydrates (g)	26.3	8.8	9.0
Minerals (mg)			
Sodium	9	2	2
Potassium	530	150	150
Calcium	14	21	19
Magnesium	25	9	9
Phosphorus	150	33	34
Iron	0.8	0.2	0.3
Zinc	0.7	0.2	0.2
Copper	0.18	0.05	0.04
Manganese	0.27	0.15	0.14

[a] Average value/100 g flesh weight.
From (8). With permission.

When injured by some fungi or damaged by slicing, the sulfur-containing amino acids in these plants are transformed immediately into volatile organosulfur compounds (Figure 16.1), and exhibit fungicidal activity and/or several physiological effects. In an intact bulb, there are two major sulfur-containing amino compounds, γ-glutamylcysteines, **1-3** and *S*-alkylcysteine sulfoxides, **4-6** (see Table 16.3, and Figure 16.2 and Figure 16.3). According to Lawson *et al.*,[9-12] and Ceci *et al.*,[13] γ-glutamylcysteines are more abundant in mature bulbs than in immature young bulbs, and rapidly decrease at sprouting, which is accompanied by increases of alliin, **4** and its related amino acids (methiin, **5** and isoalliin, **6**) in turn.[10,14] These three amino acids are termed "alliin" in this article, unless otherwise stated.

Transformation of γ-glutamylcysteines to alliin is accomplished enzymatically by γ-glutamyltranspeptidase (EC 2.3.2.2) and γ-glutamylpeptidase.[13,15-17] Since alliin bears an oxygen

TABLE 16.2   Vitamins, Fatty Acids and
Dietary Fibers in Fresh Garlic Cloves
and Onion Bulbs[a]

| | | Onion[b] | |
Compound	Garlic[b]	White	Red
Vitamins (mg)			
A	0	0	0
D	0	0	0
E	0.5	0.1	0.1
K	Tr	Tr	Tr
B1	0.19	0.03	0.03
B2	0.07	0.01	0.02
Niacin	0.7	0.1	0.1
B6	1.5	0.16	0.13
B12	0	0	0
Folic acid	0.0	0.02	0.02
Pantothenic acid	0.55	0.19	0.15
C	10	8	7
Fatty acids (mg)			
Saturated	0.18	0.01	0
Monounsaturated	0.04	Tr	0
Polyunsaturated	0.41	0.03	0.03
Dietary fibers (g)	5.7	1.6	1.7
Water soluble	3.7	0.6	0.6
Water insoluble	2.0	1.0	1.1

[a] From Standard Tables of Food Composition of
Japan. 5th Revised Edition. The Resource Council
of the Science and Technology Agency of Japan,
2001. With permission.
[b] Average value/100 g flesh weight.

on its sulfur atom, the reaction of adding an oxygen atom to
the prealliin molecule, S-alkylcysteines should be proceeded
by an oxidase.[12]

When a garlic plant is injured, alliin is immediately
transformed (in 10 sec) to alkyl thiosulfinates (allicin and its
related compounds) by the enzyme alliinase (C-S-lyase, alliin
lyase, EC 4.4.1.4).[18] The structure of allicin, **9** is shown in the
bottom of Figure 16.3. The amount of alliinase in a garlic clove
was determined to be extraordinarily high. It reaches about
10 mg/g of flesh weight, occupying as much as 15% of the total

γ-L-**Glutamyl-S-alk(en)yl cysteines** *

γ-Glutamyl-*S-trans*-1-propenylcysteine, **1**; γ-Glutamyl-*S*-allylcysteine, **2**;
γ-Glutamyl-*S*-methylcysteine, **3**; γ-Glutamyl-*S-cis*-1-propenylcysteine

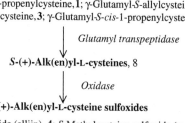

Glutamyl transpeptidase

S-(+)-Alk(en)yl-L-cysteines, 8

Oxidase

S-(+)-Alk(en)yl-L-cysteine sulfoxides

S-Allylcysteine sulfoxide (alliin), **4**; *S*-Methylcysteine sulfoxide (methiin), **5**;
S-trans-1-Propenylcysteine sulfoxide (isoalliin), **6**; Cycloalliin, **7**

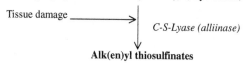

Tissue damage ⟶

C-S-Lyase (alliinase)

Alk(en)yl thiosulfinates

Allyl 2-propenethiosulfinate (allicin), **9**; Allyl methyl thiosulfinate
Allyl 1-propenyl thiosulfinate, Methyl 1-propenyl thiosulfinate
Methyl methanethiosulfinate

Spontaneous reaction

Sulfides

Diallyl disulfide, **10**; Diallyl trisulfide, **11**; Allyl methyltrisulfide, **12**;
Allyl methyl disulfide, **13**; Ajoene[†], **17**, **18**; Dithiin[†], **19**, **20**

Figure 16.1 Transformation of the sulfur compounds in garlic. The numbers after the compound's name indicate those of structures shown in Figures 16.2 to 16.5. *On the biosynthesis of this peptide, see references 21 to 23. †Both ajoene and dithiin are formed under certain conditions as mentioned in the text. Similar process proceeds in onion.

protein.[19,20] Molecular weight of alliinase was determined to be 55,000 kD.[24] Excluding carbohydrates from the molecule, it becomes 51,451 kD, which is expressed constitutively as 1.9 kb mRNA in a clove.[25,26] The alliinase requires pyridoxal 5-phosphate as a cofactor, and Mg, Mn, or Co as stimulators.[18,27] Alliin is degraded by alliinase into pyruvic acid and ammonia with the Km, 6 mM at pH 6.5, and at 33 to 37°C.[18,25,28] Alliinase is heat labile, and irreversibly inactivated at pH 3.8 or lower.[29] A pyridoxal-directed agent, hydroxylamine and amino oxyacetate were found as inhibitors for alliinase.[18,29–31]

TABLE 16.3   Sulfur and Nonsulfur Compounds That
Characterize Fresh Garlic Cloves

Compound	Amount g/100 g
Total sulfur compounds	2.3[a]
Sulfur	0.48[a]
Cysteine sulfoxides	
S-Allylcysteine sulfoxide (alliin)	1.7[a]
S-Methylcysteine sulfoxide (methiin)	0.23[a]
S-trans-1-Propenylcysteine sulfoxide (isoalliin)	0.02–0.12[9,10]
Cycloalliin	0.25[a]
γ–Glutamylcysteines	
γ–Glutamyl-S-trans-1-propenylcysteine	0.3–0.9[9,32]
γ–Glutamyl-S-cis-1-propenylcysteine	0.006–0.015[10]
γ–Glutamyl-S-allylcysteine	0.2–0.6[9,32]
γ–Glutamyl-S-methylcysteine	0.01–0.04[10]
Nonsulfur compounds	
Saponins, mostly in β-sitosterol based	0.035–0.042[33]
Sapogenin, as β-sitosterol	0.019[33]

[a] Values obtained for dry weight of garlic by Ueda *et al.* (1991)[34] are represented after multiplying with 0.65 for the flesh weight (65% moisture). The superscripts show references.

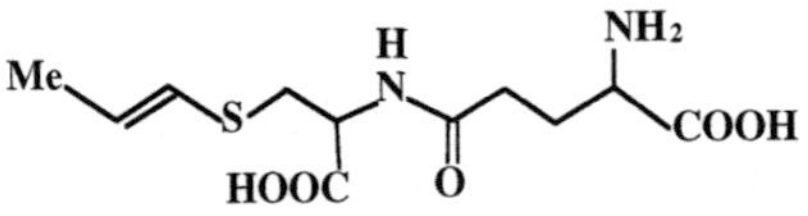

γ-Glutamyl-*S-trans*-1-propenylcysteine, **1**

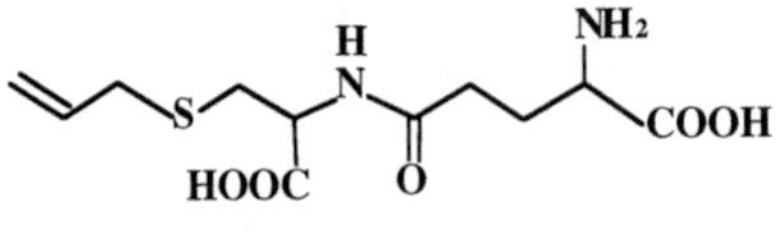

γ-Glutamyl-*S*-allylcysteine, **2**

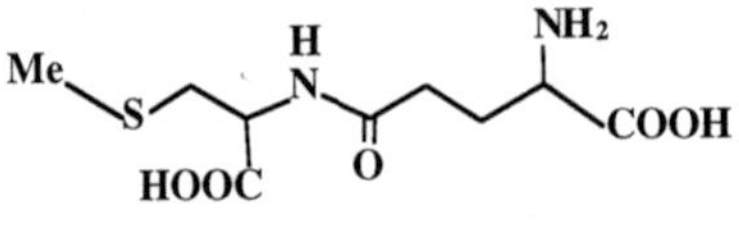

γ-Glutamyl-*S*-methylcysteine, **3**

Figure 16.2   Major γ-glutamylcysteines in garlic and onion. Me: methyl group.

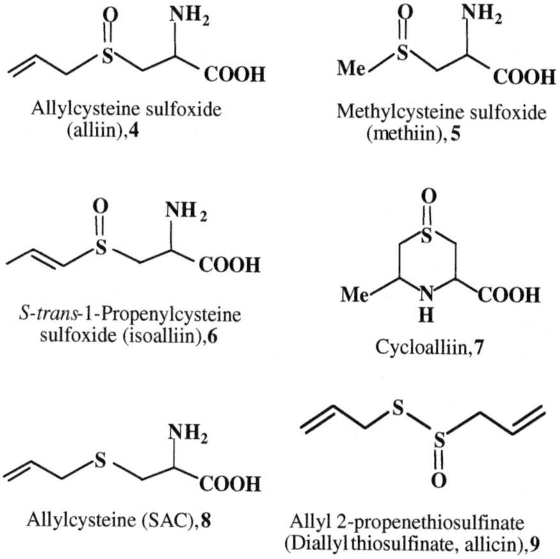

Allylcysteine sulfoxide
(alliin), 4

Methylcysteine sulfoxide
(methiin), 5

S-trans-1-Propenylcysteine
sulfoxide (isoalliin), 6

Cycloalliin, 7

Allylcysteine (SAC), 8

Allyl 2-propenethiosulfinate
(Diallyl thiosulfinate, allicin), 9

Figure 16.3 Principal *S*-alk(en)ylcysteine sulfoxides, cycloalliin, allylcysteine (SAC) and allicin (as a representative sulfinate).

Tissue damage causes binding of alliinase and alliin, which are localized separately in garlic, i.e., the former is located in bundle sheath cells scattered inside of the clove, and the latter in most cells of the clove.[19] Although alliin and its parent γ-glutamylcysteines coexist in a garlic bulb, only alliin is converted to allicin, and γ-glutamylcysteines remain as they are in the crushed bulb.[10,32,35,36] Actually, most of all garlic products, garlic tips, pastes or even dried garlic powder, have γ-glutamylcysteines, but no alliin.[29] Based on these facts, the Japan Health Food Nutrition Association (November 2001, Tokyo) selected γ-glutamyl-*S*-allyl-L-cysteine (GSAC), **2** as a maker compound to certify whether a commercial product is really made of garlic or not, and if the product is determined to have certain amounts of GSAC, the association permits the manufacturer to categorize the product as a "health food."[36] However, the physiological function of GSAC is not well known.[37]

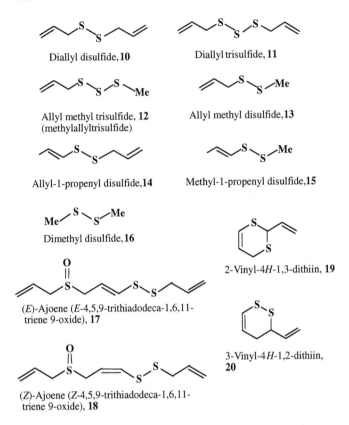

Figure 16.4   Principal alk(en)yl sulfides, and allicin-derived ajoenes and dithiins.

Allicin is converted spontaneously and quickly to alkyl sulfides. The principal structures of alkyl sulfides are shown in Figure 16.4. The bactericidal and fungicidal principle of allicin is known to be extremely unstable and reactive.[12,39] According to Lawson et al.,[12,40] the half-life of allicin is not as short when kept in water: it was about 2 days in crushed garlic, while in water it was as lengthy as 40 days. The stability of allicin decreases more in nonpolar solvents than in water, and it is temperature dependent: storing at −70°C, the half-life was 2 years.[12]

TABLE 16.4   Alk(en)yl Sulfide Detected from Garlic
Oil and Onion Oil[a]

	mg/100g flesh bulb	
	Garlic	Onion
Methyl-1-propenyl disulfide		0.82
Methyl propyl disulfide		0.28
Allyl methyl disulfide	15.7	
Diallyl disulfide	139.1	
Dimethyl trisulfide	2.3	1.34
Allyl methyl trisulfide	49.5	
Diallyl trisulfide	161.6	
Methyl propyl trisulfide		0.38
Methyl-1-propenyl trisulfide		1.11
Total content	369.0 ± 29.5	3.93 ± 0.03

Note: The two plants do not share the same compound with each
other, excluding dimethyl trisulfide.
[a] Steam-distilled garlic and onion oil samples were analyzed by
gas chromatography. The garlic oil was obtained from 'White-
roppen,' and onion oil from a yellow onion 'Sapporo-kii.'
(From T Ariga, T Seki, H Kumagai. Unpublished data, 1997. With
permission.)

Conversion of allicin to sulfides is known to proceed with
rather complicated reactions, and thus many structures of
sulfides are produced.[41,42] The garlic oil, which can be obtained
either by steam distillation or by solvent extraction from
crushed garlic bulbs, is entirely made of sulfides (see Table
16.4).[43]

Onion produces sulfides only 4 mg/100 g flesh bulb, and
the amount is barely comparable with that in a gram of garlic
clove (see Table 16.4).[44] However, once the oil is prepared, it
can be used not only for the industrial food processing, but
for laboratory studies to examine the onion function as a
counterpart of garlic oil. If there is some difference in certain
functions between the two oils, it might be due to the structural difference in their components. Onion has no allyl group
(or 2-propenyl group), but it has 1-propenyl and propyl groups,
which are not present in garlic (see Table 16.4).

Lachrymatory Factor (LF)

The LF is characteristic to onion, and its entity has been determined to be a labile compound named propanethial S-oxide, **22**. As a reason why onion yields the factor but garlic does not, it has been recognized that the formation of the nonlachrymatory 1-propenyl thiosulfinates from 1-propene-sulfenic acid (the parent of LF, **21**) proceeds spontaneously very rapidly in garlic, but slow in onion. Hence, the route for LF formation from its parent was open in onion.[14,45,46] However, the latest findings by Imai et al.[47] made it clear that for the production of LF, a novel enzyme, lachrymatory factor synthase, is specifically needed to catalyze the reaction (see Figure 16.5). It was demonstrated that the addition of the anti-LF synthase antibody onto onion slices completely inhibited formation of LF. Their findings may be useful for the development of LF-free onion.

Organoselenium Compounds in Garlic and Onion

Allium plants are known to contain organoselenium compounds with similar profile to organosulfur compounds, however the amounts can vary markedly depending upon the growth conditions. According to Block,[48] volatiles from garlic represented a gas chromatographic pattern of selenides quite similar to that of sulfides, although the quantity of the former was extremely less (1/12,000) than the latter. A general characteristic of *Allium* plants, especially garlic and onion, is their strong potential to uptake inorganic selenium, either selenate or selenite, from the soil or water culture medium, and synthesize organoselenium compounds. Kotrebai et al.[49] reported that under the same condition for selenium, enrichment onion contained 140 μg/g Se in dry sample, but garlic showed as high as 1,355 μg/g Se in dry samples in their maximum levels. Ip and Lisk also observed a similar difference between onion and garlic.[50] As a major form of Se-containing peptides, γ-glutamyl-Se-methyl selenocysteine, **26** has been determined, and from this peptide, Se-methyl selenocysteine, **27**, a potent agent for cancer prevention, may derive (see Figure 16.6). el-Bayoumy et al.[51] reported that the chemopreventive

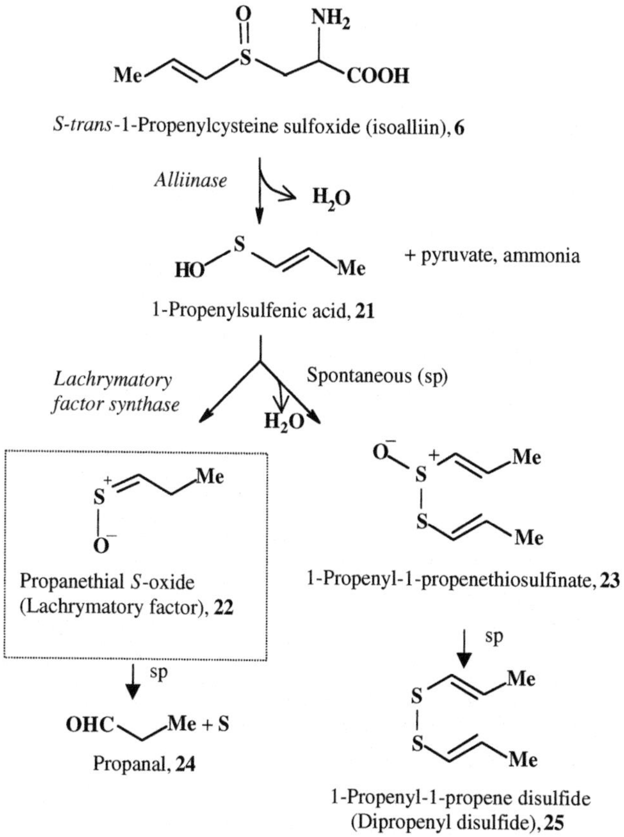

Figure 16.5   Alliinase reaction for the production of lachrymatory factor and thiosulfinate in onion. (From S Imai, N Tsuge, M Tomotake, Y Nagatome, H Sawada, T Magata, H Kumagai. An onion enzyme that makes the eyes water. *Nature* 419:685, 2002. With permission. Copyright: Macmillan Magazines Ltd.)

effect of diallyl selenide on a mammary cancer was at least 300 times more than that of diallyl sulfide. Ip and his colleagues[52–54] have also demonstrated the efficacy of these compounds through animal studies. The ability of the garlic plant to accumulate organoselenium compounds reveals potential to produce novel formulation of functional foods or production of anticancer chemicals.

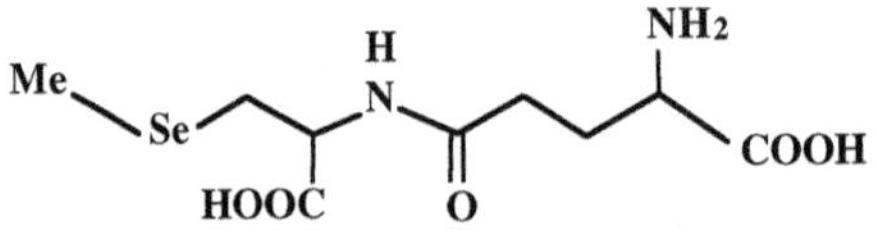

γ-Glutamyl-*Se*-methyl selenocysteine, **26**

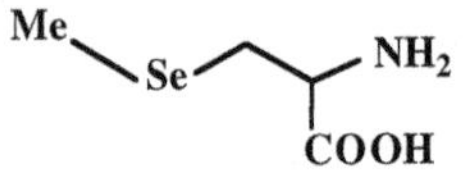

Se-Methyl selenocysteine, **27**

Figure 16.6   Principal organoselenium compounds in garlic.

PROCESS-INDUCED COMPOUNDS FROM GARLIC COMPONENTS

Ajoene and Dithiin

It has been reported that allicin is transformed not only to ally sulfides, but also to ajoene, **17, 18** and dithiin, **19, 20** when it is treated under certain conditions[45,55–58]; first of all, chopped garlic is placed at room temperature for several minutes to produce much allicin, followed by addition of alcohol or vegetable oil, then incubation at 70 to 80°C for 20 min. Thus, large amounts of ajoene and dithiin can be obtained[45,57,58]; the structures and their isomers are shown in Figure 16.4. These process-induced products, especially ajoene, have been reported to show potent antiplatelet effect.[45,55,58,59] The ajoene may presumably be formed when garlic is cooked by slicing and frying in a cooking oil.

Allylcysteine

This amino acid (*S*-allyl-L-cysteine, SAC, **8**), as described previously, is a naturally occurring compound, although its content is as low as 10 μg/g in a flesh garlic.[60] However, it was found that SAC increased markedly when crushed garlic was aged for 6 months or longer in an alcoholic solution.[61,62] The aged garlic extract (AGE) so obtained is used as a concentrated alcoholic solution or dried powder. Since AGE was first developed by Japanese manufacturers, its pharmacological function has been studied extensively by many Japanese researchers.[61–64]

Allixin

This is a phytoalexin isolated from garlic.[65] It has a structure not involving sulfur: 3-hydroxy-5-methoxy-6-methyl-2-penthyl-4*H*-pyran-4-one (see Figure 16.7).[65,66] As the name indicates, allixin is produced when garlic, especially its bulb, encounters some stress such as strong sunlight, chemicals, and microorganisms. Because it is not detected from the common garlic usually used for cooking or processing, it is necessary to treat garlic for allixin generation.[66] According to Kodera *et al.*[66] allixin was produced in the garlic clove by midsummer sunlight radiation for 3.2 h. Through this treatment, allixin was detected at 4 to 6 µg/g from both irradiated and nonirradiated sides of the clove. Their recent paper described that 3,000 to 7,400 µg/g of allixin was obtained from the surface tissue of garlic bulbs that had been placed in nets and stored in a drafty room without air conditioning for up to 2 years.[67] Although the role of allixin in a garlic plant remains unclear, its pharmaceutical application was the main objective of their study. Allixin was found to confer antioxidant activity through its phenolic hydroxyl group,[68] and inhibited DNA damage caused by aflatoxin binding. These effects may give rise to the prevention of aflatoxin-induced mutagenesis.[69] For this purpose, 75 µg/ml of allixin *in vitro* was used. As far as allixin is concerned, its pharmacological application is of interest.[70–73]

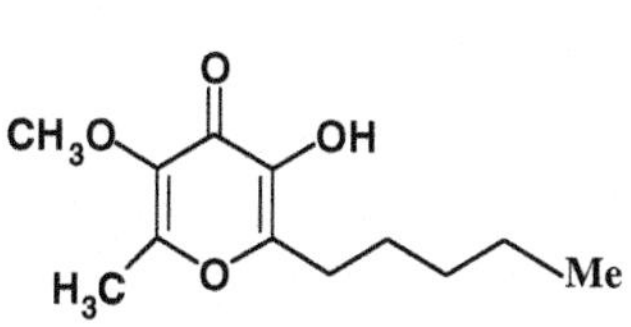

Figure 16.7 Structures of allixin.

PHYSIOLOGICAL AND NUTRITIONAL FUNCTIONS OF GARLIC AND ONION, AND THEIR HEALTH BENEFITS

In garlic and onion plants, the γ-glutamylcysteines, **1-3** and their metabolites, sulfur compounds, appear to be stored only for protecting their own plant bodies. Actually, allicin, **9** the most reactive compound generated from allylcysteine sulfoxide upon tissue damage, exhibits strong antifungal and bactericidal activities, and thus, these plants are protected from their enemies.[40,74] Even in our bodies, sulfur-containing compounds play a defensive mechanism. In addition to longtime utilization of garlic and onion, current scientific studies have demonstrated that *Allium* plants belong to the most beneficial vegetables for human health.

Physiological functions of garlic and onion together with functional principles and proposed mechanisms are listed in Table 16.5. Onion has a lesser amount of sulfur compounds as compared with garlic. However, because people consume onion much more than garlic, the amount of sulfur compounds taken from onion per person may be comparable to or more than that from garlic. For example, Japanese, Turkish, and Pakistani people consumed onion about 57, 23 and 40 times more than garlic in 2000, respectively.[4] In addition, the quantitative disadvantage in ingredients of onion can usually be overcome by using onion oil, the concentrate of functional sulfur compounds.

Antibiotic Effect

The antibiotic effects of garlic can be observed for almost all microorganisms and insects as well as viruses. For more than 4,000 years, these effects have been evaluated and garlic has been used by physicians, like Pasteur (1858) and Schweitzer (1932), for the treatment of infectious diseases.[75] It is likely that ancient people with an insect bite, wounds, or stomachache would be treated by crushed garlic as an ointment or an internal medicine. At the same time, effectiveness of garlic against food spoilage was found. It is, therefore, quite natural that they decorated tombs with garlic as a phylactery.

TABLE 16.5 Physiological and Pharmaceutical Functions of Garlic and Onion

Function	Principle	Action/mechanism
Insecticidal effects	Allicin[74] DADS[76–78] DATS[76–78]	Against mosquito, mosquito larvae, tick and flea; these are killed by garlic extract or oil[76,77] It kills plant-damaging aphids[79] Inhibition of protein synthesis,[78] or amino acid incorporation by killing symbiotic bacteria, which would supply steroids to the host[74]
Repellent effects	DADS[80] DATS[80]	Against tick, fruit fly, mosquito, and other pests. It also repels ants[81–83]
(O) Insect attractant	(O) Onion juice[82] Cut/aged onions[82] Propyl/methyl sulfides[84–88] Mercaptan[86]	Ant repellent[82,83] Onion maggot, especially female selects onion plants for oviposition Damaged onion is much more attractive than intact one for the maggot since it produces much volatile sulfides[84–88]
Antibacterial effects	Allicin[89–96] (aqueous extract or pressed garlic)	Against most of all Gram-positive/ -negative bacteria[93–94] Gram-negative, high lipids *Helicobacter pylori* is much more susceptible than Gram-positive, low lipids *Stephylococcus aureus*[91] Against antibiotic-resistant strains[92] Release of oxygen radical[93] Binding with SH-group in bacteria or their enzymes via thiosulfate structure of allicin[92–95] Inhibition of RNA synthesis[97] Bacteriophages that neutralize *E. coli*, or other bacteria were isolated from garlic plant.[98]
	(O) OAqE[99–100]	Colored varieties were less active than those of less colored.[100] Alkali treatment or heating reduces the effect.[99] (Principle would be thiosulfinates.)

TABLE 16.5 (CONTINUED) Physiological and Pharmaceutical
Functions of Garlic and Onion

Function	Principle	Action/mechanism
Antifungal effects	Allicin[40,101] DATS[102] Vinyldithiin[103] Ajoene[104] AGE[105] Garlic oil[106]	Against *Candida, Trichophyton, Aspergillus, Paracoccidioides* and other fungi[40,101,102,107] Inhibition of oxygen uptake[101] Inhibition of succinate dehydrogenase[107] Affect outer surface of cells, and reduce adhesion relating to thiol group[108,109] Destruction of cytoplasmic membrane (by ajoene)[102,107]
	(O) OAqE (thiosulfinates)[110] Ace-AMP1[111] DHBA[112]	Sporicidal rather than sporistatic[110] Against *Aspergillus niger, A. flavus, A. fumigatus,* although minimal fungicidal concentration is higher than garlic[110]
Antiprotozoal effects	Ajoene[113] Allicin[114,115] DATS[116]	Against Entamoeba, Trypanosoma, Giardia, etc.[113–117] Inhibition of phosphatidylcholine synthesis (by ajoene)[113] Inhibition of cysteine proteinases of amoeba[115]
Antiviral effects	Ajoene[118,119] DATS[120,121] DADS[123]	Against influenza A and B, polio, herpes, pneumonia viruses and HIV[118–125] Suppression of HIV replication[126] Blockade integrin-dependent process of HIV-infected cells[127] Prevention of cytolysis of infected cells (ajoene)[119] Inhibitory to retrovirus infection[128] Indirect *in vivo* effects through immune activation would be involved.
	(O) Lectins[128]	(Negative) Poliovirus did not decrease on onion[129] (Negative) An outbreak of hepatitis A was suspected to be associated with green onions.[130]

TABLE 16.5 (CONTINUED) Physiological and Pharmaceutical
Functions of Garlic and Onion

Function	Principle	Action/mechanism
Antiparasitic effects	Garlic oil[131,132] Allicin[133] DADS[134]	Anthelmintic against roundworm,[134] hookworm[135] and microfilariae[136] Against Leishmania mexicana, L. donovani,[137,138] African trypanosomes[134] and nematodes in the soil[131] Morphological alteration of mitochondrial membrane[137] Garlic may have adjuvant and prophylactic effects[93]
Antiplatelet effects	MATS[139–143] Ajoene[42,59,144–146] Dithiin[140,147] Allicin[148] Garlic oil[149–151] AGE[152-154] GAqE[148]	Inhibition of preaggregation (from discoid to pseudopods),[145] and adhesion or binding to vessels[151] Inhibition of aggregation: inhibition of cyclooxygenase, leading to poor production of $PGG_2/PGH_2/TXA_2$[154–157] Inhibition of TXB_2 synthesis[148] PGI_2 production of aorta is not inhibited *in vivo*[59,142,143] Fibrinogen receptor (glycoprotein IIb/IIIa) on platelets is hampered, so that platelet binding suppressed.[146]
	(O) Flesh onion juice[158] Onion oil[159,160] Adenosine[159,161] Allicin[159] Flavonol (quercetin)[162]	Prevention of platelet-mediated cardiovascular disorders in coronary artery-stenosed dogs[163] IC_{50} for collagen-induced platelet aggregation of OAqE was 90 mg/ml, much weaker than GAqE, 6.6 mg/ml. Adenosine and allicin inhibited aggregation without affecting arachidonic acid metabolites.[159] OAqE decreases TXB_2 in rat serum.[160] Trisulfides inhibited thromboxane synthesis.[13]

TABLE 16.5 (CONTINUED)   Physiological and Pharmaceutical
Functions of Garlic and Onion

Function	Principle	Action/mechanism
		Elevation of cyclic AMP, depending on PGI_2 production (Welsh onion).[164,165]
		Quercetin (2.5 mM) inhibited platelet aggregation in PRP by 90%.[162]
Fibrinolytic effects	Allylsulfides[150] DATS[166] Garlic oil,[167,168] powder[169,170]	Determined by clinical trials[166–168,170,171] Lower plasma fibrinogen level[167,168,172] tPA release from vessels is increased, but PAI-1 activity unchanged[169]
	(O) OAqE[158,173] Onion oil[150,174,175]	Elevation of fibrinolysis *in vitro* and *in vivo*[150,173,174]
Vasodilative effects	Dry powder[176] Adenosine[159] Ajoene[177] Garlic extract/ GAqE[178,179]	Cell membrane hyperpolarization with CA^{2+}/HPO_4^- channel closing[180] Activation of nitric oxide synthase[178] Inhibition of endothelin-1-induced contraction[179] Increase microcirculation[181]
Lipid-lowering effects	Allylsulfides[182] Raw garlic[183,184] Lyophilized garlic[185] GAqE[186,187] Cycloalliin/SCA/ γ-GSMC[188,189] DADS/DATS[189] Allicin[190]	Inhibition of hepatic cholesterol synthesis[183] Inhibition of G6PDH and malic enzyme activities[185] Increase excretion of steroids[185] Inhibition of HMG-CoA reductase, C7AHX and FAS, resulting in decrease of LDL-cholesterol[186–188] Enhance lipid catabolism by increasing noradrenaline and UCP[189,191,192]
	(O) Onion juice,[173,174] oil[150,173] MCSO[193] Dried powder[175]	High cholesterol diet-induced high cholesterol, triglyceride and lipoprotein levels were significantly lowered in animals[174,175] and human.[149–152] Decrease cholesterol in LDL-VLDL.[175]

TABLE 16.5 (CONTINUED)   Physiological and Pharmaceutical
Functions of Garlic and Onion

Function	Principle	Action/mechanism
Hypoglycemic effect	SAC/AGE[194] γ-GSMC[188,189] DADS/DATS[189] Allicin[190]	Function as insulin secretagogue[195–197] Improve aortic endothelial dysfunction in insulin-dependent model[198,199] Prevention of adrenal hypertrophy[194]
	(O) Freeze dried powder[175] MCSO[193,200]	Reverse abnormalities in albumin, urea, creatinine in diabetic rats, probably because of its hypoglycemic and hypocholesterolemic effects.[175,193,200]
Antiaging effects	Garlic extract[201]	Increase normal cell proliferation, but inhibitory to cancer cell growth[201] New insights into old remedy, a garlic[202]
Antioxidant effects	Garlic oil[203] Diallylsulfides[203] AGE[204] SAC[205–208] Allyl mercaptan[209]	Increase activities of SOD, catalase and glutathione peroxidase[203] Inhibition of 8-iso-PGF(2 alpha)[204] and iNOS productions of macrophage, but increase eNOS in endothelial cells[205] Inhibition of NF-κB activation[188,190]
Blood pressure-lowering effect	Allicin[190] Raw garlic[190] GAqE[210]	Direct relaxant effect on smooth muscle[211,212] Inhibition of prostanoid synthesis[213] Inhibition of renin-angiotensin system[214] Enhance nitric oxide system[214] Inhibition of adenosine deaminase[210] β-Adrenergic receptor blocking (suggestive)[93]
Diuretic effects	Garlic proteins[215]	Diuretic and natriuretic responses; probably mediated by a sodium pump inhibition at the sodium tubular reabsorption level of the kidney[215]

TABLE 16.5 (CONTINUED)   Physiological and Pharmaceutical
Functions of Garlic and Onion

Function	Principle	Action/mechanism
Anti-inflammatory and immuno-modulatory effects	DAS[216] GAqE[217] GEE[217,218] Protein[219]	Inhibition of cytochrome P450-mediated oxidative metabolism[216] Increase Th1-type cytokine response[138] Proliferation of lymphocytes via upregulation of IL-2 and its receptor[217,218] Hypertrophy of lymphoidsheath and lymph nodes[219] Inhibition of monocyte and T-cell interleukin production[220]
	(O) Thiosulfinates[221–223] Cepaene[221–223]	Inhibition of cyclooxygenase and lipoxygenase; "Cepaene" is more potent than thiosulfinates.[221–223] Inhibition of chemotaxis of granulocytes, thus antiasthmatic.[224,225]
Hormone secretory effects	DADS/DATS/ ACS[224–226] Allyl-mercaptan[227]	Noradrenaline secretion via beta-adrenergic action[224,225] Increase testicular testosterone, and decrease plasma corticosterone[226] Increase testosterone catabolism in culture cells[227]
Vitamin B_1 absorption	Allicin[228–230]	Allithiamine, an adduct formed between allicin and thiamine, is absorbed well by alimentary tract, and exhibits thiamine effect.[228–230]
Anticancer effects		See Table 16.6.

Abbreviations: DADS, diallyl disulfide; DATS, diallyl trisulfide; (O), subjects for onions; OAqE, onion aqueous extract; AGE, aged garlic extract; Ace-AMP1, a cationic protein with 93 amino acids and 4 disulfide bonds; DHBA, 3,4-dihydroxybenzoic acid, an oxidation product of quercetin; MATS, methyl allyl trisulfide (allyl methyl trisulfide); GAqE, garlic aqueous extract; PG/TX, prostaglandin and/or thromboxane; AMP, adenosine monophosphate; PRP, platelet-rich plasma; tPA, tissue-type plasminogen activator; PAI-1, plasminogen activator inhibitor-1; SAC, S-allylcysteine; γ-GSMC, γ-glutamyl-S-methyl-cysteine; G6PDH, glucose-6-phosphate dehydrogenase;

TABLE 16.5 (CONTINUED) Physiological and Pharmaceutical
Functions of Garlic and Onion

HMG-CoA, hepatic beta-hydroxy-beta-methylglutaryl coenzyme A; C7AHX, cholesterol 7-alpha-hydroxylase; FAS, fatty acid synthetase; UCP, uncoupling protein;
MCSO, S-methylcysteine sulfoxide; LDL, low-density lipoprotein; VLDL, very low-
density lipoprotein; SOD, super oxide dismutase; iNOS, inducible nitric oxide synthase; eNOS, endothelial nitric oxide synthase; NF-κB, nuclear factor-κB; DAS, diallyl
sulfide; GEE, garlic ethanol extract; Th1, helper T-cell 1; IL-2, interleukin-2; Cepaene,
ajoene-like sulfur compounds isolated from crushed onion; ACS, allylcysteine sulfoxide.
The superscripts show references.

Cavallito and Bailey initially carried out identification
of the bactericidal principle in 1944.[231] Alkyl thiosulfinates
are recognized as the most effective compounds exhibiting
antimicrobial activity. The studies on these effects of garlic
and related *Allium* plants that had been conducted by many
researchers up to 1994 were thoroughly documented as a
review by Fenwick and Hanley in 1985.[5]

As can be seen in Table 16.5, allicin (alkyl thiosulfinates),
9 and ajoene, **17, 18**, with a common sulfinate structure, are
responsible for the antibiotic effects. Owing to the instability
of allicin, garlic extract has usually been used as an allicin-
containing solution to examine bactericidal activity, and IC_{50}
was determined to be about 0.5 to 3 mg/ml.[232] In the aqueous
extract (about 50%, w/w, homogenate), content of allicin is
estimated to be about 1 µg/ml (corresponding to 8 to 10 µ*M*).
Ariga *et al.*[43] observed the growth inhibition of *Tricophyton
mentagrophytes* (a major fungus in the water eczema) by an
aqueous garlic extract. The extract diluted 100 times or more
with water (about 10 ng/ml allicin) showed a clear inhibition
zone, and *T. mentagrophytes* did not grow inside of the zone
for up to 7 days.

Antithrombotic Effect

Suppression of platelet aggregation is the most prominent
effect of garlic intake. Bordia and Bansal (1973)[149] described
the suppressive effect of garlic or garlic oil against platelet
aggregation in human studies. Ariga *et al.* (1981)[139] isolated a

potent inhibitory compound from garlic oil, and identified it as methyl allyl trisulfide (MATS, **12**). About 10 μM of MATS inhibited platelet aggregation induced by almost all known inducers, collagen, arachidonic acid, epinephrine, thrombin, and ADP. The target of MATS was determined to be the prostaglandin (PG) hydroperoxidase-reacting site, which is located between the reacting sites of cyclooxygenase and thromboxane synthase in the platelet's arachidonic acid cascade. Since the inhibition of this enzyme resulted in a poor production of thromboxane A_2 (TXA_2), a strong agonist for platelet aggregation, the platelet aggregation would be hampered. MATS also inhibited the aortic PG hydroperoxidase, and suppressed the synthesis of prostacyclin (PGI_2), a strong antiplatelet agent. However, MATS clearly inhibited the thrombus formation in the experimentally injured rat vessels.[142,143]

Ajoene was discovered as the most potent platelet inhibitor from an oil-macerated crushed garlic by Apitz-Castro *et al.* (1983)[55]; ajoene inhibited both of the arachidonate metabolism and the membrane G-protein level signal transduction of platelets.[144–146] Anticancer effects of ajoene are discussed in Section F: Anticancer effect of garlic and onion.

The enhanced fibrinolysis and vascular dilation have been observed primarily with garlic oil in human and animal studies.[167,179] Thus, garlic may prevent thrombus formation through multilateral mechanisms involving antithrombotic, lipid-lowering, vasodilative, and antiatherosclerotic effects.

Lipid-Lowering Effect

Lipid- and cholesterol-lowering effects have been clearly demonstrated by many investigators through their human and animal studies.[171,182–189] As shown in Table 16.5, several garlic compounds have been reported to be effective. For this purpose, the daily doses applied for animal studies are 0.1 to 0.2 mg of allyl sulfides, 1 g of garlic, or 4 mg of ether extract per kg body weight. The lipid-lowering effects of SAC, **8** and GSMC, **3** appear to be due to their degradation products, those that may be produced from the compounds soon after the absorption from the intestine.[188,189]

Recently, Lin et al.[233] reported that a single oral dose of flesh garlic homogenate to rats significantly reduced the intestinal microsomal triacylglycerol transfer protein (MTP) mRNA expression. Since MTP plays a pivotal role in the assembly and secretion of chylomicrons from intestine to the blood circulation, the reduced gene expression may be the important factor in the lipid-lowering effects of garlic.

Hypoglycemic Effects

It has been proposed that garlic compounds stimulate insulin production and prolong insulin turnover. Especially, sulfur compounds from garlic were considered to protect insulin from its inactivation with cysteine, glutathione, and albumin by blocking their SH groups.[197] Since, alliin (200 mg/kg) is known to reduce the plasma glucose level in the alloxan-induced diabetic rats, and regenerate their pancreas,[234] eating boiled garlic, which has no annoying odor, may render the antidiabetic effect.

Other Effects

Garlic and its sulfur compounds have many other beneficial effects as shown in Table 16.5, e.g., blood pressure-lowering, diuretic, anti-inflammatory, immunomodulatory, and hormone secretory effects, as well as enhancing effect on vitamin B$_1$ absorption. Anticancer effects of garlic and onion are discussed in the following section.

The recent findings by Oi *et al.*[224,226] are noticeable. They found that garlic compounds, especially allyl-containing structures, stimulate both noradrenaline secretion from nervous systems and testosterone from testis. These hormones may act to modulate lipid and protein metabolisms (enhancement of lipid catabolism[235,236] and protein anabolism[237]).

ANTICANCER EFFECTS OF GARLIC AND ONION

A number of epidemiological and experimental studies imply that garlic and onion can be considered as important functional foods for cancer prevention.[238] Fresh extract, powder, oil, and several organosulfur compounds derived from garlic and onion are reported to exhibit anticarcinogenic, antimutagenic, and antitumorigenic activities as summarized below.

1. Epidemiological studies on the cancer prevention by garlic and onion: *Allium* vegetables and their organosulfur compounds have been extensively studied on their chemopreventive effects against cancer. Epidemiological studies have provided evidence that consumption of *Allium* vegetables reduces the incidence of several cancers. Hu *et al.*[239] conducted the case-control study in northeast China, and found that the consumption of onion was inversely related to the risk of developing brain cancer. Similar studies were reported by Key *et al.*,[240] Challier *et al.*,[241] Levi *et al.*,[242] and Witte *et al.*,[243] that garlic intake significantly reduced the risk of prostate and breast cancers, and increased protective power against stomach and colorectal cancers. The recent critical review has summarized epidemiologic studies on the relationship between garlic consumption and incidence of cancers of the stomach, colon, head and neck, lung, breast and prostate which suggested preventive effects of garlic consumption on stomach and colorectal cancers.[244] Site-specific case-control studies on stomach and colorectal cancers suggest a protective effect of high intake of raw and/or cooked garlic. These chemopreventive effects of garlic are afforded by many diverse mechanisms, including inhibition of carcinogen formation, modulation of carcinogen metabolism, inhibition of mutagenesis and genotoxicity, inhibition of cell proliferation, and increase of apoptosis.[245]

2. Experimental studies on the preventive effects of garlic and onion from mutagenesis and carcinogenesis: A number of experimental studies performed for garlic and onion *in vitro* and *in vivo* support the evidences provided by epidemiologic studies. The overall anticancer effects that have been reported for garlic and onion by some hundreds of researchers can be collected largely into three groups: e.g., (1) Antimutagenic and anticarcinogenic effects, (2) Antiproliferative effects, and (3) Differentiation effects involving the apoptotic effect. These effects

of garlic and onion together with effective principles and proposed mechanisms are summarized in Table 16.6. Among the effects, prevention of mutagenesis caused by several carcinogens is deemed to be most prominent. This effect might be due to the reduction of a genotoxicity of mutagenic agents by sulfur compounds in garlic and onion through the modulation of detoxification enzyme systems.

FAMILY USE OF GARLIC

The garlic sold in markets has a moisture content of about 65%, and it can be used for a few months when stored in a refrigerator or in circulating air at ambient temperature. However, the composition and quality of garlic changes depending on the processing or cooking methods.

Sliced, Cut, or Crushed Garlic

As shown in Figure 16.1, alliin, **4** in the damaged cells, is transformed to allicin, **9** and allicin turns spontaneously into sulfides, **10-16**. In organic solvents or in cooking oils, the half-life of allicin is very short as compared to that in water, in which the half-life is estimated at up to 2 days.[29] Hence, the allicin may be recovered from garlic chopped in water. On the other hand, volatile and nonpolar sulfides produced upon slicing may effectively be trapped with cooking oils. If the sliced garlic is dried up by blowing the air or with a microwave to prepare garlic tips, both allicin and sulfides may be lost completely. However, even in this case, γ-glutamylcysteines, **1-3** will remain in the tips as mentioned above.

Boiled or Heated Garlic

Heating greatly changes the components of garlic. If an intact bulb is heated, alliin stays unchanged inside the bulb, and thus, pungent odor generation can be killed. However, if the heating is not quite sufficient to denature the alliinase, a large amount of alliin is transformed into allicin and sulfides while cooking or eating the bulb. When chopped garlic is heated, allicin and sulfides generated on the surfaces may disappear, and only a

TABLE 16.6 Anticancer Effects of Garlic and Onion

Function	Principle	Action/mechanism
Antimutagenic and anticarcinogenic effects	Garlic powder[241] DAS[242,243] DADS[242,243] DATS[245] DASO[246] DASO2[246] SAC[247] DATS/DADS (Allitridi)[248]	Reduction of nitrosoamine-initiated lesions in placenta with suppression of CYP-2E1[241,246] Prevention of nitrosoamine and AFB1-initiated hepatocarcinogenesis[242,249] Increase metabolism of AFB1 to less- or nongenotoxic AFQ1 and AFM1[249] Increase AFB1-glutathione conjugates (the phase II metabolism)[243,244,249] SAC up-regulates SOD and catalase activities in DMBA-induced carcinogenesis[247] Induction of folate receptor alpha and calcyclin expressions[248]
Antiproliferative effects	DADS[250–253] DATS[254] Ajoene[255,256] Allicin[257]	Inhibition of the growth of human tumor cell lines, in which Ca^{2+} is increased[250,254] Inhibition of growth of neoplastic cells and induce apoptosis[251,256,258,259] Cell cycle arrest into G2/M phases with depression of 34[cdc2] protein[252] Reduction of tumor volume by 69% without ill consequences of host cells[253] Transient decrease in intracellular glutathione[257]
Differentiation-inducing effects	Garlic oil and onion oil[260]	Induction of HL-60 leukemic cell differentiation into granulocytic lineage[260] Enhance the effect of all trans-retinoic acid synergistically

Abbreviations: DASO, diallyl sulfoxide; DASO2, diallyl sulfone; SAC, *S*-allylcysteine; CYP-2E1, one of the members of cytochrome P450; AF, aflatoxin; SOD, superoxide dismutase; DMBA, 7,12-dimethylbenz[a]anthracene.

small amount of alliin may remain in the pieces. Heating the chopped garlic in cooking oils, allicin, sulfides as well as alliin must be present, in the oils or within the pieces. Hence, people prefer to cook garlic with cooking oils or fatty meats.

Pickled Garlic

Pickled garlic has long been prepared for family use with vinegar, alcohol or honey as a medium. The pickled clove in any medium loses alliinase activity more or less, depending on the period of its preservation. It takes at least a month to penetrate a medium into the clove with a gradual decrease in alliinase activity. Acetic acid in vinegars (about 5%) and alcohol (20%) have been reported to be effective media in prevention of olfactory annoyance with garlic odor.[29,261] The former medium inhibits alliinase activity, and the latter converts alliin (*S*-allylcysteine sulfoxides, **4**) into the deoxygenated alliin (*S*-allylcysteine, **8**), which is no longer the substrate for alliinase. The clove pickled after slicing or being powdered has quite different ingredients from that of an intact clove. Such processing has been developed by many manufacturers to produce specialized products (see the industrial processing of garlic described next).

INDUSTRIAL PROCESSING OF GARLIC

Garlic products are sold worldwide in markets or drug stores, and recently, as OTC (over-the-counter) products which place first rank in the selling records of Germany. There are many methods for preparing garlic products, and hence, components involved are different from one product to another. However, every product as such in capsules or tablets has been sold with the only name, garlic or its supplement. This must be a problem for people using the product for their health promotion. Production of different garlic products is described next.

Aged Garlic Extracts

The aged garlic extracts (AGE) have been produced by prolonged (about 6 months) soaking of chopped garlic into 20% alcohol.[61,62,262] During the aging, γ-glutamyl-*S*-allylcysteine, **2** and γ-glutamyl-*S*-1-propenylcysteine, **1** are hydrolyzed into *S*-allylcysteine, **8** and *S*-1-propenylcysteine, respectively.[10,12] Therefore, preparation of AGE is successful for releasing functional *S*-alkenylcysteines from their parent peptides without further production of allicin and sulfides. In addition, the aging affords novel compounds *S*-allylmercaptocysteine and cystine, which are absent in garlic, although the content of total sulfur compounds decreases by about 50%.[29] On the other hand, the longtime storage produces caramelized compounds, with which the extracts are usually colored green and may cause deterioration of products. The discoloration will be mentioned in the section of onion products (onion powder).

Garlic Oil

Garlic oil is prepared by steam distillation of chopped garlic, and used for production of some sauces, pizza, cakes, ham and sausages. The major compounds of garlic oil are diallyl disulfide, **10**, diallyl trisulfide, **11** and allyl methyl trisulfide, **12**. These compounds do not change during longtime storage, especially at 4°C. However, once these compounds are separated, they may rapidly degrade by releasing sulfur atoms even at −80°C.

Garlic Powder

There are many types of garlic powder products produced in different ways.[263] Surprisingly, most powdered products have no detectable amounts of alliin and sulfides.[43] Producing such powders, manufacturers would usually take the easiest way: e.g., chopping garlic into fine pieces, followed by dehydration and pulverization. Although the ingredient of powder should be controlled to meet the formulation for a desired product, such as spiced sausages and sauces, the loss of sulfur compounds should be minimized. The only sulfur compound detected from such powders was γ-glutamyl-*S*-allylcysteine, **2**. On the contrary, some powders known as "allicin potential" contain both alliin and alliinase, in addition to γ-glutamylcysteines.[29] Preparing such a powder, the alliinase activity would be inhibited under freezing conditions throughout the production procedures. If such powder with "allicin potential" is ingested as acid-resistant capsules to avoid acid inactivation in stomach, the allicin, which must be generated in the intestine, may work somehow, although the true activity *in vivo* remains to be clarified.

ONION PRODUCTS

Because storage and processing of onion has been described elsewhere,[264] some of its major products are described in this section.

Dehydrated Onion Pieces

Cured or dried onion has a 4 to 5% moisture to allow good storage and acceptable quality. The product is processed to make powder, granules, flakes or slices, then used for the formulation of sausage, meat products, many kinds of soups and sauces as well as dressing. Although there are several important parameters to evaluate these dehydrated products, high pungency is a prime requirement. Toasted onion product is manufactured from dehydrated onion to give it a desirable flavor. For producing dehydrated products with high quality, cultivars with high content of sulfur compounds, for example, white onion, have been preferentially used. Cultivars giving good toasted product are those possessing a high reducing sugar content.[5]

Onion Powder

Onion powder is prepared either from dehydrated onion pieces or from puree. A stronger flavored product is obtained by spray drying. The powder is a uniform product of which 95% passes a sieve of 0.25 mm aperture size. This is the finest among onion products including grits, flakes, slices and rings, and used for soups, relishes, sauces, and products that do not require onion appearance and texture.[5]

On the discoloration developing during the processing of onion, many publications have described its cause, since it is of particular interest for manufacturers producing high-quality onion products.[265,266] Lukes *et al.*[266] demonstrated that in garlic puree, the contents of S-1-propenylcysteine sulfoxide, **6** were significantly correlated to the development of a green pigment. They also demonstrated that storing the puree at 23 to 28°C could prevent color development for as long as 32 days. At lower temperatures, 12°C or 3°C, it colored green upon 18 days of storage, and dark to blue-green upon 32 days of storage. These evidences clearly suggest that quick conversion of S-1-propenylcysteine sulfoxide to its metabolites (sulfinyl compounds and sulfides) by the enzymatic action of alliinase is a positive factor in preventing discoloration of garlic. The same mechanism may be adapted to onion.

Onion Oil

As garlic oil, onion oil is obtained by distillation of minced onion. Most onion oil components are generated enzymatically from their precursors such as S-1-propenylcysteine sulfoxide, **6**, S-1-propylcysteine sulfoxide, and S-methylcysteine sulfoxide, **5**. Therefore, essential oil is an incorrect term to be used for oil from onion or garlic. The minced onion is allowed to stand at ambient temperature for a few hours prior to distillation to complete the enzymatic and successive chemical reactions. The onion oil can be obtained in 0.002 to 0.03% yields as a brown-amber liquid, and collected from the bottom of a vessel adapted under a steam condenser.

The chemical composition of onion oil is confined to a series of sulfides, namely dimethyl disulfide, **16**, dipropenyl disulfide, dipropyl disulfide, and dipropenyl trisulfide. According to Fenwick and Hanley,[5] onion oil possesses (on a weight basis) 800 to 1,000 times the odor strength of fresh onion, but its commercial value may be many thousand times that of the onion. Actually, the product smells too strong, but its availability expands to many food productions because of its solubility, lack of color, and strong aroma. From functional food viewpoint, onion oil is now comparable to garlic oil.

Onion Salt

Onion salt is a mixture of onion powder and salt for use at the table or in cooking. It can be used whenever salt is required. The product is prepared with an anticaking agent (calcium stearate, 1 to 2%), and hydrogenated vegetable oil.

Pickled Onion

Small onions such as a button onion or a silver skin onion may be preserved in vinegar as pickled products. A translucent product with a desired texture is preferable. Usually, the salted onion in 10% saline solution for 24 h is transferred to a bottle, and then spiced vinegar is added. This is best eaten after 2 weeks, and it may be used within 6 months. As mentioned in the section of garlic, pickling onion in the acid solution fully retains cysteine derivatives, which may exhibit hypoglycemic and hypolipidemic activities as shown in Table 16.5.

REFERENCES

1. E Hyams. *Plants in the Service of Man: 10,000 Years of Domestication.* JM Dent & Sons, London, (1971).

2. T Tsuneyoshi, AV Nosov, Y Kajimura, S Sumi, T Eto. RFLP analysis of the mtDNA in garlic cultivars. *Jpn J Breed* 42 (bessatsu 2):164–165, 1992.

3. HI Maass, M Klaas. Infraspecific differentiation of garlic (*Allium sativum* L) by isozyme and RAPD markers. *Theor Appl Genet* 91:89–97, 1995.

4. The FAOSTAT Data Provided by Food and Agricultural Organization of the United Nations (FAO), searched on February, 2003.

5. GR Fenwick, AB Hanley. The genus Allium, part 3. *CRC Crit Rev Food Sci Nutr* 23:1–73, 1985.

6. Tokyo Central Market of Vegetables and Fruits, 2003. Personal communication.

7. KK Nielson, AW Mahoney, LS Williams, VC Rogers. X-ray fluorescence measurements of Mg, P, S, Cl, K, Ca, Mn, Fe, Cu, and Zn in fruits, vegetables, and grain products. *J Food Comp Analysis* 4:39–51, 1991.

8. Standard Tables of Food Composition in Japan, Fifth Revised Edition. The Resources Council of the Science and Technology Agency of Japan, 2001.

9. LD Lawson. Bioactive organosulfur compounds of garlic and garlic products: role in reducing blood lipids. In: Human medicinal agents from plants. Vol. ACS Symp Ser 534, AD Kinghorn and MF Balandrin, Eds. *Am Chem Soc Books,* Washington, D.C., 1993, pp 306–330.

10. LD Lawson, ZYJ Wang, BG Hughes. Gamma-glutamyl-*S*-alkyl-cysteines in garlic and other Allium species: precursors of age-dependent trans-1-propenyl thiosulfinates. *J Nat Prod* 54:436–444, 1991.

11. LD Lawson, ZYJ Wang, BG Hughes. Identification and HPLC quantitation of the sulfides and dialk(en)yl thiosulfinates in commercial garlic products. *Planta Med* 57:363–370, 1991.

12. LD Lawson, ZYJ Wang. Changes in the organosulfur compounds released from garlic during aging in water, dilute ethanol, or dilute acetic acid. *J Toxicol* 14:214, 1995.

13. LN Ceci, OA Curzio, AB Pomilio. Gamma-glutamyl transpeptidase/gamma-glutamyl peptidase in sprouted *Allium sativum*. *Phytochemistry* 31:441–444, 1992.

14. LD Lawson, SG Wood, BG Hughes. HPLC analysis of allicin and other thiosulfinates in garlic clove homogenates. *Planta Med* 57:263–270, 1991.

15. SJ Austin, S Schwimmer. L-Gamma-glutamyl peptidase activity in sprouted onion. *Enzymologia* 40:273–285, 1971.

16. S Schwimmer, SJ Austin. Enhancement of pyruvic acid release and flavor in dehydrated Allium powders by gamma glutamyl transpeptidases. *J Food Sci* 36:1081–1085, 1971.

17. S Schwimmer, SJ Austin. Gamma glutamyl transpeptidase of sprouted onion. *J Food Sci* 36:807–811, 1971.

18. M Mazelis, L Crews. Purification of the alliin lyase of garlic, *Allium sativum* L. *Biochem J* 108:725–730, 1968.

19. GS Ellmore, RS Feldberg. Alliin lyase localization in bundle sheaths of the garlic clove (*Allium sativum*). *Am J Bot* 81:89–94, 1994.

20. EJM Van Damme, K Smeets, S Terrekens, F Van Leuven, WJ Peumans. Isolation and characterization of alliinase cDNA clones from garlic (*Allium sativum* L) and related species. *Eur J Biochem* 209:751–757, 1992.

21. T Suzuki, M Sugii, T Kakimoto. Metabolic incorporation of L-valine-[C14] into S-(2-carboxypropyl) glutathione and S-(2-carboxypropyl) cysteine in garlic. *Chem Pharm Bull* 10:328–331, 1962.

22. JE Lancaster, ML Shaw. Gamma-glutamyl peptides in the biosynthesis of S-alk(en)yl-L-cysteine sulphoxides (flavour precursors) in Allium. *Phytochemistry* 28:455–460, 1989.

23. JE Lancaster, ML Shaw. Metabolism of gamma-glutamyl peptides during development, storage, and sprouting of onion bulbs. *Phytochemistry* 30:2857–2859, 1991.

24. A Rabinkov, XZ Zhu, G Grafi, G Galili, D Mirelman. Alliin lyase (alliinase) from garlic (*Allium sativum*). Biochemical characterization and cDNA cloning. *Apple Biochem Biotechnol* 48:149–171, 1994.

25. H Jansen, B Mueller, K Knobloch. Characterization of an alliin lyase preparation from garlic (*Allium sativum*). *Planta Med* 55:434–439, 1989.

26. T Ariga, H Kumagai, M Yoshikawa, H Kawakami, T Seki, H Sakurai, I Hasegawa, T Etoh, H Sumiyoshi, T Tsuneyoshi, S Sumi, K Iwai. Garlic-like but odorless plant *Allium ampeloprasum* 'Mushuu-ninniku.' *J Japan Soc Hort Sci* 71:362–369, 2002.

27. EV Goryachenkova. Enzyme in garlic which forms allycine (allyinase), a protein with phosphopyridoxal. *Chem Abst* 47:4928 (Russian), 1952.

28. A Stoll, E Sheebeck. Chemical investigation of alliin, the specific principle of garlic. *Adv Enzymol* 11:377–400, 1951.

29. LD Lawson. The composition and chemistry of garlic cloves and processed garlic. In: HP Koch, LD Lawson, Eds. *Garlic* 2nd ed. Maryland: Williams & Wilkins, 1996, pp 37–109.

30. SJ Edwards, G Britton, HA Collin. The biosynthetic pathway of the S-alk(en)yl-L-cysteine sulphoxides (flavour precursors) in species of Allium. *Plant Cell Tissue Organ Culture* 38:181–188, 1994.

31. B Muller. Garlic (*Allium sativum*): quantitative analysis of the tracer substances alliin and allicin. *Planta Med* 56:589–590, 1990.

32. M Mutsch-Eckner, O Sticher, B Meier. Reversed-phase high-performance liquid chromatography of S-alk(en)yl-L-cysteine derivatives in *Allium sativum* including the determination of (+)-S-allyl-L-cysteine sulphoxide, gamma-L-glutamyl- S-allyl-L-cysteine and gamma-L-glutamyl-S-(trans-1-propenyl)-L-cysteine. *J Chromatogr* 625:183–190, 1992.

33. MA Smoczkiewicz, D Nischke, H Wieladek. Microdetermination of steroid and triterpene saponin glycosides in various plant materials. I. Allium species. *Microchim Acta* [Vienna] II:43–53, 1982.

34. Y Ueda, H Kawajiri, N Miyamura, R Miyajima. Content of some sulfur-containing components and free amino acids in various strains of garlic. *Nippon Shokuhin Kogyo Gakkaishi (J Jpn Soc Food Sci Technol)* 38:429–434, 1991.

35. MM Eckner, O Sticher, B Meier. Reverse-phase high-performance liquid chromatography of S-alk(en)yl-L-cysteine derivatives in *Allium sativum* including the determination of (+)-S-allyl-L-cysteine sulphoxide, and gamma-L-glutamyl- S-allyl-L-cysteine and gamma-L-glutamyl-S-(trans-1-propenyl)-L-cysteine. *J Chromatogr* 625:183–190, 1992.

36. The Garlic Research Group of Japan Food and Health Association, Tokyo, 2001. Personal Communication.

37. L Liu, Y-Y Yeh. Water soluble organosulfur compounds of garlic inhibit fatty acid and triglyceride syntheses in cultured rat hepatocytes. *Lipids* 36:395–400, 2001.

38. V Sreenivasamurthy, KR Sreekantiah, DS Johar. Studies on the stability of allicin and alliin present in garlic. *J Sci Ind Res* 20C:292–295, 1961.

39. I Laakso, TS Laakso, R Hiltunen, B Muller, H Jansen, K Knobloch. Volatile garlic odor components: gas phases and adsorbed exhaled air analyses by headspace gas chromatography-mass spectrometry. *Planta Med* 55:257–261, 1989.

40. BG Hughes, LD Lawson. Antimicrobial effects of *Allium sativum* L. (garlic), *Allium ampeloprasum* (elephant garlic), and *Allium cepa* L. (onion), garlic compounds and commercial garlic supplement products. *Phytother Res* 5:154–158, 1991.

41. E Block. The Chemistry of garlic and onions. *Sci Am* 252:114–119, 1985.

42. E Block, S Ahmad, JL Catalfamo, MK Jain, R Apitz-Castro. The chemistry of alkylthiosulfinate esters. IX, Antithrombotic organosulfur compounds from garlic: structural, mechanistic, and synthetic studies. *J Am Chem Soc* 108:7045–7055, 1986.

43. T Ariga, T Seki, H Kumagai. Unpublished data, 1997.

44. T Ariga, T Seki, H Kumagai. Measurement of sulfur compounds in garlic products in Japan. 1990, unpublished data.

45. E Block, S Ahmad, MK Jain, RW Crecery, R Apitz-Castro, MR Cruz. *(E,Z)*-Ajoene: a potent antithrombotic agent from garlic. *J Am Chem Soc* 106:8295–8296, 1984.

46. LD Lawson, BG Hughes. trans-1-Propenyl thiosulfinates: new compounds in garlic homogenates. *Planta Med* 56:589, 1990.

47. S Imai, N Tsuge, M Tomotake, Y Nagatome, H Sawada, T Magata, H Kumagai. An onion enzyme that makes the eyes water. *Nature* 419:685, 2002.

48. E Block. Oraganoselenium and organosulfur phytochemicals from genus Allium plants (onion, garlic): relevance for cancer protection. In: H Ohigashi, T Osawa, J Terao, S Watanabe, T Yoshikawa, Eds. *Food Factors for Cancer Prevention.* Tokyo: Springer-Verlag, 1997, pp 215–221.

49. M Kotrebai, M Birringer, JF Tyson, E Block, PC Uden. Selenium speciation in enriched and natural samples by HPLC-ICP-MS and HPLC-ESI-MS with perfluorinated carboxylic acid ion-pairing agents. *Analyst* 125:71–78, 2000.

50. C Ip, DJ Lisk. Enrichment of selenium in allium vegetables for cancer prevention. *Carcinogenesis* 15:1881–1885, 1994.

51. K el-Bayoumy, YH Chae, P Upadhyaya, C Ip. Chemoprevention of mammary cancer by diallyl selenide, a novel organoselenium compound. *Anticancer Res* 16:2911–2915, 1996.

52. C Ip. Lessons from basic research in selenium and cancer prevention. *J Nutr* 128:1845–1854, 1998.

53. C Ip, M Birringer, E Block, M Kotrebai, JF Tyson, PC Uden, DL Lisk. Chemical speciation influences comparative activity of selenium-enriched garlic and yeast in mammary cancer prevention. *J Agric Food Chem* 48:2062–2070, 2000.

54. Y Dong, D Lisk, E Block, C Ip. Characterization of the biological activity of gamma-glutamyl-*Se*-methylselenocysteine: a novel, naturally occurring anticancer agent from garlic. *Cancer Res* 61:2923–2928, 2001.

55. R Apitz-Castro, S Cabrera, MR Cruz, E Ledezma, MK Jain. Effect of garlic extract and of three pure components isolated from it on human platelet aggregation, arachidonate metabolism. Release reaction, and platelet ultra structure. *Thromb Res* 32:155–169, 1983.

56. MH Brondnitz, JV Pascale, L Van Derslice. Flavor components of garlic extract. *J Agric Food Chem* 19:273–275, 1971.

57. B Iberl, G Winkler, K Knobloch. Products of allicin transformation: ajoenes and dithiins, characterization and their determination by HPLC. *Planta Med* 56:202–211, 1990.

58. R Apitz-Castro, JJ Badimon, L Badimon. Effect of ajoene, the major antiplatelet compound from garlic, on platelet thrombus formation. *Thromb Res* 68:145–155, 1992.

59. E Block. The organosulfur chemistry of the genus Allium: implication for organic chemistry of sulfur. *Angew Chem Ed Engl* 31:1135–1178, 1992.

60. M Sugii, T Suzuki, S Nagasawa. Isolation of (-)*S*-propenyl-L-cysteine from garlic. *Chem Pharm Bull* 11:548–549, 1963.

61. S Nakagawa, S Kasuga, H Matsuura. Prevention of liver damage by aged garlic extract and its components in mice. *Phytother Res* 3:50–53, 1989.

62. Y Hirao, I Sumioka, S Nakagami, M Yamamoto, S Hatano, S Yoshida, T Fuwa, S Nakagawa. Activation of immunoresponder cells by the protein fraction from aged garlic extract. *Phytother Res* 1:161–164, 1987.

63. J Imai, N Ide, S Nagae, T Moriguchi, H Matsuura, Y Itakura. Antioxidant and radical scavenging effects of aged garlic extract and its constituents. *Planta Med* 60:417–420, 1994.

64. H Saito. *Ninnniku-no-Kagaku (Garlic Science)*. Tokyo: Asakura-shoten, 2000, pp 274, in Japanese.

65. Y Kodera, H Matsuura, S Yoshida, T Sumida, Y Itakura, T Fuwa, H Nishino. Allixin, a stress compound from garlic. *Chem Pharm Bull* 37:1656–1658, 1989.

66. Y Kodera, M Ayabe, K Ogasawara, K Ono. Allixin induction and accumulation by light irradiation. *Chem Pharm Bull* 49:1636–1637, 2001.

67. Y Kodera, M Ayabe, K Ogasawara, S Yoshida, N Hayashi, K Ono. Allixin accumulation with long-term storage of garlic. *Chem Pharm Bull* 50:405–407, 2002.

68. N Ide, AB Nelson, BH Lau. Aged garlic extract and its constituents inhibit Cu(2+)-induced oxidative modification of low density lipoprotein. *Planta Med* 63:263–264, 1997.

69. T Yamasaki, RW Teel, BH Lau. Effect of allixin, a phytoalexin produced by garlic, on mutagenesis, DNA binding and metabolism of aflatoxin B1. *Cancer Lett* 59:89–94, 1991.

70. T Moriguchi, H Matsuura, Y Itakura, H Katsuki, H Saito, N Nishiyama. Allixin, a phytoalexin produced by garlic, and its analogues as novel exogenous substances with neurotrophic activity. *Life Sci* 61:1413–1420, 1997.

71. C Borek. Antioxidant health effects of aged garlic extract. *J Nutr* 131:1010S–1015S, 2001.

72. Y Kodera, M Ichikawa, J Yoshida, N Kashimoto, N Uda, I Sumioka, N Ide, K Ono. Pharmacokinetic study of allixin, a phytoalexin produced by garlic. *Chem Pharm Bull* 50:354–363, 2002.

73. GB Mahady, H Matsuura, SL Pendland. Allixin, a phytoalexin from garlic, inhibits the growth of *Helicobacter pylori in vitro*. *Am J Gastroenterol* 96:3454–3455, 2001.

74. SV Amonkar, EL Reeves. Mosquito control with active principle of garlic, *Allium sativum*. *J Econ Entomol* 63:1172–1175, 1970.

75. HD Reuter, HP Koch, LD Lawson. Antibacterial effects. In: HP Koch, LD Lawson, ed. Garlic 2nd edition. Maryland: Williams & Wilkins, 1996, pp 164.

76. SV Amonkar, A Banerj. Isolation and characterization of larvicidal principle of garlic. *Science* 174:1343–1344, 1971.

77. KC George, J Eapen. *In vivo* studies on the effect of garlic oil in mice. *Toxicology* 1:337–344, 1973.

78. KC George, SV Amonkar, J Eapen. Effect of garlic oil on incorporation of amino acids into proteins of Culux pipiens quinquefasciatus Say larvae. *Chem Biol Interac* 6:169–175, 1973.

79. MS Venugopal, V Narayanan. Effects of Allitin on the green peach aphid (Myzus persicae Sulzer). *Int Pest Control* 23:130–131, 1981.

80. J Kabelic. Antomikrobielle Eingenschaften des Knoblauchs. *Pharmazie* 25:266–270, 1970.

81. L Stjernberg, J Berglund. Garlic as an insect repellent. *JAMA* 284:837, 2000.

82. CJ Bierman. Insect repellant, Belgian Patent BE 896,552 (C1.AOIN), August 16, 1983, *NL Appl* 82/2,260, June 4, 1982.

83. PM Guarrera. Traditional anthelminthics, antiparasitic and repellent uses of plants in Central Italy. *J Ethnopharmacol* 68:183–192, 1999.

84. Y Ishikawa, T Ikeshoji, Y Matsumoto, M Tsutsumi, Y Mitsui. Field trapping of the onion and seed-corn files with baits of fresh and aged onion pulp. *Appl Ent Zool* 16:490–493, 1981.

85. Y Matsumoto, AJ Thornsteinsnn. Olfactory response of the larvae of the onion maggot *Hylemyia antiqua* Meigen (Diptera: Anthomyiidae) to organic sulfur compounds. *Appl Ent Zool* 3:107–111, 1968.

86. Y Matsumoto, AJ Thornsteinsnn. Effect of organic sulfur compounds on oviposition in onion maggot *Hylemyia antiqua* Meigen (Diptera: Anthomyiidae). *Appl Ent Zool* 3:5–12, 1968.

87. Y Ishikawa, Y Matsumoto, M Tsutsumi, Y Mitsui. Mixture of 2-phenylethanol and n-valeric acid, a new attractant for the onion and seed-corn flies, *Hylemya antiqua* and *H. plature* (Diptera: Anthomyiidae). *Appl Ent Zool* 19:448–455, 1984.

88. T Ikeshoji. *S*-propenylcysteine sulfoxide in exudates of onion roots and its possible decompartmentalization in root cells by bacteria into attractant of the onion maggot, *Hylemya antiqua* (Diptera: Anthomyiidae). *Appl Ent Zool* 19:159–169, 1984.

89. KS Al-Delaimy, SH Ali. Antibacterial action of vegetable extracts on the growth of pathogenic bacteria. *J Sci Food Agric* 21:110–112, 1970.

90. KN Shashikanth, SC Basappa, V Sreenivasamurthy. A comparative study of raw garlic extract and tetracycline on caecal microflora and serum proteins of albino rats. *Folia Microbiol* 29:348–352, 1984.

91. GP Sivam, JW Lampe, SR Swanzy, JD Potter. *Helicobacter pylori*—*in vitro* susceptibility to garlic (*Allium sativum*) extract. *Nutr Cancer* 27:118–121, 1997.

92. GP Sivam. Protection against *Helicobacter pylori* and other bacterial infections by garlic. *J Nutr* 131:1106S–1108S, 2001.

93. HD Reuter, HP Koch, LD Lawson. Therapeutic effects and application of garlic and its preparations. In: HP Koch, LD Lawson, Eds. *Garlic* 2nd ed. Maryland: Williams & Wilkins, 1996, pp 135–211.

94. J Kabelic, N Hejtmankova-Uhrova. The antifungal and antibacterial effects of certain drugs and other substances. *Vet Med* [Plague] 13:295–303, 1968.

95. A Rabinkov, T Miron, L Konstantinovski, M Wilchek, D Mirelman, L Weiner. The mode of action of allicin: trapping of radicals and interaction with thiol-containing proteins. *Biochim Biophys Acta* 1379:233–244, 1998.

96. N Didry, M Pinkas, L Dubreuil. Antimicrobial activity of naphtoquinones and Allium extracts combined with antibiotics. *Pharm Acta Helv* 67:148–151, 1992.

97. RS Feldberg, SC Chang, AN Kotik, M Nadler, Z Neuwirth, DC Sundstrom, NH Tompson. *In vitro* mechanism of inhibition of bacterial cell growth by allicin. *Antimicrob Agents Chemother* 32:1763–1768, 1988.

98. LM Yacobson. The isolation of bacteriophage from vegetables and fruits. *Zh Mikrobiol Epidemiol Immunobiol* 17:584–585, 1936.

99. JH Kim. Antibacterial action of onion (*Allium cepa* L.) extracts against oral pathogenic bacteria. *J Nihon Univ Sch Dent* 39:136–141, 1997.

100. EI Elnima, SA Ahmed, AG Mekkawi, JS Mossa. The antimicrobial activity of garlic and onion extracts. *Pharmazie* 38:747–748, 1983.

101. MA Adetumbi, BHS Lau. Inhibition of *in vitro* germination and spherulation of coccidioides immitis by *Allium sativum*. *Curr Microbiol* 13:73–76, 1986.

102. LE Davis, JK Shen, Y Cai. Antifungal activity in human cerebrospinal fluid and plasma after intravenous administration of *Allium sativum*. *Antimicrob Agents Chemother* 34:651–653, 1990.

103. LE Davis, JK Shen, RE Royer. *In vitro* synergism of concentrated *Allium sativum* extract and amphotericin B against *Cryptococcus neoformans. Planta Med* 60:546–549, 1994.

104. S Yoshida, S Kasuga, N Hayashi, T Ushiroguchi, H Matsuura, S Nakagawa. Antifungal activity of ajoene derived from garlic. *Appl Environ Microbiol* 53:615–617, 1987.

105. PP Tadi, RW Teel, BHS Lau. Anticandidal and anticarcinogenic potentials of garlic. *Int Clin Nutr Rev* 10:423–429, 1990.

106. NBK Murthy, SV Amonkar. Effect of a natural insecticide from garlic (*Allium sativum* L.) and its synthetic form (diallyl-disulphide) on plant pathogenic fungi. *Indian J Exp Biol* 12:208–209, 1974.

107. M Szymona. Effect of phytoncides of *Allium sativum* on the growth and respiration of some pathogenic fungi. *Acta Microbiol Pol* 1:5–23, 1952.

108. MA Ghannoum. Studies on the anticandidal mode of action of *Allium sativum* (garlic). *J Gen Microbiol* 134:2917–2924, 1989.

109. MA Ghannoum. Inhibition of candida adhesion to buccal epithelial cells by an aqueous extract of *Allium sativum* (garlic). *J Appl Bacteriol* 68:163–169, 1990.

110. MC Yin, SM Tsao. Inhibitory effect of seven Allium plants upon three Aspergillus species. *J Food Microbiol* 49:49–56, 1999.

111. S Tassin, WF Broekaert, D Marion, DP Acland, M Ptak, F Vovelle, P Sodano. Solution structure of Ace-AMP1, a potent antimicrobial protein extracted from onion seeds. Structural analogies with plant nonspecific lipid transfer proteins. *Biochemistry* 37:3623–3637, 1998.

112. U Takahama, S Hirota. Deglucosidation of quercetin glucosides to the aglycone and formation of antifungal agents by peroxidase-dependent oxidation of quercetin on browning of onion scales. *Plant Cell Physiol* 41:1021–1029, 2000.

113. D Mirelman. Inhibition of growth of Entamoeba histolytica by allicin, the active principle of garlic extract (*Allium sativum*). *J Infect Dis* 156:243–244, 1987.

114. JA Urbina, E Marchan, K Lazardi, G Visbal, R Apitz-Castro, F Gil, T Aguirre, MM Piras, R Piras. Inhibition of phosphatidylcholine biosynthesis and cell proliferation in Trypanosoma cruzi by ajoene, an antiplatelet compound isolated from garlic. *Biochem Pharmacol* 45:2381–2387, 1993.

115. S Ankri, T Miron, A Rabinkov, M Wilchek, D Mirelman. Allicin from garlic strongly inhibits cysteine proteinases and cytopathic effects of *Entamoeba histolytica*. *Antimicrob Agents Chemother* 4i:2286–2288, 1997.

116. ZR Lun, C Burri, M Menzinger, R Kaminsky. Antiparasitic activity of diallyl trisulfide (Dasuansu) on human and animal pathogenic protozoa (Trypanosoma sp. *Entamoeba histolytica, Giardia lamblia*) in vitro. *Ann Soc Belg Med Tropic* 74:51–59, 1994.

117. D Mirellman, S Veron. Knoblauch fur die Amobiasis-Therapie (Referat). *Neue Aerztliche*, No. 59, p. 10, 1987.

118. ND Weber, DO Andersen, JA North, BK Murray, LD Lawson, BG Hughes. *In vitro* virucidal effects of *Allium sativum* (garlic) extract and compounds. *Planta Med* 58:417–423, 1992.

119. R Walder, Z Kalvatchev, R Apitz-Castro. Selective *in vitro* protection of SIVagm-induced cytolysis by ajoene, [(E)-(Z)-4,5,9-trithiadodeca-1,6,11-triene-9 oxide. *Biomed Pharmacother* 52:229–235, 1998.

120. Y Tsai, LL Cole, LE Davis, SJ Lockwood, V Simmons, GC Wild. Antiviral properties of garlic: *in vitro* effects on influenza B, herpes simplex, and coxackie viruses. *Planta Med* 51:460–461, 1985.

121. Y Cai. Anticryptococcal and antiviral properties of garlic. *Cardiol Pract* 9:11, 1991.

122. P Josling. Preventing the common cold with a garlic supplement: a double-blind, placebo-controlled survey. *Adv Ther* 18:189–193, 2001.

123. S Shoji, K Furuishi, R Yanase, T Miyazaka, M Kino. Allyl compounds selectively killed human immunodeficiency virus (type-1)-infected cells. *Biochem Biophys Res Commun* 194:610–621, 1993.

124. K Nagai. Experimental studies on the preventive effect of garlic extract against infections with influenza virus. *J Jpn Assoc Infect Dis* 47:321–325, 1973. In Japanese.

125. K Nagai. Experimental studies on preventive effect of garlic extract against infections with influenza and Japanese encephalitis viruses in mice. *J Jpn Assoc Infect Dis* 47:111–115, 1973. In Japanese.

126. R Walder, Z Kalvatchev, D Garzaro, M Barrios, R Apitz-Castro. *In vitro* suppression of HIV-1 replication by ajoene [(E)-(Z)-4,5,9-trithiadodeca-1,6,11-triene-9 oxide. *Biomed Pharmacother* 51:397–403, 1997

127. AV Tatarintsev, PV Vrzhets, DE Yershuv, AA Shchegolev, AS Turgiyev, EV Karamov, GV Kornilayeva, TV Makarova, NA Fedorov, SD Varfolomeyev. The ajoene blockade of integrin-dependent processes in an HIV-infected cell system. *Vestn Ross Akad Med Nauk* 11–12:6–10, 1992. (Russian)

128. EJ Van Damme, K Smeets, I Engelborghs, H Aelbers, J Balzarini, A Pusztai, F van Leuven, IJ Goldstein, W Peumans. Cloning and characterization of the lectin cDNA clones from onion, shallot and leek. *J Plant Mol Biol* 23:365–376, 1993.

129. AS Kurdziel, N Wilkinson, S Langton, N Cook. Survival of poliovirus on soft fruit and salad vegetables. *J Food Prot* 64:706–709, 2001.

130. CM Dentinger, WA Bower, OV Nainan, SM Cotter, G Myers, LM Dubusky, S Fowler, ED Salehi, BP Bell. An outbreak of hepatitis A associated with green onions. *J Infect Dis* 183:1273–1276, 2001.

131. M Tada, Y Hiroe, S Kiyohara, S Suzuki. Nematicidal and antimicrobial constituents from *Allium grayi* Regel and *Allium fistulosum* L. var caespitosum. *Agric Biol Chem* 52:2383–2385, 1988.

132. N Pena, A Auro, H Sumano. A comparative trial of garlic, its extract and ammonium-potassium tartrate as anthelmintics in carp. *J Ethnopharmacol* 24:199–203, 1988.

133. M Araki, Y Yokota, M Kuga, S Chin, F Fujikawa, K Nakajima, H Fujii, A Tokuoka, Y Hirota. Anthelminthics. *Yakugaku Zasshi* (J Pharm Soc Japan) 72:979–982, 1952.

134. AJ Nok, PC Onyenekwe. *Allium sativum*-induced death of African trypanosomes. *Prasitol Res* 82:634–637, 1996.

135. CT Soh. The effects of natural food-preservative substances on the development and survival of intestinal helminth eggs and larvae. II. Action an *Ancylostoma duodenale* larvae. *Am J Trop Med Hyg* 9:8–10, 1960.

136. GJ Bastidas. Effect of ingested garlic on *Necator americanus* and *Ancylostoma caninum*. *Am J Trop Med Hyg* 18:920–923, 1969.

137. E Ledezma, A Jorquera, H Bendezu, J Vivas, G Prez. Antiproliferative and leishmanicidal effect of ajoene on various Leishmania species: ultrastructural study. *Parasitol Res* 88:748–753, 2002.

138. T Ghazanfan, ZM Hassan, M Ebtekar, A Ahmadiani, G Naderi, A Azar. Garlic induces a shift in cytokine pattern in *Leishmania major*-infected BALB/c mice. *Scand Immunol* 52:491, 2000.

139. T Ariga, S Oshiba, T Tamada. Platelet aggregation inhibitor in garlic. *Lancet* 1:150–151, 1981.

140. H Nishimura, T Ariga. Vinyldithiins in garlic and Japanese domestic Allium (*A. victorialis*). In: NT Huang, T Osawa, CT Ho, RT Rosen, Eds. *Food Phytochemicals for Cancer Prevention I*, ACS Symposium Series 546. Washington, D.C., 1994, pp 128–143.

141. T Ariga, A Takeda, S Teramoto, T Seki. Inhibition site of methyl allyl trisulfide: A volatile oil component of garlic, in the platelet arachidonic acid cascade. In: H Ohigashi, T Osawa, J Terao, S Watanabe, T Yoshikawa, Eds. *Food Factors for Cancer Prevention*. Tokyo: Springer-Verlag, 1997, pp 231–234.

142. T Ariga, K Tsuji, T Seki, T Moritomo, J Yamamoto. Antithrombotic and antineoplastic effects of phyto-organosulfur compounds. *BioFactors* 13:251–255, 2000.

143. T Ariga, T Seki, K Ando, S Teramoto, H Nishimura. Antiplatelet principle found in the essential oil of garlic (*Allium sativum*) and its inhibition mechanism. In: T Yano, R Matsuno, K Nakamura, Eds. *Developments in Food Engineering*. Tokyo: Blackie Academic & Professional, 1994, pp 1056–1058.

144. R Apitz-Castro, J Escalante, R Vargas, MK Jain. Ajoene, the antiplatelet principle of garlic, synergistically potentiates the antiaggregatory action of prostacyclin, forskolin, indomethacin, and dipyridamole on human platelets. *Thromb Res* 42:303–311, 1986.

145. R Apitz-Castro, E Ledezma, J Escalante, A Jorquera, FM Pinate, J Mreno-Rea, G Carrillo, O Leal. Reversible prevention of platelet activation by (*E,Z*)-4,5,9 trithiadodeca-1,6,11-triene 9-oxide (ajoene) in dog under extracorporeal circulation. *Arzneim Forsch* 38:901–904, 1988.

146. R Apitz-Castro, E Ledezma, J Escalante, MK Jain. The molecular basis of the antiplatelet action of ajoene: direct interaction with the fibrinogen receptor. *Biochem Biophys Res Commun* 141:145–150, 1986.

147. H Nishimura, CH Wijaya, A Satoh, T Ariga. Platelet aggregation inhibitory activity of vinyldithiins and their derivatives from Japanese domestic *Allium* (*A. victorialis*). In: JR Whitaker, NF Haard, cf. Shoemaker, RP Sing, Eds. *Food Health in the Pacific Rim.* Trumbull, CT: Food & Nutrition Press, 1999, pp 114–124.

148. M Thomson, T Mustafa, M Ali. Thromboxane-B2 levels in serum of rabbits receiving a single intravenous dose of aqueous extract of garlic and onion. *Prostaglandins Leukot Essent Fatty Acids* 63:217–221, 2000.

149. A Bordia, HC Bansal. Essential oil of garlic in prevention of atherosclerosis. *Lancet* 2:149–1492, 1973.

150. A Bordia, HC Bansal, SK Arora, SV Singh. Effect of essential oils of garlic and onion on alimentary hyperlipemia. *Atherosclerosis.* 21:15–19, 1975.

151. A Bordia, KD Sharma, YK Parmar, SK Verma. Protective effect of garlic oil on the changes produced by 3 weeks of fatty diet on serum cholesterol, serum triglycerides, fibrinolytic activity, and platelet adhesiveness in man. *Indian Heart J* 34:86–88, 1982.

152. M Steiner, AH Kahn, D Holbert, RI Lin. A double-blind crossover study in moderatory hypercholesterolemic men that compared the effect of aged garlic extract and placebo administration on blood lipids. *Am J Clin Nutr* 64:866–870, 1966.

153. H Kiesewetter. Effect of garlic on platelet aggregation in patients with increased risk of juvenile ischaemic attack. *Eur J Clin Pharmacol* 45:333–336, 1993.

154. M Steiner, W Li. Aged garlic extract, a modulator of cardiovascular risk factors: a dose-finding study on the effects of AGE on platelet functions. *J Nutr* 131:980S–984S, 2001.

155. AN Makheja, JY Vanderhoek, JM Bailey. Inhibition of platelet aggregation and thromboxane synthesis by onion and garlic. *Lancet* 1:781, 1979.

156. AN Makheja, JY Vanderhoek, RW Bryant, JM Bailey. Altered arachidonic acid metabolism in platelets inhibited by onion or garlic extracts. *Adv Prostaglandin Thromboxane Res* 6:309–312, 1980.

157. M Ali, M Thomson, MA Alnaqeeb, JM Al-Hassan, SH Khater, SA Gomes. Antithrombotic activity of garlic: its inhibition of the synthesis of thromboxane-B2 during infusion of arachidonic acid and collagen in rabbits. *Prostaglandins Leukot Essent Fatty Acids* 41:95–99, 1990.

158. M Ali, T Bordia, T Mustafa. Effect of raw versus boiled aqueous extract of garlic and onion on platelet aggregation. *Prostaglandins Leukot Essent Fatty Acids* 60:43–47, 1999.

159. AN Makheja, JM Bailey. Antiplatelet constituents of garlic and onion. *Agents Actions* 29:360–363, 1990.

160. T Bordia, N Mohammed, M Thomson, M Ali. An evaluation of garlic and onion as antithrombotic agents. *Prostaglandins Leukot Essent Fatty Acids* 54:183–186, 1996.

161. AN Makheja, JY Vanderhoek, JM Bailey. Effects of onion (*Allium cepa*) extract on platelet aggregation and thromboxane synthesis. *Prostaglandins Med* 2:413–424, 1979.

162. K Janssen, RP Mensink, FJ Cox, JL Harryvan, R Hovenier, PC Hollman, MB Katan. Effects of the flavonoids quercetin and apigenin on hemostasis in healthy volunteers: results from an *in vitro* and a dietary supplement study. *Am J Clin Nutr* 67:255–262, 1998.

163. WH Briggs, JD Folts, HE Osman, IL Goldman. Administration of raw onion inhibits platelet-mediated thrombosis in dogs. *J Nutr* 131:2619–2622, 2001.

164. JH Chen, HI Chen, SJ Tsai, CJ Jen. Chronic consumption of raw but not boiled Welsh onion juice inhibits rat platelet function. *J Nutr* 130:34–37, 2000.

165. KK Nagda, SK Ganeriwal, KC Nagda, AM Diwan. Effect of onion and garlic on blood coagulation and fibrinolysis *in vitro*. *Indian J Physiol Pharmacol* 27:141–145, 1983.

166. A Bordia, SK Verma, KC Srivastava. Effect of garlic (*Allium sativum*) on blood lipids, blood sugar, fibrinogen and fibrinolytic activity in patients with coronary artery disease. *Prostaglandins Leukot Essent Fatty Acids* 58:257–263, 1998.

167. A Bordia, HK Joshi, YK Sanadhya, N Bhu. Effect of essential oil of garlic on serum fibrinolytic activity in patients with coronary artery disease. *Atherosclerosis* 28:155–159, 1977.

168. A Bordia, HK Joshi. Garlic on fibrinolytic activity in cases of acute myocardial infraction part 2. *J Assoc Physicians India* 26:324–326, 1978.

169. C Legnani, M Frascaro, G Guazzaloca, S Lusovici, G Cesarano, S Coccheri. Effects of a dried garlic preparation on fibrinolysis and platelet aggregation in healthy subjects. *Arzneimittelforschung* 43:119–122, 1993.

170. IV Andrianova, VG Ionova, EG Demina, AA Shabalina, IA Karabasova, LI Liutova, TE Povorinskaia, AN Orekhov. Use of allikor for the normalization of fibrinolysis and hemostasis in patients with chronic cerebrovascular diseases. *Klin Med* (Mosk) 79:55–58. 2001. (Russian)

171. JV Gadkari, VD Joshi. Effect of ingestion of raw garlic on serum cholesterol level, clotting time and fibrinolytic activity in normal subjects. *J Postgrad Med* 37:128–131, 1991.

172. H Kiesewetter, F Jung, C Morowietz, G Pindur, M Heiden, E Wenzel. Effect of garlic on blood fluidity and fibrinolytic activity: a randomised, placebo-controlled, double-blind study. *Br J Clin Pract* 44:24–29, 1990.

173. JH Chen, HI Chen, JS Wang, SJ Tsai, CJ Jen. Effects of Welsh onion extracts on human platelet function *in vitro*. *Life Sci* 66:1571–1579, 2000.

174. A Bordia, SK Verma, AK Vyas, BL Khabya, AS Rathore, N Bhu, HK Bedi. Effect of essential oil of onion and garlic on experimental atherosclerosis in rabbits. *Atherosclerosis* 26:379–386, 1977.

175. GS Sainani, DB Desai, MN Natu, KM Katrodia, VP Valame, PG Sainani. Onion, garlic, and experimental atherosclerosis. *Jpn Heart J* 20:351–357, 1979.

176. S Wolf, M Reim. Effect of garlic on conjunctival vessels: a randomised, placebo-controlled, double-blind trial. *Br J Clin Pract* 44:36–39, 1990.

177. G Siegel, J Embden, K Wenzel, J Mironneau, G Stock. Potassium channel activation in vascular smooth muscle. In: GB Frank, Ed. *Excitation-Contraction Coupling in Skeletal, Cardiac, and Smooth Muscle*. New York: Plenum Press, 1992, pp 53–72.

178. I Das, NS Khan, SR Sootanna. Potent activation of nitric oxide synthase by garlic: a basis for its therapeutic applications. *Curr Med Res Opinion* 13:257–263, 1995.

179. S Kim-Park, DD Ku. Garlic elicits a nitric oxide-dependent relaxation and inhibits hypoxic pulmonary vasoconstriction in rats. *Clin Exp Pharm Physiol* 27:780–786, 2000.

180. SA Mirhadi, S Singh. Effect of garlic extract on *in vitro* uptake of Ca(2+) and HPO4(2–) by matrix of sheep aorta. *Indian J Exp Biol* 25:22–23, 1987.

181. J Wohlrab, D Wohlrab, WCH Marsch. Acute effect of a dried ethanol–water extract of a garlic on the microhaemovascular system of the skin. *Drug Res* 50: 606–612, 2000.

182. A Bordia, SK Verma. Effect of garlic feeding on regression of experimental atherosclerosis in rabbits. *Artery* 7:428–437, 1980.

183. K Slowing, P Ganado, M Sanz, E Ruiz, T Tejerina. Study of garlic extracts and fractions on cholesterol plasma levels and vascular reactivity in cholesterol-fed rats. *J Nutr* 131:994S–999S, 2001.

184. HG Preuss, D Clouatre, A Mohamadi, ST Jarrell. Wild garlic has a greater effect than regular garlic on blood pressure and blood chemistries of rats. *Int Urol Nephrol* 32:525–530, 2001.

185. MS Chi, ET Koh, TJ Stewart. Effects of garlic on lipid metabolism in rats fed cholesterol or lard. *J Nutr* 112:241–248, 1982.

186. AA Qureshi, N Abuirmeileh, ZZ Din, CE Elson, WC Burger. Inhibition of cholesterol and fatty acid biosynthesis in liver enzymes and chicken hepatocytes by polar fractions of garlic. *Lipids* 18:343–348, 1983.

187. AA Qureshi, ZZ Din, N Abuirmeileh, WC Burger, CE Elson. Suppression of avian hepatic lipid metabolism by solvent extracts of garlic: impact on serum lipids. *J Nutr* 113:1746–1755, 1983.

188. RK Agarwal, HA Dewar, DJ Newell, B Das. Controlled trial of the effect of cycloalliin on the fibrinolytic activity of venous blood. *Atherosclerosis* 27:347–351, 1977.

189. Y Oi, T Kawada, C Shishido, K Wada, Y Kominato, S Nishimura, T Ariga, K Iwai. Allyl-containing sulfides in garlic increase uncoupling protein content in brown adipose tissue, and noradrenaline and adrenaline secretion in rats. *J Nutr* 129:336–342, 1999.

190. A Eldayam, D Mirelman, E Peleg, M Wilchek, T Miron, A Rabinkov, S Sadetzki. The effects of allicin and enalapril in fructose-induced hyperinsulinemic hyperlipidemic hypertensive rats. *Am J Hypertens* 14:377–381, 2001.

191. Y Oi, T Kawada, K Kitamura, F Oyama, M Nitta, Y Kominato, S Nishimura, K Iwai. Garlic supplementation enhances norepinephrine secretion, growth of brown adipose tissue, and triglyceride catabolism in rats. *J Nutr Biochem* 6:250–255, 1995.

192. Y Oi, M Okamoto, M Nitta, Y Kominato, S Nishimura, T Ariga, K Iwai. Alliin and volatile sulfur-containing compounds in garlic enhance the thermogenesis by increasing norepinephrine secretion in rats. *J Nutr Biochem* 9:60–66, 1998.

193. PS Babu, K Srinivasan. Influence of dietary capsaicin and onion on the metabolic abnormalities associated with streptozotocin-induced diabetes mellitus. *Mol Cell Biochem* 175:49–57, 1997.

194. S Kasuga, M Ushijima, N Morihara, Y Itakura, Y Nakata. Effect of aged garlic extract (AGE) on hyperglycemia induced by immobilization stress in mice. *Nippon Yakurigaku Zasshi* 114:191–197, 1999. In Japanese.

195. KT Augusti, CG Sheela. Antiperoxide effect of *S*-allyl cysteine sulfoxide, an insulin secretagogue, in diabetic rats. *Experientia* 52:115–120, 1996.

196. MLW Chang, MA Johnson. Effect of garlic on carbohydrate metabolism and lipid synthesis in rats. *J Nutr* 110:931–936, 1980.

197. PT Mathew, KT Augusti. Studies on the effect of allicin (diallyl disulphide-oxide) on alloxan diabetes. Part I. Hypoglycaemic action and enhancement of serum insulin effect and glycogen synthesis. *Indian J Biochem Biophys* 10:209–212, 1973.

198. T Baluchnejadmojarad, M Roghani, H Homayounfar, M Hosseini. Beneficial effect of aqueous garlic extract on the vascular reactivity of streptozotocin-diabetic rats. *J Ethno-pharmacol* 85:139–144, 2003.

199. HD Brahmachari, KT Augusti. Effects on orally effective hypoglycaemic agents from plants on alloxan diabetes. *J Pharm Pharmacol* 14:617, 1962.

200. K Kumari, BC Mathew, KT Augusti. Antidiabetic and hypolip-idemic effects of S-methyl cysteine sulfoxide isolated from *Allium cepa* Linn. *Indian J Biochem Biophys* 32:49–54, 1995.

201. L Svendsen, SIS Rattan, BFC Clark. Testing garlic for possible antiaging effects on long-term growth characteristics, morphology and macromolecule at synthesis of human fibroblasts in culture. *J Ethnopharmacol* 43:125–133, 1994.

202. K Rahman. Garlic and aging: new insights into an old remedy. *Ageing Res Rev* 2:39–56, 2003.

203. A Helen, CR Rajasree, K Krishnakumar, KT Augusti, PL Vijayammal. Antioxidant role of oils isolated from garlic (*Allium sativum* Linn) and onion (*Allium cepa* Linn) on nicotine-induced lipid peroxidation. *Vet Human Toxicol* 41:316–319, 1999.

204. SA Dilon, GM Lowe, D Billington, K Rahman. Dietary supple-mentation with aged garlic extract reduced plasma and urine concentrations of 8-iso-prostaglandin F_{2a} in smoking and non-smoking men and women. *J Nutr* 132:168–171, 2002.

205. KM Kim, SB Chun, MS Koo, WJ Choi, TW Kim, YG Kwon, HT Chung, TR Billiar, YM Kim. Differential regulation of NO availability from macrophages and endothelial cells by the garlic component S-allyl cysteine. *Free Radical Biol Med* 30:747–756, 2001.

206. T Yamasaki, L Li, BHS Lau. Garlic compounds protect vascular endothelial cells from hydrogen peroxide-induced oxidant injury. *Phytother Res* 8:408–412, 1994.

207. Z Geng, Y Rong, BH Lau. *S*-allyl cysteine inhibits activation of nuclear factor kappa B in human T cells. *Free Radic Biol Med* 23:345–350, 1997.

208. PN Kourounakis, EA Rekka. Effect on active oxygen species of alliin and *Allium sativum* (garlic) powder. *Res Commun Chem Pathol Pharmacol* 74:249–252, 1991.

209. C Egen-Schwind, R Eckard, FH Kemper. Metabolism of garlic constituents in the isolated perfused rat liver. *Planta Med* 58:301–305, 1992.

210. MF Melzig, E Krause, S Franke. Inhibition of adenosine deaminase activity of aortic endothelial cells by extracts of garlic (*Allium sativum* L.). *Die Pharmazie* 50:359–361, 1995.

211. MB Aqel, MN Gharaibah, AS Salhab. Direct relaxant effects of garlic juice on smooth and cardiac muscles. *J Ethnopharmacol* 33:13–19, 1991.

212. Y Ozturk, S Aydin, M Kosar, KHC Baser. Endothelium-dependent and independent effects of garlic on rat aorta. *J Ethnopharmacol* 44:109–116, 1994.

213. KK Al-Qattan, I Dhan, MA Alnaqeeb, M Ali. Thromboxane-B2 prostaglandin-E2 and hypertension in the rat 2-kidney 1-clip model: a possible mechanism of the garlic-induced hypertension. *Prostaglandins Leukot Essent Fatty Acids* 64:5–10, 2001.

214. A Mohamadi, ST Jarrell, SJ Shi, NS Andrawis, A Myers, D Clouatre, HG Preuss. Effects of wild versus cultivated garlic on blood pressure and other parameters in hypertensive rats. *Heart Dis* 2:3–9, 2000.

215. CV Pantoja, NT Martin, BC Norris, CM Contreras. Purification and bioassays of a diuretic and natriuretic fraction from garlic (*Allium sativum*). *J Ethnopharmacol* 70:35–40, 2000.

216. HG Jeong, YW Lee. Protective effects of diallyl sulfide on N-nitrosodimethylamine-induced immunosuppression in mice. *Cancer Lett* 134:73–79, 1998.

217. M Colic, D Vucevic, V Kilibarda, N Radicevic, M Savic. Modulatory effects of garlic extracts on proliferation of T-lymphocytes *in vitro* stimulated with concanavalin A. *Phytomedicine* 9:117-124, 2002.

218. M Colic, M Savic. Garlic extracts stimulate proliferation of rat lymphocytes *in vitro* by increasing IL-2 and IL-4 production. *Immunopharmacol Immunotoxicol* 22:163–181, 2000.

219. T Ghazanfari, ZM Hassan, M Ebrahimi. Immunomodulatory activity of a protein isolated from garlic extract on delayed type hypersensitivity. *Intl Immunopharmacol* 2:1541–1549, 2002.

220. D Guyonnet, C Belloir, M Suschetet, MH Siess, AM Le Bon. Mechanisms of protection against aflatoxin B1 genotoxicity in rats treated by organosulfur compounds from garlic. *Carcinogenesis* 23:1335–1341, 2002.

221. CG Sheela, K Kumud, KT Augusti. Anti-diabetic effects of onion and garlic sulfoxide amino acids in rats. *Planta Med* 61:356–357, 1995.

222. W Dorsch, E Schneider, T Bayer, W Breu, H Wagner. Anti-inflammatory effects of onions: inhibition of chemotaxis of human polymorphonuclear leukocytes by thiosulfinates and cepaenes. *Int Arch Allergy Appl Immunol* 92:39–42, 1990.

223. H Wagner, W Dorsch, T Bayer, W Breu, F Willer. Antiasthmatic effects of onions: inhibition of 5-lipoxygenase and cyclooxygenase *in vitro* by thiosulfinates and "Cepaenes." *Prostaglandins Leukot Essent Fatty Acids* 39:59–62, 1990.

224. Y Oi, M Okamoto, M Nitta, Y Kominato, S Nishimura, T Ariga, K Iwai. Alliin and volatile sulfur-containing compounds in garlic enhance the thermogenesis by increasing norepinephrine secretion in rats. *J Nutr Biochem* 9:60–66, 1998.

225. A Stroll, E Seebeck. Chemical investigation of alliin, the specific principle of garlic. *Adv Enzymol* 11:377–400, 1951.

226. Y Oi, M Imafuku, C Shishido, Y Kominato, S Nishimura, K Iwai. Garlic supplementation increases testicular testosterone and decreases plasma corticosterone in rats fed a high-protein diet. *J Nutr* 131:2150–2156, 2001.

227. JT Pinto, C Qiao, J Xing, BP Suffoletto, KB Schubert, RS Rivlin, RF Huryk, DJ Bacich, WD Heston. Alteration of prostate biomarker expression and testosterone utilization in human LNCaP prostatic carcinoma cells by garlic-derived *S*-allylmercaptocysteine. *Prostate* 45:304–314, 2000.

228. M Fijiwara, H Watanabe. Allithiamine, a newly found compound of vitamin B1. *Proc Imp Acad* [Tokyo] 28:156–158, 1952.

229. K Ikeda. Studies on the nutritional value of Allium plants. LII. Nutritional significance of the ingestion of garlic (a supplement). *Bitamin* [Kyoto] 40:263–267, 1969.

230. M Fujiwara. Allithiamine and its properties. *J Nutr Sci Vitaminol* 22:57–62, 1976.

231. CJ Cavallito, JH Bailey. Allicin, the antibacterial principle of *Allium sativum*. I. Isolation, physical properties, and antibacterial action. *J Am Chem Soc* 66:1950–1951, 1944.

232. EC Delaha, VF Gragusi. Inhibition of mycobacteria by garlic extract (*Allium sativum*). *Antimicrob Agents Chemother* 27:485–486, 1985.

233. MC Lin, EJ Wang, C Lee, KT Chin, D Liu, JF Chiu, HF Kung. Garlic inhibits microsomal triglyceride transfer protein gene expression in human liver and intestinal cell lines and in rat intestine. *J Nutr* 132:1165–1168, 2002.

234. CG Sheela, KT Augusti. Antidiabetic effects of S-allyl cysteine sulphoxide isolated from garlic *Allium sativum* Linn. *Indian J Exp Biol* 30:523–526, 1992.

235. M Deseautels, J Himms-Hagen. Roles of noradrenaline and protein synthesis in the cold-induced increase in purine nucleotide binding by rat brown adipose tissue mitochondria. *Can J Biochem* 57:968–976, 1979.

236. BE Levin, J Triscari, C Sallivan. Altered sympathetic activity during development of diet-induced obesity in rats. *Am J Physiol* 244:347–355, 1983.

237. P De Feo. Hormonal regulation of human protein metabolism. *Eur J Endocrinol* 135:7–18, 1996.

238. JA Milner. Garlic: its anticarcinogenic and antitumorigenic properties. *Nutr Rev* 54:S82–86, 1996.

239. J Hu, CL Vecchia, E Negri, L Chatenoud, C Bosetti, X Jia, R Liu, G Huang, D Bi, C Wang. Diet and brain cancer in adults: a case-control study in northeast China. *Int J Cancer* 81:20–23, 1999.

240. TJ Key, PB Silcocks, GK Davey, PN Appleby, DT Bishop. A case-control study of diet and prostate cancer. *Br J Cancer* 76:678–687, 1997.

241. B Challier, JM Perarnau, JF Viel. Garlic, onion and cereal fibre as protective factors for breast cancer: a French case-control study. *Eur J Epidemiol* 14:737–747, 1998.

242. F Levi, C Pasche, CL Vecchia, F Lucchini, S Franceschi. Food groups and colorectal cancer risk. *Br J Cancer.* 79:1283–1287, 1999.

243. JS Witte, MP Longnecker, CL Bird, ER Lee, HD Frankl, RW Haile. Relation of vegetable, fruit, and grain consumption to colorectal adenomatous polyps. *Am J Epidemiol* 144:1015–1025, 1996.

244. AT Fleischauer, L Arab. Garlic and cancer: a critical review of the epidemiologic literature. *J Nutr* 131:1032S-1040S, 2001.

245. AML Bon, MH Siess. Organosulfur compounds from allium and the chemoprevention of cancer. *Drug Metabol Drug Interact* 17:51–79, 2000.

246. KA Park, S Kweon, H Choi. Anticarcinogenic effect and modification of cytochrome P450 2E1 by dietary garlic powder in diethylnitrosamine-initiated rat hepatocarcinogenesis. *J Biochem Mol Biol* 35:615–622, 2002.

247. D Haber-Mignard, M Suschetet, R Berges, P Astorg, MH Siess. Inhibition of aflatoxin B1- and N-nitrosodiethylamine-induced liver preneoplastic foci in rats fed naturally occurring allyl sulfides. *Nutr Cancer* 25:61–70, 1996.

248. K Sakamoto, LD Lawson, JA Milner. Allyl sulfides from garlic suppress the *in vitro* proliferation of human A549 lung tumor cells. *Nutr Cancer* 29:152–156, 1997.

249. D Guyonnet, C Belloir, M Suschetet, MH Siess, AML Bon. Mechanisms of protection against aflatoxin B(1) genotoxicity in rats treated by organosulfur compounds from garlic. *Carcinogenesis* 23:1335–1341, 2002.

250. CC Wu, LY Sheen, HW Chen, WW Kuo, SJ Tsai, CK Lii. Differential effects of garlic oil and its three major organosulfur components on the hepatic detoxification system in rats. *J Agric Food Chem* 50:378–383, 2002.

251. S Balasenthil, CR Ramachandran, S Nagini. S-allylcysteine, a garlic constituent, inhibits 7,12-dimethylbenz[a]anthracene-induced hamster buccal pouch carcinogenesis. *Nutr Cancer* 40:165–172, 2001.

252. S Balasenthil, KS Rao, S Nagini. Apoptosis induction by *S*-allyl-cysteine, a garlic constituent, during 7,12-dimethyl-benz[a]anthracene-induced hamster buccal pouch carcinogenesis. *Cell Biochem Funct* 20:263–268, 2002.

253. SG Sundaram, JA Milner. Diallyl disulfide inhibits the proliferation of human tumor cells in culture. *Biochim Biophys Acta* 1315:15–20, 1996.

254. CS Yang, SK Chhabra, JY Hong, TJ Smith. Mechanisms of inhibition of chemical toxicity and carcinogenesis by diallyl sulfide (DAS) and related compounds from garlic. *J Nutr* 131:1041S–1045S, 2001.

255. EK Park, KB Kwon, KI Park, BH Park, EC Jhee. Role of Ca(2+) in diallyl disulfide-induced apoptotic cell death of HCT-15 cells. *Exp Mol Med* 34:250–257, 2002.

256. K Hirsch, M Danilenko, J Giat, T Miron, A Rabinkov, M Wilchek, D Mirelman, J Levy, Y Sharoni. Effect of purified allicin, the major ingredient of freshly crushed garlic, on cancer cell proliferation. *Nutr Cancer* 38:245–254, 2000.

257. LM Knowles, JA Milner. Depressed p34cdc2 kinase activity and G2/M phase arrest induced by diallyl disulfide in HCT-15 cells. *Nutr Cancer* 30:169–174, 1998.

258. Y Li, L Yang, JT Cui, WN Li, RF Guo, YY Lu. Construction of cDNA representational difference analysis based on two cDNA libraries and identification of garlic-inducible expression genes in human gastric cancer cells. *World J Gastroenterol* (2):2, 2002.

259. M Li, JR Ciu, Y Ye, JM Min, LH Zhang, K Wang, M Gares, J Cros, M Wright, J Leung-Tack. Antitumor activity of Z-ajoene, a natural compound purified from garlic: antimitotic and microtubule-interaction properties. *Carcinogenesis* 23:573–579, 2002.

260. LM Knowles, JA Milner. Diallyl disulfide inhibits p34(cdc2) kinase activity through changes in complex formation and phosphorylation. *Carcinogenesis* 21:1129–1134, 2000.

261. R Pentz, CP Siegers. Garlic preparations: methods for quantitative and qualitative assessment of their ingredients. In: HP Koch, LD Lawson, Eds. *Garlic* 2nd ed. Maryland: Williams & Wilkins, 1996, pp 109–134.

262. H Amagase, JA Milner. Impact of various sources of garlic and their constituents on 7,12-dimethylbenz[a]anthracene binding to mammary cell DNA. *Carcinogenesis* 14:1627–1631, 1993.

263. HB Heath. *Source Book of Flavors*, AVI Publishing Co., Westport, CT, 1981, p 99.

264. GR Fenwick, AB Hanley. The genus Allium, part 1. *CRC Crit Rev Food Sci Nutr* 22:199–271, 1985.

265. JS Pruthi, LJ Singh, G Lal. Nonenzymatic browning in garlic powder during storage. *Food Sci* 9:243, 1960.

266. TM Lukes. Factors governing the greening of garlic puree. *J Food Sci* 36:1577–1588, 1986.

17

Functional Foods from Date Fruits

JIWAN S. SIDHU

Department of Family Sciences, College for
Women, Kuwait University, Kuwait

SUAD N. AL-HOOTI

Biotechnology Department, Kuwait Institute
for Scientific Research, Kuwait

INTRODUCTION

For centuries, the date fruit (*Phoenix dactylifera* L.) has occupied an important place in the diets of people in the Arab world. Interestingly, dates can be consumed or utilized for human consumption in every stage of fruit development. Usually, date fruit has been classified into four main maturity stages, i.e., *kimri, khalal, rutab,* and *tamer.*[1] At the *kimri* stage, the fruit is young, has a hard texture, is green in color, and can be used for the preparation of pickles and chutney. Depending on the cultivar, the *khalal (or bisr)* stage fruit develops a typical yellow, purplish-pink, red or yellow-scarlet color; retains a firm texture; has attained maximum size and weight; and can be used for jam, butter, dates-in-syrup, or consumed raw as fresh fruit.[2] At the *rutab* stage, half of the fruit becomes soft, darkens in color, becomes less astringent but sweeter, and can be used for jam, butter, date bars, date paste, or eaten raw. During the final or ripe stage, *tamer*, the whole fruit develops a dark brown color, a soft texture with a wrinkled appearance, the maximum total solids, the highest sweetness, and the lowest astringency. A majority of the date produce enters world trade as *tamer* fruit and is consumed at this stage.[3] *Tamer* stage fruits can also be sun dried for prolonged shelf life and storage.

The value and importance of the date palm tree and its fruits do not need any further explanation. It is mentioned and honored in the Holy Quran, and recommended by the Prophet (Peace Be upon Him). Date fruit is a rich source of carbohydrates (mainly sugars), certain vitamins, minerals, and dietary fiber.[4] In the Arab world, date fruit is cherished for its flavor and nutritive value. This plant is well suited to grow in arid regions where there are hot and dry climates with limited rainfall. Date trees are usually propagated vegetatively by offshoot, as the trees grown from seeds show high genetic heterozygosity that results in not true-to-type male and female seedlings.[5] However, these vegetative techniques are slow and have long generation times. Efforts are now being made to propagate date palm trees through newly emerging tissue culture techniques.[6-8]

Considering the importance of date palm trees and date fruits to this region, the history of date cultivation, varieties, production, marketing, chemical composition, preharvest treatments, postharvest handling of fresh fruits, nutritive value, health benefits, date-fruit-based functional foods, and future research needs are covered in this chapter.

HISTORY OF DATE CULTIVATION

The date palm (*Phoenix dactylifera* L.) belongs to the Arecaceae (or Palmae) family. Date palm has been cultivated for a long time in the semiarid and desert areas of Middle East, Pakistan, and India; in California, U.S.; in the Canary Islands; and in the northern African countries for fuel, shade, ornamentation, fiber, food, and as building material.[9] This family not only includes date palm, but also other kinds of palm trees such as oil palms, coconut palms and Washington palms. No one has documented where exactly the date palms originated, but it is suggested that they first originated in Babel, Iraq, or in Dareen, or Hofuf, Saudi Arabia, or Harqan, an island in Bahrain. From these locations, it spread to other places.[10] Date palms were introduced to Andalus by the Arabs during the seventh and eighth centuries. They were later spread throughout the deserts of the Middle East and North Africa by the Bedouin tribes of the Arab countries. It is believed that after the victory over India by Alexander the Great around 327 B.C., date palm trees were introduced to India. Date seeds were introduced to the arid areas in the U.S., namely the states of New Mexico, California, and Arizona, in 1769.[11] At present, the Middle East and North African countries are the major date-fruit producing countries in the world. However, the U.S. also produces sizable quantities of date fruits in North America, as does Spain in the European Union.

Numerous cultivars of date trees exist, but their exact number is not known. Hussain and El-Zeid[12] reported the existence of 400 cultivars, and Nixon[13] indicated the probability of 250 named varieties, but only a few dozen cultivars of economic and commercial importance. Four imported cultivars are grown commercially in the U.S., with three-fourths of the cultivated area under *Deglet Noor* cultivation and 10% under the three cultivars, *Zahidi, Khadrawy,* and *Halawy.* Knight[14] has discussed most of the important cultivars being grown commercially the world over. The *Yahidi* cultivar is hardy and drought resistant, and exhibits vigorous growth; being high yielding, it is commercially grown in Iraq. *Hallawi* (meaning sweet) bears a light-colored fruit. It is one of the leading cultivars grown in Iraq and exported from Basra. Another cultivar *Sayer* (meaning widespread), has very sweet fruit and is a hardy plant able to tolerate adverse climatic conditions. It has the highest production and is important in commercial trade. Cultivars like *Khudari, Nabbut-Al-Seif, Sullaj, Sukai, Maktumi, Sultana, Shagra, Nabtat Ali, Shbibi, Barni, Rabiaa, Safawi, Shalabi,* and *Sifri* are the major cultivars of commercial importance in Saudi Arabia.[15] *Zaghloul, Duwiki,* and *Hayani* in Egypt; *Kabkabe* and *Khustawai* in Iran; and *Barhee, Maktoom, Shalabi, Sukkari,* and *Khustawai* in Iraq are some of the other important cultivars being grown commercially.[16]

DATE PRODUCTION AND MARKETING

The total annual world production of date fruits was reported to be 5,190,000 tons during the 2000 crop year, and is expected to increase further due to the efforts being made by various countries to encourage and popularize date palm cultivation through modern techniques of plant breeding.[17] Iran is the leading producer of date fruit in the world with a total production of 930,000 tons in 2000 (Table 17.1). Egypt, Saudi Arabia, Pakistan, Algeria, Iraq, and United Arab Emirates are the other major producers of date fruits. It is expected that the recent advances in tissue culture techniques and the

TABLE 17.1 Major Date-Fruit Producing
Countries of the World (1000 tons)

Country	Production	Country	Production
World	5,190	Iraq	400
Algeria	430	Israel	10
Egypt	890	Kuwait	9
Libya	133	Qatar	17
Sudan	176	Oman	135
Tunisia	103	Spain	7
USA	20	Saudi Arabia	712
Bahrain	17	Morocco	74
China	125	UAE	318
Iran	930	Yemen	29
Mauritania	22	Pakistan	580
Chad	18	Turkey	9

(From FAO. *FAO Production Year Book: FAO Statistics Series,* Vol. 54 Rome: FAO, 2000, pp 168–169. With permission.)

improved agricultural practices will boost date production in this part of the world. Most of the fresh dates at the *khalal* and *rutab* stages of maturity are packaged in wooden or plastic crates (2 to 3 kg/crate), and offered for sale in the fruit and vegetable markets in these countries. Dates at the *tamer* stage of maturity are packaged in tin cans, plastic bags, or straw baskets, either in pressed or unpressed form and are bought by the consumers from the date markets usually located in a central place in the town. The major portion of dates is consumed at the *tamer* stage; however, a significant amount of *khalal* date fruit is also consumed.[18] Lack of a proper layout of markets, improper handling of produce, an unclean environment, inferior packaging and wrapping materials, and unhygienic storage methods are some of the pertinent problems in date marketing. Asia is the largest importer of date fruits in the world. In Asia, China and India are the major importers of date fruits. In Europe, France is the leading importer followed by Germany. The U.S., followed by Canada, is the main importer of date fruits in North America.

PHYSICO-CHEMICAL CHARACTERISTICS

Like most other fruits, dates are a rich source of mainly carbohydrates (fructose and glucose), but smaller quantities of vitamins, minerals, and other minor constituents are also present. Table 17.2 to Table 17.4 show the chemical composition of five major cultivars grown in the UAE at various stages of maturity.[4] In a majority of the cultivars, the sucrose content increased rapidly as the date fruit matured, reaching a predominant level at the *khalal* stage (42.58%), and then decreased to a nondetectable level at the *tamer* stage of maturity. When the date fruit matured from the *kimri* to the *tamer* stage, the fructose content increased approximately threefold, this accounts for the characteristic sweet taste of *tamer* date fruits. A majority of the Saudi Arabian cultivars[19] having higher concentrations of reducing sugars at the *tamer* stage are of the soft-type date fruits. In addition to the major constituent carbohydrates, date fruit also contains significant amounts of protein, crude fiber, pectin, tannins, minerals, and vitamins

TABLE 17.2   Sugar Contents of Five Date-Fruit Cultivars
Grown in UAE at Various Stages of Development (%, dry basis)

Cultivar	Maturity stage	Fructose	Glucose	Sucrose	Total sugars
Bushibal	Kimri	11.80	18.55	2.64	32.99
	Khalal	7.70	14.11	42.58	64.42
	Rutab	23.88	38.57	5.76	68.21
	Tamer	39.91	39.48	ND	79.39
Gash Gafaar	Kimri	12.54	19.23	5.40	37.17
	Khalal	9.52	16.67	35.53	61.72
	Rutab	24.98	40.99	1.16	67.13
	Tamer	39.61	39.68	ND	79.29
Gash Habash	Kimri	11.89	18.85	4.91	35.65
	Khalal	21.52	35.99	2.57	60.08
	Rutab	29.12	31.50	14.51	75.13
	Tamer	39.50	38.48	ND	77.97
Lulu	Kimri	14.55	19.57	2.16	36.28
	Khalal	23.78	37.24	6.24	67.26
	Rutab	34.58	39.44	0.91	74.93
	Tamer	39.55	38.55	ND	78.50
Shahla	Kimri	13.96	21.66	2.58	38.20
	Khalal	11.80	24.63	16.95	53.38
	Rutab	—	—	—	—
	Tamer	38.58	40.04	ND	78.62

Note: Estimations were missed. ND = Not detected
(From SN Al-Hooti, JS Sidhu, H Qabazard. Physico-chemical characteristics of
five date-fruit cultivars grown in the United Arab Emirates. *Plant Foods
Human Nutrition* 50:101–113, 1997. With permission.)

(Table 17.3 to Table 17.5). The date fruit is quite low in crude
fat, which usually ranges from 0.5% at the *kimri* stage to 0.1%
at the *tamer* stage of maturity.[4] Evidently, date fruit, like most
other fruits, cannot be considered an important source of fats
or fatty acids in our diet. The crude fiber content of date fruits
at the *kimri* stage is substantially higher (6.2 to 13.2%) than
that at the *tamer* stage (2.1 to 3.0%) of maturity.[16] The total
dietary fiber content (comprised of pectin, hemicellulose, cellulose, gums, mucilages, resistant starch, and lignin) depends
on the stage of maturity of the date fruits.[20] The total fiber
decreases as the date fruits lose their firm texture and become
soft at the *tamer* stage. Research during the last 3 decades

TABLE 17.3   Proximate Composition of Date Fruits of Different Cultivars Grown in the UAE at Various Stages of Maturity (% dry basis)

Cultivar	Maturity stage	Protein, (Nx6.25)	Fat	Ash	Pectin (as Ca pectate)	Tannins	Crude fiber
Bushibal	Kimri	6.4	0.5	3.8	3.9	2.3	12.3
	Khalal	3.4	0.2	2.8	8.4	1.3	4.8
	Rutab	2.4	0.2	2.0	9.1	0.9	3.2
	Tamer	2.3	0.1	2.0	1.4	0.4	2.5
Gash	Kimri	6.4	0.5	3.7	3.8	2.3	10.6
Gafaar	Khalal	5.3	0.3	2.8	14.3	1.4	5.0
	Rutab	2.5	0.1	2.3	10.4	0.9	3.3
	Tamer	2.4	0.1	2.0	1.3	0.4	2.7
Gash	Kimri	5.7	0.2	3.5	8.5	2.5	10.5
Habash	Khalal	2.9	0.2	3.0	9.4	1.6	5.2
	Rutab	2.4	0.1	2.3	8.5	1.2	4.3
	Tamer	2.0	0.1	1.8	1.9	0.4	2.5
Lulu	Kimri	6.3	0.7	3.9	6.2	1.8	10.9
	Khalal	3.0	0.4	2.3	7.0	1.2	4.2
	Rutab	2.6	0.3	2.0	7.7	0.8	3.4
	Tamer	2.5	0.1	1.6	1.8	0.4	2.7
Shahla	Kimri	5.5	0.2	3.8	9.4	2.5	9.6
	Khalal	3.4	0.4	2.9	10.2	1.3	6.2
	Rutab	—	—	—	—	—	—
	Tamer	2.1	0.2	1.9	1.4	0.4	2.9

Note: estimations were missed.
(From SN Al-Hooti, JS Sidhu, H Qabazard. Physico-chemical characteristics of five date-fruit cultivars grown in the United Arab Emirates. *Plant Foods Human Nutrition* 50:101–113, 1997. With permission.)

has shown that an adequate intake of dietary fiber (20 to 25 g daily) lowers the incidence of colon cancer, heart diseases, diabetes, and other diseases. Obviously, consumption of 100 g of date fruit (six to seven dates) would provide us with about 50% of the recommended daily amount of dietary fiber. The total dietary fiber of dates decreases from 13.7% at the *kimri* stage to 3.6% at the *tamer* stage of maturity.[21,22] The decreases in the pectin, hemicellulose, cellulose, and lignin contents during date-fruit ripening range from 1.6 to 0.5, from 5.3 to 1.3, from 3.4 to 1.4 and from 3.5 to 0.3%, respectively. This

TABLE 17.4   Mineral Composition of Date Fruits of Various Cultivars Grown in the UAE at Different Stages of Maturity (mg/100g edible part, dry basis)

Maturity stage	Cultivar	Macroelements					Microelements			
		Ca	P	Na	K	Mg	Fe	Zn	Cu	Mn
Kimri	Bushibal	192.7	124.7	5.2	1668.6	121.6	1.03	1.80	0.79	1.20
	Gash Gafaar	160.3	110.9	5.1	1520.8	131.5	1.03	1.55	0.63	1.39
	Gash Habash	153.3	89.5	5.2	1522.9	117.3	1.03	1.34	0.75	0.75
	Lulu	188.9	123.9	5.1	1529.8	134.1	1.02	1.63	0.71	0.90
	Shahla	145.1	104.7	5.1	1612.3	129.8	1.02	1.35	0.68	0.92
Khalal	Bushibal	84.0	57.4	8.2	1106.4	75.8	1.43	0.77	0.53	0.65
	Gash Gafaar	103.5	69.7	5.1	1137.8	71.8	1.03	1.01	0.54	0.62
	Gash Habash	104.8	67.8	5.0	938.4	107.9	0.81	0.62	0.53	0.59
	Lulu	84.5	72.7	5.1	927.2	83.1	0.71	0.64	0.51	0.51
	Shahla	95.3	84.0	9.4	1188.2	86.0	1.74	1.02	0.65	0.69
Rutab	Bushibal	85.1	67.9	3.2	744.8	95.2	0.73	0.69	0.37	0.77
	Gash Gafaar	94.9	81.8	5.3	895.5	88.6	1.26	0.88	0.47	0.57
	Gash Habash	78.6	62.5	3.3	872.3	67.4	0.94	0.49	0.35	0.33
	Lulu	58.1	59.1	1.8	910.5	60.6	1.15	0.48	0.45	0.28
	Shahla[a]	—	—	—	—	—	—	—	—	—
Tamer	Bushibal	48.8	48.8	5.1	558.9	49.2	1.83	0.46	0.27	0.44
	Gash Gafaar	56.5	60.1	3.1	592.6	48.2	1.54	0.53	0.35	0.37
	Gash Habash	50.6	55.5	1.5	573.7	50.9	1.61	0.67	0.30	0.32
	Lulu	43.2	64.7	2.2	402.8	43.6	1.38	0.29	0.28	0.31
	Shahla	48.2	68.2	2.3	652.1	53.3	2.17	0.49	0.29	0.31

[a] Rutab fruits of Shahla cultivar were not received for analysis.

(From SN Al-Hooti, JS Sidhu, H Qabazard. Physico-chemical characteristics of five date-fruit cultivars grown in the United Arab Emirates. *Plant Foods Human Nutrition* 50:101–113, 1997. With permission.)

TABLE **17.5**   Vitamin Contents of Dates and Some Other Dried
Fruits (per 100g edible parts)

Fruits	Vitamin A (IU)	Thiamin (mg)	Riboflavin (mg)	Niacin (mg)	Ascorbic acid (mg)
Dates	50	0.09	0.10	2.2	0
Raisins	44	0.01	0.19	1.1	3.2
Apricots	12,669	0.04	0.15	3.6	9.5
Prunes	1,987	0.08	0.16	1.9	3.3

(From SE Gebhardt, R Cutrufelli, RH Mathews. Composition of Foods. Agriculture Handbook No. 8-9, U.S. Department of Agriculture, Human Nutrition Information Service. Washington, D.C.: USDA, 1982, pp 97. With permission.)

shows that maximum benefit can be obtained by consuming
fresh dates (i.e., those at the *kimri, khalal,* and *rutab* stages)
rather than the fully mature *tamer* fruits. The presence of
resistant starch in the fresh dates will provide an additional
advantage as it may be prebiotic-promoting conducive conditions for the growth of desirable bifidobacteria in the lower
gastrointestinal tract.[23] Obviously, date fruits are an equally
good source of important minerals (e.g., potassium and iron).

POSTHARVEST HANDLING OF DATE FRUITS

Date fruit at the *tamer* stage of maturity has low moisture
and high sugar contents, and thus, a good storage life. Because
of its good shelf life, it is, therefore, referred to as a self-
preserving fruit. Generally, most date fruits are consumed at
the *tamer* stage of maturity, but substantial amounts are also
consumed in the perishable, immature *khalal* and *rutab*
stages. Not only are the date fruits at these immature stages
rich in dietary fiber, ascorbic acid, and β-carotene, they are
traditionally quite popular in date-growing regions.[24] However, unfortunately, date fruits at these immature stages of
maturity not only have higher moisture contents and are
susceptible to microbial spoilage, but also are available for a
very limited period during the season.

Potentially pathogenic bacteria such as *Escherichia coli,* *Staphylococcus aureus,* and *Bacillus cereus* have been identified in date fruits together with lactic acid bacteria, yeasts, *Aspergillus flavus,* and *A. parasiticus.* Fresh date fruits (at the *khalal* and *rutab* stages) are the most heavily contaminated of all samples, probably due to their higher moisture contents.[25] One approach to achieving this objective of shelf life extension is the use of antifungal agents such as potassium sorbate. The frozen storage of date fruits at the *rutab* stage is another possibility.[26]

ALLERGIC REACTIONS FROM DATE FRUITS

The occurrence of food allergies and other food sensitivities are on the increase and are affecting a significant number of people in the world today.[27] Allergic reactions to date fruits and the pollen from date palm trees in some individuals have been reported over the last few years. The antigenicity and allergenic properties of date palm pollen are more of a cultivar-specific phenomenon than a species-specific one, but are governed by the number, quantities, or both of the major allergen epitopes possessed by that variety or cultivar.[28] Generally, skin sensitivities to the pollen extracts from date trees are reported to be lower than those to the pecan in Israel.[29] Moreover, a definite relationship between the abundance of these trees in a region and incidences of skin responders to their pollen extracts exists, as sensitivity is frequent in areas rich in date and pecan plantations. Pollen pollution decreases considerably with the distance from these trees and is usually quite low at distances of approximately 100 m away from the source of this contamination. Since only male date palm and pecan trees produce pollen, their plantation near residential areas should be avoided.

   The results on the characterization of antigens and allergens of date fruits and pollen indicate that date palm fruit is a potent allergen. The sera from date fruit-allergic as well as pollen-allergic patients recognize common fruit-specific epitopes. Considerable heterogeneity in patient responses to date fruit and pollen extracts from different cultivars does exist.[30] Sera from skin prick tests (SPT), persons evaluated by ELISA (enzyme-linked immunosorbent assay) and RAST (radio allergosorbent test), and anti-IgE immunoblot experiments indicated that about 13% of the patients were SPT-positive for at least two of the extracts from the eight cultivars. As date fruits are allergenic, standardized extracts are required to diagnose and treat these allergic reactions in patients.

Date palm polypeptides share cross-reactive IgE and IgE epitopes with a number of foods implicated in the oral allergy syndrome. Most such cross-reactivities in other allergens are attributed to the presence of carbohydrate chains and profilins.[31] Date pollen can cross-react with antigens from Artemisia, birch, cultivated rye (*Secale cereale*), Timothy grass (*Phleum pratense*), Bermuda grass (*Cynodon dactylon*), and Sydney golden wattle (*Acacia longifolia*). IgE binding of the endoglycosidase-digested date-fruit extracts to an atopic serum pool is restricted to only very low molecular weight bands of 6.5 to 8 kDa. The date palm polypeptides bind IgE through glycosyl residues and share cross-reactivities with many other antigens from trees and grasses.

NUTRITIVE VALUE AND HEALTH BENEFITS

Date fruits have been so important in Arabian diets that they are called the "Fruit of Life." At the time of the breaking of the days fast in the evening during the Holy Month of Ramadan, a few date fruits are eaten. People buy their dates either once or twice annually, or in smaller batches throughout the year. If dates are available in the house, naturally the pattern of consumption is affected and one is definitely inclined to consume more of the fruit. Now a number of companies are producing cleaned, washed, and pressed *tamer* dates in attractive packages. Small amounts are also pitted, stuffed with nuts (almonds), and/or mixed with anise or fennel seeds.[32]

The nutritive value of dates is mainly measured in terms of their carbohydrate contents. *Tamer* dates are a very rich source of readily available carbohydrates in our diet, as dates (without pits) contain more than 80% total mono sugars.[33] In most date cultivars, the sucrose present at the *khalal* and *rutab* stages of maturity is almost completely converted to glucose and fructose through the action of the invertase enzyme present in date fruits. *Tamer* dates are also a rich source of total dietary fiber (about 12.97 to 13.32%, on a dry basis), both water soluble and water insoluble fractions, both having proven health benefits. Date waste dietary fiber, when fed to white albino rats for 8 weeks, significantly lowered the total cholesterol, triglycerides, and phospholipids in the livers of the rats.[34] The total serum lipids and low-density lipoprotein cholesterol decreased by 32 to 48%, while serum triglycerides and total cholesterol decreased by 23 to 35%, respectively. A large amount of dates eaten as *khalal* and *rutab*, would also supply soluble dietary fiber such as pectins (4 to 5% dry basis). Therefore, dates make significant nutritional contributions to the dietary fiber intake of human population.

In Saudi Arabia, an average person consumes about 100 g of dates every day.[32] Thus, the amount of dates consumed per person would meet 13% of their daily requirement for total energy, more than 11% of their daily requirement for iron, about 7% of their daily requirement for ascorbic acid, and 6% of their daily requirement for proteins. Some people "date lovers," consume even higher amounts of dates at the *tamer, khalal,* and *rutab* stages; thus they may obtain much higher amounts of these valuable nutrients in their diets. The traditional Moslem habit of consuming dates with milk, especially during the fasting month of Ramadan, has strong scientific logic. Besides providing energy, the vitamin C present in dates enhances the absorption of iron, thus making dates and milk a good nutritional combination in terms of iron, vitamin C, and proteins. A number of dietary constituents are known to influence the absorption of iron present in food products. Ascorbic acid, being an enhancer, would also improve the absorption of the iron present in the date fruit.[35]

Dates, at all stages of maturity, are low in fat (about 1%), but quite rich in minerals, certain B-complex vitamins, and polyphenolic compounds. Among the minerals, dates are especially rich in potassium, but at the same time, are low in sodium, thus proving to be an excellent food for persons suffering from hypertension. With respect to other minerals, dates are considered good sources of iron, copper, sulfur, and manganese; and fair sources of calcium, chloride, and magnesium. Dates also contain moderate amounts of thiamin, riboflavin, and folic acid. Considering the B-complex vitamins in relation to the calories they contain, dates are fairly good sources of these nutrients. In terms of recommended levels of thiamin, riboflavin, and nicotinic acid of 0.4, 0.6, and 6.6 mg/100 calories, the date fruits provide 0.32, 0.35, and 8.0 mg/1000 calories, respectively.[36]

Dates are also a rich source of many phytochemicals such as phenolic compounds. The astringency of *kimri*-stage date fruit is due to the presence of phenolic substances, generally known as tannins. Though many types of tannins are found in date fruits, two main groups that are thought to be mainly important in producing astringency sensations are phenolic acids and condensed tannins. The phenolic acids are comprised of cinnamic acid derivatives originating from the amino acid, phenylalanine, while condensed tannins or proanthocyanidins, are polyphenolics. Polyphenolics are produced through a complex phenylpropanoid pathway starting with the conversion of phenylalanine to cinnamic acid.[37] *Kimri*-stage date fruit contains a maximum amount of tannins, which decreases as the fruit's maturity progresses towards the *tamer* stage.[4] These phenolics are known to be strong antioxidants and prevent oxidative damage to DNA, lipids, and proteins, which may play a role in chronic diseases such as cancer and cardiovascular disease.[38] As regards nutritional contributions, *tamer* date fruits are not only good sources of sugars and dietary fiber, but they can also supply reasonable amounts of potassium, phosphorus, calcium, iron, thiamin, riboflavin, and nicotinic acid in our diet. However, consumption of *khalal* and *rutab* fruits must be encouraged, as these are better sources of some of the nutrients, especially dietary fiber, ascorbic acid, β-carotene and many phytochemicals.

ANTIOXIDANT, ANTIMUTAGENIC, AND IMMUNOSTIMULANT PROPERTIES

For the first time, the antioxidant and antimutagenic properties of date fruits have been reported.[39] Under *in vitro* conditions, dose-dependent inhibition of superoxide and hydroxyl radicals by the aqueous extract of date fruit was observed. Using riboflavin photoreduction method, an extract equivalent to 0.08 mg/ml of the date fruit was required to scavenge 50% of the superperoxide radicals. For quenching of the same amount of hydroxyl radical, an extract equivalent to 2.2 mg/ml of date fruit was needed. For the total quenching of superoxide and hydroxyl radicals, extract equivalent to 1.5 and 4.0 mg/ml of date fruit was required, respectively. Date-fruit extract also inhibited the lipid peroxidation and protein oxidation in a dose-dependent manner. To achieve 50% inhibition of lipid peroxides in a ferrous/ascorbate system, an extract of 1.9 mg/ml of date fruit was required. In time-course study of lipid peroxides, the complete inhibition of TBARS formation was obtained using an extract of 2.0 mg/ml of date fruit. In the high-ferrous/ascorbate-induction system, an extract of 2.3 mg/ml of date fruit inhibited carbonyl formation by 50% when measured by DNPH reaction method. An extract concentration of 4.0 mg/ml of date fruit completely inhibited lipid peroxide and protein carbonyl formation. In a dose-dependent manner, the date-fruit extract inhibited benzo(a)pyrene-induced mutagenicity on Salmonella tester strains TA-98 and TA-100. Date-fruit extract at levels of 3.6 mg/plate and 4.3 mg/plate was needed for 50% inhibition of His+ revertant formation in TA-98 and TA-100, respectively. These findings open up very interesting avenues of exploration of the quite potent antioxidant and antimutagenic properties of date fruits for their implications in harnessing health benefits in the form of processed functional foods in the near future.

Although as folkloric medicine, many other plants given to mothers after childbirth or to invalids are known to possess immunostimulant activity, the immunomodulatory activity of date-fruit extract has only recently been demonstrated.[40] Date-fruit extract enhanced haemagglutinating antibody titers

(HA), plaque-forming cell (PFC) counts in the spleen, and the macrophage migration index (MMI) as an indicator of cell-mediated immunity in humans. Feeding of some other plants such as *Prunus amygdalus* (almond) and *Buchanania lanzan* (*chirronji*) significantly enhanced both cell-mediated immunity (CMI) and humoral immunity in BALB/c mice as evidenced by an increase in the MMI, HA titers, and PFC counts. In comparison, the feeding of *Euryale ferox* (*tel makhana*), *Phoenix dactylifera* (*chhohara*), and *Zingiber officinale* (*sonth*) stimulated humoral immunity to a greater extent than CMI. These findings provide a scientific basis for feeding the above plant foods to mothers after childbirth and to invalids having relatively poor immune systems to improve their infection-fighting ability and overall health.

ANTIMICROBIAL PROPERTIES

The *Berhi* date-fruit extract (20%, w/v) is capable of inhibiting the growth of *Bacillus subtilis, Staphylococcus aureus, Salmonella typhi,* and *Pseudomonas aeruginosa* by about 80 to 99%.[41] Spore germination of *B. subtilis* can be inhibited completely using various concentrations of date extract. Cell elongation and depression in the cell wall of this bacterium incubated in a growth medium containing date extract can be observed in a scanning electron microscope. The growth and morphology of yeast cells can also be affected by date-fruit extract. Various concentrations of *Berhi* date extract can cause growth inhibition and a significant reduction in germ tube formation of *Candida albicans.* Even a 5% date extract produces better inhibition of *C. albicans* than the antifungal agent amphotericin B. The date extract induced a leakage of the cytoplasmic contents from the yeast cells in direct proportion to the concentration of the date extract.[42]

The effect of date-fruit extract on the ultrastructure of yeast cells can be examined by using transmission as well as scanning electron microscopes. Even a 5% extract of date fruit weakens the yeast cell wall with indications of cell distortion and in some cases a partial collapse, as observed with a scanning electron microscope.[43] Increasing the date extract concentration to 20% produced more severe damage to the yeast cell wall, leading to lysis and leakage of the cytoplasmic contents and ultimate cell death. Ultrastructural evidence showed irregularly shaped yeast cells when treated with date-fruit extract; the cell wall layers were prominently affected. The date-extract-treated cells showed losses in cell membrane integrity, aggregation of the cytoplasmic contents, and detachment of the plasmalemma from the cell wall. These multiple effects of date extract on *Candida* yeast cells may be exploited for prophylaxis to control yeast infections.

Use of date extract as a mouth rinse to control the adhesion of a *Candida* species to human buccal epithelial cells (BEC) has been explored. Adhesion of *Candida albicans,* *C. tropicalis,* and *C. kefyr* to BECs was significantly reduced (by between 25 and 52% of the control values) after both short- and long-term periods of yeast cell exposure to different concentrations of date-fruit extract.[44] Using 10% date extract rinse, a significant reduction in adherence of yeast cells to BECs can be achieved immediately or 5 to 20 min after an oral rinse. However, a pretreatment of either *Candida* or BEC, or both, with date-fruit extract gave reduced adherence; the extent of this change was the largest when both types of cells were pretreated with date extract. About 56 to 85% inhibition in germ tube formation in *Candida albicans* by date-fruit extract may be the contributory factor responsible for the effects on cell adherence. Evidently, more investigations into the identification of the bioactive constituents, their bioavailability, and the mechanisms by which they contribute to reduce the risk of cancer and other chronic diseases in humans will enhance the value of processed food products based on date fruits as medical foods, nutritional foods, hypernutritional foods, designer foods, therapeutic foods, super foods, prescriptive foods, nutraceuticals, pharmafoods, or functional foods.

FUNCTIONAL FOODS FROM DATE FRUITS

During the last 2 decades, the number of epidemiological studies that have been undertaken support an inverse relationship between the intake of fruits and vegetables and the risk of cancer and other chronic diseases in humans. All of these experimental studies show that a high intake of fruits and vegetables (at least 400 g daily) is appropriate to lower the risk of these chronic diseases. The presence of a number of phytochemicals in fruits and vegetables may, in part, explain their beneficial effects.[45] Studies in animal models and cell cultures have provided a lot of information about the possible mechanisms by which a diet rich in fruits and vegetables may reduce the risk of these chronic diseases in humans. Although date fruits are rich in some of these phytochemicals, the scientific information generated on this fruit so far is scanty.

At present, most of the date fruits are consumed directly with little or no processing, but the quantity of processed date products is growing rapidly in this region. A number such processed date products are now available in the local market throughout the year. The development of new functional foods made from date products (Figure 17.1) would not only enhance the commercial and economic value of this local crop in this region, but would also create a good market for export.

Commercially Packed Dates

Although sizeable quantities of dates are consumed at perishable immature stages (*khalal* and *rutab*), the majority of date produce is consumed in the dry *tamer* stage with moisture content of less than 20%.[4] Traditionally, dates were bulk-packed in bags or baskets without fumigation or even normal washing and offered for marketing. To meet the high quality standards expected by consumers more recently, the date producing and exporting countries have established a number of bulk-packing houses with modern facilities.

Pickles and Chutney

Pickles and chutney come in many varieties: relishes, fresh-pack pickles, brine pickles, fruit pickles, and pickles-in-oil. Date fruits at the *kimri* and *khalal* stages of maturity are suitable for making pickles and chutney. Pickles-in-oil and chutney prepared from *kimri* date fruits[46,47] resemble in tex-

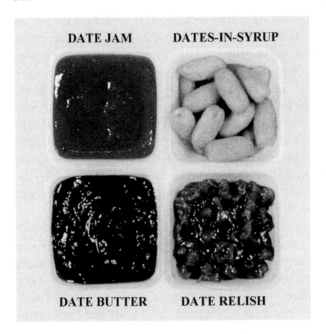

Figure 17.1   Some of the processed date fruit products.

ture and flavor other similar and extremely popular products being commercially prepared from raw mango fruit.[48] The ample amounts of sugars and other nutrients of *kimri*-stage fruit make it suitable for producing good-quality pickles and chutney. Additionally, their shape, size, and green color make them look similar to olives. Except for their lower acidity values, the sweetness and textural characteristics of *kimri*-stage date fruits are identical to those of mango fruit, and offer good potential for preparing pickles-in-oil and sweet chutney for local consumption. Pickles-in-oil are prepared using pitted, sliced *kimri* fruit with a blend of various spices, condiments, and mustard oil. Sweet chutney is prepared from peeled, pitted *kimri* fruit using various spices, condiments, and sugar. The detailed procedures for preparing pickles-in-oil and sweet chutney have been described earlier.[46,47]

Brine and salt-stock pickles are other popular products prepared from *kimri* date fruits.[49] The pickles made by both the methods were microbiologically safe as coliforms were absent, and the products had acceptable sensory quality even after 3 months of storage. The duration of the pickling process varies from prolonged fermentation for brine pickles to very limited fermentation for fresh-pack pickles or no fermentation as for mango and other fruit pickles.[48] As date pickles are popular in this region of the world, a number of date fruit pickles are now available in the local market. Most of the important factors for pickling, such as brine concentration, use of additives like sorbic acid and acetic acid, heat processing requirements, and shelf life have been studied in greater detail.[50]

Date Jam, Date Butter, and Dates-in-Syrup

Jam can be defined as a mixture of fruits and sweetening agents brought to a suitable gelled consistency, with or without other permitted ingredients. Jelly is similar to jam except that a clear fruit extract is used to obtain a transparent final product. Traditionally, jam is a self-preserved, cooked mixture of fruit and sugar (honey is often qualified as a sugar), with total soluble solids content of 68.5% or higher.[51] A good jam can be prepared when the sugar content is 65%, the pectin is 1% and the pH is about 3.0 to 3.2. Citric acid is often added if fruit does not have enough of its own. The degree of preservation is related to the water activity of the product. Mainly the sugar and pectin present in jam are responsible for the water activity. A sugar:date pulp ratio of 55:45 is usually used for jam making. Certain date-fruit cultivars, such as *Khalas, Sukkary,* and *Ruzeiz*, possess desirable chemical compositions and are highly suitable for jam making.[52] For jelly making, a clarified date juice: sugar ratio of 1:1 is used, and the finished product has total soluble solids content of around 73° Brix and a pH of 3.57, with a shelf life of up to 6 months at room temperature.[53]

Date fruits, having high sugar contents, are very suitable for jam manufacture. The *rutab*-stage date fruits have a reasonable quantity of the pectin required for jam preparation. *Tamer* fruits, having the highest sugar content, have been utilized for the preparation of date butter, which can be used in a manner similar to peanut butter. All the steps are similar to jam making, except the pH of the pulp and sugar mixture is adjusted to 4.5 to 4.7, and the total soluble solids content attained is 74 to 75°Brix. A sugar: date pulp ratio of 40:60 is normally used in date butter making. Peeled, pitted whole date fruits at the *khalal* stage of maturity are used in the preparation of dates-in-syrup.[51] Sugar syrup of 50°Brix is boiled till it is concentrated to about 75 to 80°Brix. The pH of the syrup is adjusted to between 2.8 and 3.0 using citric acid. The hot syrup is poured into glass jars containing peeled, pitted date fruits, and the jars are capped immediately. The drained weight of processed fruit is kept at a minimum of 55%. The capped jars are processed in hot water (95°C) for 30 min, then cooled to room temperature and labeled.

Date Syrup (*Dibs*)

Dates, being high in soluble sugars, are naturally an excellent raw material for the preparation of syrup for use in a variety of food products. Though the relative proportion of the various sugars may vary among different cultivars, the general procedure for extraction of date syrup is more or less the same. The *tamer* date fruits are pitted and minced and can be extracted with hot water, autoclaving, or enzymatic treatments. Extraction by autoclaving at 15 psig for 10 min with 2.5 times their weight of water seems to be the best method for dates.[54] Recently, Al-Hooti et al.[55] made use of pectinase and cellulase enzymes to obtain almost double the recovery of soluble solids than were obtained with the conventional hot water and autoclaving extraction methods. Subsequently, this date syrup extracted with pectinase and cellulase was found to be a good substitute for sucrose in bakery products.[56] Microwave heating is another alternative source that gives better uniformity in product temperature, in a comparatively shorter time. This leads to better quality and yield of syrup than is possible with the traditional heating methods.[57]

Date syrup produced by these methods can be used in a variety of food products, such as cakes, carbonated beverages, soft frozen yogurt, milk-based drinks, nutritious creamy foods, and ready-to-serve date juice beverages.[58,59] A formulated food based on date syrup, butter, hazelnuts, dried skim milk, cocoa, starch, lecithin, and baking powder yields a finished product with 6.13% protein, 19.86% fat, 47.8% total sugars, and a good amount of minerals.[60] Due to the hot weather prevailing in this part of the world for most of the year, these date-juice or syrup-based drinks have a good potential for commercial sales and marketing.

Date Paste

A number of fruits, such as apple, apricot, mango, raisin, and strawberry, are converted into paste on a commercial scale for use in baby foods, baked goods, and confectionery.[61] Date fruit is not only the richest source of sugars, but it is as nutritious as many of the above fruits in terms of certain vitamins and minerals. The production of date paste is, therefore, of particular interest to the food industry for use in various functional foods. Processing of date fruits into paste not only preserves them, but also results in reduced transportation and storage costs, since the stones (pits) removed in the process constitute about 10 to 20% of the whole fruit weight. An additional advantage is the availability of date-fruit paste for the food industry throughout the year.

For the preparation of date paste, pitted *tamer* date fruits are either soaked in hot water or steamed for a few minutes, but care must be taken to avoid unnecessary leaching of soluble sugars from the date fruit. About 3 min of steaming at 10 psig or soaking in water at 95°C for 5 to 15 sec is sufficient to soften the fruits for date manufacture. Citric acid or ascorbic acid (0.2% on a fruit basis) is desirable to lower the pH of date paste for improved shelf life and to maintain a desirable color during the storage period of 16 weeks at 5°C. The water activity (a_w) and pH of date paste prepared by this method are within the safe limits of 0.57 and 5.4, respectively.[62] Date paste can be an intermediate product in the date processing industry, when even lower grade date fruits are used for preparing this value-added product.[63] Moreover, the date paste industry is a low capital-intensive operation given that it manufactures a product with wider applications in food product development.

Date paste and date pieces can also be added to a number of food products such as baked goods and ice cream. The sugar in ice cream can easily be replaced up to 50% with date paste without adversely affecting its quality.[64] Date pieces (10%) can also be added to ice cream, but overrun is slightly lowered. Addition of 4 to 8% date paste gives marked improvements in the dough rheological properties, delays gelatinization, improves gas production and retention, prolongs the shelf life, retards staling, and improves the crumb and crust character- istics of bread.[65] Date paste added to cookies results in a higher cookie spread ratio increasing with amounts added up to the 20% level. These mono sugars are known to prevent the crys- tallization of sucrose in cookies during the cooling off period immediately after they are taken out of the baking oven.[66] This property of date paste can be used to our benefit for producing cookies with a smoother surface and higher spread ratio.

Candy and Confectionery

Date fruits are rich in carbohydrates (about 78%) but low in proteins (2 to 3%) and fat (1%), thus serving mainly as a source of calories. Turning date fruit into nearly a complete food would require supplementation with proteins, dietary fiber and fats. Recently, the trend is shifting toward the use of blends of vegetable and dairy proteins to formulate a vari- ety of functional foods, particularly candies, energy bars and confectionery. These products are particularly popular among children and adolescents. Bars made with *tamer* date pulp, sesame seeds, almonds and oat flakes have been found to be quite acceptable to consumers.[67] The average ash, fat and protein contents of 1.78, 6.09, and 7.83% in the control date bars (containing date paste and almonds) changed to 2.60, 3.90, and 9.56% in these date bars fortified with sesame seeds, almonds, skim milk powder, and rolled oats, respectively. Such fortified bars can also supply a reasonable amount of fat, protein, fiber, and minerals.

The plain date bars prepared from date fruit, almonds, coconut, groundnuts, and pistachios can also be coated with chocolate, and stored for about 6 months at 25 ± 5°C without affecting their consumer quality attributes.[68] Storage under these conditions resulted in a significant decrease in moisture, a_w, pH, and sugar content, but an increase in the Brix and pigment levels of both the plain and chocolate-coated date bars. The date cultivars, however, show a lot of variability in their suitability for confectionery making.[69] Date candy prepared using date paste, roasted groundnuts, and desiccated coconut can be plain or coated with melted chocolate, wrapped in cellophane, and stored for about 8 weeks without any loss of consumer acceptability. Use of date paste and nuts in a 60:40 ratio, and subsequently coating with chocolate gives the best sensory scores for this type of candy. Considering the nutritional significance and popularity of date fruits among a wide stratum of the population, fortified date bars have a tremendous potential for commercial production in Arabian Gulf countries.

Miscellaneous Products

A variety of candied or glace fruits are prepared from a number of fruits for use in new food product development by the food industry. The fruit is pierced for easy penetration of sugars and is also treated with calcium chloride to toughen the texture. Use of citric acid and ascorbic acid is also commonly used in the preparation of sugar syrup (about 30 to 45°Brix) required for cooking such fruit. The process of cooking fruit with sugar syrup is repeated for short intervals over a period of many days until the soluble solids content of the cooked fruit reaches 70°Brix or higher. With such high sugar contents the candied fruit becomes shelf-life stable and can be stored for many months at room temperature.[70]

Tamaroggtt is another regional product that can be made from *tamer* date fruits and *oggtt* (a fermented dried milk). It is prepared by mixing pitted, minced *tamer* dates with concentrated fermented milk in variable ratios. After inoculation with normal yogurt culture, the milk is incubated at 42°C for about 5 hours until the acidity (as lactic acid) reaches 0.6%. The fermented milk is heated with constant stirring until a paste-like consistency is obtained. Then pitted, minced dates and other ingredients like cocoa powder, sesame seeds, and salt are added as per the formulation's instructions. The product is shaped like biscuits, air or oven dried to a moisture content of about 10%, and packed in polyethylene bags.[71] The nutritive value of *tamer* date paste can be enhanced by mixing it with skim milk powder, chocolate and fruit juices such as banana, orange, pineapple, apple, grape, or strawberry, and turning it into a high-protein product, *tamarheep*.[72] *Tamarheep* is prepared by homogenizing pitted *tamer* dates and fruit juice (1:1 or 1:1.5 ratio) and then straining the mixture to obtain a uniform paste. Skim milk powder (20 to 50 parts/100 g of the above paste) and cocoa powder (1%) are added to the above mixture and dried to moisture content of about 10%. The finished product is highly nutritious and has an average of 2.77% ash, 72.4% total sugars and 18.4% protein contents (on a dry basis).

Date juice, being rich in fermentable sugars, is a good raw material for wine making. Using a suitable strain, *Saccharomyces cerevisiae* var. *ellipsoideus*, a good quality date wine with about 12% ethanol can be prepared.[73] The date wine is comparable to grape wine and has 0.35 to 0.54% acidity and 4.0 to 4.2 pH. The lower grades of date fruits are potentially economical raw material for the production of many other products such as industrial ethanol, citric acid, baker's yeast, lactic acid, and antibiotics. Date sugars can be used for the production of baker's yeast, *Saccharomyces cerevisiae*, which is an important ingredient in the preparation of yeast-raised baked products.[74] Blackstrap molasses, the usual source for baker's yeast production, can easily be replaced with date sugars without impairing yeast biomass production. Optimum concentrations of date sugar and ammonium sulfate for the highest yield of biomass are 50.0 and 2.0 mg/ml, respectively. Waste from date-fruit processing can be used for the production of lactic acid by *Lactobacillus casei* subsp. *rhamnosus*.[75] The yeast extract, which is used as a nitrogen source, is expensive and could be replaced with a mixture of ammonium sulphate and yeast extract (4:1) without adversely affecting the production efficiency, but reducing the cost of the nitrogen source by 80%. The optimum substrate concentration in terms of glucose content may be 60 g/liter for the highest yeast biomass production.

Some of the date pulp components such as soluble sugars, proteins, vitamins, and minerals can be utilized as a fermentation substrate for oxytetracycline production by some suitable mutants of *Streptomyces rimosus*.[76] After a fermentation period of 96 hours, the cell biomass is harvested, and the antibiotic is recovered. The presence of glucose, fructose, sucrose, proteins, amino acids, certain B-complex vitamins, and minerals is conducive for the synthesis of oxytetracycline by the chosen strain, *S. rimosus*. In addition to date-fruit pulp, low-grade date fruits, immature fruits, and other wastes coming out of date-fruit processing plants can also be utilized for the production of many pharmaceuticals.

By-Products from Date Processing

Apart from waste date pulp and low-grade rejected date fruits from the processing plants, date pits are another major date by-product, constituting about 10% by weight of the whole fruit. A lot of information is available about the chemical composition and utilization of date-fruit pulp, but the information on date seed (pit) composition and utilization is limited. The date seeds, being rich in oil, proteins, minerals, and fiber, are valuable raw materials for animal feeds. Rygg[77] has suggested date seeds to be a potential source of edible oil for humans, with the resulting meal being of use for the animal feeds. On average, date seeds are 6 to 7% protein, 9 to 10% fat, 1 to 2% minerals, and 20 to 24% fiber.[78] The fats from date seeds are rich in oleic acid (58.8%), and contain linoleic acid (12.8%), and palmitic acid (10.6%). The major minerals present in date seeds are potassium, phosphorus, calcium, and magnesium. The essential amino acid profile of date seed proteins is comparable with those of other oilseeds such as soybean, groundnut, cottonseed, and sesame, thus it can partially replace some of the more expensive oilseed meals in animal and poultry feed formulations. Considering the nutritional importance of the fatty acid profile, date seeds also present a possibility for the production of edible oil for humans. Date seeds do contain high amounts of cell wall materials, which can be solubilized by treating them with 4.8 to 9.6% solution of sodium hydroxide so as to improve their nutritional value as animal feeds.

Date seeds, being rich in carbohydrates, proteins, lipids, and minerals, can be utilized as a substrate to grow *Phanerochaete chrysosporium* for the production of ligninase enzyme. This enzyme is useful in solubilizing date seeds and thus improving their nutritional utility for use in animal feeds. Crushed seeds, at water content of 10 ml/g, are inoculated with *P. chrysosporium* and incubated at 30°C for 7 days. A pH of the substrate is adjusted to 4.0 for achieving the highest yield of ligninase enzyme.[79] The date seeds can also be utilized for the production of oxytetracycline by *Streptomyces rimosus*, but only after increasing the supply of carbon and nitrogen. Date seeds, as such, cannot produce appreciable amounts of this antibiotic.[80] Besides other ingredients, date seed hydrolysate and date seed lipids are good substrate for oxytetracycline synthesis by *S. rimosus*.

FUTURE RESEARCH NEEDS

The date palm tree (*Phoenix dactylifera* L.) is a very hardy plant capable of growing under extremely hot, dry and arid climates around the world. The value and importance of the date palm tree, in providing benefits to humans, does not need any further acknowledgment. Being mentioned and honored in many verses of the Holy Quran, and recommended by the Prophet (Peace Be upon Him) is ample proof that there is a lot more to the date fruit than just its sweet taste. At this stage, our knowledge about the nutritional benefits of eating date fruits is extremely limited.

The recent advances in tissue culture need to be further exploited not only to increase the fruit size, bearing capacity, and yield per hectare, but also to improve the processing and nutritional qualities, especially the kinds and amounts of phytochemicals present in the date fruit. We need to identify clearly the chemical constituents responsible for imparting the distinct flavors to date fruits from different cultivars. The availability of modern tools of chemical analysis such as gas chromatography with mass spectrophotometry would be a boon to achieving such objectives. If the date processing industry is to develop along scientific lines for global marketing, the development of suitable food standards and regulations for such processed date-fruit products is imperative. No other food standard is available except that for the packed *tamer* dates.[81] We also need to identify, characterize and estimate the various phytochemicals present in date fruits, and their bioavailability and metabolism in humans; and determine various potential antioxidants and their stability during date-fruit processing, storage and distribution. Obviously, date fruit at all stages of maturity seems to be a storehouse of these valuable phytochemicals, which can provide immense health benefits for humankind.

REFERENCES

1. F Hussein. Date Cultivars in Saudi Arabia. Ministry of Agriculture and Water, Dept. of Research and Development, Saudi Arabia, 1970, pp 33–34.

2. SN Al-Hooti, JS Sidhu, H Qabazard. Studies on the physicochemical characteristics of date fruits of five UAE cultivars at different stages of maturity. *Arab Gulf J Scient Res* 13(3):553–569, 1995.

3. MS Mikki, AH Hegazi, AA Abdel-Aziz, SM Al-Taisan. Suitability of Major Saudi Date Cultivars for Commercial Handling and Packing. Proceedings of the Second Symposium on the Date Palm in Saudi Arabia, Vol. II, Al-Hassa, 1986, pp 9–26.

4. SN Al-Hooti, JS Sidhu, H Qabazard. Physicochemical characteristics of five date-fruit cultivars grown in the United Arab Emirates. *Plant Foods Human Nutrition* 50:101–113, 1997.

5. G Toutain. Le Palmier dattier. Culture et Production. *Al Awamia* 25:81–151, 1967 (in Arabic).

6. A Zaid. Review of Date Palm (*Phoenix dactylifera* L.) Tissue Culture. Proceedings of the Second Symposium on the Date Palm in Saudi Arabia, Vol. I, Al-Hassa, 1986, pp 67–75.

7. C Sudhersan, MM AboEl-Nil, A Al-Baiz. Occurrence of direct somatic embryogenesis on the sword leaf of *in vitro* plantlets of *Phoenix dactylifera* L. cultivar *barhee. Current Science* 65(11):887–888, 1993.

8. C Sudhersan, MM AboEl-Nil, A Al-Baiz. Direct somatic embryogenesis and plantlet formation from the leaf explants of *Phoenix dactylifera* L. cultivar *barhee. J Swamy Bot Cl,* 10(1&2):37–43, 1993.

9. RW Nixon. The date palm-tree life tree in the subtropical deserts. *Econ. Bot.* 31:15-20, 1951.

10. HM Marei. Date Palm Processing and Packing in the Kingdom of Saudi Arabia. Ministry of Agriculture and Water, Riyadh, 1971, pp 5–7 (in Arabic).

11. A Al-Tayeb. History of date cultivation in the world. Al-Yawin Newspaper, Hofuf, Kingdom of Saudi Arabia. March 19, 1982, pp 18.

12. F Hussain, A El-Zeid. Studies on Physical and Chemical Characteristics of Date Varieties of Saudi Arabia. Ministry of Agriculture and Water, Kingdom of Saudi Arabia, 1975, pp 1–60. (Arabic).

13. RW Nixon. Date culture in Saudi Arabia. *Date growers' Inst Rep* 31:15–20, 1954.

14. RJ Knight Jr. Origin and world importance of tropical fruit crops. In: S Nagy, PE Shaw, Eds. *Tropical and Subtropical Fruits.* Westport, Connecticut: AVI Publishing Co, 1980, pp 1–45.

15. Anon. Study on the Development of Date Cultivation, Production, Processing and Marketing in the Kingdom of Saudi Arabia. Arab Organization for Agricultural Development, Arab League Countries, Khartoum, 1984, pp 23–27.

16. SE El-Kassas. Effect of Some Growth Regulators on the Yield Fruit Quality of *Zaghloul* Date Palm. Proceedings of the Second Symposium on the Date Palm in Saudi Arabia, Vol. I, Al-Hassa, 1986, pp 179–186.

17. FAO. *FAO Production Year Book: FAO Statistics Series,* Vol. 54 Rome: FAO, 2000, pp 168–169.

18. SN Al-Hooti, JS Sidhu, J Al-Otaibi, H Qabazard. Extension of shelf life of date fruits at the *khalal* stage of maturity. *Indian J Hort* 52(4):244–249, 1995.

19. F Hussein, S Mostafa, F El-Samirafa, A Al-Zaid. Studies on physical and chemical characteristics of 18 date cultivars grown in Saudi Arabia. *Indian J Hort* 33:107–113, 1976.

20. M El-Zoghbi. Biochemical changes in some tropical fruits during ripening. *Food Chemistry* 49:33–37, 1994.

21. O Ishrud, M Zahid, VU Ahmad, Y Pan. Isolation and structure analysis of a glucomannan from the seeds of Libyan dates. *J Agric Food Chem* 49(8):3772–3774, 2001.

22. W Al-Shahib, RJ Marshall. Dietary fiber content of dates from 13 varieties of date palm *Phoenix dactylifera*, L. *Intl J Food Sci Technol* 37:719–721, 2002.

23. DL Topping, PM Clifton. Short-chain fatty acids and human colonic function: roles of resistant starch and nonstarch polysaccharides. *Physiol Rev* 81:1031-1064, 2001.

24. KE Aidoo, RF Tester, JE Morrison, D MacFarlane. The composition and microbial quality of prepacked dates purchased in greater Glasgow. *Intl J Food Sci Technol* 31(5):433–438, 1996.

25. HK Hassan, MS Mikki, MA Al-Doori, TS Jaffar. Preservation of high-moisture dates (*Rutab*) by antimicrobial agents. *Technical Bulletin Palm & Dates Res Center* No. 2/79, pp 1–18, 1979.

26. SN Al-Hooti, JS Sidhu, H Al-Amiri, J Al-Otaibi, H Qabazard. Extension of shelf life of two UAE date-fruit varieties at *Khalal* and *Rutab* stages of maturity. *Arab Gulf J Scient Res* 15(1):99–110, 1997.

27. SL Taylor, SL Hefle. Food allergies and other food sensitivities. *Food Technol* 55(9):68–83, 2001.

28. AA Kwaasi, RS Parhar, P Tipirneni, HA Harfi, ST Al-Sedairy. Cultivar-specific epitopes in date palm (*Phoenix dactylifera* L.) pollenosis. Differential antigenic and allergenic properties of pollen from 10 cultivars. *Ant Arch Allergy Immunol* 104(3):281–290, 1994.

29. Y Waisel, N Keynan, T Gil, D Tayar, A Bezerano, A Goldberg, C Geller-Bernstein, Z Dolev, R Tamir, I Levy. Allergic response to date palm and pecan pollen in Israel. *Harefuah* 126(6):305–310, 368, 1994.

30. AA Kwaasi, HA Harfi, RS Parhar, ST Al-Sedairy, KS Collison, RC Panzani, FA Al-Mohanna. Allergy to date fruits: characterization of antigens and allergens of fruits of the date palm (*Phoenix dactylifera* L.). *Allergy* 54(12):1270–1277, 1999.

31. AA Kwaasi, HA Harfi, RS Parhar, S Saleh, KS Collison, RC Panzani, ST Al-Sedairy, FA Al-Mohanna. Cross-reactivities between date palm (*Phoenix dactylifera* L.) polypeptides and foods implicated in the oral allergy syndrome. *Allergy* 57(6):508–518, 2002.

32. MI El-Shaarawy. Dates in Saudi Diet. Proceedings of the Second Symposium on the Date Palm in Saudi Arabia, Vol. II, *Al-Hassa,* 1986, pp 35–47.

33. RM Myhara, J Karkalas, MS Taylor. The composition of maturing Omani dates. *J Sci Food Agric* 79:1345–1350, 1999.

34. EW Jwanny, MM Rashad, SA Moharib, NM El-Beih. Studies on date waste dietary fiber as hypolipidemic agent in rats. *Z Ernahrungswiss* 35(1):39–44, 1996.

35. DJ Fleming, PF Jacques, GE Dallal, KL Tucker, PWF Wilson, RJ Wood. Dietary determinants of iron stores in a free-living elderly population: the Framingham Heart Study. *Am J Clin Nutr* 67(4):722–733, 1998.

36. CE Vandercook, S Hasegawa, VP Maier. Dates. In: S Nagy, PE Shaw, Eds. *Tropical and Subtropical Fruits*. Westport, Connecticut: AVI Publishing Co, 1980, pp 506–541.

37. RM Myhara, A Al-Alawi, J Karkalas, MS Taylor. Sensory and textural changes in maturing Omani dates. *J Sci Food Agric* 80:2181–2185, 2000.

38. PCH Hollman. Evidence for health benefits of plant phenols: local or systemic effects? *J Sci Food Agric* 81:842–852, 2001.

39. PK Vayalil. Antioxidant and antimutagenic properties of aqueous extract of date fruit (*Phoenix dactylifera* L. Arecaceae). *J Agric Food Chem* 50(3):610–617, 2002.

40. A Puri, R Sahai, KL Singh, RP Saxena, JS Tandon, KC Saxena. Immunostimulant activity of dry fruits and plant materials used in Indian traditional medical system for mothers after childbirth and invalids. *J Ethnopharmacology* 71(1/2):89–92, 2000.

41. AKJ Sallal, A Ashkenani. Effect of date extract on growth and spore germination of *Bacillus subtilis*. *Microbios* 59(240/241):203–210, 1989.

42. AKJ Sallal, KHA El-Teen, S Abderrahman. Effect of date extract on growth and morphology of *Candida albicans*. *Biomed Letters* 53(211):179–184, 1996.

43. ZA Shraideh, KHA El-Teen, AKJ Sallal. Ultrastructural effects of date extract on *Candida albicans*. *Mycopathologia* 142(3):119–123, 1998.

44. KJA El-Teen. Effects of date extract on adhesion of *Candida* species to human buccal epithelial cells *in vitro*. *J Oral Pathol Med* 29(5):200–205, 2000.

45. M Gerber, MC Boutron, S Hercberg, E Riboli, A Scalbert, MH Siess. Food and cancer: state of the art about the protective effect of fruits and vegetables. *Bull Cancer* 89(3):293–312, 2002.

46. SN Al-Hooti, JS Sidhu, J Al-Otaibi, H Al-Amiri, H Qabazard. Utilization of date fruits at different maturity stages for variety pickles. *Adv Food Sci* 19(1/2):1–7, 1997.

47. SN Al-Hooti, JS Sidhu, J Al-Otaibi, H Al-Amiri, H Qabazard. Processing of some important date cultivars grown in United Arab Emirates into chutney and date relish. *J Food Process Preserv* 21:55–68, 1997.

48. SP Das-Thakur, DR Chaudhuria, SN Mitra, AN Bose. Quality aspects of processed mango products. *Indian Food Packer* 30(5):45–50, 1976.

49. AM Hamad, AK Yousif. Evaluation of Brine and Salt-Stock Pickling of Two Date Varieties in the *Kimri* Stage. Proceedings of the Second Symposium on the Date Palm in Saudi Arabia, Vol. II, Al-Hassa, 1986, pp 245–257.

50. AK Yousif, AM Hamad, WA Mirandella. Pickling of dates at the early *khalal* stage. *J Food Tech* 20:697–701, 1985.

51. SN Al-Hooti, JS Sidhu, JM Al-Saqer, H Al-Amiri, H Qabazard. Processing quality of important date cultivars grown in the United Arab Emirates for jam, butter and dates-in-syrup. *Adv Food Sci* 19(1/2):35–40, 1997.

52. AK Yousif, M Abou-Ali, A Bou-Idress. Processing Evaluation and Storability of Date Jam. Program and Abstracts of the Third Symposium on the Date Palm in Saudi Arabia, Al-Hassa, Abstract No. I-20, 1993, p 165.

53. AK Yousif, AS Alghamdi. Suitability of some date cultivars for jelly making. *J Food Sci Technol* 36(6):515–518, 1999.

54. MI El-Shaarawy, AS Mesallam, HM El-Nakhal, AN Wahdan. Studies on Extraction of Dates. Proceedings of the Second Symposium on the Date Palm in Saudi Arabia, Vol. II, Al-Hassa, 1986, pp 259–271.

55. SN Al-Hooti, JS Sidhu, JM Al-Saqer, A Al-Othman. Chemical composition and quality of date syrup as affected by pectinase/cellulase treatment. *Food Chem* 79(2):215–220, 2002.

56. JS Sidhu, JM Al-Saqer, SN Al-Hooti, A Al-Othman. Quality of pan bread as affected by replacing sucrose with date syrup produced by pectinase/cellulase enzymes. *Plant Foods for Human Nutrition* 2002, (in press).

57. IA Ali, AI Mustafa, EA Elgasim, HA Alhashem, SE Ahmed. The Use of Microwave Heating in the Production of Date Syrup (*Dibs*). Program and Abstracts of the Third Symposium on the Date Palm in Saudi Arabia, Al-Hassa, Abstract No. I-23, 1993, p 167.

58. AM Alhamdan. Rheological properties of a new nutritious dairy drink from milk and date extract concentrate (*dibs*). *Int J Food Properties* 5(1):113–126, 2002.

59. AK Yousif, AS Alghamdi, A Hamad, AI Mustafa. Processing and evaluation of a date juice-milk drink. *Egyp J Dairy Sci* 24(2):277–288, 1996.

60. I Alemzadeh, M Vossoughti, A Keshavarz, V Maghsoudi. Use of date honey in the formulation of nutritious creamy food. *Iran Agril Res* 16(2):111–117, 1997.

61. Anon. Fruit pastes enrich the summer assortment. *CCB Rev Chocolate Confect Bakery* 6(2):24–26, 1981.

62. AK Yousif, ID Morton, AI Mustafa. Studies on Date Paste. I. Storage Stability. Proceedings of the Second Symposium on the Date Palm in Saudi Arabia, Vol. II, Al-Hassa, 1986, pp 93–112.

63. MS Mikki, WF Al-Tai, ZS Hamodi. Canning of Date Pulp and *Khalal* Dates. Proceedings of the First Symposium on the Date Palm in Saudi Arabia, Al-Hassa, 1983, pp 520–532.

64. AM Hamad, HA Al-Kanhal, I Al-Shaieb. Possibility of Utilizing Date Puree and Date Pieces in the Production of Milk Frozen Desserts. Proceedings of the Second Symposium on the Date Palm in Saudi Arabia, Vol. II, Al-Hassa, 1986, pp 181–187.

65. AK Yousif, ID Morton, AI Mustafa. Functionality of date paste in bread making. *Cereal Chem* 68(1):43–47, 1991.

66. RC Hoseney. Soft wheat products. In: RC Hoseney, Ed. *Principles of Cereal Science and Technology.* St. Paul, Minnesota: AACC, 1994, pp 275–305.

67. SN Al-Hooti, JS Sidhu, J Al-Otaibi, H Al-Amiri, H Qabazard. Date bars fortified with almonds, sesame seeds, oat flakes and skim milk powder. *Plant Foods for Human Nutrition* 51:125–135, 1997.

68. AK Yousif. Processing, shelf life and evaluation of plain and chocolate-coated date bars. *Food Sci Technol Today* 8(4):243–246, 1994.

69. AK Yousif. Suitability of some date cultivars for candy making. *Tropical Science* 41(3):156–158, 2001.

70. WN Sawaya, JK Khalil, AF Al-Shalhat, AA Ismail. Processing of Glace Dates. Proceedings of the Second Symposium on the Date Palm in Saudi Arabia, Vol. II, Al-Hassa, 1986, pp 113–119.

71. IM Al-Ruqaie, H El-Nakhal. *Tamaroggtt* a New Product from Date and *Oggtt.* Proceedings of the Second Symposium on the Date Palm in Saudi Arabia, Vol. II, Al-Hassa, 1986, pp 133–141.

72. H El-Nakhal, MI El-Shaarawy, AS Mesallam. *Tamarheep* a New Product from Dates (*tamer*) with High Protein Content. Proceedings of the Second Symposium on the Date Palm in Saudi Arabia, Vol. II, Al-Hassa, 1986, pp 143–153.

73. HK Hassan, WF Al-Tai, MM Hamando. Studies on production of date wine. I. Selection of active strain for date juice fermentation. *Tech Bull Palm & Dates Res Center* 6:1–26, 1980.

74. JA Khan, KO Abulnaza, TA Kumosani, A Abou-Zeid. Utilization of Saudi date sugars in production of baker's yeast. *Bioresource Technol* 53(1):63–66, 1995.

75. H Nancib, A Nancib, A Boudjelal, C Benslimane, F Blanchard, J Boudrant. The effect of supplementation by different nitrogen sources on the production of lactic acid from date juice by *Lactobacillus casei* subsp. *rhamnosus*. *Bioresource Technol* 78(2):149–153, 2001.

76. NA Baeshin, AA Abou-zeid. Saudi Dates as Fermentation Media for Oxytetracycline Production by Some Mutants of *Streptomyces rimosus*. Program and Abstracts of the Third Symposium on the Date Palm in Saudi Arabia, Al-Hassa, Abstract No. I-7, 1993, pp 157–158.

77. GL Rygg. Date Development, Handling and Packing in the United States, Agricultural Handbook No. 482. Washington, D.C.: USDA, 1977, pp 25–41.

78. SN Al-Hooti, JS Sidhu, H Qabazard. Chemical composition of seeds of date-fruit cultivars of United Arab Emirates. *J Food Sci Technol* 35(1):44–46, 1998.

79. MA El-Nawawy, AK Abdel-Latif, MS Al-Jassir. Ligninase Production from Micromycetes. Program and Abstracts of the Third Symposium on the Date Palm in Saudi Arabia, Al-Hassa, Abstract No. I-6, 1993, p 157.

80. AA Abou-zeid, NA Baeshin. Utilization of Saudi Date Seeds in Formation of Oxytetracycline. Program and Abstracts of the Third Symposium on the Date Palm in Saudi Arabia, Al-Hassa, Abstract No. I-8, 1993, pp 158–159.

81. Anon. Kuwaiti Standards for Packed Dates. Ministry of Commerce and Industry, State of Kuwait, Standards No. 98/894, 1998, pp 1–10.

82. SE Gebhardt, R Cutrufelli, RH Mathews. Composition of Foods. Agriculture Handbook No. 8-9, U.S. Department of Agriculture, Human Nutrition Information Service. Washington, D.C.: USDA, 1982, pp 97.

18

Functional Foods and Products from Japanese Green Tea

TERUO MIYAZAWA and
KIYOTAKA NAKAGAWA
Food and Biodynamic Chemistry Laboratory,
Graduate School of Agricultural Science,
Tohoku University,
Sendai, Japan

INTRODUCTION

Discussions of food normally focus on nutritional content and flavor. However, a great deal of attention is recently being paid to the role of food in bioregulation. Foods that possess such functions are called "functional foods." Green tea, with its sweet aroma and eternally fresh taste, has been loved and consumed since its introduction to Japan centuries ago. But recent studies have scientifically proven the health-promoting effects of green tea. The healing properties of green tea are the properties of leaf constituents, especially tea catechins. Nowadays, various functional foods and/or products containing tea catechins are produced and marketed commercially in Japan as well as overseas. To become informed about the functionality of such commercial products, it is necessary to trace the scientific history of green tea. In this chapter, we briefly review the history, health benefits, and industrial applications of green tea, with particular emphasis on the functional foods and products containing tea catechins.

THE HISTORY OF GREEN TEA

According to Chinese legend, tea consumption goes back as far as 2737 BC.[1] Around that time, Sheng Nung, a legendary Emperor known as the Divine Healer, discovered the healing power in tea leaves. Since then, traditional Chinese medicine has recommended "tea" for headaches, body pains, digestion, depression, immune enhancement, detoxification, as an energizer, and to prolong life.

At the beginning of the ninth century, tea was probably introduced in Japan by Japanese Buddhist monks, returning from their studies in China.[2] In addition to learning about Chinese culture and institutions as members of diplomatic missions, they learned from Chinese Buddhist monks about the custom of drinking tea to intensify alertness during meditation. In 805, the monk Saicho returned to Japan, bringing with him the custom of drinking tea from China and encouraging its adoption among Japanese Buddhist monks.

Nowadays, drinking tea is a very ordinary event in the life of Japanese people. There are phrases that include the word "tea" (cha) in everyday use in Japan. One of the most popular is "nichijo-sahan-ji" (literally, everyday tea-meal affair), which means an ordinary event. Now, Japan is among the five major tea producers in the world (along with China, India, Indonesia, and Sri Lanka).[2]

The generic term "tea" refers to a class of beverages made from the leaves of the *Camellia sinensis* (*C. sinensis*) plant, herbal components, or a combination of both. Actual tea (green, black, white, or oolong tea) beverage is from

C. sinensis; the differences between types are dependent on processing. Green and white teas are the least processed, while black and oolong teas are fermented for a set time and then cured.[2] Of these teas, green tea is the most popular beverage in Japan. Green tea has been called the second-most consumed beverage in the world, behind water.

EFFICACY OF GREEN TEA CATECHINS IN PROMOTING HEALTH

As described above, of all the beverages that are consumed today, tea is undoubtedly the oldest and the most widely known. A Chinese proverb says "It is better to drink green tea than to take medicine." In Japan, green tea is known as the source of "Everlasting Youth and Longevity." Indeed, recent biological studies have succeeded in demonstrating that green tea has numerous positive effects on health.[3–9] Such healing properties of green tea can be ascribed to tea leaf constituents, especially polyphenol compounds.

Tea leaves contain many kinds of polyphenols, the catechins being particularly numerous. Catechins belong to the groups of compounds generically known as flavonoids, which have a C_6-C_3-C_6 carbon structure and are composed of two aromatic rings A and B (Figure 18.1). Although chemical compositions of catechins in tea leaves vary with growing conditions, season, age of the leaves, and cultivar.

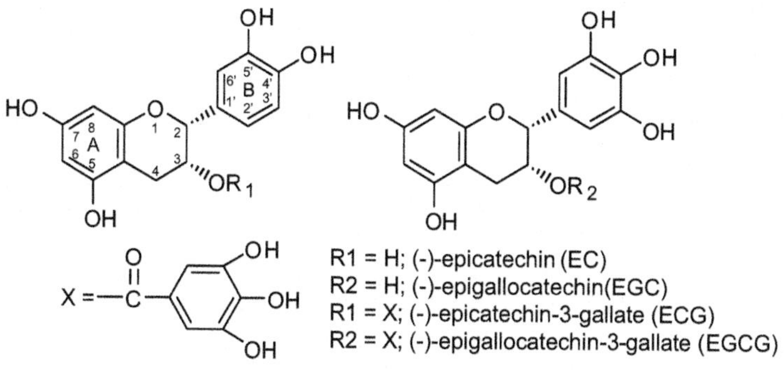

R1 = H; (-)-epicatechin (EC)
R2 = H; (-)-epigallocatechin(EGC)
R1 = X; (-)-epicatechin-3-gallate (ECG)
R2 = X; (-)-epigallocatechin-3-gallate (EGCG)

Figure 18.1   The basic structural formulas of green tea catechins.

The major green tea polyphenols are (-)-epigallocatechin-3-gallate (EGCG), (-)-epicatechin-3-gallate (ECG), (-)-gallocatechin-3-gallate (GCG), (-)-epigallocatechin (EGC), (-)-epicatechin (EC), and (+)-catechin. These catechins may account for up to 30% of the dry leaf weight. A typical cup of green tea contains between 100 to 200 mg catechins, of which 40 to 80 mg is EGCG. Among these, EGCG is believed to be the most bioactive agent. Other compounds found in tea include flavonols (quercetin, kaempferol, and rutin), xanthines (caffeine, theophylline, and theobromine), theanine (an amino acid peculiar to tea), minerals (aluminum and manganese), and trace levels of carotenoids and volatile oils. On the other hand, the term "tannins" has often been used to describe certain tea constituents. In the industrial and botanic literature, tannins are characterized as plant materials that give a blue color with ferric salts and produce leather from hides. Thus, tannins are a group of chemicals usually with large molecular weights and diverse structures. Monomeric flavonols, that is tea catechins, are precursors of condensed tannins. It might be more appropriate to use the term "tea polyphenols" or "tea flavonols," because they are quite distinct from commercial tannins and tannic acid.

It is well established that green tea catechins are good scavengers of reactive oxygen species and free radicals in both aqueous and lipophilic environments.[10–16] In fact, tea catechins are the most powerful antioxidants among the known plant polyphenols. It is reported that EGCG is 20 times more active than vitamin C, 30 times more than vitamin E, and 2 to 4 times more than butylated hydroxyanisole (BHA) or butylated hydroxytoluene (BHT).[10] The antioxidant activity of catechins increases as the number of o-dihydroxy groups increases. It was also demonstrated that catechins can act as antioxidants in synergy with tocopherols and organic acids.[17,18]

The biological (antiatherosclerotic,[19,20] anticarcinogenic,[6,21] anti-inflammatory,[22,23] and antibacterial[7,24]) activities of green tea can be attributed to the antioxidant properties of catechins. Hence, numerous studies about mechanisms underlying the beneficial properties of tea catechins against chronic diseases as well as the aging process have focused on antioxidant activity. For instance, we have previously investigated the effect of green tea catechin supplementation on the antioxidant capacity of plasma in humans by measuring plasma phosphatidylcholine hydroperoxide (PCOOH) as a specific marker of oxidative stress *in vivo.*[25-27] Plasma PCOOH levels were markedly attenuated in catechin-administrated subjects, being inversely correlated with the increase in plasma catechin levels. Our results clearly indicated that drinking green tea contributes to cardiovascular disease prevention by increasing plasma antioxidant capacity in humans. The other biological activities of tea catechins include protection from urinary tract infections, protection of the skin from the harmful effects of ultraviolet radiation, and roles in blood pressure and sugar metabolism regulation. Tea appears to be a useful means of reducing body fat and preventing obesity. In several chapters of this book, the beneficial health effects of tea catechins are noted in detail.

INDUSTRIAL APPLICATIONS OF GREEN TEA CATECHINS

Nowadays, several products of "green tea extract" are produced in Japan. They include Polyphenon (Tokyo Food Techno Co., Tokyo, Japan), SUNPHENON (Taiyo Kagaku Co., Yokkaichi, Japan), Thea-Flan (Ito-en Co., Tokyo, Japan) and others. These products are extracts of green tea, containing a high amount of tea catechins (e.g., Polyphenon 100 has more than 90% catechin purity and is caffeine-free), without any elements other than those of green tea. Such green tea extracts are utilized industrially for several purposes as described below.

The main application of green tea extract is as an antioxidant additive in edible oils and related food products. It is reported that tea catechins have potent antioxidative effects in lard, preventing its peroxidation.[28] Catechins also show antioxidativity in vegetable oils or even in oil solubilized in water.[29] They are also proven to be effective in protecting the discoloring of β-carotene.[30] Therefore, green tea catechins are widely used in several foods with high water content or in cooked products as a natural antioxidant, replacing controversial synthetic ones (i.e., BHA), owing to their water solubility and resistance to heat degradation.

Green tea extract is also used as a bacteriostatic agent. The antibacterial activity of tea catechins[7,24] has most relevance to the beverage industry. In Japan, the use of vending machines is very popular, especially by the soft drink industry. Some of the products sold in vending machines are hot (50 to 60°C). The prolonged storage of canned drinks in vending machines at these serving temperatures has caused bacterial spoilage in some cases.[31] Since tea catechins have been reported to inhibit the development and growth of bacterial spores and to reduce the heat resistance of thermophilic spores,[32] they are useful in preventing spoilage of canned drinks containing high-carbohydrate contents.

EXAMPLES OF COMMERCIAL FUNCTIONAL GREEN TEA PRODUCTS

As the effects of green tea catechins are scientifically proven, products containing the chemicals have begun to appear on the market. Commercial products from green tea catechins include supplements, drinks, and other unique products (e.g., in air filters and green tea masks). In this section, examples of commercial catechin products are described.

For easy consumption, a variety of supplements, capsules or powders containing tea catechins are marketed commercially in Japan. For instance, Catechin 100 (Tokyo Food Techno Co.) produced in capsule form contains 100 mg of catechins per capsule, and to benefit human health, a daily intake of three to nine capsules is recommended by the manufacturer. Various similar tea catechin capsules are sold in the U.S. health care market. In addition, to fortify the beneficial effects of tea catechins, other desirable components (e.g., vitamin A, C, E, and/or extracts of several plants) are often included in the catechin tablets. Such tablets are also supplied in world markets.

The beverage industry is seeking better additives to create healthy drinks. Therefore, tea catechins because of their higher antioxidative and antibacterial properties have attracted a great deal of attention as potential drink additives. The boom in catechin drinks was sparked in Japan by the release of Catechin Water (Ito-en Co.) in 1997. Catechin Water is a sports drink that contains catechins extracted from green tea. It is popular among young women concerned with health and beauty. Kirin Beverage Corp. (Tokyo, Japan) began sales of catechin drinks in 1998, and the market for catechin-containing drinks is expanding.

The Catechin Air Filter capitalizes on the remarkable effect of tea catechins in preventing influenza virus infections.[33] Matsushita Seiko (Kasugai, Japan) has developed an air purifier with a catechin filter, in which catechins work to remove bacteria and viruses from the air. An unusual product is the green tea mouth-and-nose mask. These masks (made from gauze soaked in green tea) are said to be highly effective in suppressing the spread of bacteria and in reducing odors.

Among the other products containing tea catechins are chewing gum, candies, and breath refresher capsules. In these products, catechins reduce cavities as well as deodorize breath from garlic, fish (trimethylamine), methyl mercaptan, ammonia, and tobacco smoke odor. Also marketed are catechin-containing soaps, shampoos, bath waters, skin lotion, moisture creams, cleansing powders, and cleansing packs which have the soft green color and the pleasant aroma of green tea. Toyo Tire & Rubber (Osaka, Japan) has developed catechin-containing urethane foam for use in mattresses. The catechins are said to kill bacteria in the mattress and also break down sweat to prevent odors. Various uses of catechins in medicine are now emerging. On the other hand, there are several tea products (especially for drinks) that have been approved by Japan's Ministry of Health Labor and Welfare for labeling as a "Food for Specified Health Use" (FOSHU). For instance, Kao Corp. (Tokyo, Japan) started marketing a bottled green tea product containing a large quantity of catechins in this year. The 350 ml bottled green tea, to be called Healthya, contains three to four times the amount of catechins normally found in green tea. Healthya has FOSHU approval to claim that the drink is suitable for people concerned about body fat.

OTHER CONSTITUENTS

In addition to catechins, other constituents in green tea (i.e., theanine and xanthines) are known to have biological activity. Theanine is an amino acid found only in tea plants (*Camellia sinensis*), mushrooms, and the seedlings of a few other *Camellia species* (i.e., *C. Japonica* and *C. sasanqua*). Theanine (the predominant amino acid in green tea) is 50% of the total free amino acids in the plant and between 1 to 2% of the dry weight of green tea leaves. The occurrence of theanine in tea leaves was discovered in 1950,[34] and its chemical structure was determined to be γ-ethylamino-L-glutamic acid. Theanine is an antagonist to caffeine-induced paralysis,[35] reduces blood pressure and hypertension in rats,[36] acts as a neurotransmitter in brain, can promote the synthesis of nerve growth factor in rats,[35] and enhances α-brain wave activity in humans.[35] Theanine is considered the main component responsible for the taste of green tea, which in Japanese is called "umami." Theanine is now marketed in Japan as a nutritional supplement. Tea leaf is the only presently reported natural source of theanine. However, enzymic synthesis of theanine on an industrial scale recently became possible. Theanine extracted from green tea and/or enzymatically synthesized is now added to beverages, cookies, candies, or ice cream to produce a relaxing effect and to mask the bitter taste in some foods.

Tea is a popular beverage because of its content of methyl xanthines, primarily caffeine with trace amounts of theophylline and theobromine. Tea commonly contains much less caffeine than a comparable volume of coffee (50 to 100 mg/cup *vs.* 100 to 200 mg/cup), and the amounts of other methyl xanthines present in tea are pharmacologically insignificant. Caffeine is a competitive adenosine antagonist and potent central nervous system stimulant that acts on the cortical and medullary regions of the brain, and at high doses, the spinal cord. Caffeine is also a respiratory, cardiac, and skeletal muscle stimulant, which causes coronary dilation, smooth muscle relaxation, and diuresis.

GREEN TEA IN PERSPECTIVE

As described above, green tea is now recognized as a healthy drink, a source of pharmacologically active molecules, and a functional food endowed with beneficial health properties. New products and uses are emerging from green tea. Experimental evidence of the health-promoting properties of green tea continues to increase. The factors that influence the incidence and progression of chronic diseases are becoming better defined. It is apparent that the tea is a source of phytochemicals. In human body, tea catechins are digested, absorbed and metabolized, and then exert their beneficial effects at the cellular level. In the near future, Japan will become a society with a large population of elderly people. It is expected that green tea, which contains ample amounts of catechins, will continue to serve as a familiar health-promoting tonic in the future. If you have never consumed green tea and/or catechin-containing products, why not give it a try? It may taste a little bitter at first, but the old Japanese proverb says "Good medicine is bitter."

REFERENCES

1. Gutman RL, and Ryu BH. Rediscovering tea. An exploration of the scientific literature. *HerbalGram*, 37, 33–48 (1996).

2. Y Hara. *Green Tea: Health Benefits and Applications*. Marcel Dekker Inc, New York, USA (2001).

3. Miyazawa T. Absorption, metabolism and antioxidative effects of tea catechin in humans. *Biofactors*, 13, 55–59 (2000).

4. Wang HK. The therapeutic potential of flavonoids. *Expert Opin. Investig. Drugs*, 9, 2103–2119 (2000).

5. Katiyar SK, and Elmets CA. Green tea polyphenolic antioxidants and skin photoprotection. *Int. J. Oncol.*, 18, 1307–1313 (2001).

6. Fujiki H, Suganuma M, Imai K, and Nakachi K. Green tea: cancer preventive beverage and/or drug. *Cancer Lett.*, 188, 9–13 (2002).

7. Gupta S, Saha B, and Giri AK. Comparative antimutagenic and anticlastogenic effects of green tea and black tea. *Mutat. Res.*, 512, 37–65 (2002).

8. Demeule M, Michaud-Levesque J, Annabi B, Gingras D, Boivin D, Jodoin J, Lamy S, Bertrand Y, and Beliveau R. Green tea catechins as novel antitumor and antiangiogenic compounds. *Curr. Med. Chem. Anti-Canc. Agents.*, 2, 441–463 (2002).

9. Higdon JV, and Frei B. Tea catechins and polyphenols: health effects, metabolism, and antioxidant functions. *Crit. Rev. Food Sci. Nutr.*, 43, 89–143 (2003).

10. Vinson JA, Dabbagh YA, Serry MM, and Jang J. Plant flavonoids, especially tea flavonols, are powerful antioxidants using an *in vitro* oxidation model for heart disease. *J. Agric. Food Chem.*, 43, 2800–2802 (1995).

11. Cao G, Sofic E, and Prior R. Antioxidant capacity of tea and common vegetables. *J. Agric. Food Chem.*, 44, 3426–3431 (1996).

12. He Y, and Shahidi F. Antioxidant activity of green tea and its catechins in a fish meat model system. *J. Agric. Food Chem.*, 45, 4262–4266 (1997).

13. Hirayama O, Takagi M, Hukumoto K, and Katoh S. Evaluation of antioxidant activity by chemiluminescence. *Anal. Biochem.*, 247, 237–241 (1997).

14. Wiseman SA, Balentine DA, and Frei B. Antioxidants in tea. *Crit. Rev. Food Sci. Nutr.*, 37, 705–718 (1997).

15. Sawai Y, and Sakata K. NMR analytical approach to clarify the antioxidative molecular mechanism of catechins using 1,1-diphenyl-2-picrylhydrasyl. *J. Agric. Food Chem.*, 46, 111–114 (1998).

16. Yokozawa T, Dong E, Nakagawa T, Kashiwagi H, Nakagawa H, Takeuchi S, and Chung HY. *In vitro* and *in vivo* studies on the radical-scavenging activity of tea. *J. Agric. Food Chem.*, 46, 2143–2150 (1998).

17. Hara Y, Luo SJ, Wickremashinghe RL, and Yamanishi T. Uses and benefits of tea. *Food Rev. Intl.*, 11, 527–542 (1995).

18. Antony JIX, and Shankaranaryana ML. Polyphenols of green tea. *Int. Food Ingred.*, 5, 47–50 (1997).

19. Riemersma RA, Rice-Evans CA, Tyrrell RM, Clifford MN, and Lean ME. Tea flavonoids and cardiovascular health. *QJM.*, 94, 277–282 (2001).

20. Kris-Etherton PM, and Keen CL. Evidence that the antioxidant flavonoids in tea and cocoa are beneficial for cardiovascular health. *Curr. Opin. Lipidol.*, 13, 41–49 (2002).

21. Lin JK. Cancer chemoprevention by tea polyphenols through modulating signal transduction pathways. *Arch. Pharm. Res.*, 25, 561–571 (2002).

22. Katiyar SK, Matsui MS, Elmets CA, and Mukhtar H. Polyphenolic antioxidant (-)-epigallocatechin-3-gallate from green tea reduces UVB-induced inflammatory responses and infiltration of leukocytes in human skin. *Photochem. Photobiol.*, 69, 148–153 (1999).

23. Ahmed S, Rahman A, Hasnain A, Lalonde M, Goldberg VM, and Haqqi TM. Green tea polyphenol epigallocatechin-3-gallate inhibits the IL-1β-induced activity and expression of cyclooxygenase-2 and nitric oxide synthase-2 in human chondrocytes. *Free Radic. Biol. Med.*, 33, 1097–1105 (2002).

24. Hara Y, and Ishigami T. Antibacterial activities of tea polyphenols against foodborne pathogenic bacteria [in Japanese]. *Nippon Shokuhin Kogyo Gakkaishi*, 36, 996–999 (1989).

25. Nakagawa K, and Miyazawa T. Chemiluminescence-high performance liquid chromatographic determination of tea catechin, (-)-epigallocatechin-3-gallate, at picomole levels in rat and human plasma. *Anal. Biochem.*, 248, 41–49 (1997).

26. Nakagawa K, Okuda S, and Miyazawa T. Dose-dependent incorporation of tea catechins, (-)-epigallocatechin-3-gallate and (-)-epigallocatechin, into human plasma. *Biosci. Biotech. Biochem.*, 61, 1981–1985 (1997).

27. Nakagawa K, Ninomiya M, Okubo T, Aoi N, Juneja LR, Kim M, Yamanaka K, and Miyazawa T. Tea catechin supplementation increases antioxidant capacity and prevents phospholipid hydroperoxidation in plasma of humans. *J. Agric. Food Chem.*, 47, 3967–3973 (1999).

28. Matsuzaki T, and Hara Y. Antioxidative action of tea leaf catechins. *Nippon Nogeikagaku Kaishi*, 59, 129–134 (1985).

29. Hara Y. Process for the Production of a Natural Antioxidant Obtained from Tea Leaves. U.S. Patent No., 4,673,530 (1987).

30. Unten L, Koketsu M, and Kim M. Antidiscoloring activity of green tea polyphenols on b-carotene. *J. Agric. Food Chem.*, 45, 2009–2012 (1997).

31. Nakayama A, and Samo S. Evidence of "flat sour" spoilage by obligate anaerobes in marketed canned drinks. *Bull. Japan Soc. Sci. Fish*, 46, 1117–1123 (1980).

32. Sakanaka S, Juneja LR, and Taniguchi M. Antimicrobial effects of green tea polyphenols on thermophilic spore-forming bacteria, *J. Biosci. Bioengineer.*, 90, 81–85 (2000).

33. Okubo T, and Juneja LR. Effects of green tea polyphenols on human intestinal microflora. In: Yamamoto T, Juneja LR, Chu DC, and Kim M. Eds., *Chemistry and Applications of Green Tea*, CRC Press, Boca Raton, New York, USA, pp. 109–121 (1997).

34. Sakato Y. The chemical constituents of tea: III. A new amide theanine. *J. Agri. Chem. Soc.*, 23, 262 (1950).

35. Chu DC, Kobayashi K, Juneja LR, and Yamamoto T. Theanine — its synthesis, isolation, and physiological activity. In: Yamamoto T, Juneja LR, Chu DC, and Kim M. Eds., *Chemistry and Applications of Green Tea*, CRC Press, Boca Raton, New York, USA, pp. 129–135 (1997).

36. Yokogoshi H, Kato Y, Sagesaka-Mitane Y, Takihara-Matsuura T, Kakuda T, and Takeuchi N. Reduction effect of theanine on blood pressure and brain 5-hydroxyindoles in spontaneously hypertensive rats. *Biosci. Biotech. Biochem.*, 59, 615–618 (1995).

19

Miso as a Functional Food

W. J. MULLIN

Agriculture and Agri-Food Canada,
Ottawa, Ontario, Canada

INTRODUCTION

Miso can be defined in broad terms as a fermented soybean product. It has a pleasant unique fresh sweet fragrance and has been described as having a predominantly salty, "nutty," or "meaty" flavor. Japan is by far the largest producer and consumer where in the 1990s the annual production was approximately 560,000 tons,[1] but has decreased to 48,000 tons in 2002. It is the key ingredient in miso soup, which is a staple part of Japanese cuisine, and has many other culinary uses particularly in marinades and in sauces and as a flavor enhancer. The production and consumption of miso is steeped in tradition; no history of Japan is complete without mentioning miso. The history of its development into the product as it is known today has been well reported by many scholars including Dr. Hideo Ebine,[2,3] who was a former director of the Central Miso Research Institute at the Japanese Federation of Miso Manufacturers Cooperatives in Tokyo. Another excellent history is included in The Book of Miso by Shurtleff and Aoyagi[4] who explore the development through the ages beginning with the first mention of soy sauce and chiang in China going back to the Chou dynasty of 722 to 481 B.C.

Miso is one of the many foods initially developed by ancient civilizations to provide items of high nutritional value that could be used during periods of the year when fresh fruits and vegetables were scarce or unavailable. It was equally important that the foods should be simple to prepare and to store. The basic methods of food preservation, such as dehydration and the use of alcohol and salt, were already well known. As far as miso is concerned the preparation was labor intensive but the whole process of making miso was a family affair and part of the social fabric of family life. Frequently there also was a need to feed vast armies on the move requiring preserved foods, which had to be easily transportable as well as being easy to transform into a meal. Miso is one such food that provides a highly nutritious component to the diet.

Although the origins of miso are lost in time it has survived in essentially its current form for at least a thousand years. It is generally acknowledged that it is derived from the Chinese "chiang," which is a similar fermented soyfood but relies on added spices to make a much more pungent paste. Miso does not contain anything else besides cooked soybeans and a rice, barley, or soybean "koji," plus salt, water, and bacterial cultures. There are several types of miso, which are distinguished according to location of manufacture, relative proportion of ingredients, and the length of time for fermentation, all of which influence the color, texture, and organoleptic properties of the final product. There are also some regional variants of miso such as tofu-misozuke which is produced in the Fukuoku prefecture of southern Japan.[5,6] The main ingredient of all miso is soybeans, which are soaked, cooked, and mashed before mixing with other ingredients and a special *koji*. There are three basic koji types, made from either rice or barley or soybeans, which have been fermented in a separate process before being added to the main mixture. The koji provides a source of oligosaccharides and enzymes for the final fermentation.

Miso is highly regarded in Japan as an important nutritional mainstay of the diet. The preparation of miso has evolved over many centuries[3] though the basic production principles have remained the same and it is still referred to as a brewing process. Where you find miso being made you will frequently find saki (rice wine) production too and inevitably tamari and shoyu, more commonly known as soy sauce. These industries all use the same basic fermentation processes and, more importantly, the same basic technology and expertise to maintain and produce the bacterial cultures. The vast majority of miso producers belong to the local prefecture cooperative and to the national cooperative in Tokyo.[1] These organizations are very active in providing technical and sales support to their members and assistance in sourcing soybeans and rice from outside Japan. In spite of the influx of Western cuisine and fast foods, and although there have been radical changes in production and marketing since the 1950s, miso has maintained a stable market share and value in Japanese food stores.

RAW MATERIALS

Soybeans

To make high quality miso it is essential to have the highest quality soybeans. The most sought after soybeans for miso production are still provided from domestic production in Japan, but imports from China and other producing countries are required to make up the shortfall. Consequently Japanese importers are constantly searching for the best available soybeans for miso production. The qualities should include large size, preferably in the range of 20 grams per 100 seeds, this gives a greater cotyledon to seed coat ratio than is normally found in oilseed varieties. The seed coat (testa) should be intact and shatter resistant under the normal stresses of harvest, cleaning and transportation, it should be very light in color with a white or near white hilum. The seeds should have the capacity to absorb water to increase weight ideally by a factor of 2.0 or more. The preferred varieties should not be genetically modified. Of the macro nutrients in soybeans the carbohydrate content is the most critical, particularly the oligosaccharides. The proteins and oil content do not seem to have as much influence on the final product as the sugars. The standard methods for measuring the soluble sugars are based on extraction using mild acid conditions and a colorimetric determination of the concentration. More specific chromatographic methods have been developed, which provide detailed information of the individual sugars, but there have been no studies on how the balance of these sugars affect the organoleptic properties of the final product. Only about 10% of the total annual soybean requirements for making miso in Japan are from domestic producers, the rest are imported from China and more recently from Canada, the U.S. and South America. Significant changes in quality may occur during storage at port facilities and during shipment due to fluctuations of temperature and humidity. This means that additional basic tests are required when the shipment reaches its final destination.

Rice

The production of rice in Japan is maintained at a high level by government control. Once again the highest quality is called for in the production of *koji*. Japan is currently self sufficient in rice production and home grown rice is preferred for *koji* preparation. In years where there are shortages, most of the rice for *koji* production is imported from Thailand where high quality short grain, nonglutinous rice *Oryzae sativa*, var. japonica is produced. The rice is polished to remove the outer so-called bran layers before fermentation into the *koji*.

Barley

High quality barley with a high protein content is used to make *koji*. The barley is milled, or "pearled," to remove the outer bran layers and some of the endosperm resulting in a yield of about 70% of the total weight of the original grain. The barley retains its "crease," which is still quite visible in the finished "rough" miso types.

Salt

Sea salt is most commonly used in miso manufacture as it reputedly provides some of the more subtle flavors to the final product. The less expensive salt, refined from rock salt deposits, is also used though the inclusion of too much iron has to be monitored.

Water

The purity of the water is critical to the production of miso; spring water is preferred if possible but treated water supplies are also used. In modern miso production plants the mineral inclusions are carefully monitored to prevent any variation in organoleptic properties and color density, and any disruption of the health of bacterial cultures used for fermentation.

Bacterial Cultures

In the traditional or "farmhouse" production the fermentation is started by using a small portion of the last batch of miso as the source of bacterial cultures. Although this perpetuates the distinctive organoleptic properties to the product it also increases the probability of introducing unwanted bacterial strains that might have developed inadvertently, causing problems with "off" flavors. In modern miso production, in the small-scale batch-type system as well as in large-scale, pure cultures are used, though in some facilities a small amount of mature miso is also added after it has been established that there are no unwanted bacterial strains present. There are two basic specialized cultures used for fermentation in miso production. Both are halophilic since the normal salt content is about 12% in the finished product. The standard yeasts are *Zygosaccharomyces rouxii* and *Candida versalitis*, and the lactic acid-producing bacteria is *Tetragenococcus halophilus*. These yeasts and bacteria are produced commercially by the local prefecture cooperatives and in the laboratories of the larger miso producers where quality control and a consistent product are of utmost importance. A large amount of research is devoted to improving and maintaining the purity of the bacterial cultures.

Koji Starter

The preparation of the *koji* starter, or "*tane-koji,*" is an industry itself. The starter is a preparation from the spores of *Aspergillus oryzae* on cooked brown rice that has some added ash from hard wood to provide trace minerals. This innocculum is used in the preparation of large quantities of *koji* prior to mixing with the other ingredients for miso fermentation. There are many different strains of *A. oryzae* used in making the *koji* which will impart variations to the miso fermentation and consequently the distinctive flavors.

PROCESSING TECHNOLOGY

Going back in history the preparation of miso was a family affair that took place after the fall harvests. This scenario was common to many cultures where food preservation had to be done in times of plenty, which were followed by less plentiful food supplies due to seasonal changes of the climate. Gradual changes have taken place as the day-to-day living habits have adjusted to greater urbanization and to advances in the food preservation and distribution systems. The use of technological advances and the economics of large size manufacturing plants have completely altered food manufacturing and distribution patterns. Although there are still some miso makers using traditional methods in rural areas for personal use, and also some commercial miso makers using the old traditional methods for niche markets, the vast majority is made in modern processing facilities under strict hygiene controls of ingredients and cultures from start to finish.

After the raw materials have been acquired the process of making miso begins first with the preparation of the koji. For rice or barley koji the polished grain is soaked, steamed and after cooling it is laid in trays of bamboo or perforated stainless steel 5 to 10 cm deep and inoculated with the tane-koji. The fermentation takes place at a controlled temperature, usually 30 to 35°C, and with controlled humidity. Clumps are formed as the fermentation progresses, which are periodically either manually or mechanically broken up. In the large processing facilities the scale is increased to capacities of several tons. Enclosed automated systems with modern mechanical devices reduce the probability of contamination from airborne bacteria. Within an enclosed unit the circular perforated steel "beds," several meters in diameter, are mechanically filled to a depth of several centimeters with cooked grain then inoculated with the tane-koji. Temperatures and humidity are closely monitored and controlled. The developing koji is raked to prevent the build up of clumps, and after the process has completed the finished koji is transferred to temporary storage or directly to the mixing facility. Koji is usually made on a demand basis but it can be stored for a short time in the moist state and longer if it is dried. The way the koji is prepared and, more importantly the choice of *A. oryzae* strain, has a very direct effect on the properties of the miso. The subtle differences in the resulting rice or barley koji will influence the way that the enzymes degrade the soybean proteins, carbohydrates, and lipids contributing to the distinctive organoleptic properties of the final product.

Once the koji has been prepared and the yeast and lactic acid cultures have been activated and increased in a culture medium then the miso fermentation can begin. The soybeans are mechanically cleaned to remove any foreign matter, especially stones, then soaked at room temperature for up to 24 h. After draining, the soybeans are cooked with steam using a continuous rotary cooker for large-scale production. In smaller batch processes the soybeans are boiled in water at atmospheric pressure to begin with and then for a short time at about 1.1 atmos. The method of cooking has some influence on the color development of the miso, steaming results in less color development than boiling. The cooking temperature and duration of cooking affects the denaturation of protein which becomes an important factor in the enzymatic degradation from koji enzymes. It affects the texture as well as the digestibility. The precise processing parameters are generally developed by each miso maker and will vary for each type of miso, the details are rarely published. The variation in quality according to the soybean variety, storage time, and storage conditions is an important factor in deciding the cooking time. Most processors will do a small test cook to set the soaking and cooking parameters before the production run.

Using the old traditional methods the batches could be as small as 500 g balls of miso laid in woven bamboo trays or as large as a 5- to 10-ton capacity circular vat made of cedar planks bound with bamboo hoops and topped with heavy smooth boulders to aid in expressing fermentation gasses. The fermenters would be housed in protective sheds to shield the miso from the extreme effects of rain, snow, or sun but allowed to be at ambient temperatures, which would mean fast fermentation in summer and slow in winter. The modern stainless steel, plastic-lined batch fermenters contain up to 20 tons each and are kept at constant temperature, usually about 30°C for the entire fermentation period.

The salt content of miso is ~13% by weight, which allows it to be displayed at ambient temperatures for retail. It is recommended that the unopened package can be stored safely for up to a year. High temperature, exposure to air and light may cause a deepening of color at the surface and slight changes in taste. After opening the miso should be refrigerated and can be stored safely for several months. Flexible packages of miso from single use 10 g size to larger 0.5 to 1.0 kg packages are commonly available. In the larger grocery and specialty food stores and markets in Japan miso is sold from large display tubs; 20 or more different *misos* may be on display and available for purchase.

ORGANOLEPTIC PROPERTIES

The flavors, textures, and aromas of miso, collectively known as the organoleptic properties, are extremely complex.[7,8] The regional preferences of the three main types, rice, barley or soybean miso, are well defined though this does not mean to say that where one type is favored the others can not be found. Rice miso is by far the most common, accounting for about 80% of total national production. Barley miso, the second most popular, comes from the southwest of Japan, Kyushu, Shikoku, Kanto, and Chugoku districts. Soybean miso is only produced in a small area of the Tokai-Hokuriku district, southern Japan, in the prefectures of Aich, Mie, and Gifu. The soybean miso is the most expensive, partly due to the small production but more importantly, because of the longer time needed for the fermentation to reach completion. Within a miso type there are regional differences that distinguish one miso from another and each takes on the name of that district. The color of the miso ranges from a pale yellow or buff to a deeper red of rice and barley miso to the black of soybean miso. The color development of the finished product is the result of a number of parameters including the inherent color of the cotyledons and/or seed coat, whether the soybeans were boiled or steamed, the total cooking time, the ratio of koji to soybeans, and exposure to light after it is packaged for sale. The fermentation time is a critical factor in the production of monosaccharides, oligosaccharides, amino acids, and peptides that would react together and produce color compounds.[9] "Smooth" types are ground to a smooth paste before retailing whereas "rough" types are sold without further processing after the completion of the fermentation. In rough miso a small amount of the rice or barley grain is still visible though it hardly affects the texture at all. Smooth types are generally preferred to rough but each type has its own following.

As with other fermented food and beverage products the unique nuances in miso are difficult to describe, but the most subtle are associated with the flavors and aromas. In Japan miso is frequently finished by adding 2% ethanol, which effectively arrests the fermentation and produces the predominating aroma of fresh miso. Fresh miso has a pleasant, light and sweet odor, which is not at all strong, and an appearance that is somewhat shiny on the surface termed "brightness" by the producers. The flavor is dominated by the salinity, but there

TABLE 19.1   Flavor and aroma compounds found in *Miso*

Alcohols	Esters	Acids
Ethanol	Ethyl propionate	Pentanoic
n-Propanol	Ethyl palmitate	Hexanoic
iso-Propanol	Ethyl stearate	
n-Butanol	Ethyl oleate	Aromatics
iso-Butanol	Ethyl linolate	Benzaldehyde
iso-Amyl alcohol	Ethyl linoleate	2-Phenyl-2-butenal
2,3-Butanediol	Methyl hexadecanoate	Benzenemethanol
Hexanol		4-Ethylphenol
1-Octen-3-ol		4-Ethylguaicol
		2-methoxy-4-vinylphenol

Furanones
3-Methyl-2(5H)-furanone
4-Hydroxy-5-methyl 3-(2H)-furanone (HMF)
4-Hydroxy-2,5-dimethyl-3(2H)-furanone (HDMF)
4Hydroxy-2(or5)ehtyl-5(or2)-methyl3(2H) furanone (HEMF)

are other distinct flavors that blend together to give each type of miso its overall unique organoleptic character.

The blandness of soybeans and nonfermented soy foods is totally changed by the fermentation process and the addition of *koji* and salt. Under the essentially anaerobic conditions a wide range of volatile compounds is produced via the fermentation bacteria and the enzymes associated with the koji.[7,8] Some of the flavor and aroma compounds that have been found in miso are listed in Table 19.1 but this is by no-means a complete list. Several normal and substituted short-chain alcohols have been isolated as well as a variety of long- and short-chain fatty acids and esters. Volatile aromatics as well as furanones have also been detected. The concentration of these compounds is a function of the type of miso and the length of time that the fermentation has been allowed to progress and thus there is more flavor and aroma to the more mature miso. In addition the different strains or sources of the fermentation bacteria are a significant influence on the quantities of these fermentation products too.

NUTRITIONAL PROPERTIES

Miso has been an integral part of the diet in Japan for centuries and although the availability of alternative foods has become more and more common since the beginning of the 20th century the basic diet has remained the same. There is still a predominance of rice, fish, wheat, and soy-based products at each meal presented with great pride in a truly artistic fashion. The general health of the population is testament to the superior diet and food supply with a much lower incidence of common Western diseases such as bowel and bladder cancers, lower heart disease, and hypertension, which can all be linked to lifestyle. Since miso is such a major component of the diet it is thought that it has a direct influence on health in general. Paradoxically the high-salt content of miso does not seem to be a factor in hypertension in Japan. It is a very complex mixture of nutrients and it is often very difficult to define the specific cause-effect relationship on health. It is obviously a source of carbohydrates, protein, and lipids but it has health-giving properties beyond providing energy and maintaining a healthy body. The earlier anecdotal evidence of the health-giving properties of miso have been confirmed and enhanced by more sophisticated modern investigations using specific extracts in model systems. There still remains a large amount of evidence pointing to the synergistic multi-component effects of miso on health. One of the most interesting modern uses of miso was its use in treating victims of radiation sickness from exposure after the atomic bomb explosions in Japan in 1945.[10] The overall health-giving properties were recognized and exploited.

The nutritional properties are derived from all the components of the miso. The diverse mixture of macronutrients, including proteins, lipids, and carbohydrates, as well as the micronutrients, including the phytochemicals, are transformed by the *koji* enzymes and the bacterial fermentation.[11,12] Some of the soybean proteins are hydrolysed by the proteases that are present in the koji resulting in the production of more easily digestible amino acids and peptides.[12] The enzyme activity depends on the source, i.e., manufacturing conditions and the *A. oryzae,* a kind of fungi called yellow koji, used in the preparation of the koji. One of the consequences of proteolysis is the production of peptides that induce a reduction in hypertension. The peptides inhibit effects of the angiotensin I converting enzyme (ACE).[13] Soy sauce extracts have been shown to contain alkaloidal components that have antiplatelet activity[14] and it can be assumed that miso will do the same since soy sauce is produced in a process similar to miso fermentation.

With the addition of rice or barley *koji* to the fermentation mix the monosaccharides and oligosaccharides are initially significantly increased but as the fermentation progresses they are reduced.[15,16] The soybean galactose-containing oligosaccharides are almost eliminated, which in turn greatly reduces the problems associated with subsequent fermentation and gas formation in the colon. Initially glucose is increased due to the glucosidase from the *koji* but it is then fermented by the halophilic yeasts. Dietary fiber is reduced from the initial high-prefermentation levels but it is still significant in the final product.[17]

The effects of soy phytochemicals on heart disease[18] are well documented, but it appears that the isoflavones and their derivatives are just as important in the reduction of the incidence of cancers.[19] Extracts of miso have been used in studies to determine the chemoprevention effects on induced colonic tumors and gastric tumors in experimental animals.[20,21] In another study the chemoprevention of *N*-nitroso-*N*-methyl-urea-induced mammary cancers by soyfoods and by miso alone and in combination with tamoxifen has been studied[22–24] as well as suppressive effects on *Salmonella typhimurium*-induced mutagens.[25,26] The antioxidant activity of isoflavones is greatly enhanced by the miso fermentation process, which appears to be due to the production of *o*-dihydroxyisoflavones and is most effective in soybean miso compared to rice and barley miso.[27–30] There is also a synergistic effect with tocopherol and ascorbic acid.[31] Extracts have shown inhibitory effects on isolated lines of cancer cells in model systems.[32–34]

STANDARDS AND REGULATIONS

The basic standards for miso production are covered by Japanese government regulations from the Ministry of Agriculture, Forestry, and Fisheries. These standards are strictly enforced and monitored in cooperation with the national and prefecture miso cooperatives.

The system of miso cooperatives in Japan is highly organized and an essential part of the national miso making community. The cooperatives are organized into 51 unions throughout Japan, these are grouped together into 8 national blocs on geographical lines. The blocs come together as the national organization known as Japan Federation of Miso Manufacturers Cooperatives (JFMMC) with headquarters in Tokyo. The JFMMC represents about 1,400 miso manufacturers of all sizes. All producers are eligible to become members of the local organization, and the vast majority take advantage of the membership, which can provide many benefits. The services at the local level depend on the size of the cooperatives but they can all call on assistance through the bloc association and then through the national JFMMC. The JFMMC operates the Central Miso Research Institute, which conducts basic research, tests new domestic and imported varieties, troubleshoots processing problems, and prepares and distributes cultures for the fermentation. The JFMMC is very active in public relations and publicity and also sponsors research into the health benefits and nutritional value of using miso. It publishes an annual report, which includes news on research as well as detailed production figures.

There are annual miso competitions at the local and national level that keep miso in the spotlight and do much to maintain and improve the quality of miso. These are usually scheduled to take place toward the end of the year with the national JFMMC competition held last in early November in Tokyo as a climax to the production year. The competitions are keenly contested with several hundreds of entries and are widely reported in the national and local media. Judging is a very demanding technique and it takes several years to become proficient enough to be appointed to a panel. The competition misos are submitted weeks in advance and even before being accepted tests are applied to make sure the samples are made according to specifications and no "foreign" ingredients have been included. There are divisions for each type of miso at the local and national levels and in Tokyo new products based on the traditional fermentations are exhibited. Winning a class can be very beneficial to the promotional aspects of marketing. New uses for miso are also on display with leading chefs preparing dishes that are provided to the members of JFMMC, competition participants, and to the media. A surprising range of foods have been developed using miso as an ingredient including sauces, marinades, salad dressings, and such unlikely products as hard candies.

REFERENCES

1. Japan Federation of Miso Manufacturers Cooperatives, Annual Report, 2000.

2. H. Ebine. Miso. In: N.R. Reddy, M.D. Pierson, D.K. Salunkhe, Eds. *Legume-Based Fermented Foods*, CRC Press, Boca Raton, FL, pp 48–49.

3. H. Ebine. Industrialization of Japanese Miso Fermentation. In: KH Steinkrause, Ed. *Industrialization of Indigenous Foods*. New York and Basel: Marcel Dekker, 1989, pp. 90–125.

4. W. Shurtleff, A. Aoyogi. A History of Miso. In: *The Book of Miso*. Berkeley: Ten Speed Press, 1983, pp. 214–241.

5. J. Funaki, M. Yano, H. Hayabuchi, S. Aria. Proteolysis of tofu-misozuke during ripening. Studies on tofu-misozuke prepared in Fukuoka Prefecture. *Nippon Shokuhin Kagaku Kaishi* 43:546–551, 1996.

6. J. Funaki, M. Yano, T. Misaka, K. Abe, S. Aria. Purification and characterization of a neutral protease that contributes to the unique flavor and texture of tofu-misozuke. *J. Food Biochem.* 21:191–202, 1997.

7. E. Sugawara, Y. Yonekura. Comparison of aroma components in five types of miso. *Nippon Shokuhin Kogaku Kaishi* 45:323–329, 1998.

8. Y. Hayashida, K. Nishimura, J.C. Slaughter. The importance of the furanones HDMF and HEMF in the flavour profile of Japanese barley miso and their production during fermentation. *J. Sci Food Agric.* 78:88–94, 1998.

9. S. Hondo and T. Mochizuki. Effects of saccharides on the formation of texture and color of miso. *Nippon Shokuhin Kogyo Gakkaishi* 26:509–513, 1979.

10. A. Ito, T. Gotoh, N. Fujimoto. Chemoprevention of cancers by miso and isoflavones. *J. Toxicol. Pathol.* 11:79–84, 1998.

11. T. Mochizuki, H. Yasuhira, S. Hondo, I. Ouchi, K. Rokugawa, K. Itoga. Studies on the changes of several components during miso making. *Fermentation Technol. Today.* 663–668, 1972.

12. S. Nikkuni, N. Okada, H. Itoh. Effect of soybean cooking and temperature on the texture and protein digestibility of miso. *J. Food Sci.* 53:445–449, 1988.

13. S. Nikkuni, N. Okada, H. Itoh. Effect of soybean cooking temperature on the texture and protein digestibility of miso. *J. Food Sci.* 53:445–449, 1988.

14. T. Teranaka, M. Ezawa, J. Matsuyama, H. Ebine, I. Kiyosawa. Inhibitory effects of extracts from rice-koji miso, barley-koji miso and soybean-koji miso on the activity of Angiotensin 1 converting enzyme. *Nippon Nogeikagaku Kaishi* 60:1163–1169, 1995.

15. H. Tsuchiya, M. Sato, I. Wantanabe. Antiplatelet activity of soy sauce as functional seasoning. *J. Agri. Food Chem.* 47:4167–4174, 1999.

16. S. Hondo, T. Mochizuki. Polysaccharides in soybean-steamed waste water and miso. *Nippon Shokuhin Kogyo Gakkaishi.* 26:461–468, 1979.

17. S. Hondo, T. Mochizuki. Free sugars in miso. *Nippon Shokuhin Kogyo Gakkaishi.* 26:469–474, 1979.

18. E. Takeyama, M. Fukushima, A. Kawarada, S. Okamoto. Dietary fibre contents of soybean and soybean foods. *Nippon Shokuhin Kogyo Gakkaishi* 33:263–269, 1986.

19. A.L. Lichtenstein. Soy protein, isoflavones and cardiovascular disease risk. *J. Nutr.* 128:1589–1592, 1998.

20. A. Ito, T. Gotoh, N. Fujimoto. Chemoprevention of cancers by miso and isoflavones. *J. Toxicol. Pathol.* 11:79–84, 1998.

21. Y. Masaoka, H. Wantanabe, O. Katoh, A. Ito, K. Dohi. Effects of miso and NaCl on the development of colonic aberrant crypt foci induced by azomethane in F344 rats. *Nutr. Cancer.* 32:25–28, 1998.

22. H. Wantanabe, T. Uesaka, S. Kido, Y. Ishimura, K. Shiraki, K. Kuramoto, S. Hirata, S. Shoji, O. Katoh, N. Fujimoto. Influence of concomitant miso or NaCl treatment on induction of gastric tumors by N-methyl-N^1-nitro-N-nitrosoguanidine in rats. *Oncology Reports,* 6:989–993, 1999.

23. T. Gotoh, K. Yamada, A. Ito, H. Yin, T. Kataoka, K. Dohi. Chemoprevention of N-nitroso-N-methylurea induced rat mammary cancer by miso and tamoxifen, alone and in combination. *Japan J. Cancer Res.* 89:487–495, 1998.

24. T. Gotoh, K. Yamada, A. Ito, H. Yin, T. Kataoka, K. Dohi. Chemoprevention of N-nitroso-N-methylurea-induced mammary carcinogenesis by soyfoods or biochanin A. *Japan J. Cancer Res.* 89:137–142, 1998.

25. J.E. Baggott, T. Ha, W.H. Vaughn, M.M. Juliana, J.M. Hardin, J. Grubbs. Effect of miso (Japanese soybean paste) and NaCl on DMBA-induced rat mammary tumors. *Nutr. Cancer* 14:103–109, 1990.

26. I. Kiyosawa, J. Matsuyama, C. Aria, T. Setoguchi. Suppressive effects of the methanol extracts from soybean products on SOS response of *Salmonella typhimurium* induced by mutagens and their contents of isoflavones. *Nippon Shokuhin Kagaku Kogaku Kaishi.* 42:835–842, 1995.

27. I. Kiyosawa, W. Miura, T. Sato, M. Yonenaga, H. Ebine. Suppressive effects of the methanol extracts from miso on SOS response of *Salmonella typhimurium* induced by mutagens and their isoflavone contents. *Food Sci. Technol. Int.* 2:181–182, 1996.

28. H. Esaki, R. Wantanabe, H. Masuda, T. Osawa, S. Kawakishi. Formations and changes of o-dihydroxyisoflavones during the production of soybean pastes. *Nippon Shokuhin Kagaku Kaishi.* 48:189–195, 2001.

29. H. Esaki, S. Kawaishi, T. Inoue, T. Osawa. Potent antioxidative o-dihydroxyflavones in soybean pastes and their antioxidative activities. *Nippon Shokuhin Kagaku Kogaku Kaishi* 48:51–57, 2001.

30. H. Esaki, H. Onozaki, Y. Morimitsu, S. Kawakishi, T. Osawa. Potent antioxidative isoflavones from soybeans fermented with *Aspergillus saitoi. Biosci. Biotechnol. Biochem.* 62:740–746, 1998.

31. N. Yamaguchi, S. Akatsuka. Antioxidant preparations from non-salted miso. *Nippon Shokuhin Kogyo Gakkaishi.* 31:278–280, 1984.

32. H. Esaki, H. Onozaki, Y. Morimitsu, S. Kawakishi, T. Osawa. Potent antioxidative isoflavones isolated from soybeans fermented with *Aspergillus saitoi. Biosci. Biotechnol. Biochem.* 62:740–746, 1998.

33. Y. Miyama, H. Shinmoto, M. Kobori, T. Tsushida, K. Shinohara. Inhibitory effect of extracts of miso on melanogenesis. *Nippon Shokuhin Kogaku Kogaku Kaishi.* 43:712–715, 1996.

34. A. Hirota, S. Taki, S. Kawaii, M. Yano, N. Abe. 1,1-Diphenyl-2-picrylhydrazyl radical-scavenging compounds from soybean miso and antiproliferative activity of isoflavones from soybean miso toward the cancer cell lines. *Biosci. Biotechnol. Biochem.* 64:1038–1040, 2000.

20

Fermented Soybean Products as Functional Foods: Functional Properties of Doenjang (Fermented Soybean Paste)

KUN-YOUNG PARK and KEUN-OK JUNG

Pusan National University, Busan, Korea

INTRODUCTION

Korea has a long history of eating fermented soybean foods. The historical record indicates that the cultivation of soybeans originated in Manchuria, which was part of Korea in ancient times.[1] Doenjang (Korean fermented soy paste) is a common fermented soybean food that was developed in Korea along with other processed soybean foods. Kanjang is a fermented soy sauce that is obtained during the doenjang preparation.

Historically, soybeans and processed soybean foods have been the main protein sources in the Korean diet. Doenjang is a traditional food that is both a rich protein source as well as a seasoning for enhancing the taste of foods. Doenjang has also been used as a folk medicine for emergency treatments such as removing toxins from insect and snake bites or for stopping bleeding, etc. The medicinal functions of doenjang were first described in Dongeuibogam (1613 A.D.), which was a popular traditional Korean medical text.

The functional characteristics of doenjang and other traditional Korean soybean foods have not been reported in the scientific literature until recent years, although they have a long history of use. On the other hand, some investigators have suspected that doenjang might cause stomach cancer because of possible contamination with aflatoxigenic fungi such as Aspergillus flavus in the fermented soybeans (meju), producing aflatoxin, a known potent carcinogen.[2]

The possible aflatoxin contamination from aflatoxigenic mold during doenjang fermentation has been studied.[3,4] Although aflatoxins can be produced during meju fermentation if the meju is inoculated with *Asp. parasiticus*, the toxins produced are mostly degraded and removed during the manufacture of doenjang.[3-9] On the other hand, doenjang exhibits strong antimutagenic activity against various mutagens/carcinogens, and anticancer activities in *in vitro* and *in vivo* experimental systems.[10-21] Doenjang also exhibits other functionalities such as antioxidative effects[22-33] and a reduction in the incidence of cardiovascular diseases.[34-43]

HISTORY OF KOREAN SOYBEAN FERMENTED FOODS

The cultivation of soybeans has been traced back to the ancient Kokuryo dynasty of Korea in what is now southern Manchuria in China. It is believed that soybeans and soybean foods have been consumed in Korea since the third century B.C.[44] The word, Jang, which is a Chinese character meaning fermented soybean in Korean, originated in China in Jure about the second century B.C., but Jang meant fermented animal meat in Chinese. Samkukji, written by a Chinese in the third century A.D., reported that Koreans prepared fermented foods, especially from soybeans, very well and praised the fermentation skills of the Koreans.[45] The fermentation technology of the Koreans was so advanced that they shared the techniques with neighboring countries. The Chinese agricultural technology book, Jeminyosul (A.D. 530 to 550) written by a governor, Maeeunsa, stated that shi (bacteria-fermented soybean) in Korea migrated to China and Japan.

The words, Jang (mold-fermented soybean) and shi, were shown in a February article by King Sinmoon in his third year (683 A.D.) during the Silla dynasty; it had also been reported that soybeans were cultivated in A.D. 1C (99 A.D.). Daeboyulryong (701 A.D.) mentioned the words jang, shi and maljang that described soybean products, and Jungchangwonmoonseu (739 A.D.) also mentioned maljang. *Donga*, written by Shinjungbaesuk in Japan, indicated that maljang was imported from Korye (the old name of Korea) and renamed it miso.[46] The Japanese Jang miso seems to have further been developed into the Japanese traditional miso using rice-soybean meju instead of only soybean maljang.

During the Korye dynasty, the name of maljang changed to maeyjo and then to meju, the maljang soaked in brine in a clay pot and ripened; the solid sediment was called doenjang (soy paste) and the watery part was called kanjang (soy sauce). Dongeubogam (A.D. 1613), written by Hurjun, which is a famous Korean medicine book, described how to make medicinal doenjang using soybeans and how to fix soured doenjang. Jungbosanlimkyungje written by Yojungim (A.D. 1760) introduced 45 different processing methods of soybean foods, describing how many days fermentation for making jang, selection of water, salt quality, how to handle the pottery, fixing jang with an off-taste, etc. Kyuhapchongseo written by Madam Lee (1759 to 1824) wrote and described the proper preparation methods for various jangs in great detail.

Commercial production of fermented soybean products (jang in Korean) started around 1930 A.D. Japanese-built jang factories in Korea during the occupation (1909 to 1945 A.D.) to supply soybean products for the Japanese in Japan. After liberation from Japan in 1945, Koreans took over the factories. Since the Korean war (1950 to 1953), military personnel and people living in large cities have mostly consumed commercial fermented soybean products due to shortages of fermentation space and the time required for the preparation of the Korean traditional soybean products. However, traditional soybean fermented products are still prepared by families living in rural areas. Recently, the large factories have begun producing traditionally fermented Korean soybean products as well as manufacturing the modified Japanese-style soybean products.

MANUFACTURING METHODS AND CHARACTERISTICS OF DOENJANG FERMENTATION

There are two ways of preparing traditional doenjang, traditional homemade doenjang, and commercially made doenjang. The standardized traditionally prepared doenjang process is shown in Figure 20.1.[7,47] Meju preparation is the first step in making doenjang. Meju is a naturally fermented soybean block and the main ingredient for making doenjang. The meju microorganisms are sources of enzymes in the fermentation of soybeans. The macromolecules; protein, carbohydrates, and fats, of the soybeans are degraded into small molecules of peptides, amino acids, oligosaccharide, sugars, free fatty acids and various processed phytochemicals from the soybeans during the fermentation of meju and the further fermentation to doenjang.

The standardized process for making doenjang is as follows; soybeans are soaked in water for 12 h at 15°C and cooked for 4 h at 100°C or autoclaved ($1kg/cm^2$) for 30 min. The cooked soybeans are crushed and molded ($(8 - 12) \times (12 - 18) \times (15 - 25)$ cm in size) into an $8 \times 12 \times 20$ cm block. The soybean blocks are dried for 3 days in the air, tied up with rice straw and then traditionally hung at the edge of an eave for about 1 to 2 months to initiate the natural fermentation. The *Bacillus sp.* in the inner part of the meju, and molds and yeasts on the outside are involved in the fermentation. The mejus are washed with water and the mold, straw, and dirt brushed from the surface. The meju is then placed in an earthen jar, and

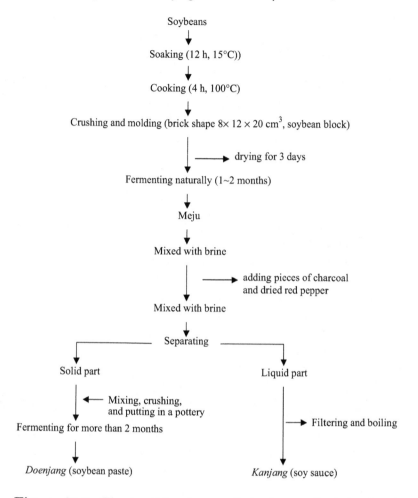

Figure 20.1   The traditional manufacturing method for doenjang and kanjang.

brine is added at ratios of meju, salt and water of 18.4 ± 4.4:14.6 ± 2.1:67.0 ± 4.4 (approx. 1:1:4 w/w/w).[47] Dried red peppers and pieces of charcoal (6 and 4 pieces, respectively) are added to the mixture and fermented for 2 to 3 months in the clay pot. During this period, the amino acids and sugars formed by the enzymatic hydrolysis of the soy proteins and carbohydrates exude out to the brine where they undergo the

Malliard reaction giving a dark brown color to the products. The halophilic yeasts, such as *Sacchromyces rouxii*, grow in the mixture and produce alcohols and other organic compounds, which give the flavor to the soy sauce.[1]

The meju-brine mixture is then filtered. The liquid part of the mixture is kanjang (soy sauce), which can be boiled to destroy microorganisms and remove off-flavors. The crushed solid part is packed into another clay pot, additional natural salt is spread on top to prevent microbial contaminations, and fermentation allowed to continue for an additional 2 to 6 months to make doenjang. The doenjang can also be prepared without separating the liquid part; when this is done less water is added to the meju and salt mixture and then it is fermented without separating out the soy sauce.

Soybeans naturally contain *Bacillus* sp., especially *Bacillus subtilis*. Various molds such as *Rhizopus, Mucor, Penicillium*, and *Aspergillus* sp. are also involved in meju fermentation. About 20 different molds have been detected in naturally fermented meju samples collected from various regions of Korea. However, *Asp. oryzae, Penicillium* sp, *Mucor* sp. and *Rhizopus* sp. are the predominant fungi.[48] Among the eight kinds of bacteria isolated, *Bac. subtilis* was the predominant species detected. The ratio of Bacillus sp. to molds is about 100:1, thus the major microorganism for meju fermentation are *Bacillus* sp. During ripening of meju in brine, yeasts such as *Saccharomyces* sp. and lactic acid bacteria are involved along with the molds and the *Bacillus* sp.[49] During the ripening of soybean paste, again *Bacillus* sp. are involved. *Bacillus* sp. can survive the long heat treatment of soybeans[50] and become the major microorganism during doenjang fermentation.

FUNCTIONAL PROPERTIES OF DOENJANG

Nutritional and Functional Components in Soybean and Doenjang

Soybeans used to be called "animal meat produced in the field" in Korea. The soybean typically contains 9.7% moisture, 36.2% protein, 17.8% lipid, 25.7% carbohydrate, 5.0% fibers, and 5.6% ash.[51] Depending on the variety, soybeans contain 33 to 47% protein, 11 to 23% lipid, 21 to 28% carbohydrates, and lesser amounts of other functional phytochemicals.

When 15 traditionally prepared doenjang samples were analyzed, they averaged 54.7% water, 13.8% crude protein, 8.0% crude lipid, 14.4 ml titratable acidity, and 11.8% salt. The level of free amino acid was 3.81% of which glutamic acid (25%) was the highest, but leucine, alanine, histidine, lysine, proline, and valine were also found in relatively high concentrations. Lactic acid was the most abundant organic acid, and acetic, malic, citric, and oxalic acids were also detected. Linoleic acid (52.2%), oleic acid (20.7%), and linolenic acid (8.7%) were the most abundant unsaturated fatty acids, comprising 81.6% of the total fatty acids.[52] Doenjang also contained oligosaccharides and 3.1% dietary fiber. The digestibility of raw soybean is 55%, cooked soybean 65%, but doenjang is 85%.[53]

The functional compounds found in soybeans and doenjang include the following: protease inhibitor (Bowman-Birk inhibitor), peptides, and amino acids from proteins; oligosaccharides of raffinose and stachyose and dietary fibers from carbohydrates; fatty acids (oleic, linoleic, and linolenic acids), vitamin E, and lecithin from the lipid fraction; the minerals P, Ca, Mg, S; and phytochemicals such as phytic acid, saponins, squalene, sterols (β-sitosterol), isoflavones (daidzein, genistein), phenolic acids (syringic, vanillic, chlorogenic, ferulic, cinnamic acids, etc), lignan, carotenoids (lutein, α-, and β-carotenes), etc. Some of the above compounds can be obtained by eating soybeans and some are formed during the fermentation of doenjang; many exhibit antimutagenic, anticarcinogenic, antiaging, antioxidative, antiarteriosclerosis, and other beneficial properties.

Safety of Doenjang

Since doenjang is traditionally manufactured by natural fermentation, Crane et al.[2] suspected that doenjang might be contaminated with aflatoxins as a result of contamination with aflatoxigenic molds during the fermentation. They indicated that the high incidence of stomach cancer in Koreans was probably due to the consumption of the aflatoxin contaminated doenjang. Aflatoxins, especially aflatoxin B_1 (AFB_1) are very potent carcinogens and are produced by *Asp. flavus* and *Asp. parasiticus* as secondary metabolites during storage of foods such as peanuts and corn.[54]

In order to confirm whether aflatoxin contamination is possible during the meju and doenjang fermentation, experiments were carried out by inoculating meju with *A. parasiticus*. Aflatoxins were produced in the inoculated meju as shown in Figure 20.2. AFG_1 production was high, though it degraded quickly, whereas AFB_1 synthesis was low during the 4 weeks of fermentation. After 2 months of fermentation in brine, aflatoxin content decreased to 10 to 20% of the original level. Almost 100% of aflatoxins produced in the fermented meju were degraded after 3 months of ripening in the brine.[3,4]

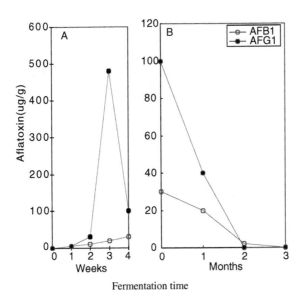

Figure 20.2   Aflatoxin B_1 and G_1 production by *A. parasiticus* with *A. oryzae* and *B. subtilis* during the fermentation of meju block (A) made from var. Jangyeop soybean and degradation of aflatoxins during ripening of the fermented meju in brine (B) for 3 months.

Thus if the ripening period is long enough, most of the aflatoxins that might be produced during fermentation can be degraded. It has been suggested aflatoxins are degraded by ammonia that is formed during the fermentation of the proteins in the soybeans, as well as by the pieces of charcoals added to the mixture, light, competitive growths of various microorganisms (mixed culture condition), melanoidin formation during the fermentation, etc.[3-7] The main fermenting microorganism, *B. subtilis* during doenjang fermentation significantly inhibits the growth of *Asp. flavus* and *Asp. parasiticus*.[4,8,9] Soybean and meju were very poor substrates for aflatoxin production.[55]

There was considerable toxin degradation that occurred during fermentation, even when *A. parasiticus* was intentionally inoculated into the soybean block. Thus, even though aflatoxin contamination is possible under natural conditions, the levels are very low and the toxins would probably be degraded during the fermentation process, therefore doenjang is essentially safe from aflatoxin contamination.[7]

Fermented soybean paste also exhibited antimicrobial activity against other pathogenic microorganisms, including: *B. cereus, E. coli, Listeria monocytogenes, Staph. aureus, Strep. faecalis, E. coli 0157:H7, B. subtilis* and *Sal. typhimurium*. However, nonfermented soy paste did not show antimicrobial activities against *B. cereus, E. coli, Listeria monocytogenes* and *Strep. faecalis*. The antimicrobial compounds in the fermented soy paste appear to be peptides, 4-hydroxy benzoic acid and benzoic acid formed during the fermentation,[56] which supports the antibiotic properties attributed to doenjang as a traditional folk medicine.

Antimutagenic Activity of Doenjang

The effect of doenjang on aflatoxin B_1 (AFB_1)-induced mutagenicity has been evaluated along with other fermented soybean foods and rice.[14,21] Strong antimutagenic activity toward AFB_1 was observed by treatment with a methanol extract of doenjang. The AFB_1-mediated mutagenesis was completely inhibited at the level of 25mg/plate of the traditional doenjang extract.

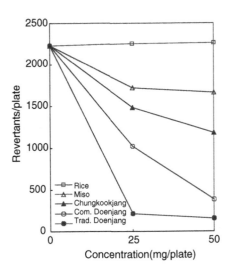

Figure 20.3   Effect of methanol extract of various soybean-fermented foods on the mutagenesis of aflatoxin B_1 in *Salmonella typhimurium* TA100 strain.

Commercial doenjang, chungkookjang and miso also exhibited antimutagenic activities, however, the traditional doenjang showed the highest antimutagenic activity (Figure 20.3).

At the same concentration, 64 to 66% and 39 to 53% of the AFB_1-induced mutagenesis was blocked when the methanol extracts of raw and cooked soybeans were added to the system, respectively.[21] Cooked soybeans inhibited mutagenicity less than did raw soybeans, probably due to the destruction of the trypsin inhibitor, which has chemopreventive effects,[57,58] by heat treatment, but the fermented soybeans (doenjang) were the most effective ($p < .05$). The higher antimutagenic activity shown by the doenjang is apparently the result of some end products of soybean fermentation. The lower antimutagenic activities of miso and chungkookjang as compared with that of doenjang are probably due to the smaller portion of soybeans used and shorter fermentation period, respectively.[59] Our experiment confirmed the fact that the mutagenicity of AFB_1 was inhibited in the presence of doenjang extract under several experimental conditions.[10,13–16,18,20,21]

Meju, fermented soybean block, exhibited an antimutagenic effect toward AFB_1. We have recently demonstrated that soybean grain meju and soybean-crushed brick-shaped meju show 87 and 77% antimutagenic effects against AFB_1 respectively, in *Salmonella typhimurium* TA100, while the cooked soybeans only showed 31% antimutagenicity (unpublished results). When rats were fed AFB_1 with meju they did not develop tumors, however, rats fed AFB_1 without meju did develop tumors. Thus meju appears to also have chemopreventive effects.

Traditional doenjang contains about 10 to 12% of NaCl, but the NaCl was extracted from the doenjang during the methanol extraction. NaCl itself did not show mutagenicity, but NaCl with carcinogen-enhanced mutagenic and carcinogenic effects.[60] It is reported that NaCl plays a comutagenic/cocarcinogenic role in the presence of MNNG (*N*-methyl-*N*-nitro-*N*-nitroguanidine).[61] However, the antimutagenic activities of the doenjang used in our studies were so strong that the comutagenic effect of the NaCl was blocked. This indicates that even if the AFB_1 is present as a rare contaminant in doenjang, the mutagenicity induced by AFB_1 can be completely blocked.[7]

The antimutagenic effects of doenjang were not limited to AFB_1, but were also exhibited toward other mutagens/carcinogens such as the indirect mutagens, benzo(a)pyrene (BaP) and N, N-dimethylnitrosamine (DMN); and the direct mutagens, MNNG and 4-nitroquinoline-1-oxide (4-NQO). In the case of direct mutagens, MNNG and 4-NQO-induced mutagenesis were completely inhibited in the *S. typhimurium* TA100 strain at the level of 25 mg/plate of the doenjang extract. BaP and DMN-induced mutagenesis were also blocked by 85 and 98%, respectively; although AFB_1 mutagenesis was completely blocked at the same level. Thus, it can be concluded that doenjang extract has strong antimutagenic activities, not only on AFB_1, but also toward other selected mutagens/carcinogens.[14]

Our experiments confirmed that the mutagenicity of AFB_1 is inhibited by methanol extracts of doenjang in various experimental systems.[10,13–16,18,20,21] The dichloromethane and ethyl acetate fractions of the extract also significantly reduced mutagenicities induced by AFB_1 as well as MNNG in *S. typhimurium* TA100 strain and in the SOS chromotest.[20] β-sitosterol glucoside and linoleic acid were identified as active antimutagenic compounds in the dichloromethane fraction by means of silica gel chromatography and NMR; genistein was also identified as a major compound in the ethyl acetate fraction by means of thin layer chromatography and HPLC.[20]

The amount of genistein was increased in fermented soy foods compared to nonfermented soy foods, which was attributed to the cleavage of the -glycosyl bond in genistin by microorganisms during fermentation.[62] We determined the contents of genistein, genistin and β-sitosterol glucoside in doenjang and soybean extracts and their fractions by HPLC.[10] The amount of β-sitosterol glucoside was not changed by fermentation. However, the concentration of genistein, which is more bioactive than genistin,[63] was significantly increased by fermentation, although the genistin level in doenjang decreased.[63]

Other experiments also indicate that soybeans contain higher levels of genistin, though the concentrations decreased, genistein increased significantly to 230 to 510 µg/g during doenjang fermentation.[27,64–66] Therefore, it is thought that the difference in the antimutagenicity between doenjang and soybeans is due to the increase in fermented active end products such as genistein in doenjang.

Trypsin inhibitor, genistein, soya saponin, α-tocopherol, β-sitosterol, and linoleic acid, among others in doenjang are all believed to be active compounds that possess antimutagenic effects.[10] All active compounds exerted a strong antimutagenic effect against AFB_1. Genistein and linoleic acid exhibited the strongest activity among them (Table 20.1). The active compounds also exhibited antimutagenic effect against the direct mutagen, MNNG in the Ames test, genistein and linoleic acid again showed the highest antimutagenic effect against MNNG of the active compounds in the SOS chromotest. Genistein and linoleic acid were suggested as the major active compounds found in doenjang.[20]

TABLE 20.1   Effect of Various Active Compounds (1.25 mg/plate) on the Mutagenicity Induced by Aflatoxin B_1(AFB$_1$, 0.75•/plate) in *Salmonella typhimurium* TA100[10]

| | AFB$_1$ | |
Sample	Revertant/plate	Inhibition rate (%)
Spontaneous	104 ± 2	
Control (AFB$_1$)	1218 ± 9[a]	
Genistein	307 ± 31[c]	82
Genistin	579 ± 37[b]	57
Linoleic acid	324 ± 28[c]	80
α-Tocopherol	786 ± 104[b]	39
β-Sitosterol glucoside	549 ± 99[b]	60
Soyasaponin	621 ± 142[b]	54
Trypsin inhibitor	646 ± 87[b]	51

[a-c] Means with the different letters beside data are significantly different at the 0.01 level of significance as determined by Duncan's multiple range test.

The *Drosophila* wing spot test was used to investigate the antimutagenic effect of genistein on the somatic mutagenicity induced by aflatoxin B_1 (AFB$_1$) (Table 20.2). Mutagen alone or mutagen with genistein were administered to the heterozygous (*mwh*/+) third instar larvae by feeding, and somatic cell mutations were detected in adult fly wing hairs. Genistein alone did not show any mutagenicity at the feeding concentrations of 1 to 4mg/mL in the test system. Genistein significantly inhibited the mutagenicity induced by AFB$_1$, especially at the lower feeding levels, small *mwh* spots that arise mostly from chromosome deletion and nondisjunction were more strongly suppressed by genistein than were the large *mwh* spots from chromosomal recombination. These results indicate that genistein inhibits AFB$_1$–induced mutagenicity in the Drosophila *in vivo* system.[20,67,68]

Linoleic acid (LA) is one of the active compounds found in doenjang. LA (0.005 to 2.5 mg/plate) did not show any toxicity or mutagenicity in the presence or absence of the S9 fraction in *S. typhimurium* TA100 strain. LA showed strong

TABLE 20.2   Effect of Genistein on the Mutagenicity
Induced by Aflatoxin $B_1(AFB_1)$ in the *Drosophila
melanogaster* Wing Spot System (*mwh*/+)

Exposure dose (mg/ml)		Frequency per wing (Number) of Single spots		No. of wings scored
AFB_1	Genistein	Small mwh	Large mwh	
0	0	0.05 (6) [0][1]	0.00 (0) [0]	120
0.5	0	0.65 (39) [100]	0.433 (26) [100]	60
0.5	1	0.033 (2)* [–3]	0.00 (0)[a] [0]	60
0.5	2	0.067 (4)* [3]	0.00 (0)* [0]	60

* Significantly different from the control at the $p < 0.05$ level.

TABLE 20.3   Effect of Linoleic Acid (LA) on the
Mutagenesis Against Aflatoxin $B_1(AFB_1, >1$/plate)
in *Salmonella typhimurium* TA100

Mutagen+LA(mg/plate)	Revertants/plate	Inhibition rate(%)
Spontaneous	106 ± 13	
AFB_1 (1µg/plate)	1017 ± 69	
AFB_1+LA 0.005	708 ± 14	34
0.05	533 ± 25	53
0.25	185 ± 6	91
0.5	125 ± 2	98
2.5	112 ± 3	99

antimutagenic effects against AFB_1 at levels between 0.005
and 2.5 mg/plate (Table 20.3). At 0.25mg of LA/plate, AFB_1
mutagenesis was inhibited by 91%, and increased to 98% at
0.5 mg/plate. A similar inhibitory effect has been observed
with the increased concentrations of LA added to other
mutagens.[69] LA again showed antimutagenic activities in the

Drosophila wing spot test. LA significantly inhibited both small mwh and large mwh spot mutations.[20] Hayatsu et al.[70] demonstrated that LA and oleic acid inhibited mutagenicities induced by AFB1, Trp-p-1, and BaP. Lim et al.[71] reported that linolenic acid effectively inhibits carcinogen/mutagen-induced mutagenesis. Strong inhibitory effects of linolenic acid were demonstrated using the Ames test and SOS chromotest against AFB_1, MNNG, and 4-NQO. Thus the unsaturated free fatty acids, mainly linoleic acid, formed during the fermentation resulted in an increased antimutagenicity of the doenjang.

ANTICANCER EFFECT OF DOENJANG

Doenjang considerably decreased the cytotoxicity and transformation induced by 3-methylcholanthrene (MCA) in C3H10T1/2 cells, protecting cells from the harmful effects of the carcinogen and inhibiting cell transformation by the carcinogen. The type II and type III foci formation from the MCA treatment was significantly decreased when doenjang extract was added to the system (Table 20.4). The type II and type III foci formation was correlated with 50 and 85% tumor formation in C3H mice.[72] The total foci numbers of type II and type III were decreased significantly (79 to 90%) by the doenjang treatment, indicating doenjang could prevent MCA-mediated carcinogenicity.

TABLE 20.4   Inhibitory Effect of the Methanol Extract from Doenjang on the Transformation of C3H/10T1/2 Cells Treated with 3-Methylchoanthrene(MCA, 10•/•)

Carcinogen sample (•/assay)	Total number			
	Type I foci	Type II foci	Type III foci	Type II and III foci
MCA (Control)	4.7 ± 0.5	10.3 ± 1.3	12.0 ± 2.2	22.3
MCA + Doenjang 10	8.9 ± 1.7	2.7 ± 0.9	2.0 ± 0.8	4.7
+ Doenjang 50	7.3 ± 1.2	3.3 ± 0.5	0.3 ± 0.5	3.3
+ Doenjang 100	4.7 ± 0.9	2.8 ± 0.5	0.3 ± 0.5	2.3

Note: The ratio of soybean and flour when preparing the doenjang was 7:3.

The growth of several human cancer cells such as AGS gastric adenocarcinoma cells, Hep 3B hepatocellular cancer cells, HT-29 colon cancer cells, and MG-63 osteosarcoma cells was significantly reduced by treatment with the methanol extract of doenjang.[11,12,20] The doenjang extract decreased the [3H] thymidine incorporation in AGS and Hep 3B cancer cells.[20] Choi et al.[73] also reported that doenjang fermented with Bacillus strain, PM3, significantly decreased the growth of human HepG2 cancer cells.

Though the methanol extract of doenjang greatly inhibited the growth of HT-29 human colon cancer cells after 6 days of incubation, the hexane fraction from the methanol extract showed the highest inhibitory effect on the growth of HT-29 cells.[11] Choi et al.[74] also indicated that the hexane fraction from methanol extracts of doenjang suppressed various tumor cells. The doenjang hexane fraction (DHF) was evaluated for its effects on cell cycle progression in the human breast carcinoma MCF-7 cells.[75] DHF induced a G1-phase arrest of the cell cycle in MCF-7 cells, which correlated with the accumulation of the hypophosphorylated form of the retinoblastoma protein (pRB) and enhanced association of pRB with the transcription factor E2F-1. As shown in Figure 20.4, the expres-

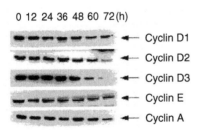

Figure 20.4 Western blot analysis of cyclins from MCF-7 cells after increasing periods of exposure to the doenjang hexane fraction (DHF). Cells were treated with 50/ml DHF for the time indicated. Total cell lysates were prepared and Western blot analysis was performed using antibodies to D-type cyclins (D1, D2 and D3), cyclin E and cyclin A and ECL detection.

sion of D-type cyclins was decreased by DHF in a time-dependent manner, but DHF did not affect the levels of cyclin E and cyclin A. However, the activity of Cdk2 and cyclin E-associated kinase was decreased in a time-dependent manner. The tumor suppressor, p53, and Cdk inhibitor, p21, which is a known downstream effector of p53; and association of p21 with Cdk2 were markedly induced in DHF-treated cells (Figure 20.5). Thus, DHF inhibited cancer cell growth by inducing an inhibition of pRB phosphorylation, decreasing expression of D-type cyclins and increasing of Cdk inhibitor p21 that appears to be responsible for the observed G1 arrest.[75]

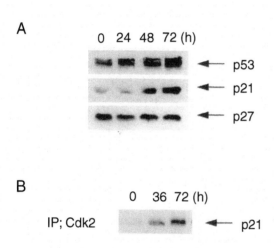

Figure 20.5 Induction of Cdk inhibitor p21 and association of p21 with Cdk2 by DHF in MCF-7 cells. (A) Cells were incubated with 50/ml DHF for the time indicated and total cell lysates were subjected to Western blot analysis using anti-p53, p21 and p27 antibody, and ECL detection. (B) Total cell lysates from untreated control cells and cells exposed to DHF for 36 and 72 h were immunoprecipitated with anti-Cdk2 antibody. Immune complexes were separated on 12% SDS-polyacrylamide gels, transferred to nitrocellulose membrane and incubated with anti-p21 antibody followed by ECL detection.

Various free fatty acids are isolated from the hexane fraction, especially linoleic acid and oleic acid. Linoleic acid (LA) was the main free fatty acid found in the hexane fraction of the doenjang methanol extract, and was identified as an active compound.[21] Our studies also indicated that LA decreased the growth of various human cancer cells[11,20] and of transplanted sarcoma-180 ascites tumor cells in Balb/c mice, and delayed the tumorigenesis process.[76–78] LA also enhanced the phagocytic activity and natural killer cell activity *in vitro* and *in vivo*,[79–81] and has been shown to modulate the T subset immune cells as a biological response modifier in sarcoma 180-transplanted mice.[82] LA also showed a capacity to differentiate Caco-2 cells[83] and F9 teratocarcinoma cells.[20]

We have already reported[84] that genistein induced a G2/M arrest in human MCF-7 and MDA-MB-231 breast cancer cells by inhibiting cdc2 and cdk2 kinase activities and decreasing cyclin B1 expression. We also demonstrated that genistein induces p21 in human PC-3-M prostate carcinoma cells, which inhibits the threshold kinase activities of cdks and associated cyclins, leading to a G2/M arrest in the cell cycle progression.[85,86] Genistein is known to inhibit angiogenesis and metastasis *in vitro*,[87,88] and reduce tumorigenesis and carcinogenesis *in vivo*.[89,90] Genistein induced differentiation of F9 tetracarcinoma cells.[20] Hong[91] suggested that the low incidence of prostate cancer in Asian men from China, Japan, and Korea is probably due to the genistein in soy foods.[91] Fotsis et al.[92] concluded that genistein may have important applications for the treatment of solid tumors and angiogenic diseases.

It has been reported that soy peptides derived from soy paste, soy hydrolyzates, and soy sauce exhibited anticancer activities in several cancer cell lines. Lee[93] reported that peptide fractions from doenjang that contained hydrophobic amino acids such as Gly, Al, Pro, Val, Leu, Ile, and Phe exhibited anticancer effects on SNUF-12 and SWF-12 human colon cancer cell lines.

Soy saponin[94] and soy Bowman-Birk inhibitor (BBI) are the active anticancer compounds from soybeans and soybean

TABLE 20.5 Effect of Hexane, Methanol and Boiling Extracts from Doenjang on Antitumor Activities and Life Span of Mice Transplanted with Sarcoma-180 Cells to Balb/c Mice

Sample	Tumor weight (g)	Survival time (day)
S-180 + PBS	3.28 ± 0.29[a]	20.8 ± 3.6
+ Hexane ext.	1.18 ± 0.15(64.0)[b]	32.9 ± 3.7(58.2)
+ Methanol ext.	0.68 ± 0.29(79.3)	34.6 ± 0.8(66.3)
+ Boiling ext.	1.65 ± 0.18(49.7)	28.9 ± 2.7(38.9)

[a] Values are mean± SD of 10 mice.
[b] The values in parentheses are the inhibition rate (%).

products. Anticancer activities of soy saponin and soy BBI from doenjang have not yet been investigated, and thus further research is needed to elucidate the active roles of these compounds and those generated during doenjang fermentation.

The solid tumor growth from transplanted sarcoma-180 cells in Balb/C mice were greatly inhibited by doenjang.[18] When 5 mg/kg of the hexane, methanol and hot water extracts from doenjang were administered to the Balb/c mice, the tumor growth was inhibited by 64.0, 79.3, and 49.7%, respectively, and the life span (prolongation effects) of the mice increased by 58.2, 66.3, and 38.9%, respectively (Table 20.5), demonstrating that the methanol extract was most effective. Spleen index, a marker for immunological activity, increased in the mice administrated the doenjang extracts compared to the control group. The phagocytic activity of the mice was increased approximately twofold by treatment with the hexane extract. The nitroblue tetrazolium (NBT) reduction rate in the peritoneal phagocytic cells of the mice was also increased about 3 times (45 vs. 16%) when treated with hexane extract.[18] Kim et al.[17] found that methanol extracts of traditional doenjang, miso (Maruseng, Co.), and cooked soybeans extended the life span of Balb/c mice with transplanted sarcoma-180 cells by 68, 41, and 11%, respectively, demonstrating that doenjang was the most effective.[17]

Lipid peroxide content of the liver was increased by the injection of the mice with sarcoma-180 cells, but decreased significantly with doenjang treatment ($p < .05$). Hepatic glutathione S-transferase activity and glutathione content decreased in mice transplanted with sarcoma-180 cells, but the activity and the content were increased by treatment with doenjang extracts in the mice, respectively. Thus doenjang extracts exhibited a possible protective effect on sarcoma cell-induced hepatotoxicity in mice.[19]

INCREASED CHEMOPREVENTIVE EFFECT OF DOENJANG

Prolonging the fermentation period in production of doenjang was found to increase its chemopreventive effects.[59] A 2-year fermented doenjang showed higher antimutagenic, anticancer, antitumor, and antimetastasis activities compared to 3-month and 6-month fermented doenjangs. The 2-year-old doenjang had increased browning products with significantly more redness and yellowish colors, but the lightness was decreased ($p < .05$). It had been reported that browning products from doenjang and kanjang can reduce the mutagenicity of various mutagens.[6,95,96] The antimutagen protection against aflatoxin B_1, and anticancer effects against AGS human gastric cancer cells and HT-29 human colon cancer cells were greatly increased when doenjang was aged for 2 years, as well as a strong induction of apoptosis in AGS cells (data not shown). The 2-year-old doenjang caused a 2- to 3-fold increase in antitumor effects on sarcoma-180 injected mice and anti-metastatic effects by colon 26-M 3.1 cells in mice, compared to the 3 or 6 months fermented doenjang.

Salt is another important factor that affects the chemopreventive activity of doenjang.[97] Pure salt is used for making doenjang commercially. However, natural salt, chunil salt, from salt ponds in the West Sea of Korea is traditionally used to make doenjang. Buddhist monks developed a traditional bamboo salt prepared in their temples. Natural salt is packed into a bamboo container plugged with mud on both sides, and then heated 1 to 9 times to more than 1300°C. The bamboo salt has a higher mineral content than the original natural salt. Because of presence of the bamboo and mud, the salt is believed to turn into a healthy salt during burning, due to chemical and physical changes. Bamboo salt is used as a folk medicine for the treatment of cancer in Korea. Doenjang made with bamboo salt had a significantly increased anticlastogenic effect in MMC-induced mice using an *in vivo* supravital staining micronucleus assay. The antimetastatic effect on colon 26-M 3.1 cells in mice was the lowest (31 to 57% inhibition) when doenjang was prepared with purified salt, but 9 times higher when bamboo salt was used in the doenjang, exhibiting the highest (97 to 98% inhibition) antimetastatic activity. Doenjang made with 1× heat treated bamboo salt also supported the high rate of antimetastatic effects (75 to 85% inhibition) (data not shown). Further studies are needed to elucidate the mechanism of the enhanced activity. Increased chemopreventive effects from adding ginger, garlic, Japanese apricot[98,99], sea tangle[100], and mushrooms[99] to doenjang were studied. Doenjang made with ginger, 5% sea tangle, *Phellinus linteus,* and *Ganoderma lucidum* exhibited higher antimutagenic and anticancer effects than the control doenjang.

Antioxidant Effects

Antioxidative activity of doenjang was reported by Cheigh et al.[22] and Lee at al.[23] Both lipid soluble and water soluble extracts from doenjang exhibited strong antioxidative activities in ground cooked meat, ground cooked fish, and in a linoleic acid mixture reaction system.[22] Soy paste (water soluble extracts) protected ground cooked meat (GCM) from lipid oxidation during 5 weeks of storage at 6°C as shown in Figure 20.6. The TBA value of the control, GCM (10 g), increased over the incubation period, but GCM with the soy paste did not. Salt increased the TBA value in GCM. The addition of salt (1.2 g) to GCM (10 g) significantly increased the TBA value, whereas adding doenjang (10 g) to the GCM (10 g) did not increase the TBA value even though it contained 12% salt. This result indicated that doenjang inhibited lipid oxidation

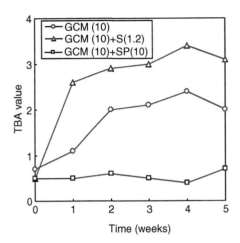

Figure 20.6   Changes in thiobarbituric acid (TBA) values of ground cooked meat (GCM) with the addition of soybean paste (SP) and salt (S) during storage for 5 weeks at 6°C.

in GCM even with a salt content that is known to cause oxidation, and thus doenjang appears to contain potent antioxidative compounds. The methanol extract from freeze dried, deffated doenjang inhibited lipoxygenase activity, and increased metal ($FeCl_3$) chelation and free radical scavenging activities.[24] It was suggested that phenolic compounds, such as isoflavonoids (genistein, daidzein, etc.) and phenolic acids (chlorgenic acid, caffeic acid, etc.), saponins, vitamin E, and browning products such as peptides and amino acids that form during doenjang fermentation and from soybeans are the antioxidative compounds found in doenjang.[25,26]

Kim et al.[29] reported that both phenolic acid and isoflavone fractions from methanol extracts of defatted meju and doenjang showed similar antioxidative effects against the oxidation of linoleic acid. Phenolic acid fractions contain vanillic, chlorogenic, p-coumalic, ferulic, and caffeic acids, of which caffeic acid content accounts for more than 70%. However, Lee and Cheigh[28] indicated that the major phenolic acid (acidic phenolics) in doenjang was syringic acid (48%). The levels of *p*-coumaric acid and ferulic acid decreased, but syringic acid, vanillic acid and *p*-hydroxybenzoic acid concentrations increased during doenjang fermentation. The neutral phenolics (isoflavonoids) were fractionated, revealing that the concentration of genistin in doenjang dramatically decreased, but the genistein concentration increased significantly.[27] Antioxidative activities of phenolic compounds markedly increased, especially the neutral phenolics, during 60 days of doenjang fermentation, with a simultaneous significant decrease in peroxide value during auto-oxidation at 50°C for 3 days. This is probably due to the high levels of genistein formed during the fermentation. The phenolic compounds in doenjang significantly decreased the peroxide value (Figure 20.7) and conjugated diene formation of linoleic acid compared to those from soybeans. Thus the increase in the aglycone of genistein concentrations in doenjang, formed during the fermentation, caused the increase in the antioxidative activity.[101]

Both hydrophilic[30] and lipophilic[31] brown pigments produced during doenjang fermentation and ripening have demonstrated antioxidative activities. The antioxidative activity increases as the absorbance at 400 nm (yellowish color) increases. Cheigh et al.[31] also indicated that browning products (Malliard browning reaction products) from soy sauce show a considerable antioxidative activity, inhibiting the formation of peroxides and conjugated dienoic acids.[25]

Free amino acids such as histidine, tryptophan, tyrosine, and methionine exhibit strong antioxidative activities.[25] Doenjang contains high levels of glutamic acid, lysine, tryptophan, histidine, and tyrosine.[26] Thus much of the antioxidative effect of doenjang may be attributed to these free amino acids formed during fermentation. Muramoto[27] reported that His-containing peptides such as His-His show antioxidative activity and act as metal-ion chelates, active-oxygen quenchers, and hydroxy radical scavengers.

Soy saponin extracts and saponins inhibit lipid peroxidation.[102] Sung[94] indicated that soy saponins effectively reduced malondialdehyde formation, and that the activity was significantly higher than that of ascorbate or α-tocopherol in

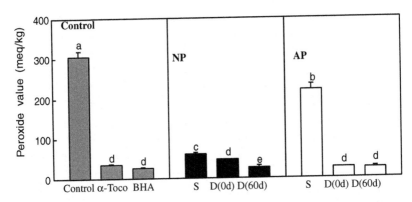

Figure 20.7 Changes in the peroxide values of linoleic acid with the addition (0.08%) of NPS, NPD(0d), NPD(60d), APS, APD(0d), APD(60d) during auto-oxidation at 50°C for 3 days. NP, neutral phenolics; AP, acidic phenolics; S, soybean; D, doenjang. [a~e]The different letters are significantly different at the 0.05 level of significance as determined by Duncan's multiple range test (n = 3).

Hep G 2 cells. Soy saponins increased both the glutathione peroxide and glutathione S-transferase levels.

Reduced Cardiovascular Diseases

Asian populations with high soybean consumption have lower CHD (coronary heart disease) rates than Western populations, which have a low intake of soybeans and soybean products.[103] CHD is a leading cause of death, accounting for 50% of all cardiac death in the U.S.[104] Recently, soybeans and soybean foods with isoflavones have been recommended as part of a heart healthy diet for Americans. High soybean intakes reduce the incidence of cardiovascular diseases.[103,105] The amino acid composition of soy protein, unsaturated fatty acids, dietary fiber, isoflavones, saponins, and plant sterols in soybeans confer potential cardiovascular protection.[106] Nondigestible protein, phytic acid, and peptides from fermented soybeans also promote healthy cardiovascular function.[107]

Fibrinolytic Activity

Fermented soybean products such as chungkookjang, natto, and doenjang show fibrinolytic activity. Kim[34] isolated a Bacillus species that secreted strong fibrinolytic enzymes from various kinds of doenjang samples. The fibrinolytic activity of the traditional doenjang samples was 10 times higher than that of the commercial doenjang samples. Bacterial strains showing the fibrinolytic activity were identified as *Bac. amyloliquefaciens, Bac. Pantotheticus,* and *B. subtilis.* The culture media and the doenjang containing the above strains revealed high levels of fibrinolytic activity. The fibrinolytic enzyme activity of the Bacillus species from doenjang was 3 to 4 times higher than that of the Bacillus species from chungkookjang and natto. The enzyme purified from *Bac. amyloliquefaciens* was pH and heat stable, and had more than 130 unit/mg specific fibrolytic enzyme activity.

The peptide products from doenjang fermentation showed antithrombotic activity.[35] Although doenjang extract and its peptides fractions all showed antithrombotic activity, fraction number 16 showed the highest activity. The major residues in the peptides of the fraction were histidine, arginine, and alanine; with highest concentrations of alanine (40.5%) of the free amino acids.

Antihypertensive Effects

ACE (Angiotensin converting enzyme) inhibitors from foods can assist hypertensive patients in controlling high blood pressure, without side effects. ACE inhibitory activity was found in fermented soybean products such as soy sauce, tempeh, natto, douchi, miso, and doenjang.[108] Peptide fermentation products are the active compounds that inhibit ACE activity. From 18 samples of meju, doenjang, chungkookjang, and natto, Hwang[36] isolated *Bacillus subtilis* SCB-3 from meju from Soonchang Kun (county), and identified it as the strain with the highest ACE inhibitory activity. Changes in composition and ACE inhibitory activity during fermentation of doenjang by *B. subtilis* were examined for 90 days. The

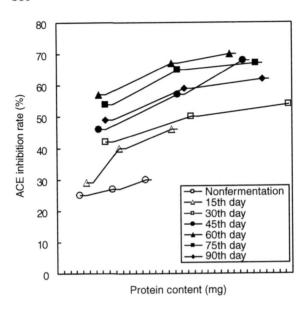

Figure 20.8   Comparison of the ACE inhibition rate according to protein contents of each extract during doenjang fermentation at 30°C by *Bacillus subtillus* SCB-3. (From JH Hwang. Angiotensin I converting enzyme inhibitory effect of doenjang fermented by *B. subtilis* SCB-3 isolated from meju, Korean traditional food. *J Korean Soc Food Sci Nutr* 26:775–783, 1997. With permission.)

maximum protease activity of the doenjang was reached after 60 days of fermentation at 30°C, at which point the ACE inhibition rate was also the highest (Figure 20.8). Protease activity increased during the first 60 days of fermentation, and then decreased during the remainder of the 90 days. This was positively correlated with increased concentrations of amino nitrogen and production of ACE inhibitory peptides. The protein content was higher in doenjang fermented more than 60 days and ACE activity increased in parallel with the increased protein content and protease activity. The short fermentation period resulted in a low ACE inhibitory activity, but long-fermented doenjang produced ACE inhibitory peptides as part of the protease activity. The fermented doenjang had increased ACE inhibitory effects whereas nonfermented did not. This might be due to the fermentation products of peptides.[37] Various peptides from doenjang were isolated and identified as ACE inhibitors. Suh et al.[38] reported that ACE inhibitor was isolated and the purified inhibitor showed a competitive inhibition with ACE. The amino acid profile of the peptides consisted of alanine, phenylalanine, leucine, glutamic acid, glycine, serine, and aspartic acid. The highest concentrations of amino acids were: 55.5% alanine, 39.9% phenylalanine, and 2.1% leucine.

Another active ACE inhibitory peptide was isolated from traditional doenjang[39] and the active compound was identified as a dipeptide, Arg-pro with ACE IC_{50} of 92 μM. Shin et al.[40] also isolated ACE-inhibiting peptide from doenjang through ultrafiltration and ion exchange prep-HPLC. The F-5 fraction was again divided into five fractions by ion exchange prep-HPLC, and all fractions showed high ACE inhibitory activity (IC50 = 2.5 to 8.3 mL), however, the F53 fraction exhibited the highest ACE inhibition. As shown in Table 20.6, the main amino acid component in the peptides was histidine with small amounts of other amino acids. His-His-Leu was also isolated from doenjang as the peptide with the highest ACE inhibitory activity and antihypertensive effect *in vitro* and *in vivo*.[41,42] Ser-Try was the peptide with the highest ACE inhibitory activity in miso.[109]

In several human trials, soy protein isolate with isoflavones have been shown to lower diastolic and systolic blood pressure.[103] High levels of dietary isoflavones significantly decreased diastolic blood pressure in women, but had no effect in men.

Reduced Serum Cholesterol Level

Isoflavones in doenjang can reduce serum cholesterol concentration, and reduce oxidation of LDL cholesterol due to their polyphenolic structures[110] and consequently reduce the risk of cardiovascular diseases. Sugano et al.[111] reported a remarkable reduction in serum cholesterol concentrations and increased excretion of steroids into feces after rats were administered the insoluble, nondigestible peptide fraction

TABLE 20.6 Amino Acid Composition of the F53 Fraction
from Soybean Paste

Amino acid	Concentration, n mole/ml		
	Total amino acids (mole%)	Free amino acids (mole%)	Peptide amino acids (mole%)
Asp	88 (1.5)	ND[a]	88 (2.1)
Glu	171 (2.8)	ND	171 (4.1)
Ser	82 (1.4)	ND	82 (2.0)
Gly	272 (4.5)	ND	272 (6.6)
His	4504 (74.2)	1681 (86.6)	2823 (68.5)
Arg	25 (0.4)	ND	25 (0.6)
Thr	128 (2.1)	ND	128 (3.1)
Ala	ND	ND	ND
Pro	ND	ND	ND
Tyr	ND	ND	ND
Val	107 (1.8)	NA[b]	107 (2.6)
Met	ND	ND	ND
Cys	ND	ND	ND
Ile	164 (2.7)	ND	164 (4.0)
Leu	ND	ND	ND
Phe	478 (7.9)	261 (13.4)	217 (5.3)
Lys	45 (0.7)	ND	45 (1.1)
Total	6064 (100.0)	1942 (100.0)	4122 (100.0)

[a] Not Detected
[b] Not Available
(From ZI Shin, CW Ahn, HS Nam, HJ Lee, HJ Lee, TH Moon. Fractionation of angiotensin converting enzyme (ACE) inhibitory peptides from soybean paste. *Korean J Food Sci Technol* 27:230–234, 1995. With permission.)

from processing soybeans. In addition to lowering cholesterol, it has been reported that isoflavones, via estrogenic effect (high affinity for the β-estrogen receptor), inhibit LDL oxidation, enhance vascular reactivity, and inhibit platelet aggregation. Phytoesterols can decrease cholesterol absorption. MUFA and PUFA in doenjang might also reduce serum cholesterol levels, LDL-c, etc.[112] Han et al.[43] indicated that increased hydrophobic amino acid concentrations lower serum cholesterol in Wistar rats. More research is needed to clarify the preventive role of other active compounds from doenjang on cardiovascular diseases.

Other Possible Functions of Doenjang

Doenjang may also have antiobesity activity,[113,114] accelerate recovery from fatigue[114] and ameliorate symptoms of menopause,[115] due to the high content of isoflavones and peptides generated during the fermentation process.

The OH groups from phenolic compounds, produced in high amounts in doenjang, can combine with phospholipase or pancreatic lipase, and thus reduce their enzyme activities. The degradation of the phospholipid portion of the lipid decreased as did the hydrolysis of triglyceride, and the undigested lipids were excreted in the feces.[113] Dietary fiber also binds undigested lipids and removes them. Soybean peptides were also reported to reduce body fat[114] by increasing thermogenesis, inhibiting the reduction in basal metabolism caused by a low-calorie diet, and eliminating adipose tissue more rapidly. Wakamiya[114] also indicated that soybean peptides can accelerate recovery from fatigue induced by concussion stress on mice dose dependently. Epidemiological data suggest that 25% of Asian women experience hot flashes, whereas 70 to 80% in Western menopausal women. Kurzer[115] reported that isoflavones can improve hot flash and vaginal dryness scores due to the menopausal symptoms (54 vs. 35% for hot flashes and 60 vs. 27% for vaginal dryness). These results need to be confirmed and detailed study of these functionalities investigated in studies using doenjang as the isoflavone source.

B. subtilis is one of the primary microorganisms in doenjang fermentation. *B. subtilis* is considered to be a probiotic[116–120] like lactic acid bacteria. *B. subtilis* preparations are sold as probiotics in most European countries,[116] where they are used for oral bacteriotherapy and bacterioprophylaxis of gastrointestinal disorders, many of which lead to diarrhea. Intake of *B. subtilis* is thought to restore the normal microbial flora following extensive antibiotic use or illness.[117] *B. subtilis* also acts as an immunostimulatory agent in a variety of diseases[118] and as an *in vitro* and *in vivo* stimulant of secretory immunoglobulin A. Pinchuk et al.[119] reported that *B. subtilis* 3 secreted antibiotics and inhibited *Helicobacter pylori* activity and showed antagonistic properties against species of the family Enterobacteriaceae. *B. subtilis* 3 produced antibiotics, one of which, amicoumacin A, exhibited anti-inflammatory properties and antibacterial activity against *Helicobacter pyroli*.

Vaseeharan and Ramasamy[121] recently demonstrated significant *in vitro* and *in vivo* antagonistic effects of Bacillus against the pathogenic vibrios. *B. subtilis* from doenjang inhibited growth of aflatoxigenic molds, *Asp. flavus* and *Asp. Parasiticus*.[8,9] Further study is needed to confirm the probiotic activities of *B. subtilis* and other microorganisms from doenjang.

CONCLUSION

Doenjang has long been consumed by Koreans who manufactured it at home by natural fermentation. Though the profound taste could be produced by fermentation, maintaining the consistent quality, and manipulating increased quality and functionality was difficult. Thus additional standardized manufacturing methods are also needed.

Korean ancestors traditionally prepared doenjang along with kanjang by fermentation. The active functional compounds such as genestein from genistin, free fatty acids such as linoleic acid, oleic acid, and linolenic acid from soy lipid, amino acids or peptides from soy protein, browning products, etc. can be formed during fermentation.

It is also necessary to develop methods for making different types of meju under controlled conditions for industry. The appropriate microorganism strains should be used for making meju. *B. subtilis, B. licheniformis, Asp. oryzae, Rhizopus sp., Mucor sp.*, etc. have been isolated as major strains, and thus should be included in select starter cultures of single or mixed strains for meju fermentation that show better taste, functionality, and fermentation patterns. Studies of probiotic functions of the starter cultures, especially for *Bacillus sp.*, and *Aspergillus oryzae, Rhizopus sp., Mucor sp.*, etc. are also needed.

The selection of the main ingredients, such as soybean, water, and salt are also very important factors for increasing functionality of doenjang along with better fermentation patterns and nutrition of doenjang. Manufacturing doenjang and kanjang together, as was the traditional way, as well as doenjang without kanjang from the meju are also necessary in order to examine their effects on taste, nutrition, and functionality of doenjang. The traditional manufacturing method should be kept, but modified methods with scientifically validated protocols need to be developed for homemade and factory-made doenjang.

REFERENCES

1. CH Lee. *Fermentation Technology in Korea.* Seoul: Korea University Press, 2001, pp 70-81.

2. PS Crane, SU Rhee, DJ Seel. Experience with 1079 cases of cancer of the stomach seen in Korea from 1962 to 1968. *Am J Surgery* 120:747–751, 1970.

3. KY Park, KB Lee, LB Bullerman. Aflatoxin production by Aspergillus parasiticus and its stability during the manufacture of Korean soy paste (doenjang) and soy sauce (kanjang) by traditional method. *J Food Prot* 51:938–945, 1988.

4. KB Lee. Aflatoxin Production by Aspergillus Parasiticus During the Manufacturing of Korean Soy Paste (doenjang) and Soy Sauce by Traditional Method. MS thesis, Pusan National University, Busan, 1987.

5. KY Park, ES Lee. Effect of ammonia and pH on the degradation of aflatoxin B_1 during the storage of Korean soy sauce (kanjang). *J Korean Soc Food Nutr* 18:115–122, 1989.

6. KY Park, ES Lee, SH Moon, HS Cheigh. Effects of browning products and charcoal on the degradation of aflatoxin B_1 in Korean soy sauce (kanjang) and its model system. *Korean J Food Sci Technol* 21:419–424, 1989.

7. KY Park. Destruction of aflatoxins during the manufacture of doenjang by traditional method, and cancer-preventive effects of doenjang. *J Korean Assoc Cancer Prev* 2:27–37, 1997.

8. KJ Kang, JH Jeong, JI Cho. Inhibition of aflatoxin-producing fungi with antifungal compound produced by Bacillus subtilis. *J Fd Hyg Safety* 15:122–127, 2000.

9. JG Kim, WS Roh. Changes of aflatoxins during the ripening of Korean soy paste and soy sauce and the characteristics of the changes-Part I. Effect of Bacillus subtilis on the growth and aflatoxin production of Aspergillus parasiticus. *J Fd Hyg Safety* 13:313–317, 1998.

10. KY Park, KO Jung, SH Rhee, YH Choi. Antimutagenic effects of doenjang (Korean fermented soy paste) and its active compounds. *Mutat Res* 523–524:43–53, 2003.

11. KY Park, JM Lee, SH Moon, KO Jung. Inhibitory effect of doenjang (fermented Korean soy paste) extracts and linoleic acid on the growth of human cancer cell lines. *J Food Sci Nutr* 5:114–118, 2000.

12. SY Lim, KY Park, SH Rhee. Anticancer effect of doenjang in *in vitro* sulforhodamine B(SRB) assay. *J Korean Soc Food Sci Nutr* 28:240–245, 1999.

13. KY Park, SH Moon, HS Baik, HS Cheigh. Antimutagenic effect of doenjang (Korean fermented soy paste) toward aflatoxin. *J Korean Soc Food Nutr* 19:156–162, 1990.

14. KY Park, SH Moon, HS Cheigh, HS Baik. Antimutagenic effects of doenjang (Korean soy paste). *J Food Sci Nutr* 1:151–158, 1996.

15. KY Park, SH Moon, SH Rhee. Antimutagenic effect of doenjang (Korean soy paste)- Inhibitory effect of doenjang stew and soup on the mutagenicity induced by aflatoxin B_1. *Environ Mut Carcino* 14:145–152, 1994.

16. KY Park, SY Lim, SH Rhee. Antimutagenic and anticarcinogenic effects of doenjang. *J Korean Assoc Cancer Prev* 1:99–107, 1997.

17. MK Kim, SH Moon, JW Choi, KY Park. The effect of doenjang (Korean soy paste) on the liver enzyme activities of the sarcoma-180 cell transplanted mice. *J Food Sci Nutr* 4:260–264, 1999.

18. KY Park, MH Son, SH Moon, KH Kim. Cancer-preventive effects of doenjang *in vitro* and *in vivo*. 1. Antimutagenic and *in vivo* antitumor effect of doenjang. *J Korean Assoc Cancer Prev* 4:68–78, 1999.

19. MH Son, SH Moon, JW Choi, KY Park. Cancer-preventive effects of doenjang *in vitro* and *in vivo*. 2. Effect of doenjang extracts on the changes of serum and liver enzyme activities in sarcoma-180 transplanted mice. *J Korean Assoc Cancer Prev* 4:143–154, 1999.

20. SY Lim. Studies on the Antimutagenic and Anticancer Activities of Doenjang. PhD dissertation, Pusan National University, Busan, 1997.

21. SH Moon. Antimutagenic Effect of Doenjang (Korean Soy Paste). MS thesis, Pusan National University, Busan, 1990.

22 HS Cheigh, KS Park, GS Moon, KY Park. Antioxidative characteristics of fermented soybean paste and its extracts on the lipid oxidation. *J Korean Soc Food Nutr* 19:163–167, 1990.

23. JH Lee, MH Kim, SS Im. Antioxidative materials in domestic meju and doenjang 1. Lipid oxidation and browning during fermentation of meju and doenjang. *J Korean Soc Food Nutr* 20:148–155, 1991.

24. JS Lee, HS Cheigh. Antioxidative characteristics of isolated crude phenolics from soybean-fermented foods (doenjang). *J Korean Soc Food Sci Nutr* 26:376–382, 1997.

25. GS Moon, HS Cheigh. Antioxidative characteristics of soybean sauce in lipid oxidation process. *Korean J Food Sci Technol* 19:537–542, 1987.

26. EA Bae, GS Moon. A study on the antioxidative activities of Korean soybeans. *J Korean Soc Food Sci Nutr* 26:203–208, 1997.

27. K Muramoto. Functional Properties of Proteolytic Hydrolysates of Soy Proteins. Proceedings of International Symposium on Soybean Peptides and Human Health, Seoul, 1998, pp 33–39.

28. JS Lee, HS Cheigh. Composition and antioxidative characteristics of phenolic fractions isolated from soybean-fermented food. *J Korean Soc Food Sci Nutr* 26:383–389, 1997.

29. MH Kim, SS Im, YB Yoo, GE Kim, JH Lee. Antioxidative materials in domestic meju and doenjang. 4. Separation of phenolic compounds and their antioxidative activity. *J Korean Soc Food Nutr* 23:792–798, 1994.

30. JH Lee, MH Kim, SS Im, SH Kim, GE Kim. Antioxidative materials in domestic meju and doenjang 3. Separation of hydrophilic brown pigment and their antioxidant activity. *J Korean Soc Food Nutr* 23:604–613, 1994.

31. MH Kim, SS Im, SH Kim, GE Kim, JH Lee. Antioxidative materials in domestic meju and doenjang. 2. Separation of lipophilic brown pigment and their antioxidative activity. *J Korean Soc Food Nutr* 23:251–260, 1994.

32. HS Cheigh, JS Lee, GS Moon, KY Park. Antioxidative activity of browning products fractionated from fermented soybean sauce. *J Korean Soc Food Nutr* 22:565–569, 1993.

33. Y Yoshiki, K Okubo. Active oxygen scavenging activity of DDMP(2,3-dihydro-2,5-dihydroxy-6-methyl-4H-pyran-4-one) saponin in soybean seed. *Biosci Biotech Biochem* 59:1556–1557, 1995.

34. SH Kim. New trends of studying on potential activities of doenjang – fibrinolytic activity. *Korea Soybean Digest* 15:8–15, 1998.

35. DH Shon, KA Lee, SH Kim, CW Ahn, HS Nam, HJ Lee, JI Shin. Screening of antithrombotic peptides from soybean paste by the microplate method. *Korean J Food Sci Technol* 28:684–688, 1996.

36. JH Hwang. Angiotensin I converting enzyme inhibitory effect of doenjang fermented by B. subtilis SCB-3 isolated from meju, Korean traditional food. *J Korean Soc Food Sci Nutr* 26:775–783, 1997.

37. JH Hwang. Studies on Angiotensin Converting Enzyme Inhibitory Peptide of Doenjang Fermented by B. subtilis SCB-3 Isolated from meju, Korean traditional food. PhD dissertation, Korea University, Seoul, 1996.

38. HJ Suh, DB Suh, SH Chung, JH Whang, HJ Sung, HC Yang. Purification of ACE inhibitor from soybean paste. *Agric Chem Biotechnol* 37:441–446, 1994.

39. SH Kim, YJ Lee, DY Kwon. Isolation of angiotensin-converting enzyme inhibitor from doenjang. *Korean J Food Sci Technol* 31:848–854, 1999.

40. ZI Shin, CW Ahn, HS Nam, HJ Lee, HJ Lee, TH Moon. Fractionation of angiotensin-converting enzyme (ACE) inhibitory peptides from soybean paste. *Korean J Food Sci Technol* 27:230–234, 1995.

41. ZI Shin, CW Ahn, HS Nam, HJ Lee, HJ Lee. Angiotensin I-Converting Enzyme Inhibitor Derived from Fermented Soybean Paste and Enzymatic Soybean Hydrolyzate. Proceedings of IUFost'96 Regional Symposium, Seoul, 1996, p 265.

42. R Yu, SA Park, DK Chung, HS Nam, JL Shin. Effect of soybean hydrolysate on hypertension in spontaneously hydrolysate on hypertension in spontaneously hypertensive rats. *J Korean Soc Food Sci Nutr* 25:1031–1036, 1996.

43. ES Han, HJ Lee, DH Shon. Effect of amino acid composition and average hydrophobicity of soybean peptides on the concentration of serum cholesterol in rats. *Korean J Food Sci Technol* 25:552–557, 1993.

44. *Science of Soybeans.* Seoul: Daekwang Publishing Co., 1999. pp 1–14.

45. SB Kim. *Cultura; History of Korea Food Life.* Seoul: Kwang moon kak, 1997, pp 146–260.

46. 35th Chronicle of Korea Soy Sauce Industrial Cooperative. Seoul: Korea Soy Sauce Industrial Cooperative, 1997, pp 27–32.

47. KY Park, KM Hwang, KO Jung, KB Lee. Studies on the standardization of doenjang(Korean soybean paste) 1. Standardization of manufacturing method of doenjang by literatures. *J Korean Soc Food Sci Nutr* 31:343–350, 2002.

48. MC Kim. The microbiological studies on Korean Jang. *J Gyeongsang Natl Univ* 15:1–26, 1976.

49. JY Yu. Characteristics of Meju and Microorganisms for Traditional Soybean Fermented Foods. The First Symposium and Expo for Soybean-Fermented Foods. P 31–87. The Research Institute of Soybean-Fermentation Foods. Yeungnam Univ, May 1, 1998, Kyungsan, Korea.

50. KS Choi. Isolation, Termo-Tolerance and Sterilization of Bacillus sp. Bacteria in Korean Traditional Soy Sauce (kanjang). Proceedings of the 5th Symposium and Expo for Soybean Fermented Foods, Kyungsan, 2002, pp 83–117.

51. *Food Composition Table.* 6th Revised ed. Suwon: National Rural Living Science Institute, RDA, 2001, pp 70–71.

52. SK Park, KI Seo, SH Choi, JS Moon, YH Lee. Quality assessment of commercial doenjang prepared by traditional method. *J Korean Soc Food Sci Nutr* 29: 211–217, 2000.

53. JH Lee. *65 Doenjang dishes for good health.* Seoul: Lees Com, 2003, p 10.

54. KY Park. Aflatoxin: Factors affecting aflatoxin production. *J Korean Soc Food Nutr* 13:117–126, 1984.

55. KY Park, LB Bullerman. Effects of substrate and temperature on aflatoxin production by Aspergillus parasiticus and Aspergillus flavus. *J Food Prot* 46:178–184, 1983.

56. SD Yi, JS Yang, JH Jeong, CK Sung, MJ Oh. Antimicrobial activities of soybean paste extracts. *J Korean Soc Food Sci Nutr* 28:1230–1238, 1999.

57. J Yavelow, TH Finlay, AR Kennedy, W Troll. Bowman-Birk soybean protease inhibitor as an anticarcinogen. *Cancer Res* (Suppl.) 43:2454S–2459S, 1983.

58   HG Weed, RB McGandy, AR Kennedy. Protection against dimethylhydrazine-induced adenomatous tumors of the mouse colon by the dietary addition of an extract of soybeans containing the Bowman-Birk protease inhibitor. *Carcinogenesis* 6:1239–1241, 1985.

59. SY Park. The Manufacture of Cancer Preventive Doenjang and Anticancer Effects of Doenjang. MS thesis, Pusan National University, Busan, 2003.

60. M Takahashi, T Kokubo, F Furukawa, Y Kurokawa, M Tatematsu, Y Hayashi. Effect of high salt diet on rat gastric carcinogenesis induced by N-methyl-N-nitro-N-nitrosoguanidine. *Gann* 74:28–34. 1983.

61. SH Kim, KY Park, MJ Suh. Comutagenic effect of sodium chloride in the Salmonella/Mammalian microsome assay. *Foods Biotechnol* 4:264–267, 1995.

62. M Fukutake, M Takahashi, K Ishida, H Kawamura, T Sugimura, K Wakabayashi. Quantification of genistein and genistin in soybeans and soybean products. *Food Chem Toxicol* 34:457–461, 1996.

63. M Onozawa, K Fukuda, M Ohtani, H Akaza, T Sugimura, K Wakabayashi. Effects of soybean isoflavones on cell growth and apoptosis of the human prostatic cancer cell line LNCaP. *Jpn J Clin Oncol* 28:360–363, 1998.

64. JS Kim, S Yoon. Isoflavone contents and -glucosidase activities of soybeans, meju and doenjang. *Korean J Food Sci Technol* 31:1405–1409, 1999.

65. MH Lee, YH Park, HS Oh, TS Kwak. Isoflavone content in soybean and its processed products. *Korean J Food Sci Technol* 34:365–369, 2002.

66. JH Jeong, JS Kim, SD Lee, SH Choi, MJ Oh. Studies on the contents of free amino acids, organic acids and isoflavones in commercial soybean paste. *J Korean Soc Food Sci Nutr* 27:10–15, 1998.

67. HS Yun, KY Park, WH Lee. Antimutagenic effects of genistein in Drosophila somatic mutation assaying system. *J Korean Assoc Cancer Prev* 5:135–143, 2000.

68. HS Yun, MA Yoo, KY Park, WH Lee. Antimutagenic effect of genistein toward environmental mutagen. *J Korean Environ Sci Soc* 8:569–574, 1999.

69. SY Lim, SH Rhee, KY Park. Antimutagenic effects of linoleic acid. *J Food Sci Nutr* 2:29–34, 1997.

70. H Hayastsu, S Arimoto, K Togawa, M Makita. Inhibitory effect of the ether extract of human feces on activities of mutagens: inhibition by oleic and linoleic acids. *Mutat Res* 81:287–293, 1981.

71. SY Lim, SH Rhee, KY Park. Inhibitory effect of linolenic acid on the mutagens-induced mutagenicities in Ames assay system and SOS chromotest. *Korean J Life Sci* 5:121–125, 1995.

72. CA Reznikoff, JS Bertram, DW Brakow, C Heidelberger. Quantitative and qualitative studies of chemical transformation of cloned C3H mouse embryo cells sensitive to post confluence inhibition of cell division. *Cancer Res* 33:3239–3249, 1973.

73. MR Choi, HS Lim, YJ Chung, EJ Yoo, JK Kim. Selective cytotoxic effects of doenjang (Korean soybean paste) fermented with Bacillus strains on human liver cell lines. *J Microbiol Biotechnol* 9:504–508, 1999.

74. SY Choi, MJ Cheigh, JJ Lee, HJ Kim, SS Hong, KS Chung, BK Lee. Growth suppression effect of traditional fermented soybean paste(doenjang) on the various tumor cells. *J Korean Soc Food Sci Nutr* 28:458–463, 1999.

75. YH Choi, BT Choi, WH Lee, SH Rhee, KY Park. Doenjang hexane fraction-induced G1 arrest is associated with the inhibition of pRB phosphorylation and induction of Cdk inhibitor p21 in human breast carcinoma MCF-7 cells. *Oncology Rep* 8:1091–1096, 2001.

76. TH Rhew, SM Park, HY Chung, KY Park, JC Hah. Antitumorigenic activities of linoleic acid detected by in situ hybridization on transplanted tumors in mice. *J Korean Cancer Assoc* 24:493–503, 1992.

77. JC Hah, TH Rhew, ES Choe, HS Yaung, KY Park. Antitumor effect of linoleic acid against sarcoma-180 detected by the use of protein A-gold complex in mice. *J Korean Cancer Assoc* 24:783–789, 1992.

78. YP Zhu, ZW Su, CH Li. Growth inhibition effects of oleic acid, linoleic acid and their methyl esters on transplanted tumors in mice. *J Natl Cancer Inst* 81:1302–1306, 1989.

79. JH Jeong, KH Kim, MW Chang, SD Lee, JK Seo. The influence of linoleic acid and ursolic acid on mouse peritoneal macrophage activity. *Korean J Immunol* 15:53–60, 1993.

80. KH Kim, MW Chang, KY Park, TH Rhew, YI Sunwoo. Effects of linoleic acid, ursolic acid, phytol, and small water dropwort extract on the phagocyte of mice. *Environ Mut Carcino* 13: 135–144, 1993.

81. KH Kim, MW Chang, YI Sunwoo. Effects of linoleic acid and ursolic acid on the natural killer cell activity in the mice. *J Kosin Medical College Semiannual* 8:25–45, 1992.

82. KH Kim, MW Chang, EY Kim, KY Park, YI Sunwoo. Effects of linoleic acid(LA) and ursolic acid(UA) on the T subset in the sarcoma 180-transplanted mice. *J Korean Soc Microbiol* 28:239–249, 1993.

83. X Llor, E Pons, A Roca, M Alvarez, J Mane, F Fernandez-Banares, MA Gassull. The effects of fish oil, olive oil, oleic acid and linoleic acid on colorectal neoplastic processes. *Clin Nutr* 22:71–79, 2003.

84. YH Choi, L Zhang, WH Lee, KY Park. Genistein-induced G2/M arrest is associated with the inhibition of cyclin B1 and the induction of p21 in human breast carcinoma cells. *Int. J Oncology* 13:391–396, 1998.

85. YH Choi, WH Lee, KY Park, L Zhang. p53-independent induction of p21 (WAF1/CIP1), reduction of cyclin B1 and G2/M arrest by the isoflavone genistein in human prostate carcinoma cells. *Jpn J Cancer Res* 91:164–173, 2000.

86. YH Choi, SJ Lee, M Kim, L Zhang, WH Lee, KY Park. Genistein induces inhibition of cell proliferation and programmed cell death in the human cancer cell lines. *J Korean Cancer Assoc* 30:800–808, 1998.

87. T Fotsis, M Pepper, H Adlercreutz, G Fleishmann, T Hase, R Montesano, L Schweigerer. Genistein, a dietary-derived inhibitor of *in vitro* angiogenesis. *Proc Natl Acad Sci USA* 90:2690–2694, 1993.

88. S Rauth, J Kichina, A Green. Inhibition of growth and induction of differentiation of metastatic melanoma cells *in vitro* by genistein: chemosensitivity is regulated by cellular p53. *Br J Cancer* 75: 1559–1566, 1997.

89. VE Steele, MA Pereira, CC Sigman, GJ Kelloff. Cancer chemoprevention agent development strategies for genistein. *J Nutr* 125: 713s–716s, 1995.

90. Q Cai, H Wei. Effect of dietary genistein on antioxidant enzyme activities in Sencar mice. *Nutr Cancer* 25:1–7, 1996.

91. SJ Hong. Health Potential of Genistein Against Prostate Diseases. Proceedings of International Symposium Soybean and Human Health, Seoul, 2000, pp 13-24.

92. T Fotsis, M Pepper, H Adlercreutz, T Hase, R Montesano, L Schweigerer. Genistein, a dietary ingested isoflavonoid, inhibits cell proliferation and *in vitro* angiogenesis. *J Nutr* 125:790s–797s, 1995.

93. HJ Lee. Anticancer Activity of Soybean Peptides. Proceedings of International Symposium on Soybean Peptides and Human Health, Seoul, 1998, pp 13–21.

94. MK Sung. Soy Saponins as a Cancer Prevention Agent. Proceedings of International Symposium on Soybean and Human Health, Seoul, 2000, pp 47–63.

95. PC Billings, P Newberne, AR Kennedy. Protease inhibitor suppression of colon and anal gland carcinogenesis induced by dimethylhydrazine. *Carcinogenesis* 11:1083–1086, 1990.

96. H Witschi, AR Kennedy. Modulation of lung tumor development in mice with the soybean-derived Bowman–Birk protease inhibitor. *Carcinogenesis* 10:2275–2277, 1989.

97. JO Ha, KY Park. Comparison of mineral contents and external structure of various salts. *J Korean Soc Food Sci Nutr* 27:413–418, 1998.

98. KI Lee, RJ Moon, SJ Lee, KY Park. The quality assessment of doenjang added with Japanese apricot, garlic and ginger, and samjang. *Korean J Soc Food Cookery Sci* 17:472–477, 2001.

99. SJ Lee. Studies on the Enhancement of Cancer Preventive Effects of Doenjang. MS thesis, Pusan National University, Busan, Korea, 2002.

100. CB Choi, EY Lee, DS Lee, SS Ham. Antimutagenic and anticancer effects of ethanol extract from Korean traditional doenjang added sea tangle. *J Korean Soc Food Sci Nutr* 31:322–328, 2002.

101. MC Hsieh, TL Graham. Partial purification and characterization of a soybean -glucosidase with high specific activity toward isoflavone conjugates. *Phytochemistry.* 58:995–1005, 2001.

102. HS Jeon. Effect of soybean saponins and major antioxidants on aflatoxin B1-induced mutagenicity and DNA adduct formation. MS thesis, Sookmyung Women's University, Seoul, 1998.

103. R Beaglenole. International trends in coronary heart disease mortality, morbidity and risk factors. *Epidemiol Rev* 12:1–15, 1990.

104. D Krummel. Nutrition in cardiovascular disease. In: LK Mahan, S Escott – Stump, 9th ed. *Food, nutrition and diet therapy.* Philadelphia: WB Saunders Co, 1996, pp 509–551.

105. JW Erdman. Soy protein and cardiovascular disease: A statement for health care professionals from the Nutrition Committee of the A.H.A. *Circulation* 102:2555–2559, 2000.

106. MS Anthony. Soy protein and coronary heart disease. Proceedings of 2nd International Symposium on Soybean and Human Health, Seoul, 2002, pp 35–43.

107. SW Chung, MA Choi, JS Park, KS Kim, DK Chung, HS Nam, ZI Shin, R Yu. Effect of dietary soybean hydrolysate on plasma lipid profiles, select biochemical indexes, and histopathological changes in spontaneously hypertensive rats. *Korean J Food Sci Technol* 31:1101–1108, 1999.

108. J Wu, W Ding, RE Aludo, AD Muir. Antihypertensive Effect of Fermented Soybean Products. Proceedings of 2nd International Symposium on Soybean and Human Health, Seoul, 2002, pp 113–133.

109. A Takahama, A Iwashita, M Matsuzawa, H Takahashi. Antihypertensive peptides derived from fermented paste-miso. INFORM 4(Abstract NN5):p 525, 1993.

110. WC Kendall. Effect of Soy Protein Food on Low-Density Lipoprotein Oxidation and Ex vivo Sex Hormone Receptor Activity — a Controlled Crossover Trial. Proceedings of International Symposium on Soybean & Human Health, Seoul, 2000, pp 89-96.

111. M Sugano, Y Yamada, K Yoshida, Y Hashimoto, T Matsuo, M Kimoto. The hypocholesterolemic action of the undigested fraction of soybean protein in rats. *Atherosclerosis* 72:115–122, 1988.

112. Nydahl MC, Gustafsson IB, Vessby B. Lipid-lowering diets enriched with monounsaturated or polyunsaturated fatty acids but low in saturated fatty acids have similar effects on serum lipid concentrations in hyperlipidemic patients. *Am J Clin Nutr.* 9:115–22, 1994.

113. SK Noh. Functional action of flavonoids for treatment of obesity. *Food Indust Nutr* 7:27–29, 2002.

114. T Wakamiya. The Physiological Function of Soybean Peptides. Proceedings of International Symposium on Soybean Peptides and Human Health. Seoul, 1998, pp 1–9.

115. MS Kurzer. Isoflavones and Menopausal Health. Proceedings of International Symposium on Soybean and Human Health. Seoul, 2000, pp 29–43.

116. DH Green, PR Wakeley, A Page, A Barnes, L Baccigalupi, E Ricca, SM Cutting. Characterization of two Bacillus probiotic. *Appl Environ Microbiol* 65:4288–4291, 1999.

117. P Mazza. The use of Bacillus subtilis as an antidiarrhoeal microorganism. *Boll Chim Farm* 133:3–18, 1994.

118. PL Meroni, R Palmieri, W Barcellini, G De Bartolo, C Zanussi. Effect of long-term treatment with B. subtilis on the frequency of urinary tract infections in older patients. *Chemioterapia* 2:142–144, 1983.

119. IV Pinchuk, P Bressollier, B Verneuil, B Fenet, IB Sorokulova, F Megraud, MC Urdaci. In vitro anti-Helicobacter pylori activity of probiotic strain Bacillus subtilis 3 is due to secretion of antibiotics. *Antimicrobial Agents Chemotherapy.* 45:3156–3161, 2001.

120. IB Sorokulova, DK Kirik, IV Pinchuk. Probiotics against Campylobacter pathogens. *J Travel Med* 4:167–170, 1997.

121. B Vaseeharan, P Ramasamy. Control of pathogenic Vibrio spp. by Bacillus subtilis BT 23, a possible probiotic treatment for black tiger shrimp Penaeus monodon. *Lett Appl Microbiol* 36:83–87, 2003.

21

Conventional and Emerging Food Processing Technologies for Asian Functional Food Production

HAO FENG and SCOTT A. MORRIS

Department of Food Science and Human
Nutrition, University of Illinois at
Urbana-Champaign, Urbana, IL, USA

INTRODUCTION

Traditionally, Asian foods are produced on a small scale with manual operations. The product is usually consumed locally. In the last few decades, mass production of prepared foods with modern equipment and technologies has started to play an increasingly important role in the production of many Asian food products, including foods with health claims. Conventional food processing technologies are primarily aimed at prevention of spoilage, and retention of nutrients and sensory attributes.[1] The production of functional foods, however, also requires maximizing the retention of biologically active components that are usually heat sensitive and/or susceptible to oxidative reactions.

The mass production of a food product involves separate and distinct steps called *unit operations*. Unit operations deal mainly with the transfer of energy, as well as the changes of food materials primarily by physical means but also by physical or biochemical means.[2] In the production of Kikkoman soy sauce (Figure 21.1), for example, several manufacturing steps can be identified. First, equal amounts of autoclaved soybeans and roasted and milled wheat are mixed together. By enriching the mixture with a seed culture of *aspergillus sojae*, a dry mash called *koji* is obtained. Brine composed of salt and water is added to and mixed with the *koji* to form a mash, which is called *moromi*. Fermentation of *moromi* in large tanks generates the flavor, sweetness, taste, and aroma typical to Kikkoman soy sauce. Following fermentation, the *moromi* is wrapped in cloth and pressed to separate the pure raw soy sauce. The raw soy sauce is filtered and pasteurized. The finished soy sauce is then bottled.

The production of an Asian food product may involve several different unit operations, as shown in Figure 21.1. In a specific processing step, however, one can also use different unit operations to perform the same task. For instance, soy sauce pasteurization can be achieved with a high temperature short time (HTST) heat exchanger, or with a nonthermal processing method, such as high pressure processing (HPP) or pulsed electric field (PEF). The selection of a unit operations combination for the production of an Asian functional food is determined by factors such as economics and effects on product quality or retention of key biologically active components. A good understanding of the operational principles of the conventional and emerging food processing methods, as well as their effects on product quality, including the bioactivity of key components, is essential to the development and production of an Asian functional food.

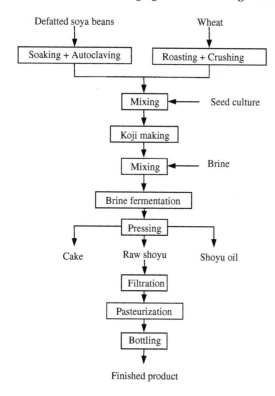

Figure 21.1 The processing steps involved in the manufacture of Kikkoman soy sauce.

DEGRADATION KINETICS OF KEY COMPONENTS

Kinetic Models

The effect of a processing method on quality changes of an Asian food product can be described with a selected kinetic model. There are several factors affecting the rate of degradation of a food component during processing and storage. Generally recognized factors include temperature, moisture content/water activity, and residence/heating time. Other parameters, such as pressure, oxygen, pH, composition (fats, sugar, protein, etc.), coreactant level, presence of trace metals and other catalysts, and light intensity, may also play an important role in certain circumstances.[3] Mathematical models have been developed to describe food quality changes. Even thought reaction pathways leading to quality degradation in foods are complicated, the kinetics of the quality changes, however, can be described by the following relationship:

$$\frac{d[A]_i}{dt} = -k_i [A]_i^n \qquad (21.1)$$

where [A] is the quality attribute for a selected quality index, k is a rate constant which is process and product specific, n is the order of the reaction, and the subscripts i = I, II etc., represent different stages a product will undergo, such as processing and storage in which different kinetic parameters are involved. In Equation 21.1, [A] can be the concentration of a bioactive component, pigment, or soluble solid; the enzyme activity; the microbial count; or a physical attribute such as porosity or rehydration capacity.

Many quality degradation reactions in food products exhibit either zero- (n = 0) or first- (n = 1) order kinetic behavior. A zero-order or pseudo-zero-order reaction was widely observed for overall quality of frozen foods, nonenzymatic browning, and chlorophyll loss.[4-8] A first-order reaction behavior was reported for ascorbic acid degradation,[9] thiosulfinate loss,[7] color loss,[10] microbial destruction,[11] texture loss,[8] and enzyme inactivation.[9] It is important to note that the degradation kinetics for components that determine product quality, such as flavor, texture or functional ingredient integrity, are usually different from those describing microbial inactivation. It is this difference that has led to the development of new processing methods that expose products to high temperatures for very short times, inactivating the microbes while minimizing quality loss.

Effect of Environmental Parameters

Factors affecting product quality can be included in the degradation kinetics equation by a rate constant which can either lump various effects into a single expression or isolate different factors using separate correlations. In the former case, the expression may take the form of

$$k_i = f\left(\text{temperature, moisture, time,}\atop O_2, \text{pH, composition, etc.}\right) \tag{21.2}$$

For the latter, a variety of correlations is available in the literature. The temperature dependence of quality decay kinetics, for example, is usually described by an Arrhenius-type equation[12]:

$$k_i = k_{0T} \exp\left(-\frac{E_a}{RT}\right) \tag{21.3}$$

where k_{0T} is an Arrhenius factor for temperature dependence, E_A is activation energy, R is universal gas constant, and T is absolute temperature. The moisture dependence can be written as[13]:

$$k_i = k_{0X} X^{-m} \tag{21.4}$$

where k_{0X} is a coefficient, X is moisture content, and m is a constant. The moisture and temperature dependence can also be taken into account in Equation (21.3) by relating the Arrhenius factor and/or the activation energy to moisture content and/or temperature using regression methods. Luyben et al.[9] considered the moisture dependency of both the Arrhenius factor and the activation energy by:

$$E_a = A + B \cdot \exp(-CX)$$
$$\ln k_o = D + E \cdot \exp(-CX) \tag{21.5}$$

where A, B, C, D, and E are constant and k_o is the Arrhenius factor. Values for parameters in Equation (21.5) of selected enzymatic reactions were given by Luyben et al.[9] and are listed in Table 21.1.

Nonenzymatic browning is a common reaction during processing and storage of many foods that causes pigment and nutrient losses. The activation energy values for nonenzymatic

TABLE 21.1 Parameters Defined in Equation (21.5) for Selected Enzymatic Reactions

Parameter	Catalase	Lipase	Alkaline phosphatase
A	2.585×10^5	3.898	4.832×10^5
B	-2.057×10^5	1.237×10^5	-4.832×10^5
C	3.699	4.880	11.366
D	86.27	-9.743	164.62
E	-85.95	38.509	-190.53

(Adapted from KChAM Luyben, JK Liou, S Bruin. Enzyme degradation during drying processes. In: P. Linko, J. Larinkari, Eds. *Food Process Eng*, Vol. 2. Applied Science Publishers Ltd., London, 1979, pp 192–209. With permission.)

browning of selected foods are tabulated in Table 21.2. Note that activation energy increases when water activity decreases.[17,21] As a result, the activation energy during storage of a processed product is higher compared with the activation energy during processing. This is because water tends to decrease the temperature sensitivity of the reaction as moisture concentration increases.[22]

SELECTED UNIT OPERATIONS IN THE PRODUCTION OF ASIAN FUNCTIONAL FOODS — CONVENTIONAL METHODS

Fermentation

Fermentation is one of the oldest forms of food preservation used by humans. It is a biochemical process in which the metabolism of microorganisms is carried out under anaerobic conditions, resulting in the production of desirable foods or beverages that are more stable, palatable, and nutritious. Fermentation can take place with a single culture, with a culture mixture, or by natural cultures in the case of indigenous fermented foods. Principal fermentation reactions involved in foods include lactic acid, propionic acid, citric acid, alcoholic, butyric acid, and acetic acid, as well as gassy fermentations.[23] Fermented foods can be classified into seven

TABLE 21.2 Activation Energy for Nonenzymatic Browning in Selected Foods

Food	a_w	X (g H_2O/g solid)	Temperature (°C)	E_a (kcal/mole)	Reference
Apricot (in storage)	NA	0.24	22–49	26	(14)
Potato (during drying)	NA	0.05–1.1	40–80	37–25	(15)
Raisin	0.6–0.8	NA	21	24	(16)
Cabbage (in storage)	0.01–0.62	NA	37	40–28	(17)
Carrot (during drying)	~0.15–0.75	NA	60–90	~125–80	(18)
Carrot (during drying)	NA	0.03–0.33	60–90	92–47	(19)
Onion (during drying)	0.29–0.95	NA	40–80	139–121	(20)

categories on the basis of the type of substrate, i.e., meat, fish/shellfish, dairy, cereals, roots crops, legumes, and vegetables/fruits. Fermentation offers several advantages: (1) enrichment of diet through development of a diversity of flavors, aromas, and textures in food substrates; (2) preservation of food through lactic acid, alcoholic, acetic acid, and alkaline fermentations; (3) improvement of nutritional values by enrichment with microbial protein, amino acids, lipids and vitamins; (4) detoxification during food fermentation; and (5) a decrease in cooking times and fuel requirements.[24,25] In modern food and beverage fermentation operations, fermentors with volumes up to 250 m³ with different configurations have been used to optimize and economize the operation. On the other hand, many Asian foods are indigenous naturally

TABLE 21.3   Selected Fermented Asian Foods

Type of product	Substrate(s)	Main Region(s)	Microorganisms	Fermentation
Bagoong	Fish, shrimps or oysters	East and Southeast Asia	Autolytic enzymes and bacteria	Proteolytic
Paak	Various fish and rice	East and Southeast Asia	Lactic acid bacteria	Lactic
Tapé ketan	Rice	Indonesia	Molds and yeasts	Alcoholic/ lactic
Soy sauce	Soya beans and wheat	China and Japan	Molds, yeast and bacteria	Proteolytic/ lactic
Tempe	Soya beans and other legumes	Indonesia	Molds	Proteolytic
Natto	Soya beans	Japan	B. subtilis	Proteolytic
Nata	Coconut or fruit juices	Philippines	Acetic acid bacteria	Acetic
Kimchi	Napa cabbage	Korea	Lactic acid bacteria	Lactic

(Adapted from T Hosoi, K Kiuchi. Natto-A food made by fermenting cooked soybeans with Bacillus subtilis (natto). In: ER Farnworth, Ed. *Handbook of Fermented Functional Foods*, New York: CRC Press, 2003, pp 227–245 and S Arai. Global view on functional foods: Asian perspectives. *Br J Nutrition* 88 Suppl:139–143, 2002. With permission.)

fermented products and they are traditionally produced with small-scale manual operations.

Fermentation plays an important role in the production of Asian foods. Table 21.3 lists selected Asian foods produced with fermentation. This is also reflected in the production of many Asian functional foods. *Natto*, a Japanese FOSHU (food for specified health use) product is made by soybean fermentation. It has a long history of use in many Asian countries other than Japan as well. Recently, a mutant of *Bacillus natto* was developed that can produce a high quantity of menaquinone-7 (vitamin K2). The *natto* produced using this mutant is thus expected to reduce the risk of osteoporosis.[26] With advances in genetic engineering, much progress has been made in food fermentation technology. Nevertheless, many indigenous fermented Asian foods and their processes are still not well known. Moreover, for those fermentation processes with well-known principles in laboratory testing, the scale up to industrial scale may pose a challenge. Some fermented Asian foods are facing problems with relatively high microbial load and short shelf life. Introduction of emerging technologies to postprocess fermented Asian functional foods will provide a possible approach to secure food safety and extend shelf life.[28]

Drying

Spray Drying

Spray drying is the most widely used industrial process involving particle formation and drying. It is highly suited for the continuous production of dry foods in either powder, granulate, or agglomerate form from liquid feedstocks such as solutions, emulsions, slurry, and puree. Spray drying involves the atomization of the feed into a spray of droplets and exposure of the droplets to hot air in a drying chamber. The sprays are produced by either rotary (wheel) or nozzle atomizers. Evaporation of moisture from the droplets and formation of dry particles proceed under controlled temperature and airflow conditions. Powder is discharged continuously from the drying chamber. Typical, a spray drying system consists of a feed pump, atomizer, air heater, air disperser, drying chamber, and subsystems for exhaust air cleaning and powder recovery. The drying can be divided into three stages: an initial period in which droplet temperature increases to the wet bulb temperature; a second period in which a concentration gradient builds up in the drop and water activity at the surface decreases; and a third period in which internal diffusion is the only limiting factor. The large surface area created by the fine droplets facilitates high heat and mass transfer in the hot air and droplet interface and results in fast drying. The residence time of the product is in the range of 5 to 120 seconds. There is a critical moisture content below which the droplet surface becomes impermeable to aroma compounds, thus preventing losses in flavor.

Spray drying is also the most common method of microencapsulation that may be used as an economical means to protect a bioactive component in a food product from being exposed to an oxidative environment.[29] One example is the spray drying of emulsions for protection of limonene in orange oil against oxidation using various matrices.[30] Microencapsulation with spray drying was tested to protect β-carotene in comparison with drum- and freeze drying methods. The cost of encapsulation with spray drying is considerably lower than freeze drying encapsulation.[29] Spray drying has found numerous applications in the production of Asian foods. Soy milk has been spray dried to reduce bulk volume and extend shelf life. Spray dried soymilks are sometimes sold with a coagulant to make instant homemade tofu.[31]

Freeze Drying

Freeze drying is a process of removing water from a product by sublimation and desorption. A freeze drying system consists of (a) a drying chamber with temperature controlled shelves (heating plates), (b) a condenser to trap water vapor removed from the product, (c) a cooling system to supply refrigerant to the condenser, and (d) a vacuum system to reduce the pressure in the chamber and condenser to facilitate the drying process. A freeze drying cycle has three phases: freezing, primary drying, and secondary drying. During the freezing phase, the goal is to freeze the mobile water of the product. Significant super-cooling may be encountered, so the product temperature may have to be much lower than the actual freezing point of the solution before freezing occurs. In the primary drying phase, the chamber pressure is reduced, and heat is applied to the product to cause the frozen water to sublime. The water vapor is collected on the surface of a condenser. At the end of primary drying, the product temperature will rise asymptotically toward the shelf temperature. This and several other methods may be used to detect the endpoint of primary drying. Since there is no mobile water in the product at the end of primary drying, the shelf temperature may be increased without causing melting. Therefore, the temperature is increased to remove bound water until the residual water content falls to the range required for optimum product stability. This phase is referred to as secondary drying, and is usually performed at the maximum vacuum the dryer can achieve. One needs to be careful not to increase product temperatures too fast so as not to exceed the glass transition of some products. The length of the secondary drying phase is determined by the nature of the product. For products rich in protein and peptides, water is required to maintain secondary and tertiary structures. If this water is removed, the product may be denatured and lose some or all of its activity. In this case, the water content must be carefully controlled.

It usually takes 14 to 18 hours to freeze dry a 1.27 cm (0.5 in) thick slab of food that initially contains 75 to 85% water. For example, for a coffee extract containing 74% water (0.5 inch thick), 60% of the drying time was used for sublimation and 40% for desorption (secondary drying). Drying time is proportional to L^2, where L is the thickness of the slab. Since in freeze drying the product remains frozen or at low temperatures most of the time, little heat damage to heat-sensitive or bioactive components occurs. In addition, because the removal of ice crystals leaves a porous, honeycomb-type structure, the product has a high rehydration capacity. Freeze drying is a slow and expensive process and is most suitable for drying of high-value products. There have not been many applications of freeze drying in the production of Asian functional foods, although there has been some laboratory testing.[32] However, the production of nutraceuticals from Asian herbs will become a growing area for freeze drying applications.

Refractance Window™ Drying

The Refractance Window™ (RW) drying system is a novel drying method developed by MCD Technologies, Inc. (Tacoma, WA). It utilizes circulating water at 95 to 97°C as a means of carrying thermal energy to a food product to be dried. The moist product to be dried is placed on the upper surface of the dryer's conveyor belt, a thin sheet of transparent plastic moving over the surface of hot water in a shallow trough. The unused heat in the water is recycled (Figure 21.2). The dried products are moved over a cold water trough before being scraped off the belt. The residence time of the product on the drying belt is typically 3 to 5 min.[33] The drying mechanisms involved in RW drying have been postulated by MCD Technologies in the following manner: Heat from the circulating hot water is conducted to the belt and then into the moist product to be dried. Additionally, infrared rays in the water are transmitted directly through the plastic film into the moist product, speeding the rapid but gentle drying process. As the product dries, the conveyor belt/wet product interface or the refractance window closes. This causes the majority of infrared radiation to be bent back into the water at the water side of the plastic film surface, leaving only conducted heat as the drying means. At this stage, however, since the product is relatively dry and thus possesses low thermal conductivity, the heat conducted through the product is also reduced. This protects the product from excess heating and curtails color and flavor degradation. RW drying, while maintaining a relatively low product temperature, also protects products from oxidization.

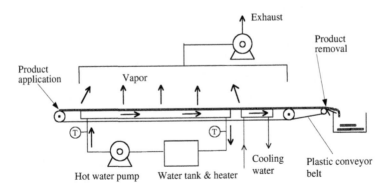

Figure 21.2 Schematic of a pilot-scale Refractance Window™ drying system.

TABLE 21.4   Carotene Losses in Carrots Among Control and Samples Dried with Drum, Freeze, and Refractance Window™ Drying Methods

Sample	Total carotene Loss (%)	α-carotene Loss (%)	β-carotene Loss (%)
Control	0.0	0.0	0.0
Drum dried	56.0 ± 1.2	55.0 ± 1.1	57.0 ± 1.3
RW[a] dried	8.7 ± 2.0	7.4 ± 2.2	9.9 ± 1.8
Freeze dried	4.0 ± 3.6	2.4 ± 3.7	5.4 ± 3.5

[a] Refractance Window™
(From BI Abonyi, H Feng, J Tang, CG Edwards, BP Chew, DS Mattinson, J K Fellman. Quality Retention in Strawberries and Carrots Dried with Refractance Window™ System. *J Food Sci* 67(3):1052–1056, 2002. With permission.)

The exact mechanism associated with RW drying is not fully understood. Nevertheless, experiments have been conducted to examine quality retention in food products dried with this novel drying method. Table 21.4 shows the total, α-, and β-carotene retention in carrot samples dried by three methods. The retention of carotenes in the RW-dried samples is comparable to that of freeze-dried. Since RW drying is a relatively simple drying method, the cost of RW drying is lower compared to freeze drying.

Osmotic Dehydration

Osmotic dehydration involves immersing plant or animal tissue in highly concentrated osmotic solutions so that a water migration from the tissue into the solution takes place, which is generally attributed to the existence of selective-permeable cell membranes. Water migration from the tissue to the solution is accompanied by a solute transfer from the solution into the tissue and a leaching out of some of the tissue's own solutes into the solution. The osmotic solutions can be a binary or ternary aqueous solution of mono-, di- and polysaccharides, inorganic salts, ethanol, and polyols. The driving force for water removal is the chemical potential between the solution and the intercellular fluid. The removal of water is mainly by diffusion and capillary flow, whereas solute uptake or leaching is controlled only by diffusion.[34]

Osmotic treatment results in a high- or intermediate-moisture product that usually needs subsequent processing to ensure a stable product. Despite the fact that the technique has been used in various ways for centuries, it is only since the late 1960s that osmotic dehydration has been recognized in a broader context and studied in combination with convective drying, vacuum drying, freeze drying, sun drying, pasteurizing, canning, and freezing. Extensive experiments have demonstrated that osmotic treatment can improve product quality. The quality improvement is ascribed to the direct formulation effect, which may increase the sugar-to-acid ratio and improve the texture and stability of the pigments during drying and storage.[35] The relatively low treatment temperature (30 to 50°C) also contributes to less quality loss.

Osmotic dehydration is a less energy-intensive process mainly because it does not deal with phase transition. The product is processed in a liquid phase that generally gives good heat- and mass-transfer coefficients. A characteristic of the osmotic process is the formation of a superficial concentration solute layer that reduces the mass transfer rate, but it limits solute impregnation and reduces the loss of water soluble solutes, such as ascorbic acid or fructose.[35] To improve the mass transfer, agitation of the osmotic solution, as well as vacuum osmotic dehydration, has been used.

In a drying test to compare the effect of different drying methods on the retention of lycopene activity, Shi[36] found that osmotic-vacuum dried samples have a significantly higher lycopene retention compared to hot air drying. Recently, the concept of osmotic impregnation has been expanded to the development of new functional foods. Vacuum impregnation has been studied especially for producing fortified functional fresh foods.[37,38] The purpose is to allow physiologically active components (PAC) in a solution to enter a porous food matrix during the impregnation process. The solution can contain PAC, a_w or pH depressors, antimicrobials, etc. Gras et al.[38]

tested calcium fortification of vegetables by vacuum impregnation and found that calcium impregnation takes place in the intercellular spaces of eggplant and xylem of carrots.

Extrusion

Extrusion is a process in which a highly viscous material is forced through a small opening to obtain a desired shape, which involves several unit operations, including mixing, kneading, cooking, shaping, and forming. Extruders can be classified according to the method of operation (cold forming, low-pressure forming, and high-pressure forming) and the method of construction (single- or twin-screw extruders). A food extrusion process involves complicated thermal, mechanical, and biochemical processes in which food powders turn from a glassy state into a rubbery semisolid melt, and back to a solid state after rapid cooling when moisture is flashed off at the discharge end, with accompanying biochemical reactions such as starch gelatinization, protein denaturation, and starch dextrinization.

In cold extrusion, pasta or meat products are mixed and shaped at ambient temperature. Low-pressure extrusion, with cooking and forming at temperatures below 100°C, is used to produce licorice, fish pastes, surimi, and pet foods.[39] In extrusion cooking, food is heated above 100°C. It is hence regarded as a high-temperature short time (HTST) process. Due to high temperature (up to 180°C), high pressure (up to 2,000 psi), relatively high shear rate (10 to 200 s^{-1}), and short residence time (5 s to 3 min), an extrusion cooking process plasticizes starch and protein biopolymers and inactivates microbes and enzymes.

There are two types of continuous extruders: single- and twin-screw (double) extruders. In a single-screw extruder, a screw rotates in a grooved cylindrical barrel. Food flows in a helical motion along the length of the barrel through a channel between the screw and the barrel, and is driven by friction between the food material and the inner barrel surface. A single screw can be divided into three sections. In the first section, the screw usually has deep flights to collect feed materials into the extruder. The second section is referred to as a kneading zone, where the depth of the screw is decreased to initiate compression and cooking. The final section of the screw is a melt zone, where the product reaches the outlet die and a high local pressure build-up and a high temperature coexist. Single-screw extruders have lower capital and operating costs and require less skill to operate than twin-screw extruders. A twin-screw extruder consists of two screws and can be divided into corotating and counterrotating types according to the direction of rotation, and can be either intermeshing (overlapping within the barrel) or not intermeshing.[40] Generally, counterrotating screws act like a positive displacement pump and have a C-shaped chamber in one side of the barrel. The corotating intermeshing screws that have a self-wiping function are the most commonly used in food processing operations. The advantages of the twin-screw extruders include that the throughput is independent of feedrate and that it can handle oily, sticky or very wet materials. The main limitations of the twin-screw extruders are the relatively high capital and maintenance costs.

It is important to understand the effect of operating parameters on functional components in an extrusion application. One of the primary reactions that occur in extruders is gelatinization of starches. Due to a combination of moisture, heat and mechanical forces, starch granules experience hydration and disintegration into small oligosaccharide units. The high temperature and the presence of protein and reducing sugars facilitate nonenzymatic browning due to a Maillard reaction, which will adversely affect the color, flavor, and nutrition of the final product. It was reported that at an extruder temperature of 154°C, there was a 95% retention of thiamin and little loss of riboflavin, pyridoxine, niacin or folic acid in cereals. However, losses of ascorbic acid and β-carotene were up to 50% and 90%.[41] Low temperatures and low concentrations of sugars, however, resulted in an increase in protein digestibility. Destruction of antinutritional components in soya products also improved the nutritional value of textured vegetable proteins.[39] In tests with a fiber-rich preparation using extrusion cooking, it was found that extrusion can moderately improve the hydration properties of pea hull brans, sugar beets, or lemon fibers.[42] Extrusion of wheat or rye flour under normal conditions did not change the total amount of dietary fiber. However, under drastic conditions, an increase in the amount of total fiber and a significant conversion of insoluble to soluble fractions has been found.[43] Extrusion has been used to produce a fortified ready-to-eat rice breakfast with added vitamins, minerals, and flavor compounds.[32]

Extraction

Extraction is a process of separating specific components from solid or liquid foods by means of an immiscible solvent. It can be classified into liquid-liquid, solid-liquid (or leaching), and supercritical fluid extractions. The history of extraction can be dated back to 2737 B.C., when the Chinese emperor Shen-Nung first brewed tea with boiling water.[44] Nowadays, extraction has evolved into a widely used unit operation with numerous food applications, including applications in the production of foods rich in bioactive components. Table 21.5 lists selected food applications for solid-liquid extraction. Extraction has several advantages over other separation methods because it is less energy intensive than distillation and evaporation and it can be used to separate azeotropic mixtures that are difficult to separate with distillation. Extraction deals with relatively low temperatures, so it is suitable for the separation of heat-sensitive food products.

In extraction operations, the selection of solvent is critical. Solvents used for food and biomaterials should be nontoxic, with a high capacity, high distribution coefficient, high selectivity for the solute, and low miscibility with the feed. Solvents should also be easily recoverable, stable and inert, nonflammable and nonexplosive, environmentally safe, and inexpensive.[44] It is not easy to find a solvent that can satisfy all the criteria for a specific product. The Food and Agriculture Organization of the United Nations (FAO) recommended 59 solvents for use in food processing.[47] The generally considered safe solvents for food applications include water, CO_2, alcohols (ethanol, propanol, isopropanol), acetone, methyl ethyl ketone,

TABLE 21.5 Examples of Solvent Extraction in Food Applications

Source	Solute	Solvent
Animal pancreas	Insulin	Acidic alcohol
Dry tea leaves	Tea solutes	Water
Fermentation broth	Penicillin	Amyl acetate, water
Fish scraps	Fish oil	Hexane, butanol
Fruit pomace	Pectin	Dilute acids
Green coffee beans	Caffeine	Methylene chloride, water, supercritical CO_2
Hop flowers	Hop solutes	Methylene chloride, supercritical CO_2
Oilseeds	Vegetable oils	Hexane
Orange peels	Orange essential oil	Water
Roasted coffee	Coffee solubles	Water
Vanilla beans	Vanilla	65% ethanol

(Adapted from RR Segado, Extraction techniques for food processing. In: RK Singh, SS Riviz, Eds. Bioseparation Processes in Foods, New York: Marcel Dekker, Inc., pp 17–57, 1995; RW Rousseau, Handbook of Separation Process Technology, New York: John Wiley, 1998[45]; and A Chiralt, J Martínex-Monzó, T Cháfer, P Fito, Limonene from citrus. In: J Shi, G Mazza, M Le Maguer, Eds. Functional Foods, Biochemical and Processing Aspects, Vol 2, New York: CRC Press, pp 169–188, 2002.[46] With permission.)

some dilute acid and basic solutions, aqueous solutions of nontoxic salts, hexane and other noncyclic alkanes (heptane, pentane, propane, and butane), some esters, and vegetable oils.

Extraction has been used to enrich SDG, a plant lignan in flaxseeds with ethanol as a solvent.[48] Solvent extraction finds application in obtaining anthocyanin pigments from grape pomace. Enocyanin, a deeply colored extract from red grape pomace, has been commercially produced in Italy since 1879. In this process, SO_2 is used to assist the extraction and to protect the pigments from oxidation and microbial spoilage. Extraction of crushed grape pomace with a mixture of ethyl acetate and water yielded phenolic compounds displaying antioxidant activities comparable to BHT in Rancimat tests.[49] The essential oil of yellow mustard is obtained by solvent extraction of the press cake. For components that are prone to oxidation and heat, supercritical CO_2 extraction is a good choice since it can provide high recovery (95%) compared to cold pressing. A drawback of supercritical fluid extraction is its relatively high production cost. Soybean oil, used primarily for cooking and frying in most Asian countries, is produced by an expander-solvent extraction method.[50]

Thermal Processing

Thermal processing, one of the oldest means of assuring the safety of packaged moist foods, has maintained a dominant presence in the Asian food industry. Although energy intensive, it is relatively simple, robust and, if properly managed, will assure a product that is safe and wholesome. The earliest thermally processed foods produced by Nicholas Appert in France in the early 1800s were simple glass bottles filled with food and placed in boiling water. This development was mimicked in Great Britain using hand-formed metal cans. High-pressure steam retorts and mass-produced metal cans followed, as well as the development of a good understanding of both the microbiology of pathogens and the physics of heat transfer. Because of the large amount of fuel required to produce the heat in thermal processing, efforts have been made to increase the efficiency of the basic process by modifying the process, the energy source, the package or some combination of these, but the majority of shelf-stable moist foods is still processed in a manner similar to that developed by Appert. Current industry estimates indicate that although metal cans and glass jars will remain a fixture on the grocery shelves for some time to come, new production capacity is being targeted at newer types of process-package combinations, such as retortable pouches and aseptic packaging.

Thermal Processing

Thermal processing seeks to treat the food in a closed container with enough heat to achieve a state of "industrial sterility," where pathogens have either been killed or deactivated to below a predetermined statistical level, typically one organism per several million containers. It is worth noting that this does not necessarily mean that the food product is completely devoid of pathogens but it does mean that the product is unlikely to develop any harmful pathogenic activity prior to use. In order to thermally process a food product, it is necessary to understand the nature of heat conduction in the product, and the rate at which a target organism of concern is killed.

UHT and HTST Processing

With thermal processing in closed containers came concerns about the food product's nutritional and flavor quality being degraded by the long cooking times that are necessary to achieve microbial inactivation. Since microbial inactivation is a function of both time and temperature, as is degradation of nutrients and flavor compounds, researchers surmised that it might be possible to "flash cook" foods at high temperatures for very short periods of time, particularly liquid foods, so that they would be microbially safe, either pasteurized or made shelf stable, with a much lower level of nutrient and flavor loss. The methods currently used exploit the ability to outpace the quality degradation kinetics of the product with a fast microbial inactivation. The additional benefit of this is that thermal energy is not wasted in overcooking the product in large containers, but is applied only to the extent needed for microbial inactivation.

Both ultra high temperature (UHT) and high-temperature-short-time (HTST) treatments rely on similar types of thermal treatment, differing only by the times and temperatures used in processing. The food product is passed through a heat exchanger that offers a large surface area and small section thickness, which reduces the amount of time that the product needs to be treated to minimize overcooking. The product is then transferred to a sterile container which is then sealed. By doing this, the finished package is not cooked in a retort, and there is little or no energy spent in overcooking the product. This treatment is extended into unit operations that operate under filtered air and sterile closures (i.e., aseptic packaging).

Applications in Asian Functional Food Production

The previous background on thermal processing can be used to design a thermal process for an Asian functional food. Traditionally, cook values are used to examine the overall quality of the thermal processed foods. In functional food applications, a good estimation of the degradation kinetics of the bioactive components is important. Some bioactive components (such as the pigments in fruit juices) are temperature sensitive, while others such as fiber are relatively inert and can often be considered to be stable. Further complications arise because the modes of action of many functional ingredients are poorly understood or characterized. Thus, although it may be possible to preserve the apparent integrity of a particular ingredient, it may be difficult to ensure that the functionality that is associated with that particular ingredient is maintained. Nowadays, canning, HTST pasteurization, and aseptic packaging have become a common practice in the production of many Asian food products, including numerous functional foods and nutraceuticals.

SELECTED UNIT OPERATIONS IN THE PRODUCTION OF ASIAN FUNCTIONAL FOODS — EMERGING TECHNOLOGIES

Alternate Thermal Processing Methods

Microwave Heating

Microwave ovens have become a common appliance in most industrialized countries. By definition, microwaves are electromagnetic waves in the frequency range of 300 to 300,000 MHz, corresponding to wavelengths from 1 m to 1 mm. Heating foods by microwave radiation is achieved both by the absorption of microwaves that is fixed at either 2,450 MHz or 915 MHz by rotation of dipoles in water and translation of the ionic components of the food. The ability of a material to be heated in an electromagnetic field is termed "lossiness." To quantify the "lossiness," a complex dimensionless number, the permittivity

ε^* is often used. The real component, the dielectric constant ε', expresses the ability to store energy in the material. The imaginary component, $\varepsilon,"$ represents the energy losses and is called the dielectric loss factor. Ingredients in foods with different dielectric properties will hence be heated at different speeds. Additionally, the temperature and material phase of the material has a profound effect on the heating behavior of the materials, particularly if it is frozen. Foods that are thawed or warm will generally heat faster than those that are frozen or cool.

In microwave ovens a number of standing wave patterns can be generated as a result of multiple reflections at the metal cavity walls. This will result in nonuniform heating, an inherent problem associated with microwave (MW) heating.[51] If a food is exposed to MW irradiation, it will begin to convert the MW energy into heat more and more efficiently as the temperature increases at locations where more MW energy is focused due to nonuniform distribution, causing a condition known as "runaway heating." To overcome the nonuniformity problem, two common features have been widely used in the oven design: a "stirrer," commonly a slow-turning set of reflectors built into the roof of the cavity, and a turntable to rotate the food as the energy field is shifted by the stirrer. In industrial microwave treatment, the same problem may be present, but for many processing applications, it may be possible to incorporate design features into the cavity as well as move the product around to allow different parts of the product to receive nearly equal amount of MW irradiation over a period of time. Generally, the fast and volumetric heating facilitated by microwave irradiation allows the design of a thermal process with short treatment time, and hence provides better protection of the bioactive components.

RF Heating

Radio frequency (RF) heating is similar to microwave heating in that food components will convert an oscillating electromagnetic field to heat as a function of composition and temperature (although the temperature effects are less pronounced with RF heating), but the system runs at much lower frequencies. RF heating involves applying a high-voltage AC signal to a set of parallel electrodes set up as a capacitor. The medium to be heated is sandwiched between the electrodes. RF has been used as a rapid heating method in baking and postdrying snack foods, thawing frozen foods, blanching vegetables, heating packaged bread, etc.[52]

Ohmic Heating

Ohmic heating is a special case of the above two types of alternate thermal sources, since it runs at very low frequencies and can be run in direct current mode. Rather than responding to lossiness of the materials, heating is dominated by the electrical resistance of a material. In principle, the ohmic heating effect is similar to that obtained with microwaves. However, the major advantage of ohmic compared to microwave heating is that the depth of penetration is not limited. The process allows food products to be heated to sterilization temperature in less than 90 s, and hence improves product quality and gives a better retention of bioactive components. Ohmic heating works well for the processing of various high- and low-acid Asian food products, which may be filled aseptically for ambient temperature storage and distribution.[53]

Nonthermal Technologies

High-Pressure Processing (HPP)

High-pressure processing (HPP) is also known as high hydrostatic pressure (HHP) processing or ultra-high pressure (UHP) processing, wherein a food is treated at elevated pressures of 200 to 1,000 MPa. Process temperatures during HPP treatment can be specified from below 0°C to above 100°C. Treatment time can range from a millisecond pulse to over 30 min. Hite et al.[54] in 1914 first reported the effects of HPP on food microorganisms by subjecting milk to pressures of 650 MPa and obtaining a reduction in the viable numbers of microbes. For the last 15 years, because of consumer demand for high quality, minimally processed foods with fresh characteristics and no additives, HPP has gained in popularity with the food industry. Advantages of HPP over traditional thermal processing include reduced process times; minimal heat damage; uniform and instantaneous treatment; retention of freshness, flavor, texture, and color; no vitamin C loss; minimal undesirable functionality alterations; and capacity to treat packaged foods. HHP-treated foods may have improved or imparted functional properties of food constituents.

The basic principle that explains the action of HPP treatment is the *Le Chatelier principle,* which states that any reaction, conformational change, or phase transition accompanied by a volume deduction will be favored at high pressure while reactions involving an increase in volume will be inhibited.[55] Thus HPP affects any phenomenon in food systems in which a volume change is involved and favors phenomena that result in a volume decrease. Other principles that govern HPP include the *isostatic principle,* which implies that the transmittance of pressure is uniform and instantaneous, as well as the *microscopic ordering principle,* which implies that at constant temperature, an increase in pressure increases the degree of ordering of the molecules of a substance.[56] HPP has been demonstrated to be very effective in the inactivation of vegetative cells of bacteria. It is also used to inactivate yeasts, molds, and spores with various degrees of success. Enzymes related to food quality vary in their barosensitivity. Some of them can be inactivated at room temperature by pressures of a few hundred MPa, whereas others can withstand 1000 MPa.[57]

The microbial inactivation of HPP is attributed to the pressure-induced collapse of intracellular vacuoles, and damage to cell walls and cytoplasmic membranes. Knorr[58] proposed that the disruption of metabolic processes caused by the effects of high pressure on cellular enzymes might be another inactivation mechanism. Pressure used in the HPP of foods appears to affect only noncovalent bonds (i.e., ionic, hydrogen and hydrophobic bonds), leaving covalent bonds intact. This permits the destruction of microbial activity without affecting food molecules that contribute to the flavor and nutrition of the food.

The application of pressure influences biochemical reactions since most of these reactions are associated with volume changes. Hoover et al.[59] reported that pressure affects reaction systems in two ways: by reducing the available molecular space and by increasing interchain reactions. As a result, reactions involved with the formation of hydrogen bonds are favored by high pressure, since such bonding leads to a decrease in volume. Cheftel[60] demonstrated that pressures above 100 to 200 MPa often cause the dissociation of oligometric structures into subunits, partial unfolding and denaturation of monomeric structures, and protein gelation if protein concentration and pressure are high enough. The effect of HPP on soy protein functionality has been investigated by several research groups.[61–63] HPP is also studied for its capacity to reduce microbial load and extend shelf life for kimchi.[28] In Japan, HPP has been successfully used in the production of niche and high quality jams and fruit juices.

Pulsed Electric Field (PEF)

Pulsed electric field (PEF) processing involves the application of high-voltage pulses for a few microseconds to food placed or flowing between two electrodes. The concept of PEF is relatively simple. Electric energy at a low power level is collected over an extended period and stored in a capacitor. The collected energy is then discharged almost instantaneously at very high levels of power. The generation of pulsed electric fields requires two major devices: a pulsed power supply and a treatment chamber that converts the pulsed voltage into pulsed electric fields. If the electric field intensity between the electrodes reaches the range of 12 to 35 kV/cm, a pronounced lethal effect on microorganisms can be observed. The precise mechanisms by which microorganisms are destroyed by PEF are not well understood, but some postulations have been brought forth to describe the inactivation actions. Zimmermann et al.[64] proposed that pores are formed in cell membranes when the applied electric field causes the electrical potential of the membrane to exceed the natural level of 1 V. The pores then cause swelling and rupturing of the cells. The induced oxidation and reduction reactions within the cell structure that disturb metabolic processes may be another cause of inactivation.[65] The highly reactive free radicals produced from food components due to electrolysis reactions of food components in the electric field can also have a bactericidal effect.

The advantages of PEF include good bactericidal effect; good color, flavor and nutrient retention; no evidence of toxicity; and short process time. The limitations of PEF may arise from its low enzyme and spore inactivation capacity; difficulty in treating food with relatively high conductivity; suitability for liquid foods only; and possible electrolysis. In foods, large molecules that carry charges, such as protein, ionic polysaccharides, polar lipids, and molecules containing double bonds or sulphydryl groups, will undergo modifications when exposed to PEF.[66] Foods with a relatively high protein content experience deposition on the anode of a PEF system due to protein aggregation. The most common application of PEF has been focused on food preservation and product quality aims, including extending the shelf life of orange juice, apple juice, milk, and liquid eggs.[67] Shelf-life studies show that PEF treatment can extend the refrigerated shelf life of fresh citrus juice to beyond 60 days.[68] PEF is also studied for its potential as a novel membrane permeabilization method in the fields of medicine and biosciences.[69] It has been found that PEF can increase the permeability of plant cell membranes so that the yield of juice in extraction operations is increased.[70] In tests with the yeast *Phaffia rhodozyma*, Kim et al.[71] recorded a 98% increase in electropermeabilization in carotenoid pigments extraction, showing the potential to use PEF in bioactive component extraction.

Irradiation

Ionizing radiation includes γ-rays from cobalt 60 or cesium 137, and to a lesser extent, X-rays generated in electrically driven machines, as well as electron beams.[72] Irradiated foods are required to carry a radura symbol and the words *treated by irradiation* or *treated with radiation*. The main advantages of irradiation include little or no heating of food, ability to treat packaged and frozen foods, low energy requirement, automatic operation, and low operational costs. The major limitations are the high capital cost and the public resistance due to fear of induced radioactivity in processed foods.

When food is irradiated, most of the radiation passes through the food without being absorbed. The small amount that is absorbed destroys any insects on grains, produce or spices; extends shelf life; and prevents fruits and vegetables from ripening too fast. Higher doses can kill foodborne pathogens that can contaminate foods and cause foodborne diseases. Food irradiation is a nonthermal treatment that achieves its effects without raising the food's temperature significantly, thus minimizing nutrient losses and changes in food texture, color, and flavor.

The effects of irradiation on food proteins present in a food matrix are dose and product dependent. Generally, peptide linkages are not attacked, and the main effects are concentrated around sulfur linkages and hydrogen bonds.[73] The radicals formed are mostly immobile and are prone to recombination rather than to reaction with other food components. At commercial dose levels, irradiation has little or no effect on the digestibility of proteins or composition of essential amino acids. Enzymes are not affected by irradiation. In some products, however, irradiation can alter the functionality of proteins. In irradiated eggs, for example, loss of viscosity in the white and off-flavor development in the yolk were observed. Irradiation can break high molecular weight carbohydrates into small units. Carbohydrates may react with hydroxyl radicals to form ketones, aldehydes, or acids, leading to a drop in pH. Starch may degrade to dextrins, maltose, and glucose, resulting in a decrease in the viscosity in solution. Irradiation does not change in the degree of utilization of the carbohydrate and hence there is no reduction in nutritional value. Lipid oxidation can be triggered by irradiation. Highly unsaturated fats are more readily oxidized than less unsaturated fats. This process can be slowed by eliminating oxygen by vacuum or modified atmosphere. Most vitamins are not affected severely by irradiation up to 10 kGy. Water soluble vitamins vary in their sensitivity to irradiation. Reported sensitivity is in the order of thiamin > ascorbic acid > pyridoxine > riboflavin > folic acid > cobalamin > nicotinic acid.[39] Among the fat soluble vitamins, vitamin E is the most sensitive. Therefore, foods with high vitamin E content, such as oils and dairy products, are not suitable for irradiation due to off-flavor generation. There are few studies on the effect of irradiation on the bioactivity of food components. Ayed et al.[74] used gamma irradiation to extend the shelf life of grape pomace and found improved anthocyanin yields.

Ultraviolet (UV) and Pulsed Light

Ultraviolet (UV) radiation for the purpose of food preservation utilizes the bactericidal effect of light in the wavelengths of 200 to 320 nm. Commercial UV light processing involves the use of mercury lamps, which generate 90% of their energy at a wavelength of 253.7 nm. UV irradiation is safe, environmentally friendly, energy saving, and less expensive compared to conventional bactericidal treatments. To achieve microbial inactivation, the UV radiant exposure must be at least 400 J/m^2. Light in the pulsed form can also be used to inactivate foodborne pathogens, which include pulsed ultraviolet light (PUV) and pulsed light. The latter involves intense and short duration pulses of broad-spectrum 'white' light ranging from a UV wavelength of 200 nm to near infrared regions of 1000 nm with peak emissions between 400 to 500 nm. In a pulsed light system, a capacitor stores electrical energy to generate rapidly released high-intensity high-power pulses. Such pulses are used to generate intense flashes of broad-spectrum light by electrically ionizing a xenon gas lamp. In operation, light flash duration is typically about 300 μs and the light intensity can be as high as 20,000 times that of sunlight. The materials to be sterilized are typically exposed to 1 to 20 flashes per second.

The bactericidal effect of UV light is due to the DNA absorption that causes cross-linking between neighboring pyrimidine nucleoside bases (thymine and cytosine) in the same DNA strand.[75] Once the threshold of cross-linking has been exceeded, the number of cross-linkings is beyond repair, and cell death occurs[75]. In pulsed light treatment, pulses of light induce photochemical and photothermal reactions in foods. The UV component causes photochemical changes, whereas visual and infrared lights cause photothermal effects. Inactivation occurs by several mechanisms, including chemical modification and cleavage of the DNA. Conventional UV treatment primarily affects DNA by mechanisms that are reversible under certain experimental conditions. Pulsed light, however, is able to produce extensive irreversible damage to DNA, proteins, and other macromolecules.[76]

Microbial inactivation with either UV or broad-spectrum light is basically a surface treatment. The penetration depth of light in many food materials is limited, especially in opaque foods or on irregular surfaces. Therefore, light treatments are mainly used for inactivation of liquid foods with low solid content and surface sterilization of packaging materials. Nevertheless, UV light has been successfully used in the fruit and vegetable juice industry to pasteurize juices and to achieve a five-log reduction. The CiderSure UV processor developed by OESCO Inc. (Conway, MA, U.S.) has found applications in apple juice and apple cider pasteurization. It is a technique especially designed for small juice producers. Light Process™ (California Day-Fresh Foods, Glendora, CA, USA) is a UV cold pasteurization technique used to process carrot juice. Juices treated with Light Process are reported to have high nutrient retention, good color, fresh-like taste, and retention of natural enzymes. In certain circumstances, UV treatment may compromise the quality of a food product by introducing off-flavors, such as rancidity, tallowiness, fishiness, cardboard flavor, and oxidized flavor.[77] UV irradiation may provide Asian functional food and nutraceutical producers an inexpensive alternative to pasteurize liquid products to better protect the bioactive components.

Ultrasound

Ultrasound is energy generated by sound waves of 20,000 or more vibrations per second. It can be divided into two categories, low-intensity and high-intensity ultrasounds. Low-intensity ultrasound is used as a nondestructive analytical method while high-intensity or power ultrasound finds applications in emulsion generation, dispersion of aggregated materials, drying, and modification and control of the crystallization process.[78] The bactericidal effect of ultrasonic waves has long been observed. Recent studies of ultrasound reported increased inactivation rates when bacterial spores were simultaneously exposed to ultrasonic waves and heat (thermoultrasonication).[79] Ultrasound in combination with pressure treatment was also studied to achieve a higher inactivation rate.[80]

Several theories have been proposed to describe the inactivation mechanism of ultrasound. When ultrasonic waves pass through a liquid consisting of alternate rarefaction and compressions, bubbles or cavities can be formed if the amplitude of the waves is high enough. This phenomenon is known as cavitation. The collapse of the bubbles creates high pressure that causes cell membranes to disrupt and the cell wall to break down.[81] Application of ultrasound to a liquid can also lead to the formation of OH– and H+ species and hydrogen peroxide.[82] These species also have important bactericidal properties. When ultrasound is combined with heat and pressure, a synergistic effect has been reported. This synergistic effect was attributed to the disruption of the bacterial spore cortex, which resulted in protoplast rehydration and loss of heat resistance.[83] In the case of sonication assisted by elevated pressure, the increase in inactivation rate was probably due to an increase in bubble implosion intensity, as postulated by Pagán et al.[80]

Besides food preservation studies, ultrasound has also been tested in acoustic assisted unit operations in drying, extraction, crystallization, thawing, freezing, and cutting. Ultrasound food cutters have been used in the food industry for cutting various products and are especially suited to glutinous Asian food products. Ultrasonically assisted extraction of herbs and Chinese plants for the production of helicid, bernerine, and bergenin has resulted in purer products in a shorter time.[84] An industrial-scale ultrasonic reactor dedicated to the solvent extraction of different herbs has been installed in the PLAFAR factory in Brasov, Romania.[85]

REFERENCES

1. G Mazza, BD Oomah. *Functional Foods: Processing Technologies for Retention of Biologically Active Compounds.* Dallas: AIChE CofE, 1999, pp 1–6.

2. CJ Ceankkoplis. *Transport Processes and Unit Operations.* 2nd ed. Boston: Allyn and Bacon, Inc., 1983, pp 1–2.

3. YC Lee, JR Kirk, CL Bedford, DR Heldman. Kinetics and computer simulation of ascorbic acid stability of tomato juice as function of temperature, pH and metal catalyst. *J Food Sci* 42:640–644, 1977.

4. M Karel, JTR Nickerson. Effects of relative humidity, air, and vacuum on browning of dehydrated orange juice. *Food Technol* 18:1214–1218, 1964.

5. I Saguy, S Mizrahi, R Villota, M Karel. Accelerated method for determining the kinetic model of quality deterioration during dehydration. *J Food Sci* 43:1861–1864, 1978.

6. S Resnick, G Chirife. Effect of moisture content and temperature on some aspects of nonenzymatic browning in dehydrated apple. *J Food Sci* 44:601–605, 1979.

7. CM Samaniego-Esguerra, IF Boag, GL Robertson. Kinetics of quality deterioration in dried onion and green beans as a function of temperature and water activity, Lebensm.-Wiss. *u-Technol* 24:53–58, 1991.

8. PS Taoukis, TP Labuza, IS Saguy. Kinetics of food deterioration and shelf-life prediction. In: KJ Valentas, E Rotstein, RP Singh Eds. *Handbook of Food Engineering Practice.* CRC Press, New York, 1997, pp 361–403.

9. KChAM Luyben, JK Liou, S Bruin. Enzyme degradation during drying processes. In: P Linko, J Larinkari, Eds. *Food Process Engineering*, Vol. 2. London: Applied Science Publishers Ltd., 1979, pp 192–209.

10. H Chou, WM Breene. Oxidation decoloration of b-carotene in low-moisture model systems. *J Food Sci* 37:66–68, 1972.

11. S Bruin, KChAM Luyben. Drying of food materials: a review of recent developments. In: AS Mujumdar, Ed. *Advances in Drying.* New York: Hemisphere Pub, 1980, pp 155–215.

12. I Saguy, M Karel. Modeling of quality deterioration during food processing and storage. *Food Technol* Feb:78–85, 1980.

13. C Strumillo, J Adamiec. Energy and quality aspects of food drying. *Drying Technol* 14(2):423–448, 1996.

14. ER Stadtman, HA Barker, VA Haas, G Mackinney. Studies on the storage of dried fruit. III. The influence of temperature on the deterioration. *Ind Eng Chem* 38:324–329, 1946.

15. CE Hendel, VG Silveira, WO Harrington. Rates of nonenzymatic browning of white potato during dehydration. *Food Technol* Sep:433–438, 1955.

16. MJ Copley, MJWB Van Arsdel. Food Dehydration, vol 2. Products and Technology. Connecticut: AVI Pub. Co. 1964.

17. S Mizrahi, TP Labuza, M Karel. Computer-aided predictions of extent of browning in dehydrated cabbage. *J Food Sci* 35:799–803, 1970.

18. K Eicher, R Laible, W Wolf. The influence of water content and temperature on the formation of Maillard reaction intermediates during drying of plant products. In: D Simato, JL Multon Eds. *Properties of Water in Foods.* Dordrecht, The Netherlands: Martinus Nijhoff Publishers, 1985.

19. K Müller, W Bauer. *Detection and Kinetics of Chemical Reaction During Drying of Foods.* D MacCarthy, Ed. Concentration and drying of foods. New York: Elsevier Applied Sci. Publ., 1986.

20. RS Rapusas, RH Driscoll. Thermophysical properties of fresh and dried white onion slices. *J Food Eng* 24:49–164, 1995.

21. CE Hendel, VG Silveira, WO Harrington. Rates of nonenzymatic browning of white potato during dehydration. *Food Technol* Sep:433–438, 1955.

22. TP Labuza, M Saltmarch. The nonenzymatic browning reaction as affected by water in foods. In: LB Rockland, GF Stewart, Eds. *Water Activities: Influence on Food Quality*, New York: Academic Press, 1981.

23. DYC Fung. Food fermentation. In: FJ Francis, Ed. *Encyclopedia of Food Science and Technology*, 2nd ed. New York: John Wiley & Sons, 2000.

24. KH Steinkraus. Introduction to indigenous fermented foods. In: KH Steinkraus, Ed. *Handbook of Indigenous Fermented Foods,* New York: Marcel Dekker, Inc, 1996.

25. L Leistner. Fermented foods. In: R Macrae, RK Robinson, MJ Sadler, Eds. *Encyclopedia of Food Science Food Technology and Nutrition*, London: Academic Press, 2000.

26. T Hosoi, K Kiuchi. Natto-A food made by fermenting cooked soybeans with Bacillus subtilis (natto). In: ER Farnworth, Ed. *Handbook of Fermented Functional Foods*, New York: CRC Press, 2003, pp 227–245.

27. S Arai. Global view on functional foods: Asian perspectives. *Br J Nutr* 88 Suppl:139–143, 2002.

28. JW Lee, DS Cha, KT Hwang, HJ Park. Effects of CO_2 absorbent and high-pressure treatment on the shelf life of packaged kimchi products. *Int J Food Sci Technol* 38:519–524, 2003.

29. SA Desobry, FM Netto, TP Labuza. Comparison of spray-drying, drum drying and freeze-drying for b-carotene encapsulation and preservation. *J Food Sci* 62:1158–1162, 1997.

30. M Minor, F Weinbreck, de CG Kruif. Innovations in encapsulation of food ingredients. *Innovations in Food Technol* 76–77, May 2002.

31. K Liu. Oriental soyfoods. In: CYW Ang, K Liu, YW Huang, Eds. *Asian Foods, Science and Technology*, Lancaster, PA: Technomic Publishing Co., Inc., 1999.

32. BS Luh. Rice products. In: CYW Ang, K Liu, YW Huang, Eds. *Asian Foods, Science and Technology*, Lancaster, PA: Technomic Publishing Co., Inc., 1999

33. BI Abonyi, H Feng, J Tang, J., CG Edwards, BP Chew, DS Mattinson, J K Fellman. Quality retention in strawberries and carrots dried with Refractance WindowTM System. *J Food Sci* 67(3):1052–1056, 2002.

34. MS Rahman, CO Perera, Drying and food preservation. In: MS Rahman. Ed. *Handbook of Food Preservation*, New York: Marcel Dekker, Inc., 1999, pp 173–216.

35. AL Raoult-Wack. Recent advances in the osmotic dehydration of foods. *Trends in Food Sci Technol* 5:255–260, 1994.

36. J Shi, M LeMaguer, Y Kakuda, A Liptay, F Niekamp. Lycopene degradation and isomerization in tomatoes during dehydration. *Food Res Int* 32:1521, 1999.

37. P Fito, A Chiralt, N Betoret, ML Gras, M Cháfer, J Martínez-Monzó, A Andrés, D Vidal. Vacuum impregnation and osmotic dehydration in matrix engineering: application in functional fresh food development. *J Food Eng* 49:175–183, 2001.

38. ML Gras, D Vidal, N Betoret, A Chiralt, P Fito. Calcium fortification of vegetables by vacuum impregnation: Interactions with cellular matrix. *J Food Eng* 56:279–284, 2003.

39. PJ Fellows. *Food Processing Technology.* 2nd ed. Cambridge: Woodhead Publishing Limited, 2000, pp 294–308.

40. DR Heldman, RW Hartel. *Principles of Food Processing*, Gaithersburg: Aspen 1998, pp 253–283.

41. JM Harper. Food extrusion. *CRC Crit Rev Food Sci Nur*, February:155–215, 1979.

42. MC Ralet, JF Thibault, VG Della. Influence of extrusion cooking on the physico-chemical properties of wheat bran. *J Cereal Sci* 11, pp 249–259, 1990.

43. I Bjorck, M Nylan, NG Asp. Extrusion cooking and dietary fiber: effects on dietary fiber content and on degradation in the rat intestinal tract. *Cereal Chem* 61:174–179, 1984.

44. RR Segado. Extraction techniques for food processing. In: RK Singh, SS Riviz, Eds. *Bioseparation Processes in Foods,* New York: Marcel Dekker, Inc., pp 17–57, 1995.

45. RW Rousseau. *Handbook of Separation Process Technology*, New York: John Wiley, 1998.

46. A Chiralt, J Martínex-Monzó, T Cháfer, P Fito. Limonene from citrus. In: J Shi, G Mazza, M Le Maguer, Eds. *Functional Foods, Biochemical and Processing Aspects*, Vol 2, New York: CRC Press, pp 169–188, 2002.

47. FAO/WHO. General Requirements. Vol. 1. In: Codex Alimentarius. Food and Agriculture Organization, World Health Organization. Rome: United Nations, 1992.

48. DJA Jenkins. Incorporation of flaxseed or flaxseed components into cereal foods. In: SC Cunnane, LU Thompson, Eds. *Flaxseed in Human Nutrition*, Champaign: AOCS Press, 1995, pp 281–294.

49. F Bonilla, M Mayen, J Merida, M Medina. Extraction of phenolic compounds from red grape marc for use as food lipid antioxidants. *Food Chem* 66:209–215, 1999.

50. YW Huang, CY Huang. 1999. Traditional Chinese functional foods. In: CYW Ang, K Liu, YW Huang, Eds. *Asian Foods, Science and Technology*, Lancaster, PA: Technomic Publishing Co., Inc.

51. T Ohlsson, N Bengtsson. Microwave technology and foods. *Adv Food Nutr Res* 43:65–140, 2001.

52. Y Zhao, B Flugstad, JW Park, JH Wells. Using capacitive (radio frequency) dielectric heating in food processing and preservation — A review. *J Food Processing Eng* 23:25–55, 2000.

53. Anon. Commercial development of ohmic heating garners 1996 industrial achievement award. *Food Technol* Sept:114–115, 1996.

54. BH Hite, NJ Giddings, CE Weakly. The Effects of Pressure on Certain Microorganisms Encountered in the Preservation of Fruits and Vegetables, Bull. 146 W. Va. Univ. Agric. Exp. Sta., Morgantown, 1914, p 1.

55. E Palou, A Lopez-Malo, GV Barbosa-Canovas, BG Swanson. High-pressure treatment in food preservation. In: MS Rahman, Ed. Handbook of Food Preservation, New York: Marcel Dekker, Inc., 1999, pp 533–576.

56. K Heremans. From living systems to biomolecules. In: C Balny, R Hayashi, K Hermans, P Masson, Eds. *High Pressure and Biotechnology*, London: John Libbey and Co. Ltd., 1992, pp 37–44.

57. MP Cano, A Hernandez, B De Ancos. High-pressure and temperature effects on enzyme inactivation in strawberry and orange products. *J Food Sci* 62:85–88, 1997.

58. D Knorr. Hydrostatic pressure treatment of food: microbiology. In: GW Gould, Ed. *New Methods of Food Preservation*. London: Blackie Academic and Professional, 1995, pp 159–175.

59. DG Hoover, C Metrick, AM Papineau, DF Farkas, D Knorr. Biological effects of high hydrostatic pressure on food microorganisms. *Food Technol* 43:99, 1989.

60. JC Cheftel. High-pressure, microbial inactivation and food preservation. *Food Sci Technol Int* 1:75, 1995.

61. N Kajiyama, S Isobe, K Uemura, A Noguchi. Changes of soy protein under ultrahigh hydraulic pressure. *Int J Food Sci and Technol* 30:147–158, 1995.

62. E Molina, AB Defaye, DA Ledward. Soy protein pressure-induced gels. *Food Hydrocolloids* 16:625–632, 2002.

63. H Zhang, L Li, E Tatsumi, S Kotwal. Influence of high pressure on conformational changes of soybean glycinin. *Innovative Food Sci Emerging Technol.* 4:269–275, 2003.

64. U Zimmermann, G Pilwat, F Riemann, Dielectric breakdown on cell membranes. *Biophys J* 14:881–889, 1974.

65. SE Gilliland, ML Speck. Mechanism of the bactericidal action produced by electrohydraulic shock. *Appli Microbiol* 15:1038–1044, 1967.

66. L Barsotti, JC Cheftel. Food processing by pulsed electric field, II: biological aspects. *Food Rev Int* 15:181–213, 1999.

67. PC Vasavada, D Heperkan. Non-thermal alternative processing technologies for the control of spoilage bacteria in fruit juices and fruit-based drinks. *Food Safety* Feb/March: 8–13, 46–48, 2002.

68. CE Morris. FDA regs spur nonthermal R&D. *Food Eng* July/August: 61–68, 2000.

69. D Knorr, A Angersbach, MN Eshtiaghi, V Heinz, DU Lee. Processing concepts based on high-intensity electric field pulses. *Trends Food Sci Technol* 12:129–135, 2001.

70. MI Bazhal, NI Lebovka, E Vorobiev. Pulsed electric field treatment of apple tissue during compression for juice extraction. *J Food Eng* 50:129–139, 2001.

71. NH Kim, JK Shin, HY Cho, YR Pyun. Effects of high-voltage pulsed electric fields on the extraction of carotenoid from Phaffia rhodozyma. *Korean J Food Sci Technol* 31:720–726, 1999.

72. MS Rahman. Irradiation preservation of foods. In: MS Rahman, Ed. *Handbook of Food Preservation*. New York: Marcel Dekker, Inc., 1999, pp 397–419.

73. LL Dock, JD Floros. Thermal and nonthermal preservation methods. In: MK Schmidl, TP Labuza, Eds. *Essentials of Functional Foods*. Gaithersburg: Aspen, 2000, pp 49–87.

74. N Ayed, HL Liu, M Lacroix. Improvement of anthocyanin yield and shelf-life extension of grape pomace by gamma irradiation. *Food Res Int* 32:539–543, 1999.

75. R Miller, W Jeffrey, D Mitchell, M Elasri. Bacterial responses to ultraviolet light. *Am Soc Microbiol* 8:535–541, 1999.

76. Anon. Pulsed Light Technology. In: Kinetics of Microbial Inactivation for Alternative Food Processing Technologies. U.S. Food and Drug Administration, June 2000.

77. BE Ellickson, V Hasenzahl. Use of light-screening agent for retarding oxidation of process cheese. *Food Technol* 12:577, 1958.

78. JW Povey, TJ Mason. *Ultrasound in Food Processing.* Glasgow: Blackie, 1998.

79. JA Ordóñez, MA Aguilera, ML García, B Sanz. Effect of combined ultrasonic and heat treatment (thermoultrasonication) on the survival of a strain of Stapgylococcus aureus. *J Dairy Sci* 54:61–67, 1987.

80. R Pagán, P Mañas, A Palop, FJ Sala. Resistance of heat-shocked cells of Listeria monocytogenes to manosonication and manothermosonication. *Letters Applied Microbiol* 28:71–75, 1999.

81. G Scherba, RM Weigel, JR O'Brien. Quantitative assessment of the germicidal efficiency of ultrasonic energy. *Appl Environ Microbiol* 57:2079–2084, 1991.

82. KS Suslick. Homogenous sonochemistry. In: KS Suslick, Ed. *Ultrasound. Its Chemical, Physical and Biological Effects*, New York: VCH, 1988.

83. J Raso, P Mañas, R Pagán, FJ Sala. Influence of different factors on the output power transferred into medium by ultrasound. *Ultrasonics Sonochem* 5:157–162, 1998.

84. TJ Mason, L Paniwnyk, JP Lorimer. The uses of ultrasound in food technology. *Ultrason Sonochem* 3:S253–S260, 1996.

85. http://www.plafarbu.com

Index